C0-AXH-704

A Monograph on
POLYCLAD TURBELLARIA

Stephen Prudhoe, O.B.E.

Department of Zoology
British Museum (Natural History)

BRITISH MUSEUM (NATURAL HISTORY)

OXFORD UNIVERSITY PRESS

First published 1985 by
British Museum (Natural History)
Cromwell Road, London SW7 5BD
and
Oxford University Press, Walton Street, Oxford OX2 6DP

Oxford New York Toronto
Melbourne Auckland
Kuala Lumpur Singapore Hong Kong Tokyo
Delhi Bombay Calcutta Madras Karachi
Nairobi Dar es Salaam Cape Town
and associated companies in
Beirut Berlin Ibadan Mexico City Nicosia

Published in the United States
by Oxford University Press, New York

ISBN 0 19 858518 7 OUP edition
ISBN 0 565 00977 X BM (NH) edition

British Library Cataloguing in Publication Data

Prudhoe, Stephen
A monograph on the polyclad turbellaria.
1. Polycladida
I. Title
595.1'23 QL391.P7

Printed in Great Britain by Henry Ling Ltd., at the Dorset Press, Dorchester, Dorset

Contents

Preface

Polyclad turbellarians are interesting marine flatworms, but they have received relatively little notice from students of marine life. In fact, there exists only one monograph dealing with the group as a whole. It was prepared by a former member of the Naples Marine Station, Dr Anton Lang, and published in 1884. Since then, few zoologists have directed serious attention towards these worms, and principally they are: P. Hallez (1878–1913); F. F. Laidlaw (1902–6); Sixten Bock (1913–31); Kojiro Kato (1934–68); Libbie H. Hyman (1938–60) and Ernst Marcus and his wife Eveline du Bois-Reymond Marcus (1947–68). Much of the information gathered by these authors is taxonomic, based primarily on studies of the copulatory complexes in serial section, an exercise that requires patience and an understanding of functional morphology in these worms.

In recent years there has been a notable increase in underwater exploration by marine zoologists and others, and students are now frequently meeting with polyclads, especially in the warmer seas. As a result, there is a desire for information on the identity and natural history of these animals. Such information may, unfortunately, be obtained only from specialists, who are now almost non-existent. Therefore, it seems that a well-nigh unexplored field of marine animals awaits serious and sustained investigation.

Herein is an attempt to review available information and provide a basis for students who might wish to undertake a serious study of the Polycladida. The book is divided into six sections: (i) a general account of structure, natural history, classification and technique; (ii) a classification of families and genera, with definitions, differential keys and diagnostic line-drawings of type-species, where possible; (iii) a bibliography of various taxa; (iv) an alphabetical list of specific names, with indications of their nomenclatoral history; (v) a subject-index; and (vi) a list of references.

The compilation of this book has been effected intermittently during the past thirty-five or so years, and although every effort has been made to guard against blunders, imperfections will, no doubt, be noticed. The author is, therefore, grateful to his colleagues, Dr David I. Gibson and Rodney A. Bray for reading through the general account and offering helpful suggestions. Thanks are also due to Charles Hussey, who has given unstinted help with problems concerning references, and to David Cooper for preparing excellent serial-sections of these worms at various times. Grateful acknowledgement is made to those authors whose original drawings have been utilized. Finally, the author wishes to respectfully mention the late Dr F. F. Laidlaw and the late Dr L. H. Hyman, both of whom gave much advice and encouragement at the outset of this task.

Introduction to the Polyclad Turbellaria

EXTERNAL FEATURES The body of a polyclad is unsegmented, bilaterally symmetrical, dorso-ventrally flattened, somewhat leaf-like, and varying in outline from broadly oval (*Planocera*) to elongate or ribbon-like (*Cestoplana*). The smallest known adult form (*Stylostomum notulata*) may be only 1 mm long, whereas the largest (*Discoplana gigas*) may reach a length of more than 15 cm, but polyclads generally range between 10 mm and 60 mm in length. The creeping or ventral surface is flattened, usually without pigmentation, whereas the upper or dorsal surface is slightly convex and often brownish or greyish. Some larger forms, however, particularly those living in the Indo-west Pacific waters, are frequently brilliantly coloured and strikingly patterned. Many forms are pellucid or translucent. At or near the anterior end of the body there may be a pair of tentacles, these being ear-like projections or mere folds of the anterior margin. A pair of dorsal processes, often retractile, may be situated posteriorly to the anterior margin, near the cerebral organ (Fig. 1A). The dorsal surface is generally smooth, but it occasionally bears numerous papillae (*Thysanozoon*), tubercles (*Ommatoplana*) or slender processes (*Hoploplana*). In the genus *Enantia*, there are well developed cuticular spines placed near the lateral and posterior margins of the body, while *Styloplanocera fasciata* is provided with a marginal row of tiny sensory papillae round the body.

Photoreceptors, usually referred to as 'eyes', may be seen as small dark specks lying beneath the dorsal epidermis (Fig. 1). They are variably distributed in the anterior or cephalic region, and sometimes in the marginal or submarginal zones of the body.

The mouth (Fig. 1) lies in the mid-ventral line of the body, more often near the middle, but is sometimes placed either more anteriorly (*Pseudoceros*) or posteriorly (*Stylochus*). In the region of the mouth, the median line on the dorsal surface of the body is often raised into a longitudinal ridge by an underlying pharynx (Fig. 1). In pellucid and translucent forms, the typical dendritic ramifications of the intestine are plainly visible, and when the intestinal trunk and its branches contain food and are seen through the body wall, the worm resembles a leaf, with the trunk resembling the mid-rib and its branches the veins (Fig. 1).

The male and female copulatory organs are separate and usually appear, more especially on the ventral side, as whitish areas almost invariably lying posteriorly to the pharynx and opening to the exterior either individually or via a common pore in the median line on the ventral surface. Several forms possess a well defined ventral sucker (a glandulo-muscular adhesive organ), a shallow adhesive depression or an epidermal pad on the ventral surface, acting presumably as an organ of attachment. When present, the ventral sucker (Fig. 1B) lies more or less centrally in the median line and posteriorly to the female genital pore. The adhesive depression, however, lies at or near the posterior end of the body, except in a few acotyleans where it is situated between the genital pores.

BODY-WALL This comprises a single superficial epithelium or epidermis investing a basement layer or membrane supported by layers of muscle fibres enclosing a connective tissue or parenchyma in which lie various organ systems. The *epidermis* consists of a layer of ciliated cells, among which are scattered several types of gland cells and many sensory cells.

Ventrally, the ciliated cells are columnar or cuboidal, whereas dorsally they are invariably columnar and usually taller, and there is sometimes a considerable difference in this respect. Each of these cells possesses several cilia, which are better developed ventrally than dorsally, being relatively longer and thicker. At the base of each cell there is a mass of fibres by which the cell is attached to the basement layer.

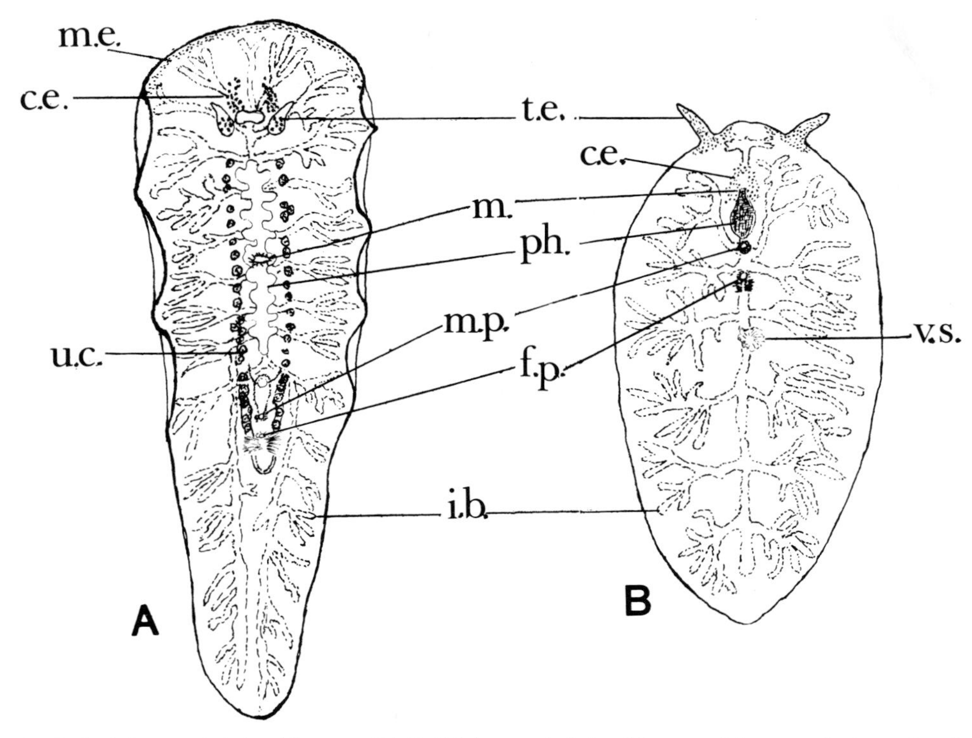

Fig. 1 **A**, Acotylean, ventral view (diagrammatic); **B**, Cotylean, ventral view (diagrammatic). **c.e.**, cerebral eyes; **f.p.**, female pore; **i.b.**, intestinal branch; **m**, mouth; **m.e.**, marginal eyes; **m.p.**, male pore; **ph.**, pharynx; **t.**, tentacles with eyes; **u.c.**, uterine canal; **v.s.**, ventral sucker.

Glandular cells are especially numerous in the dorsal epidermis, particularly in the anterior region, and in many species in the marginal zones ventrally. They rarely occur in the tentacles. These cells are pyriform or oval, open externally through short narrow necks and contain a rounded nucleus, possessing a nucleolus, lying at the base of each cell. Although the several kinds of gland-cells are similar in appearance they differ considerably in the nature of their secretions, the function and chemistry of which little is known. These gland cells may be divided into two groups: (*a*) cyanophilic glands that are stained blue by haematoxylin and whose secretions are viscous fluids; and (*b*) eosinophilic glands that are stained pinkish or red by eosin and whose secretions are formed into bodies of definite shape. According to the nature of their products, these cells are known as rhabdite cells, pseudorhabdite cells, slime cells, pigment cells and so on.

Rhabdite cells are present in all known polyclads, except, according to Graff (1878), in *Stylochoplana tarda.* They usually occur in large numbers, especially in the dorsal epidermis. There are, however, a few species in which such cells are subepidermal. These cells produce solid, cylindrical or fusiform, refringent bodies known as rhabdites or hyaloids, which are disposed vertically above the nucleus within the cell. Lang (1884) has described the early development of a rhabdite in *Thysanozoon brocchii.* A single cell may contain one or more rhabdites, all of which are more or less of uniform size; the size varies greatly among different species. In *Meixneria furva,* for instance, they measure only 3 μm in length, whereas in *Acerotisa typhla* they attain a length of 25 μm. The purpose of rhabdites in the epidermis has not yet been satisfactorily explained, but when a polyclad is disturbed or irritated rhabdites are discharged in great numbers. They appear to form a slimy or gelatinous coating over the body, presumably for protection. Moreover, this covering might be noxious, or even toxic, to a predator, although there is no firm evidence of this. It has, however, been noticed that rhabdites occurring among tissues of prey in the gut of a polyclad have been those of the predator, suggesting that rhabdites are used in the capture of prey.

Pseudorhabdite cells, slime-rod cells or calyciform cells frequently occur in large numbers in both the dorsal and the ventral epidermis, especially when rhabdites are scarce. Their eosinophilic secretions are moulded into granular blocks or plates, irregular in shape, refringent and slightly tinted. These blocks are arranged in superimposed layers within the cell. As the blocks exude through the cell openings, being viscous, they adhere to one another to form a sticky mass. They have been thought to assist in holding captured prey and in sealing wounds in the body wall.

In some forms, mucoid or adhesive cells are present in the marginal and submarginal zones of the ventral epidermis. They secrete a mucoid substance which probably enables the worm to maintain a firm hold on any hard but moist surface. The ability to do this is important to those forms much exposed to the action of surf.

Pigment cells occur principally in the dorsal epidermis and, rarely, in the ventral body wall. These flask-shaped cells possess a large vacuole enclosed in a thin wall of cytoplasm. The vacuole contains a tinted transparent fluid in which lie granules of pigment. Pigment in the epidermis is usually absent in the tentacular and cerebral regions.

Other kinds of gland cells are sometimes present in the epidermis, but their functions are even more obscure than those of the commoner types.

Epidermal cells of all kinds frequently sink into the parenchyma, dorsally and ventrally, and it is possible that they might be regarded as unicellular glands that discharge their secretions to the exterior by means of long ductules passing through the body wall.

Sensory cells occur mainly in the marginal zones of the body and sparsely in the dorsal epidermis. The outer surface of each sensory cell is concave, the concavity holding the base of a long flagellum or a tactile bristle. A nerve fibre connects the cell to a neurone lying in the subepidermal nerve net.

The genus *Enantia* bears lateral 'thorns' on the dorsal surface in the marginal regions of the body. Graff (1890) has shown that each thorn arises from cuticular material secreted by a number of epidermal cells raised up as a papilla.

Nematocysts or 'stinging cells' and similar structures (Martin, 1914) are sometimes found in the epidermis and the parenchyma, mainly in the dorsal and lateral fields of the body, and sometimes in the gastrodermis and lumen of the gut branches. For a time, they were thought to have developed in the tissues of polyclads, but it is now clear that nematocysts in these worms originate from ingested coelenterates, especially from the polyps of hydroids. It is thought that the pharyngeal secretions of the polyclad prevent the explosion of the nematocysts before the prey is devoured. The nematocysts are borne by amoebocytes from the gut to the epidermis, where they are ready to discharge their cnidocils against prey or an attacker, for they retain their ability to be expelled by mechanical stimulation. In the parenchyma of *Anonymus virilis*, dorsally to the intestinal branches, there appear to be individual groups or batteries of nematocysts and various other microscopic weapons. These lie in pathways to the dorsal epidermis, in which they wait for use. Polyclads appear to be unaffected by the poison released by cnidocils exploding in the tissues of their prey, for the secretions of the digestive system seem to render the polyclad immune against tissues so contaminated.

Basement layer, lamella, or membrane is a lamellated or stratified non-cellular layer of hyaline tissue supporting the epidermis and serving as an attachment for the muscle layers lying beneath. It is relatively thick in the middle region of the body, becoming thinner towards the margins. Contrary to the opinion of earlier writers, the basement membrane is not nucleated. Pedersen (1966) has described the histological and histochemical features of this 'subepidermal membrane' in *Discocelides langi*.

Musculature. This consists of several subepidermal layers of fibres, orientated in different directions, which are mainly responsible for the general movements of the worms. They give rise to numerous fibres passing dorso-ventrally through the parenchyma or attaching themselves to various organs. The fibres are unstriated. The musculature is usually more strongly developed ventrally than dorsally, and at the middle of the body is thicker than towards the margins. There is variation in the number of layers of fibres in the body wall. Frequently there are six layers in the ventral musculature and five in the dorsal. In some forms, however, only two or three layers in each set have been described. Dorsally, the common arrangement of the layers is: (i) an exceedingly thin outer transverse or circular layer of muscle fibres lying immediately beneath the basement membrane; (ii) a longitudinal layer; (iii) and (iv) a double layer of diagonally disposed muscles, the fibres of the two sets intersecting and running in opposite directions to form a single layer of criss-cross muscles; and (v) a well developed inner transverse layer. The common arrangement of the ventral layers of muscles is similar, except that the inner transverse or circular layer lies between the two diagonal layers and an additional innermost longitudinal layer occurs internally to the diagonal layers. This ventral inner longitudinal layer is often the most powerful of the muscle layers. Occasionally, as in *Limnostylochus annandalei*, the layer of diagonal fibres is absent.

Papillae. In a few genera, the dorsal surface of the body is raised into many papillae or tubercles. In some species of *Thysanozoon* the papillae contain appendages of the intestinal branches. The purpose of these superficial structures is not known, but possibly they are organs of flotation and allow the worms to be distributed quickly by inshore and off-shore water currents.

Parenchyma This is of mesodermal original and is a loose tissue enclosed by the subepidermal musculature. It forms a connective tissue, which supports the internal organs and endows the body with a considerable degree of flexibility. Histologically, the parenchyma is chiefly composed of polygonal nucleated cells with long irregular processes or filaments that unite to form a delicate, irregular, honeycomb tissue. The alveoli of this tissue are filled with a gelatinous and slightly granular matrix, in which lie various kinds of cells, including sunken epidermal cells, and unicellular gland cells associated with the digestive and reproductive systems. The parenchyma also contains wandering cells or amoebocytes assisting in growth and in the regeneration of lost or injured parts. Pigment granules may also occur in the parenchyma, especially among the Cotylea. It seems possible that some of these granules may be coloured excretory products, for while they are plentiful in the parenchyma in well nourished worms, they are sparse in starving individuals. The histological and histochemical features of the parenchyma in *Discocoelides langi* have been described by Pedersen (1966).

ORGANS OF ATTACHMENT These structures are situated ventrally in the median line and comprise a true sucker, an adhesive disc or a genital sucker.

The *true sucker* is a protuberance on the ventral surface of the body. It consists of a modified epithelium covering a very thin basement membrane and a thick muscular lamella differentiated from the parenchyma and originating from the longitudinal muscles of the body wall and from the dorso-ventral muscles serving as retractor muscles. Such suckers are known only among many cotyleans and lie more or less in the middle region of the ventral surface, posteriorly to the female genital pore. The purpose of the ventral sucker is not known, but it might be used as a holdfast against the strong action of waves.

The *adhesive disc* is a shallow depression or a thin pad constructed similarly to a sucker, but not differentiated from the parenchyma. This structure lies ventrally, near the posterior end of the body in boniniid and cestoplanid polyclads.

The *genital sucker* occurs between the male and female genital pores in a few species. It is a depression in the ventral wall of the body covered by a thickening of the body musculature through which pass dorso-ventral and diagonal retractor muscles.

NERVOUS SYSTEM This system in polyclads is bilaterally symmetrical and characterized by the differentiation of a brain or cerebral organ and paired nerve cords, linked by transverse commissures to form a nerve plexus situated immediately below the ventral and dorsal subepidermal musculatures. The cerebral organ lies in the median line between the anterior margin of the body and the pharynx, seldom occurring posteriorly to the mouth (*Oligocladus*). Typically, it is either more or less globular or transversely elliptical, usually with an indentation on the anterior and posterior margins, thereby being superficially bilobed. On the anterior margin of each lobe there is a mass of ganglion cells or extra-capsular ganglia ('Körnerhaufen' of German authors), from which nerves pass to the eyes, to the tentacles and to the anterior margin of the body. The cerebral organ consists of a membraneous capsule enclosing a mass of nerve fibres surrounded by ganglion cells of several types. From the central core of fibres up to ten pairs of nerve cords extend anteriorly, laterally and posteriorly in the dorsal and ventral parenchyma. The anteriorly directed nerves leave the cerebral organ and extend over the cephalic region, where they connect with various sense organs. Three or more pairs of well developed longitudinal nerve cords emerge from the cerebral organ and proceed laterally and posteriorly. Of these, a conspicuous ventral pair, passing alongside the pharynx, are possibly motor nerves, whereas a dorsal pair, extending through the sublateral fields of the body, are possibly sensory. Sometimes the longitudinal nerve cords bifurcate soon after leaving the cerebral organ. The nerve cords give off transverse connectives that anastomose at various places to form a dorsal and a ventral synaptic nerve plexus attached to the subepithelial musculature by branching nerve fibres. This plexus is more strongly developed ventrally than dorsally.

In some forms the longitudinal, dorsal and ventral nerve cords arise from a plexus of nerves around the cerebral organ. The nervous system of *Theama evelinae* is less complex, as transverse connections appear to exist only between the main longitudinal nerve cord and the submarginal nerve in each lateral field of the body. In *Boninia mirabilis* Bock (1923*b*) describes a nerve sheath embedded in the ventral musculature and containing a pair of longitudinal thickenings, possible representing the motor nerves in the typical system. It therefore seems that variations in the pattern of the nervous system might have taxonomic importance.

SENSE ORGANS In polyclads the sense organs are represented by a ciliated anterior marginal groove, photoreceptors or eyes, tactile and chemosensory receptors.

The *marginal or auricular groove* occurs subventrally on the anterior margin of the body. It often appears as a shallow, whitish, furrow lined with an epithelium carrying bristles, as well as cilia that are longer than those on the rest of the body. The precise function of this furrow appears to be uncertain, but it is probably comparable with the 'ciliated pit' of triclad turbellarians, which is innervated and chemosensory.

Tentacles, when present, are paired and lie near the cerebral organ or on the anterior margin of the body, but in the acotylean genus *Styloplanocera* a row of small marginal innervated tentacle-like tufts occurs dorsally round the body. The tentacles near the cerebral organ are known as nuchal tentacles, and those on the anterior margin as marginal tentacles. The former may be muscular and retractile, usually without pigment, whereas the latter are mere folds or lappet-like outgrowths on the dorso-anterior margin and may carry some pigment. In preserved specimens, the nuchal tentacles may be retracted to such a degree that their presence is indicated merely by a pair of shallow unpigmented depressions in, or slight elevations or bosses on, the dorsal surface. Tentacles appear to be sensitive to touch and water movement.

Pigmented photoreceptors, 'eyes' or 'eye-spots' occur in the parenchyma, mainly in the anterior third of the body. They may be observed in cleared specimens as small black or brownish spots, usually arranged in definite groups: (*a*) cerebral eyes overlie or border the cerebral organ; (*b*) tentacular eyes, larger and darker than the cerebral eyes, lie within or beneath the tentacles, alongside the cerebral organ, or mingle with the cerebral eyes when tentacles are absent, but they rarely lie as deep in the parenchyma as the cerebral eyes; (*c*) frontal eyes scattered anteriorly, often fanwise, from the cerebral organ; and (*d*) marginal or submarginal eyes, sometimes forming a complete series round the body. In some forms (*Latocestus*) the eyes are scattered fanwise over the cephalic region without being differentiated into particular groups and are referred to as cerebro-frontal eyes. Generally, polyclads living in deep water have smaller eyes than those living in or near the tidal zones. Furthermore, mud-loving and deep water acotyleans have marginal eyes, whereas those forms frequenting shallow waters and tidal zones are often without them. In a few species eyes have not been observed. The eyes have the structure of pigment-cup ocelli. Each consists of a cup or bowl of pigment cells surrounding refractive rods connected with photosensitive or retinal cells lying at the opening of the pigment-cup and communicating with the cerebral organ by means of sensory or optic nerves. Eyes are sensitive only to light coming towards the open end of the pigment-cup, thus the cups in an individual animal are disposed for the perception of light from any direction. Generally, the cerebral eyes see light from above and below, and the tentacular eyes from the front, behind and sides. A few polyclads appear to be positively phototactic, but the majority are negatively phototactic. Usually the epidermis in the ocular regions has no pigment.

Lanfranchi *et al.* (1981) found both epidermal and cerebral eyes in the pelagic larvae of *Thysanozoon brocchii* and *Stylochus mediterraneus*, whereas in the young of *Notoplana alcinoi* only cerebral eyes occur. The epidermal eye is a cup-shaped pigmented cell with its cavity filled with lamellae of ciliary origin. A small photosensitive cell covers the opening of the cup. This eye lies between the epidermis and the basement membrane, but later sinks into the parenchyma and coverts to a cerebral eye. The cerebral eye consists of a cup lined with ciliated pigment cells and two types of photosensitive cells. These eyes lie in the parenchyma near the cerebral organ. As the pelagic larva develops so the cilia of the pigmented cells disappear.

Special sensory cells sensitive to touch (*tangoreceptors*) lie immediately below the subepidermal musculature, more especially on or near the margins of the body. They are numerous in the marginal tentacles of euryleptid and pseudocerotid polyclads, and occur in bundles in the apical regions of the dorsal papillae of *Thysanozoon*. Their endings project externally through the epidermis as stiff tactile bristles and flagella, often disposed in groups or tufts.

Chemosensory and rheotactic receptors are widely distributed over the body, thus, respectively, enabling the worm to detect food and changes in the flow of water over the surface of the body.

DIGESTIVE SYSTEM This shows a remarkable constancy in design and is highly characteristic of the order. It consists of a mouth, a muscular pharynx lying in a pharyngeal chamber and a ramified intestine.

The mouth is a simple aperture in the ventral wall of the body situated in the median line, frequently in the middle region. It sometimes occurs closely posterior to the cerebral organ (*Eurylepta*), anteriorly to the cerebral organ (*Oligocladus*), or very occasionally in the posterior region of the body (*Latocestus*). It is enclosed by a sphincter. As a rule, the mouth is distinctly separated from, and anterior to, the genital pores. Exceptionally it lies together with the genital pores in an invagination of the ventral body wall—the 'genito-buccal atrium' (*Miroplana*)—or, the mouth and the male pore form a common aperture (*Stylostomum*). The mouth opens directly into a spacious elongate chamber containing a protrusible muscular pharynx. This chamber—*the pharyngeal chamber*, pharyngeal sheath or buccal cavity—is longitudinally disposed in the median line, frequently in the middle third of the body length. It often appears to possess a number of lateral pockets of varying depth arranged more or

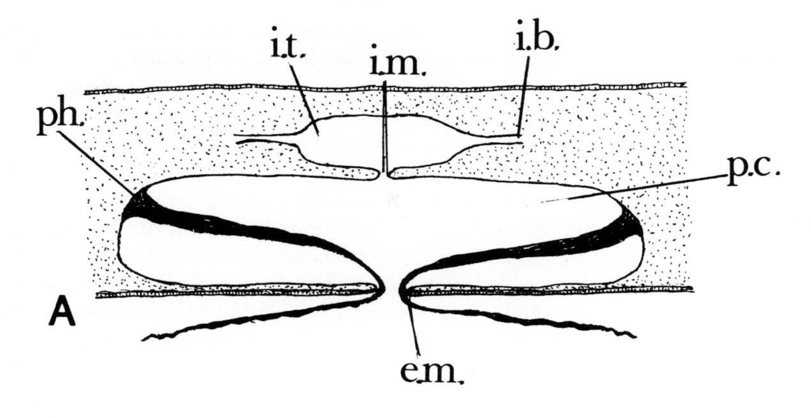

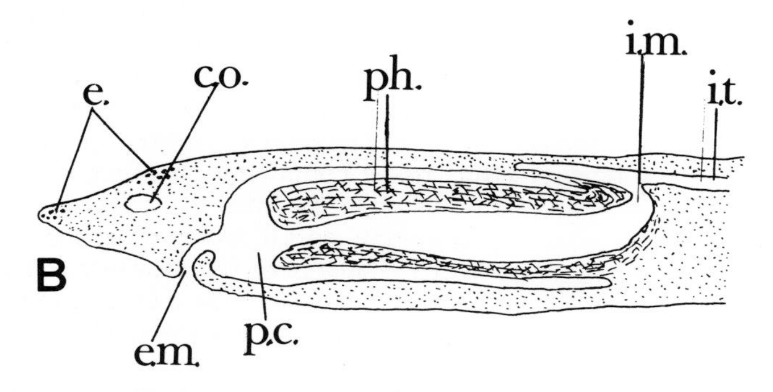

Fig. 2 Pharyngeal complexes (diagrammatic). **A**, transverse section with 'ruffled' pharynx; **B**, longitudinal section with tubular pharynx. **c.o.**, cerebral organ; **e.**, eyes; **e.m.**, external mouth; **i.b.**, intestinal branch; **i.m.**, internal mouth; **i.t.**, intestinal trunk; **p.c.**, pharyngeal chamber; **ph.**, pharynx.

less in pairs. The ventral wall of the chamber is the ventral wall of the body, but dorsally the chamber is coated with a thin layer of coarse muscle fibres and lined with a low ciliated epithelium, through which pass ducts of gland cells lying in the parenchyma. In some forms (*Cryptophallus* and *Oligocladus*) the hinder end of the pharyngeal chamber is provided with a distinct appendix, which holds no part of the pharynx.

The plicate pharynx, when quiescent within its chamber, may appear to be 'ruffled', collar- or ring-like, campanulate or tubular. The ruffled pharynx is commonly met with in acotylean polyclads (Fig. 1A). This condition is due to the pharynx being attached to the upper marginal walls of the pharyngeal chamber, including the lateral pockets. This type of pharynx is therefore a highly muscular fold of the chamber and its pockets, thereby dividing the chamber into an upper (dorsal) and a lower (ventral) compartment, both of which are in communication with each other by the space formed by the free margin of the pharyngeal fold (Fig. 2A). In this instance, the mouth opens into the central or the posterior region of the ventral pharyngeal compartment. In some species of the family Pseudocerotidae, when the pharynx lies within its chamber it appears ring- or collar-like, but still divides the chamber into an upper and a lower compartment. In this instance, the pharyngeal chamber is without lateral pockets. The campanulate and tubular pharynges (Fig. 2B) are folds arising from the posterior wall of the chamber and directed anteriorly (Euryleptidae and Prosthiostomidae).

They invariably lie immediately posterior to the cerebral organ. Here there are no lateral pockets in the wall of the pharyngeal chamber and the mouth occurs at the anterior end of the chamber.

When a polyclad feeds, the free margin of the pharynx is thrust out through the mouth and expands into a membraneous discoid structure with frilled margins. It is endowed with a powerful musculature and is capable of expanding to such an extent that prey as large as, or even larger than, the predator may be readily engulfed.

Histologically, the pharynx is variable. Corrêa (1949) has reported as many as fifteen layers of tissue in the ruffled pharynx of *Zygantroplana henriettae,* although nine layers are more common among polyclads with this type of pharynx. Generally, the dorsal and ventral walls are invested with a low epithelium, which may be ciliated and in some instances appears to secrete a thin cuticle or membrane. The epithelium of each wall is usually supported by an exceedingly thin basement membrane, beneath which lies a layer of longitudinal or retractor muscles (*i.e.,* fibres running from the base of the pharynx to its free margins) investing a layer of circular muscles. The relative position of these two muscle layers is not constant throughout the group, for their arrangement in one wall is sometimes reversed in the other. In the ruffled pharynx there is a centrally disposed lamella of circular muscles, on one or both sides of which lies a well developed nerve plexus arising from the ventral nerve cords of the body and with neurones sometimes situated outside the plexus. The tubular and bell-shaped pharynges may have a central nerve plexus instead of a central layer of muscles. Between the subepithelial musculatures and the central lamella of muscles or the central nerve plexus of the pharynx, lies a loose parenchymatous tissue containing cyanophilic and eosinophilic gland cells. The eosinophilic cells open chiefly in the marginal regions of the pharynx, whereas the cyanophilic cells open on both the dorsal and the ventral surfaces of the pharynx. Radial muscle fibres extend through the pharynx from the outer to the inner wall.

According to Laidlaw (1902), the outer and inner walls of the tubular pharynx of *Prosthiostomum* are clothed with a much flattened epithelium. The outer epithelium covers a narrow layer of longitudinal muscle fibres, succeed by a wider circular layer investing a second longitudinal layer regarded as the retractor muscle layer. Neurones lie among the fibres of the retractor muscle layer. The inner epithelium covers a thick layer of circular muscle fibres investing a narrower longitudinal layer. Between the inner and outer musculatures is a parenchyme containing numerous gland cells. Many radial muscle fibres are gathered into distinct bundles extending from one epithelium to the other.

The inner or intestinal mouth, sometimes referred to as the diaphragm, is situated in the roof of the dorsal compartment of the pharyngeal chamber in forms with a ruffled pharynx, and at the posterior end of the chamber in forms with a tubular pharynx. This mouth is furnished with a sphincter and opens into the gut or intestinal system consisting of a longitudinal trunk with paired lateral branches. It seems that the external mouth often opens into the middle region of the pharyngeal chamber, and the intestinal mouth is situated almost directly above it. The general pattern of variation in their positions is related to the fact that as the outer mouth wanders posteriorly so that inner mouth moves anteriorly, and conversely the anterior movement of the former leads to a posterior shifting of the latter. An extreme of these movements is found in the genus *Cestoplana,* in which the external mouth opens into the posterior region of the pharyngeal chamber and the intestinal mouth into the anterior region. In forms with a tubular or campanulate pharynx, the external mouth is always at the anterior end of the chamber, and the intestinal mouth at the posterior end.

The intestinal trunk, 'main-gut' or gastrovascular cavity appears to function as a stomach and extends along the median line, varying considerably in length. It lies dorsally to the pharynx, but in some cotyleans it runs posteriorly from the pharyngeal chamber. It possesses four or more pairs of lateral branches, and often a forwardly directed branch reaching to the cerebral organ or beyond. Sometimes the forwardly directed branch trifurcates posteriorly to the cerebral organ, the median limb passing over the cerebral organ and the two lateral limbs passing alongside the organ. The intestinal branches ramify repeatedly towards the periphery of the body, where they usually terminate blindly. In some forms, the intestinal branches anastomose to form an intricate network of canals. In the genus *Thysanozoon* diverticula from the intestinal branches may enter papillae on the dorsal surface of the body. In the genus *Diplopharyngeata*, the intestinal system consists of an intestinal trunk with many pairs of lateral branches which are said to be without secondary branches. The musculature of the intestinal system is variable in its development. Often distinct sphincters are set as regular intervals along the branches, which thereby have a moniliform appearance. These sphincters appear to be responsible for peristalsis of the gut branches.

The single layered columnar epithelium or gastrodermis lining the main gut and its branches contains innumerable gland cells, mainly of two kinds, the phagocytic cells and the granular claviform Minot's cells. The gastrodermis may also be ciliated. Minot's cells are often more abundant in the main gut than in its branches, or *vice versa*. They have been thought to secrete products regarded as precursors of digestive enzymes, but this is uncertain as some polyclads appear to be without them, particularly during a period of starvation. Correa (1949), however, after investigating this question holds the opinion that Minot's cells probably do not produce digestive enzymes. Initially, food undergoes extracellular digestion in the lumen of the main gut and its branches. Correa also found that in *Zygantroplana henriettae* the gland cells of the branches absorb food particles, lose their boundaries and complete digestion intracellularly, after which the boundaries of the gland cells reappear, and the gastrodermis assumes its original structure.

Usually polyclads are without any form of anal opening and unassimilated material from the gut branches is accumulated in the intestinal trunk and discharged through the mouth by the pharynx. On the other hand, in *Oligocladus sanguinolentus* and a few other species there is a dorsal anal opening at the hinder end of the intestinal trunk, in *Yungia* the gut branches have dorsal appendages opening through small pores scattered over the dorsal surface of the body, and in *Cycloporus* the branches terminate in small vesicles situated marginally and opening dorsally.

EXCRETORY SYSTEM This system in polyclads is little known, for it is difficult to detect, even in compressed living material, and various attempts to trace it in its entirety seem to have been fruitless. Although parts of it have been observed in young *Leptoplana*, in *Cestoplana* and in *Cryptophallus sondaicus*, only in *Thysanozoon brocchii* has much of the system been made out with any degree of certainty.

On the scanty evidence available, the excretory system appears, as in other groups of platyhelminths, to be composed of three parts: (i) bulbous flame cells opening into; (ii) capillary vessels (probably efferent ducts of the flame cells) leading to; (iii) large longitudinal tubules. The flame cells lie in the parenchyma and contain a lumen filled with a fluid into which a number of long cilia project and oscillate. These cells open into the proximal ends and along the course of capillaries lined with cilia. The capillaries open into a number of non-ciliated longitudinal tubules which appear as sinuous canals running the length of the body in the marginal regions. They give off branches connecting the canals with the dorsal (*Thysanozoon*) or the ventral (*Crytopophallus*) wall of the body and open to the exterior through tiny nephridiopores. This system is, in fact, a protonephridial one, for its tubes are open externally and closed internally by flame cells. In addition to an excretory function, osmoregulation is probably one of the chief functions of the system.

RESPIRATORY AND CIRCULATORY SYSTEMS In the absence of such systems in polyclads, oxygen is probably passed to the tissues in two ways. Firstly, the spontaneous waving of the epidermal cilia produces a flow of water over the body, thereby allowing the animal to obtain part of its oxygen requirement and give off carbon dioxide by diffusion through the epidemis. Secondly, peristalsis of the digestive system, together with oscillation of the cilia of the intestinal lining, produces a rough circulation of water entering with food which possibly aids the transport of dissolved oxygen to the internal organs.

REPRODUCTIVE SYSTEM Sexual dimorphism is not known among polyclads, which are protandrous hermaphrodites, except for a few forms (*Discoplana gigas*) that appear to be protogynous. Though the young adult may function primarily as a male animal and the ageing adult as a female, the periods of maturation of sperm and ova overlap. There is no direct evidence of self-fertilization, nor has asexual reproduction been described.

Typically, the *testes* are small compact follicles widely distributed in the ventral parenchyma, sometimes extending between the gut branches into the dorsal parenchyma, and infrequently they may be disposed more or less in two sublateral longitudinal bands. In some forms, however, they lie entirely among the ovaries in the dorsal parenchyma. The testes are numerous in the young adult, but as the female organs become functional, so the testes degenerate and decrease in number. The spermatozoa produced by polyclads have been classified into three types: filiform, flagellate and flanged. Kato (1940) found the spermatozoa of eight species of Japanese acotyleans to be of the filiform type, measuring 4 μm to 18 μm long according to species. The thin tunicae propriae of the testes give rise to fine sperm ductules, seminal capillaries or *vasa efferentia*, which unite to form a delicate network opening at various points into a pair of spermiducal vesicles, seminal canals or *vasa deferentia*.

These latter canals are thin walled and sometimes have a moniliform appearance. They are symmetrically disposed laterally to the pharynx, or to the median line posteriorly to the pharynx, and with maturity they become convoluted and much swollen with sperm. Posteriorly to the pharynx, the vasa deferentia turn medially to open into the male copulatory complex. In some forms each vas deferens bifurcates to form an inner limb extending to the complex and an outer limb terminating shortly after the bifurcation, or continuing posteriorly to unite with the outer limb of the opposite side to form a loop behind the female copulatory complex. On the other hand, in the genus *Planocera* the vasa deferentia appear to arise at a level posteriorly to the female genital pore and extend anteriorly to the hinder region of the pharynx, where they turn inwardly towards the median line. In many species, the medially directed portions of the vasa deferentia lead directly into a seminal vesicle, whereas in other species the limbs unite to form a canal which has been described as an *ejaculatory duct*, a vasa deferentia communis, or a common sperm duct opening into a seminal vesicle, a prostatic organ or an intromittent organ. Hyman (1953*a*) considers it proper to use the term ejaculatory duct to embrace 'the entire terminal part of the male duct from the proximal end of the seminal vesicle to the tip of the penis or everted cirrus'. Bock (1913) appears to have a similar opinion of the ejaculatory duct in its widest sense, but in its strictest sense it is a canal uniting the seminal vesicle with the prostatic organ (*Stylochoplana*), with the prostatic duct (*Stylochus*) or with the penis-papilla (*Euplana*). It is here considered, mainly in agreement with earlier workers, that the ejaculatory duct arises from the union of the vasa deferentia or from the proximal end of the seminal vesicle and extends to the male pore. Along its course it may be successively modified histologically into a seminal vesicle, a prostatic organ, an intromittent organ and a male antrum, which together constitute the male copulatory complex or apparatus. In a new plehniid genus, *Nephtheaplana*, the ejaculatory duct is said to run from a seminal vesicle, by-passing the prostatic organ and opening independently on the ventral surface of the body. Occasionally, a seminal vesicle and, or, a prostatic organ are absent. The distal portions of the vasa deferentia are sometimes bulbous and coated with a thick musculature and represent accessory seminal vesicles, false seminal vesicles or *spermiducal bulbs*, and appear to function as storage organs for sperm. The primary *seminal vesicle* is rounded, oval, pyriform or arcuate, often with a thin layer of nucleated cells investing a thick muscular wall. Among members of the family Stylochidae, the musculature of a pair of spermiducal bulbs and the seminal vesicle is interwoven in such a manner as to form a tripartite or anchor-shaped seminal vesicle. In many forms, the ejaculatory duct runs directly from the seminal vesicle to the penis-papilla, and in this instance the *prostatic organ* or pars prostatica is either free or absent. When free or independent, the prostatic organ lies in the parenchyma, usually either dorsally or ventrally to the ejaculatory duct, into which it opens by a short canal–*the prostatic canal* (Stylochidae). The independent prostatic organ might be regarded as a modified diverticulum of the ejaculatory duct, a vestige or perhaps a rudiment of which may be seen in the genus *Enterogonia.* When the prostatic canal lies directly between the seminal vesicle and the penis-papilla (intromittent organ), and sperm passes through its lumen during mating, it is said to be 'interpolated' or 'intercalated' (Leptoplanidae). Although usually distinctly separated, the prostatic organ and intromittent organ are occasionally enclosed in a thick muscular sheath. As a rule, the prostatic organ is a large muscular structure lined with a ciliated glandular epithelium producing a granular eosinophilic material that mingles with sperm during mating. The precise function of the prostatic organ is not known, but its secretions are probably important in maintaining the viability of the sperm. In some forms, the epithelial lining of the prostatic organ may be so tall as to contain longitudinal chambers running parallel with the lumen, or be thrown into deep radial folds. In the parenchyma around the prostatic organ often lie numerous extra-capsular gland cells, the efferent ducts of which pass through the musculature of the organ and open into its lumen (*Stylochus*). In forms without a distinct prostatic organ, a portion of the ejaculatory duct has an epithelium containing eosinophilic gland cells (*Pericelis*), or such cells lie in the parenchyma and their efferent ducts pass through the wall of the ejaculatory duct and discharge their secretions into its lumen (*Chromyella*). A few species have been described without a prostatic organ, but in these species it is very likely that a portion of the ejaculatory duct, just before it enters the intromittent organ, is invested with extracapsular cells functioning as prostatic glands, as in *Chromyella,* yet such glands may not be apparent until the male copulatory complex is fully developed. It is worthy of note that sperm is not stored in the prostatic organ, even though sperm might very occasionally be seen in its lumen. From this organ, the ejaculatory duct leads into the male antrum, usually through a muscular penis-papilla disposed in the roof of the antrum. This latter portion of the ejaculatory duct has been called the ductus communis, the ductus masculinus communis or penis canal. In forms possessing an independent prostatic organ, a prostatic canal runs from the organ to unite with the ejaculatory duct, which in this region is also known as the ductus communis, leading into a *penis-papilla.* This papilla is protrusible and when very small it may be tipped with a hard tubular stylet, probably derived from cuticular material secreted by gland cells lying in the walls of the papilla. In the families Planoceridae and Gnesiocerotidae there is no penis-papilla, but this region of the ejaculatory duct is modified into an intromittent organ by forming a spacious chamber containing a long cuticularized papilla or lined with spines. This structure is able to turn itself inside out to form a long protuberance or spiny cirrus for sexual union. The male antrum varies in development, for it may be a short narrow canal, or a spacious chamber lined with a ciliated epithelium. In some species, the male antrum is divided into a narrow inner chamber and a spacious outer one. The inner chamber, known as the *penis-pocket,* encloses the penis-papilla or penial stylet and opens into the outer chamber through a conical protuberance resembling a small penis-papilla and known as a *penis-sheath,* which appears to act as a guide for the papilla or stylet during copulation. The *antrum masculinum* opens ventrally through an aperture in the median line, generally posterior to the pharynx. In the genus *Stylostomum,* however, the antrum may open into the pharyngeal chamber near the mouth, and in a few species of other genera the male and female apertures unite to form a common gonopore. In other species, these apertures lie adjacent to each other, and in other instances the apertures are situated in a common depression in the body wall.

A single male copulatory complex is usually present, and this almost invariably lies posteriorly to the pharynx and anteriorly to the female complex. The duplication and multiplication of the male complex is not uncommon. In some species of the family Pseudocerotidae and of the genus *Cestoplana* the male complex is duplicated, whereas the cotylean *Anonymus virilis* has several such complexes disposed in a longitudinal row in each sublateral field of the body. Members of some families (Discocelididae and Boniniidae) possess many musculo-glandular organs, usually known as accessory prostatic organs or *prostatoids* ('Apidiorgane' of Bock, 1927*a*), lying in the parenchyma and opening into the male antrum. Each prostatoid is a pyriform structure with a thick muscular wall, through which pass the efferent ducts of extracapsular gland cells investing the prostatoid and whose granular eosinophilic secretions are discharged through the free end of the prostatoid. The narrow ends of the prostatoids may be armed with a hard cuticular stylet. The precise function of these pyriform organs is not known, but they are usually regarded as modified prostatic organs, which may or may not be connected with the ejaculatory duct or the vasa deferentia. Histologically, they closely resemble the adenodactyls of triclad turbellarians, which have been thought to be stimulators in copulation.

The male copulatory complex tends to lie anteriorly to the male genital pore in acotylean polyclads, whereas the complex tends to lie vertically over, or posteriorly to, the male pore among cotyleans.

The ovaries are rounded follicles, usually larger and less numerous than the testes, and lie principally in the dorsal parenchyma, though they may mingle with the testes between the gut branches in the ventral parenchyma. Occasionally, the ovaries are disposed in two lateral longitudinal bands. They produce endolecithal eggs, for each ovary possesses a germinative zone and a sterile zone. Oöcytes develop in the generative zone and yolk is formed in the sterile zone, as independent yolk glands or vitellaria are absent, so that each egg contains an oöcyte and its own yolk. According to Bock (1913) the germinal layer lies dorsally in the ovaries of many acotyleans, whereas in most cotyleans it lies ventrally. Usually, each ovarian follicle is bounded by a thin tunica propria, but in *Chromoplana bella* the tunic appears to be invested with a thick layer of muscle fibres. The tunica gives off delicate canals or *oviducts* lined with a ciliated epithelium. These canals form an anastomosing system lying ventrally to the follicles and opening at various points into a pair of longitudinal *uterine canals* situated laterally to the pharynx and, or, the intestinal trunk and ventrally to the intestinal branches. In *Polyposthia similis*, however, the oviducts on each side of the body form a single tube lying between the pharynx and the male copulatory complex, and in *Nymphozoon bayeri* the multiple vaginae open into a single uterine sac. Certain members of the family Chromoplanidae appear not to develop uterine canals, as is the case in the family Theamatidae. The uterine canals are thin walled, spacious, lined with a glandular epithelium and act as repositories for eggs. In the genus *Opisthogenia*, the uterus consists of interwoven canals in the lateral fields of the body, and in some species of the genus *Stylochus* the uterus may consist of a branched system of canals in the regions usually occupied by the normally unbranched canals. In some cotyleans and a few acotyleans, the uterine canals possess lateral vesicles which appear to function as sperm reservoirs and, or, fertilization chambers. On the other hand, some degenerate spermatozoa and egg fragments have been found in the uterine vesicles of *Boninia*, suggesting that the vesicles might also function as receptacles for the disposal of unwanted reproductive materials, which are ingested by the epithelium lining the vesicles. In *Oligocladus sanguinolentus*, however vesicles are only indirectly connected with the uterine canals by means of the oviducts. Sometimes the anterior region of the uterine canals is lined with a tall ciliated secretory epithelium, which probably has a nutritive function and keeps spermatozoa active. In fact, spermatozoa often accumulate in large numbers in this region. The posterior ends of the uterine canals open individually into the proximal end of the vagina, or by a common canal into the vagina interna.

The vagina is lined with a ciliated epithelium and may be divided into three sections: *the vagina interna*; the *vagina media*; and the *vagina externa* (antrum femininum). The vagina interna sometimes terminates, after receiving the uterine canals, in a vesicle lined with a tall secretory epithelium and coated with a feeble musculature. This vesicle may be bulbous, elongate, crescentic or Y-shaped, and it has been given various names, such as *Lang's vesicle*, female glandular organ, female accessory organ, receptaculum seminis and others, but Lang's vesicle has been most commonly used. This organ undoubtedly functions as a receptaculum seminis until the spermatozoa pass into the uterine canals, where the eggs are fertilized and retained for a time. Very occasionally Lang's vesicle may open on the ventral surface of the body (*Nonatoma* and *Ancoratheca*). When Lang's vesicle is absent sperm passes directly into the uterine canals, where it is stored. After fertilization surplus sperm and other reproductive materials are probably digested and absorbed by the epithelium lining the vesicle. Between Lang's vesicle and the entrance of the uterine canals, the vagina has been referred as the stalk, duct or canal of Lang's vesicle, which may have a moniliform appearance owing to sphincters disposed at regular intervals along its course. In a few instances, when the vesicle is absent, the inner end of the vagina interna receives the posterior ends of the uterine canals, and, after this union, it continues as a *ductus vaginalis* forming a ventrally-directed loop opening into the vagina externa or to the exterior by an aperture situated posteriorly to the female genital pore. In *Pentaplana*, the inner end of the vagina interna bifurcates into a pair of vaginal ducts, each duct giving off four or five appendages, one of which opens ventrally, and copulation is said to occur through the opening. So far, the ductus vaginalis is known only among acotyleans.

A further modification of the vagina interna is the *genito-intestinal canal*, which connects the vagina with the intestinal system. This canal occurs in both the Acotylea and the Cotylea. In the former group it is a continuation of the inner end of the vagina interna, whereas in the latter group there are one or two ducts connecting the uterine canals with the intestine.

The functions of the ductus vaginalis and the genito-intestinal canal are not yet understood, but it is very likely that they are used for the disposal of unused reproductive materials. In the case of the ductus vaginalis and Lang's vesicle with external apertures, it seems possible that they also function as copulatory organs and comparable with the copulatory canals of proseriate turbellarians. In *Discostylochus parcus* and *Enterogonia orbicularis*, which possess a genito-intestinal canal, Bock (1925*a* & *b*) found entire eggs and egg fragments in the lumen of the intestinal trunk and its branches, and he suggested that the surplus eggs may play a casual rôle in the nourishment of the worms. Bock (1927*b*) implies that the canal or stalk of Lang's vesicle, the genito-intestinal canal and the ductus vaginalis are, in acotyleans, modifications of the vagina interna. It is perhaps worth considering that, because among turbellarians adjacent structures tend to fuse, the ductus vaginalis arose from an adhesion and then a coalescence of the wall of Lang's vesicle with the ventral wall of the body, and a similar process occurred in a union of the vesicle with the intestinal trunk or one of its branches to form a genito-intestinal canal.

The *vagina media* or cement duct is that section of the vagina into which open innumerable 'shell' glands ('Kittdrüsen' of German authors) lying in the surrounding parenchyma, and usually form an alate mass on either side of the female copulatory complex, when viewed dorsally or ventrally. This region of the vagina is better known as the 'shell' chamber, since it is here that fertilized eggs are coated with a gelatinous material when egg laying occurs. The epithelium of this chamber is sometimes thrown into a spiral fold of four or five turns.

The *vagina externa* or antrum femininum extends from the 'shell' chamber to a genital pore or to a genital atrium, both of which lie ventrally in the median line. In a few species a vagina externa is absent.

When the male and female genital pores are separated, the former lies anteriorly to the latter, but in one or two genera the positions are reversed. The entire vagina is coated with a thin inner layer of circular muscle fibres and an outer longitudinal layer of fibres and lined with a ciliated epithelium. In some instances, however, the epithelium of the stalk of Lang's vesicle appears to have a glandular function, and sometimes the vagina externa is provided with an exceedingly well-developed muscular wall and is termed the *vagina bulbosa.* Occasionally, a bulbous muscular structure opens into the vagina externa anteriorly and is the *bursa copulatrix.* Multiplication of the female copulatory complex is rare, but it is well shown in the genus *Nymphozoon.* Contrary to the usual arrangement, the female copulatory organs lie anteriorly to the male in the cotylean genera *Opisthogenia* and *Anonymus* and in some members of the acotylean family Polyposthiidae.

REPRODUCTION AND DEVELOPMENT Polyclads are hermaphroditic and engage in reciprocal mating or cross-fertilization. Hypodermic impregnation or perforation of the body wall has been noted, when packets of spermatozoa (?spermatophores) have apparently been injected into the body tissues of other individuals of the same species. The spermatozoa presumably migrate through the parenchyma to meet with and fertilize eggs in the uterine canals. This appears to occur shortly before spawning, and Lang (1884) found evidence of this in *Cryptocelis alba* and *Prostheceraeus albocinctus.* Haswell (1907*b*) detected a site of hypodermic impregnation in *Echinoplana celerrima.* Likewise, Bock (1913) found evidence of similar behaviour in *Hoploplana grubei* and *Prostheceraeus vittatus,* and Poulter (1975) in *Prosthiostomum katoi.* Verrill (1895) found numerous specimens of *Stylochus ellipticus* having 'spermatophores' implanted in various regions of the dorsal parenchyma. Nevertheless, genuine spermatophores are apparently produced by polyclads, because claviform spermatophores have been found attached to the female genital pore of *Gnesioceros sargassicola* (*see* du B. R. Marcus & Marcus, 1968). It is generally assumed that polyclads, although hermaphroditic, are not self-fertilizing. Yet self-fertilization does seem to be possible among those species whose female and male genital pores are closely approximate or open together into a common antrum, thus providing the intromittent organ with easy access into the vagina of the same individual. Self-fertilization is known to occur among other groups of platyhelminths.

After copulation, which appears to happen at any time and not to be governed by any of the rhythms of nature, the spermatozoa are stored temporarily in Lang's vesicle, should this organ be present, or in the uterine canals. Fertilization happens in the vagina interna or in the uterine canals, usually just before egg laying commences.

Asexual reproduction is not known among polyclads.

It is generally thought that polyclads breed between early and late summer, but this may not always be the case. For instance, Remane (1929) shows that in Keil Bay *Notoplana atomata* is mature in the autumn and lays its eggs during December and January, and according to Girard (1850*a*) this species spawns in January and February in New England. Moreover, the present writer has examined specimens of *N. atomata* from the western and eastern coasts of North America and found the uterine canals to be packed with mature eggs from August to December. In British waters, this species deposits its eggs from August to December. Yang (1974) states that *Stylochus ijimai* spawns between May and October in Korea, with the peak period between July and September. How often an individual of this species lays eggs during this period is not stated, but as many as 132 000 eggs may be laid at one time. That a polyclad may spawn only during a certain period, but have ripe spermatozoa available throughout the year, has been shown by the investigations of Thum (1970, 1974). He found specimens of *Notoplana acticola* with mature testes and spermatozoa available throughout the year, particularly in autumn and winter. The female reproductive phase became active only in spring. Individuals with mature eggs and sperm were most frequent in late spring and summer. Copulation occurred during late summer and autumn, when spawning took place. Pearse & Wharton (1938) found that *Stylochus inimicus* laid between 7000 and 21 000 eggs per individual during one summer month. They also reported and figured brooding by an individual of this species which was seen covering with its body for several days a mass of eggs laid by several individuals. Rzhepishevskji (1979) mentions

Stylochus pilidium showing similar parental care. If these worms were actually brooding, presumably they continue doing so until the larvae hatch, but this suggestion requires confirmation.

The eggs, of which there are usually more than one in a single thin capsule, are laid in spiral or somewhat zig-zag chains or ribbons, or in plate-like masses covered with a sticky gelatinous substance derived partly from special glands lying in the ventral parenchyma and partly from the 'shell' glands opening into the vagina, thereby enabling the eggs to be fastened firmly to stones, shells, algae and other surfaces. An exception to this practise is found in *Discoplana takewakii* which is said to deposit its eggs in a cocoon in the genital bursa of a Japanese brittle-star. The cocoon is irregular in shape, measures about 4 mm by 2 mm and contains about 1000 eggs.

It is not known whether the worms die after egg laying or whether their gonads are reconstituted for another cycle of reproductive activity.

The rate of development of polyclad eggs is influenced by water temperature, salinity and probably other factors, and it is known that the more yolk an egg contains the more slowly the embryo develops.

The karyology of polyclads has been little studied, in fact, it has been directed only to the chromosome numbers of about 30 species. The haploid numbers of these species so far ranges from 2 to 16. Several of these species belong to the family Leptoplanidae, the chromosome numbers of which appear to be so variable as to be of no systematic importance at family level.

The egg is endolecithal, inasmuch as yolk is contained in the egg cytoplasm. Embryonic development is unmodified. The zygote undergoes holoblastic, unequal, spiral cleavage, similar to that of acoelous turbellarians, nemertines, polychaetes and some molluscs; yet the later development of the polyclad embryo differs greatly. The eggs may be in various stages of cleavage when laid. It is not intended here to give an account of the embryonic development, for such information has been given by Hyman (1951), and is readily available in textbooks on embryology or on invertebrate zoology. Moreover, there appear to be differences in interpretation by various authors of the segmentation of the fourth quartet cells, particularly 4d. Kato (1940), however, has given a detailed account of the early development of some Japanese polyclads from the laying of eggs to the hatching of larvae. He also mentions these interpretative problems.

The development of polyclads is either direct or indirect. The direct type involves no metamorphosis of the hatched larva and is so far known only among acotyleans. The embryo on hatching is flattened. Its epidemis is ciliated and often bears a variable number of sensory hairs or sensillae, surrounded by patches of long cilia, arranged symmetrically on the margins of the body. It has two or three pairs of eyes, and a mouth, which may be closed during the embryonic and larval stage, is formed in the ventral median line behind the middle of the body. Usually, the embryo has an elementary pharynx and a main gut with lateral branches, and resembles a miniature adult, without reproductive organs, when it escapes from the egg capsule. This juvenile form is frequently found among inshore plankton.

The indirect form of development involves a metamorphosis of the free-swimming larva, as found in the Cotylea and some Acotylea. When the embryo is ready to emerge from its capsule, it is ciliated, ovoid, and bears four, eight or ten posteriorly-directed lobes or processes, filled with parenchyma and bearing a fringe of long cilia. These lobes are widely distributed, for in the eight-lobed form there is a broad, median, antero-ventral lobe, opposed dorsally by a much smaller weakly-developed lobe, and behind the former there is a pair of ventral lobes, bordered by a pair of lateral lobes. Finally, a further pair of lobes is formed at the posterior end of the body. Between the ventral lobes lies a mouth, leading into an intestinal cavity through a narrow stomodaeum invested with mesodermal tissue and representing the primordium of the pharynx. No intestinal branches are apparent. There is also a long cilium or a tuft of cilia at each end of the body, and a cerebral organ giving rise to central and lateral nerves, together with two small groups of dorsal eyes in the anterior region. According to Lang (1884), the protonephridia are derived from two posterior ectodermal cellular bands, but Kato (1940) failed to find such bands in his study of the development of some Japanese polyclads. Between the cerebral organ and the anterior extremity of the body lies a columnar frontal organ, consisting of stellate cells and interstitial material, and derived from an invagination of the ectoderm. The anterior cilia arise from the frontal organ, which soon degenerates to leave no trace in the adult worm.

Upon hatching, the larva, known as *Müller's larva*, thought to represent a modified trochophore, becomes planktonic–the distribution phase of its life cycle–and after a few days the ciliated lobes are absorbed into the body, the cilia disappear and the body loses its plumpness and becomes a dorso-ventrally flattened, antero-posteriorly elongate organism. Dawydoff (1940) found ten-lobed types of Müller's larvae, which are relatively large, variably pigmented, and abundant in plankton off the coast of Vietnam. The additional pair of lobes is situated on the dorsal surface of the larva, posteriorly to the median lobe, although the position is not constant, apparently varying according to species. Dawydoff proposed the generic name *Lobophora* [preoccupied] for the octolobate and the decalobate forms of Müller's larvae, but Dawydoff's name can have no validity other than that of a larval group-name.

Another type of planktonic larva involved in an indirect life cycle is *Götte's larva*. This larva closely resembles Müller's larva in structure. It is characterized by the possession of only four ciliated lobes on the surface of the body. These are indicated by a low protuberance on the dorsal surface, a broad antero-ventral or pre-oral lobe and a pair of large ventral lobes. The frontal organ is not as strongly developed as in Müller's larva.

So far, Götte's larva is known only in the development of certain species of the acotylean genera *Stylochus* and *Notoplana*, but other species of these genera have a direct life cycle. Lang (1884) considered that Götte's larva, although resembling the pilidium larva of nemertines, was a transitional stage of the eight-lobed Müller's larva, but subsequent evidence (Kato, 1940) points to it being an independent larval form. Its metamorphosis appears to be similar to that of Müller's larva.

According to Kato (1940), in *Planocera reticulata* the embryo develops in the egg capsule into an 'intra-capsular Müller's larva', undergoes metamorphosis within the capsule, hatches as a larval form resembling the adult, and represents an intermediate type of development between the direct and the indirect. The larva differs from the typical Müller's larva in its flattened oval body, in the absence of a dorsal lobe, in all seven lobes being disposed on the ventral surface of the body, and in the narrow intestinal cavity provided with well-developed lateral branches. Upon hatching, the young worm creeps about on the sea-bed and is never free-swimming. The egg capsule may contain more than one embryo, and if any die, their yolk material and tissues serve as food for the survivors which then develop into giant larvae.

It now appears that polyclad larvae which undergo metamorphosis are quadrilobate, heptalobate, octolobate or decalobate.

Precocious development occurs in *Graffizoon lobatum* Heath which has the form of Müller's larva in which male and female reproductive organs have developed. This has been regarded as a neotenic euryleptid, but Heath (1928) concludes that it is an instance of paedogenesis. However, it is not yet known whether the form described by Heath will pass through the normal metamorphosis of a euryleptid or reproduce in normal form. A further instance of precocious development was recorded by Marcus (1950), who found Müller's larva of *Cycloporus gabriellae* possessing a male reproductive system, but no indication of female organs. He states 'that the larval sexual stage is not a definitive one, but will pass through normal metamorphosis from an immature juvenile to the full-grown worm that becomes mature again. This phenomenon is known as dissogony' (found in Ctenophora and some nereid polychaetes.) He likewise considers the development of *Graffizoon lobatum* as an instance of dissogony.

Both Götte's and Müller's larvae are strongly attracted towards light, thus appear in surface waters and thereby have an excellent chance of being widely distributed by surface currents and winds. The function of the epidermal lobes in these planktonic larvae has yet to be determined, but they possibly enable the larva to maintain an equilibrium in surface waters. According to Pearse & Wharton (1938), the pelagic Götte's larva of *Stylochus* 'appeared to be strongly positively phototropic, slightly negatively rheotropic and rather unresponsive to gravity and temperature variations'.

In young polyclads, the early stages in the development of sex organs may be seen as strings of cells. It is not unusual to find specimens with the male copulatory complex well advanced in development, whereas the female complex is indicated merely by a chain of cells. If food is plentiful juvenile polyclads grow to maturity in from six to eight weeks.

Available evidence suggests that all cotyleans have an indirect life cycle, in which Müller's larva undergoes metamorphosis. It is, however, known that certain groups of marine invertebrates generally have an indirect life cycle, but in Antarctic waters some of their representatives have a direct life cycle in which the young are much like the adults. It would, therefore, be interesting to know whether or not members of the cotylean subfamily Laidlawiinae have a direct life cycle in Antarctica.

ECOLOGY The Polycladida are almost exclusively marine, the known exceptions being species of the genus *Limnostylochus*, individuals of which appear to prefer the fresh or brackish waters of Assam and Borneo. Nevertheless, individuals of inherently salt-water species do occasionally wander into brackish waters. These flatworms are primarily inhabitants of the sublittoral zone extending in depth to the edge of the continental shelf, but some species enter the intertidal zone with the incoming tide in search of food and often get left in pools and in damp situations under stones and seaweeds by the ebb-tide. Other species, however, seem rarely to venture into the intertidal zone. A few have been found in the mesopelagic zone, for *'Stylochus' crassus* Verrill was dredged on dark green mud in about 2000 metres off the New England coast of the U.S.A., and *Plehnia arctica* (Plehn) has been dredged on grey mud at 1275 metres off Jan Mayen Island in the Arctic Ocean. The latter species is eyeless and has also been found at 12 metres. Between tide marks, polyclads may be found living in mud, under stones and rocks, in empty mollusc shells, hiding among seaweeds, and in crevices and holes in rock pools, emerging from their hiding places when touched by the rising tide and shielded from direct light. From this, it may be accepted that polyclads are more active at night than in daylight. They also dwell in cavities of sponges and corals, especially on reefs, where their flattened bodies allow them to creep into narrow crevices in search of food or find protection from predators. Some may often be found in beds of barnacles and bivalves between the tidemarks.

Many species have individual preferences for a particular substrate. Some stylochids occur only in areas with a muddy bottom or a mixture of mud and sand, other forms prefer to live among certain kinds of seaweeds. Generally, polyclads avoid fine arenaceous sea-floors, but some very small forms may be constituents of an interstitial fauna in areas of coarse sand, more especially in tropical and subtropical waters.

Relatively few species are pelagic as adults. Their very thin, horizontally-flattened bodies and their body fluid being lighter than sea water give them the buoyancy to drift among plankton. They have also been found among floating masses of *Sargassum* weed, or attached to driftwood and other flotsam in tropical and warm-temperate waters. They have been taken in depths ranging from a few metres to over 1000 metres.

Many associate with other types of invertebrates, particularly bivalve molluscs. Some species have, so far, only been found living in shells inhabited by hermit crabs, sometimes devouring the eggs of the crab, but also feeding on hydroids attached to the inner surface of the occupied shell. Several instances are known of where in one locality a species of polyclad usually associates with another type of invertebrate, whereas in a different locality the same species forms no similar association. This has been shown by Hyman (1953*a*) who records that *Notoplana sanjuania* mostly associates with the crab *Paralithodes camtschatica* and was 'found either on the crab's back or inside barnacles growing on the crab' in Pavlov Bay, Alaska, but 'no similar association has been reported for the Puget Sound specimens' of this species. There are many records of species of polyclads apparently preferring to associate with particular invertebrates, and a good instance of this is shown by the occurrence of *Ceratoplana colobocentroti* underneath the sea-urchin *Colobocentrotus stratus* on Krakatau I., Indonesia, and in the Hawaiian Is, two localities roughly 8000 km apart.

Several forms live in the branchial chamber or pallial groove of molluscs, and Wheeler (1894) opened about 100 individuals of *Buscyon* and found one to six specimens of *Hoploplana inquilina* in the branchial cavity of nearly every snail. *Notoplana patellarum* frequently occurs in the mantle cavity of the limpet *Patella oculus* in South Africa. It seems possible that while hiding in this way the polyclad acquires some of its oxygen requirement from water circulating through the mantle cavity. Likewise, those polyclads living in shells occupied by hermit crabs may find that the crab's respiratory current keeps the inside of the shell aerated. *Stylochus ellipticus* has been found not only under rocks and among seaweeds, but also in burrows of nemertines and shrimps, and in shells of live oysters and barnacles, thus hinting that its associations are mainly for protection and the avoidance of direct light. Further, *Taenioplana teredini* and *Stylochoplana affinis* are known only from the empty burrows of *Teredo* in which they lay eggs.

Another type of association is shown by *Stylochoplana parasitica* which Kato (1935*a*) regards as parasitic in the pallial groove of chitons. When removed from the chiton *Liolophura japonica*, the worms died after a few days. The reason for this is not known, but it is possible that the worms became adapted to a high oxygen intake from water passing over the gills of the chiton. When removed from the 'host' and placed in an artificial environment the oxygen available was insufficient and the worms died. A somewhat similar association was recorded by Hallez (1984), who found that *Leptoplana tremellaris*, living amid and feeding on colonies of the ascidian *Botryllus*, died within 24 hours after being removed from the colonies; whereas other worms of the same species not found among *Botryllus* lived for several weeks, some for several months, under laboratory conditions and without the presence of *Botryllus*. Likewise, Crozier (1917) found a species of *Pseudoceros* feeding on the ascidian *Ectinascidia turbinata*, but when removed and placed with another ascidian, *Ascidia atra*, it refused to feed and died within two days. Kato (1935*b*) describes *Discoplana takawakii* as a parasite of an ophiuran and simply states that the worm lives in the genital bursa and 'feeds on gonads attached to the bursa, since no germ cells were found in the infested bursa'. Further, the worm is said to deposit a light brownish cocoon containing about 1000 eggs in the bursa. While it is implied that the worm castrates its 'host', to regard *D. takewakii* as a parasite of ophiurans requires a much closer investigation. Again, Jokiel & Townsley (1974) give a biological account of a new species of *Prosthiostomum*, which they consider to be an 'obligate ectoparasitic symbiont', feeding only on the polyps of a genus of stony corals (*Montipora*).

To sum up, it seems that a case for parasitism by polyclads is not established with any degree of certainty, for although some polyclads may seem partial to tissues of certain prey and may ultimately become addicted to those tissues, it does not follow that they are parasitic. Excluding the predatory aspect, polyclads living in association with other marine invertebrates may be regarded purely as commensals having unilateral or mutual dependence. Moreover, it should not be overlooked that most polyclads are negatively phototactic, and those living in or near the tidal zone tend to hide themselves in cavities or chambers of various marine invertebrates, but in deeper water this tendency appears to be much less evident. Sometimes many individuals of a species are found together under a rock or other hiding place, and the species is then thought to be gregarious. This is not necessarily so, for often they seemingly gather in an area where food is briefly plentiful, or in a suitable hiding place in an area where such a facility is limited. They might also gather in areas where temperatures and salinity are at one time suitable for reproduction.

Like other groups of tubellarians, polyclads are carnivorous, devouring other types of worms, small crustaceans, molluscs and ascidians, and may even be cannibalistic, for Marcus (1950) found a specimen of *Acerotisa* [= *Stylostomum*] and a *Prosthiostomum* pharynx in the intestinal trunk of *Prosthiostomum gilvum.* Some species appear to feed on algae, but not exclusively. Some polyclads seem to have a predilection of certain types or species of prey. The presence of potential prey is detected either by water disturbance, caused by the prey, or by chemoreception. The worm seizes its victim with its pharynx and entangles it with mucus. There is evidence that nematocysts or similar structures may be used in the capture of prey. They are also scavengers and readily feed on dead or injured animals, the body fluids of which are detected by the anterior sensory groove of the worm. Marcus (1952) records the presence of fish blood in the intestine of *Cycloporus gabriellae.* Polyclads likewise feed on sedentary animals, such as alcyonarians, polyzoans and ascidians, particularly the latter. The marginal regions of the ventral surface of polyclads are amply supplied with epidermal cells producing an adhesive substance. When feeding on encrusted colonial animals the margins of the worm become firmly attached to the surface of the encrustation and inside the area covered by the worm, the free margin of the pharynx extrudes, withdraws individual zooids from the colony or discharges proteolytic secretions over the entrapped portion of the colony to macerate the zooids for ingestion. The softened tissue is sucked into the intestinal branches, where digestion is completed. Another form of extracorporeal feeding occurs when a large prey is engulfed by the pharynx and the secretions from the margins and inner surface of the pharynx induce a degree of digestion outside the body. This, together with pressure exerted by the pharyngeal muscles, breaks up the prey, which is then sucked into the intestinal trunk and its branches to be fully digested, the gastrodermal cells absorbing the products. They also feed similarly on gastropod molluscs, into the shell of which the pharynx may be protracted to envelop the body of the snail within. Poulter (1975) observed the tubular pharynx of a prosthiostomid to protrude and nip off a fragment of tissue from the margin of the body of another polyclad and ingest it.

From time to time reports have appeared of polyclads causing considerable damage to commercial oyster beds. In fact, members of the genus *Stylochus* are frequently found in the shells of oysters, mussels and barnacles. That certain of these members have a tendency to associate with oysters is shown by their occurrence among oyster colonies in wire baskets suspended from wharfs, etc. Nevertheless, statements accusing polyclads of the wholesale destruction of oyster beds are misleading, as no direct evidence has been given on whether diseased or healthy oysters are attacked, whether polyclads make no distinction, or whether they are scavenging on the remains of oysters left by known and confirmed predators. Stead (1907) points to the destruction of oyster beds by *Notoplana australis* in Australia, but he did mention the abundance of a polychaete predator and an oyster-drill among the oysters. *Stylochus* specimens have been said to tear off small pieces of tissue as an oyster moves within its shell. It is, however, known that *Stylochus* will devour eggs and spat of oysters, but whether they do so in sufficient enough numbers to effect the economy of oysters is questionable, although claims have been made of heavy destruction, particularly in the Japanese oyster industry (Kato, 1940). Pearse & Wharton (1938) investigating the 'oyster-leech' or 'wafer' in Florida found that *Stylochus inimicus* (= *frontalis*) attacked and devoured oysters, but they concluded that high mortality among oysters is probably never due to a single cause, but to a number of unfavourable factors. From some investigations it seems that oyster beds become infested with *Stylochus* when temperatures and salinities are high, but there has been no direct mention of the effect of these conditions on the oysters.

Hyman (1944*a*) records *Taenioplana teredini* only from empty burrows of *Teredo* in Hawaiian waters, and she thinks 'that the polyclad feeds on *Teredo* and may constitute an important enemy of the pest'. Edmonson (1945) produces experimental evidence to support the claim that this worm preys on shipworms.

Much of our knowledge on the feeding habits and prey of polyclads deals with benthic forms. Information on the food of pelagic polyclads is negligible, but Graff (1892*b*) records *Planocera simrothi* occurring in a shell among a swarm of an oceanic mollusc (*Janthina*). Moreover, in the same species of polyclad, fragments believed to be of siphonophores were found in the intestinal trunk.

Polyclads can endure protracted fasting, and a stylochid has been known to have lived 185 days without food. During such periods the body gradually becomes smaller.

Just as little is known of the feeding habits of polyclads, so less is known of their predators. It is unlikely that they are attacked and eaten, when adult, by most other predators, because their epidermal secretions appear to be distasteful, if not poisonous. Laidlaw (1902) records a large *Thysanozoon*(?) secreting an enormous amount of mucus which killed other animals in the same container, and Arndt (1943) found that extracts of polyclads injected into the heart and coelom of guinea-pigs and frogs produced toxic symptoms, sometimes with lethal results. Since triclads have epidermal gland cells similar to those of polyclads, it is worthy of note that Moseley (1874) reported an instance of where the tip of the tongue was applied to the body surface of a land planarian and produced 'a feeling of unpleasant tingling and it was accompanied with light swelling'. Some polyclads are beautifully coloured and impressively patterned, mainly in black, red and yellow, and because of this it seems possible that such worms are looked upon as unpalatable by possible predators. This suggestion seems to be partly supported by the behaviour of the young of certain fishes, which appear to simulate polyclads in movement and colour, as well as in undulating gently and fluttering slowly away when approached (Randall & Emery, 1971). The degree of protection with which polyclads are furnished probably places them at a high level in the food chain. It is very likely that their greatest casualties occur during their larval and juvenile stages, when the potency of their epidermal secretions might not be as strong as it is in the adult.

Riser (1974) states that *Taenioplana teredini*, an inhabitant of teredinid burrows, is avidly eaten by a nereid polychaete, which also lives in shipworm burrows. Poulter (1975) found that a species of *Prosthiostomum* was eaten by puffer fish apparently without ill effects.

MOVEMENT/LOCOMOTION Polyclads have two modes of locomotion: creeping and swimming, generally in a forward direction. The surface over which a polyclad travels appears to have some influence on its rate of movement, for those living on a muddy bottom are often sluggish, whereas those living among algae or on a hard bottom are more active. Nevertheless, the much flattened body allows the worm to flow over any moist surface. The majority of these animals are negatively phototactic and therefore move about at night or in areas well shaded from direct light, but some highly-coloured forms, such as pseudocerotids, may be seen swimming or creeping in areas exposed to direct sunlight. Locomotion tends to give the body a more elongate outline than that shown when resting.

Creeping is a slow gliding motion, for which these worms have been likened to pieces of living film or gelatin flowing gently over any surface in search of food. This form of locomotion is accomplished either by ciliary action producing a current from anterior to posterior, together with a slight muscular rippling of the vental surface. A slightly quicker method of creeping is effected by undulations generating a succession of shallow waves passing posteriorly along the margins of the body. The waves of both sides move sometimes alternately (ditaxic), sometimes simultaneously (monotaxic). Some forms also occasionally creep leech-like by extending and fixing the anterior region of the body and drawing the remainder forward, but sometimes these forms show no inclination to swim or float as do many other polyclads. A few forms also creep by expanding and contracting the lobes of an everted pharynx. When creeping, the anterior region of the body is usually raised to allow the sensitive marginal furrow to test its surroundings.

Swimming gracefully and freely is undertaken mainly by the larger elliptical forms, and is accomplished by the expanded margins of the body undulating delicately and rhythmically, similarly to the rippling of the lateral fins of a cuttlefish. From this mode of swimming, such polyclads have been called 'scarf'- or 'skirt'-dancers. A method for swimming faster is produced by expanding the cephalic region of the body into two lateral 'wings' that are flapped rapidly over the dorsal and ventral sides in a manner similar to that of a swimming ray. Elongate forms tend to swim rapidly by undulating the whole body in a dorso-ventral, wave-like, motion similar to that of a swimming leach. Swimming is effected with ease and rapidity but is of short duration.

Cheng & Lewin (1975) record *Gnesioceros sargassicola* floating in large numbers among *Sargassum* weed in the Gulf of California. Some of these worms were seen swimming in a forward direction by waving the lateral margins of the anterior region of the body. Some were also seen to be 'attached to weed only by the posterior end, while the rest of the body swayed freely in the water'.

Apparently, pseudocerotids are very good swimmers. Levetzow (1943) notes that *Thysanozoon brocchii* moves quite rapidly, and that the worms might cover fair distances on calm nights. Such nocturnal migrations appear to be well known to Neapolitan fishermen. There is, however, the probability that the distances covered by specimens of this species are due, at least in part, to the papillae on the dorsal surface of the body which give the worms a degree of buoyancy, thus allowing them to drift passively or with little movement in water currents.

Many, if not all, polyclads have the ability to float and glide clinging to the surface of the water with their ventral sides uppermost. The movement of larvae from the direct type of development is similar to that of the adult. In the indirect type the planktonic larva revolves in various ways in its early life, but when older it rotates around its long axis, the anterior end being uppermost.

COLORATION In life polyclads usually appear whitish, drab or light brown, some are, however, brightly coloured and strikingly marked or banded, while others are so pellucid or translucent as to be almost invisible, except for the various internal organs seen as whitish areas. Most of the beautifully-coloured species belong to the Cotylea and are mainly found in tropical and warm-temperate waters, which are clear or much less turbid than waters beyond these areas, where polyclads are usually of dull appearance. In many respects, the highly coloured pseudocerotids resemble nudibranch molluscs in beauty of colouring and gracefulness of movement.

Colour in these worms springs from several sources. Superficial coloration arises from special cells lying in the dorsal epidermis and more frequently encountered among the cotyleans. Such cells are, however, not entirely responsible for pigmentation, as this may also be partly due to granules of pigment lying on either side of the basement membrane, in the musculature of the body wall or in the parenchyma, or to coloured materials in the excretory system or even to coloured bodies, such as zooxanthellae in tissues of coral, lying in the branches of the intestinal trunk and awaiting digestion. Tissues of prey undergoing digestion within the gut branches may also contribute towards the ground coloration. For instance, Crozier (1917) found that three sets of specimens of a species of *Pseudoceros* after feeding on ascidians of different colour had each adopted the colour of their prey. Moreover, in one instance, he found that the colour of the prey occurred not only in the digestive system of its polyclad predator, but also in its epidermis. *Cycloporus papillosus* is often yellow, green, red or black according to the colour of the ascidian upon which it has been feeding. Similarly, Levetzow (1943) reports that there are two colour varieties of *Thysanozoon brocchii* living in the Bay of Naples and mainly feeding on *Ciona intestinalis*, one being the light brown colour of its prey, the other violet, harmonizing with the colour of the *Botryllus* encrusting its fellow ascidian. Bock (1927*a*) records a specimen of *Apidioplana* having adopted the coloration of the gorgonian it had fed upon. Marcus (1947) mentions that, according to the algae among which they live, specimens of *Stylochoplana aulica* adopt the colour of the algae, but since the worms are probably not vegetarians, this comparable coloration is likely to have arisen from prey that had fed upon the weed. Similarly, some pelagic polyclads living among *Sargassum* having a brown and white coloration resembling that of most other animals associating with the weed.

According to Lang (1884), yellow, orange and red pigments usually occur in the dorsal epidermal pigment-cells, whereas pigments in the parenchyma are usually grey, brown or black. Variations in the predominance of various colours in a species sometimes appears to be influenced by the condition of the animal, for a well-nourished worm may be deeply coloured, whereas in an undernourished worm the colour may be weak and its markings much less apparent.

Nothing is known of the chemical properties of the pigments found in polyclads.

It is not yet known whether the brilliant coloration and distinctive markings together represent a defensive attribute, particularly as a warning of not being palatable. Furthermore, except in instances of worms living on encrustations of colonial animals, it is not known whether a highly-coloured worm moving around in search of food harmonizes with the nature of the bottom of its environment.

It seems, therefore, that coloration in polyclads may be natural or acquired from ingested prey.

REGENERATION Compared with the freshwater triclad turbellarians, polyclads have low powers of regeneration. Apparently, fragments containing the cerebral organ will regenerate into complete animals, whereas fragments from behind the cerebral organ do not regenerate anteriorly, but do so posteriorly. Moreover, post-cerebral fragments are able to crawl. It is not unusual to see the hinder half of a polyclad with the appearance of having undergone regeneration, but showing no trace of copulatory complexes. One often meets a specimen with a colourless or whitish patch of tissue in the marginal zone of the body, suggesting that replacement of a lost portion has taken place. In fact, Collingwood (1876) records an instance of a large wedge-shaped wound in the middle of the dorsal ridge of the body of *Pseudoceros hancockanus* which had closed up within 24 hours.

PARASITES Bacteria-like organisms have been recorded in the gut of polyclads, but their function has not been determined. These worms also appear to be common hosts of sporozoan Protozoa, the cysts of which are frequently encountered in the parenchyma. Gregarines often appear in large numbers in the gut branches, and *Ophiodina discocelides* is a common parasite of *Discocelis tigrina.* Orthonectids are also known to live in polyclads. Bock (1913) records the presence of very young cestodes, with four suckers and some indication of proglottid formation in the gut of *Stylochus orientalis.* These young cestodes were probably tetraphyllidean plerocercoids, the adults of which are commonly found in the spiral valve of elasmobranchs. Moreover, trematode metacercariae have been found free in the gut, or encapsulated in the musculature of polyclads, and Graff (1892*b*) records, but does not describe, a mature trematode in the gut branches of a planktonic polyclad, *Planocera simrothi.* The presence of larvae of parasitic platyhelminths in the gut and tissues implies that these hosts are intermediates in the life cycle of certain parasites that reach maturity in fishes, but it is likely that such infestations in polyclads have been acquired accidentally. Nematodes have also been found in the digestive system, but it was never stated whether they were parasites, or free-living forms accidentally ingested with food. It is possible, however, that small polyclads might prey on free-living nematodes and ingest them whole.

DISTRIBUTION *General remarks.* Very little is known of the effects of the chemical and physical properties of sea water on the distribution of polyclads. Nevertheless, as with other groups of marine invertebrates, their distribution is probably influenced by temperature, by salinity, by the degree of light penetration, by the nature of inshore and offshore water currents, as well as by certain ecological factors and climatic conditions. Of these influences, temperature is likely to be the major environmental element determining the range of a species, which may be stenothermal or eurythermal. The unevenness of coast lines probably has some influence on distribution, for the physical properties of waters in estuaries, bays and harbours differ somewhat from those of waters in the open sea. Being generally negatively phototactic, polyclads seek to hide from direct light, so if a stretch of sandy shore, for instance, is without rocks, stones or other objects under which to hide, the worms will avoid it. That the distribution of some species is limited by physical factors is shown by the boreal species *Notoplana atomata*, for the southern limit of its range occurs at about 45°N in the eastern Atlantic, about 41°N along the eastern coast of North America and about 47°N along the western coast.

A few groups seem to have a preference for certain regions—the euryleptid subfamily Laidlawiinae for the Antarctic, the families Latocestidae and Pseudocerotidae for tropical and subtropical regions–whereas other groups favour no particular region, but do avoid the polar latitudes. A few species appear to be widely distributed, but none are cosmopolitan.

It is implicit that species with a direct life cycle and young having little pelagic life are not so widely dispersed as those species with an indirect life cycle and a larva with a planktonic existence. But this is not necessarily so, because the varying physical conditions of the seas seem to restrict many of the latter species to certain regions, as will be exemplified below.

The greatest number and diversity of genera and species, although perhaps not in the number of individuals of a species, occur in tropical zones, especially among coral reefs that offer an abundance of food, as well as holes and crevices in which to hide from direct light and possible predators. Over 900 species of polyclads have been described and over half of these were found between latitudes 30°N and 30°S and about two-thirds of these in the Indo-west Pacific region. This region, apparently part of the Tertiary Tethys Sea, might have been the centre of the distribution of modern polyclads. There is a high degree of speciation among polyclads in tropical and subtropical regions, and this may be due to an ability to cope adequately with habitats constantly subjected to changes effected by the many climatic factors prevailing in these regions.

Clearly, polyclads are essentially animals of warm waters, and the number of species recorded so far gradually declines towards the colder regions. In fact, only about five species are known to occur in the Antarctic and a similar number in the Arctic. That temperature has much influence on the distribution of these worms is further indicated by the fact that coasts along which cold-water currents flow harbour definitely fewer species than coasts washed by warmer waters.

It also seems likely that the polyclad fauna of the southern hemisphere is not as varied as that of the northern, because the coastal regions of the former are not so vast as those of the latter.

The specific components of the polyclad fauna of an area, particularly an island, may occasionally change for some reason or other. For instance, Hyman (1939*d*) found that polyclads collected in the Bermudas in 1935 were broadly different from those collected in the same area by Verrill in 1898 and 1901. Miss Hyman, however, suspected that the change was probably brought about by pollution 'of the waters by chemicals released from a shipwreck'. It should also be mentioned that there was a sudden lowering of sea temperature in Bermuda in 1901, destroying much sea life among the islands.

Unfortunately, insufficient is known of the biology of polyclads to judge how they deal with the varying demands of the marine environment, but since they undoubtedly belong to an ancient group of animals, it may be accepted that they cope satisfactorily. Simonette &

Delle Cave (1978) have described a fossil from the Middle Cambrian thought to be that of a polyclad.

Vertical distribution. Polyclads are mainly inhabitants of coastal waters, ranging from high-water mark to depths reaching to the edge of the continental shelf. A few species have been found at deeper levels, and others are entirely pelagic. Very little is known of their vertical distribution, but it does seem that many species, although normally inhabiting shallow sublittoral waters, frequently wander into the intertidal zone, sometimes in search of food, sometimes to spawn. Other species, however, prefer to remain below the low-water mark and seldom, if ever, venture into the intertidal zone.

These flatworms may also have their vertical distribution partially controlled by their preference for certain ecological factors. Some species prefer to live among green seaweeds, others among brown weeds, and others among red weeds: thus forms having such preferences may be allotted levels matching the zonation of the weeds. Others of these worms choose particular types of sea floors, such as mud, sand and mud, shells, gravel and so on, whereas others prefer to settle among colonies of encrusting animals.

The effect of temperature is shown, on the one hand, by the interesting eurythermal species *Notoplana atomata,* for it extends from the intertidal zone down to a depth of 200 metres in the coastal waters of northwestern Europe, and during the months of July and August along the Atlantic coast of North America from Newfoundland to Massachusetts, it has been found in intertidal pools with temperatures ranging from 9·5°C to 24°C. The stenothermal species *Discocelides langi* and *Cryptocelides loveni* are, on the other hand, known only from well below the low-water mark in the cold-temperate or boreal region of the north-east Atlantic, where temperatures are less variable.

As mentioned above, many polyclads are negatively phototactic, thus their vertical movement may also be influenced by the strength of light passing into the water.

As a matter of interest, Bock (1913) states that he found *Leptoplana tremellaris* only in the littoral zone of the west coast of Scandinavia, but in other parts of Scandinavia it has been collected at depths to 10 metres. In the English Channel, the species is found in the intertidal zone and to depths of 60 or 70 metres, and in the Mediterranean to 100 metres. It is known, however, that it moves into the intertidal zone to breed. Thum (1974) likewise found some indication of the sexual activity of *Notoplana acticola* being associated with the position of the worm at varying levels in the tidal zone; genuine hermaphrodites were generally more numerous in the lower tidal zone. Moreover, it appears that species commonly found below the low-tide level in the Gulf of Mexico are usually found in the intertidal zone further north along the Atlantic coast of North America.

Most oceanic polyclads are epipelagic, but Palombi (19242b) records forms from the mesopelagic of the Caribbean.

Horizontal or geographical distribution. Information on the geographical distribution or zoogeography of polyclad species is relatively scarce. In fact, the only countries where they have been studied to any extent are the Mediterranean (Lang, 1884), Scandinavia (Bock, 1913), the Atlantic coasts of Canada and the U.S.A. (Hyman, 1940*a*), the Pacific coasts of Canada and the U.S.A. (Hyman, 1953*a*), the Brazilian coast (Marcus and du Bois-Reymond Marcus, 1947–1958) and Japan (Kato, 1934–1944). Furthermore, not a great deal of attention has been directed towards their systematics and to the degree of morphological variation within individual species.

Since there is a lack of information on the polyclad faunas in many parts of the world, it is difficult to recognize zoogeographical areas of distribution, except perhaps to support the four tropical shelf-regions: the Indo-west Pacific, the eastern Pacific, the western Atlantic and the eastern Atlantic (Briggs, 1974). Generally, species found in one region appear to be different from those occurring in others. The four biotic regions have been thought to have emerged around the early Cretaceous, and zoogeographers have divided each of these regions into provinces, each province possessing a fauna partly endemic to it.

There is, however, sufficient information available on the polyclad fauna of the North Atlantic Ocean to recognize patterns of zoogeography. From these patterns, it is possible to determine an eastern North Atlantic cold-temperate or boreal province extending from northern Norway to the English Channel, southward of which a warm-temperate or Lusitanean province extends to the Mediterranean and Black Sea, the coasts of Morocco and the Cape Verde Is. Likewise, there appears along the Atlantic coasts of Canada and the U.S.A. to be a boreal province extending from Baffin Bay southward to the neighbourhood of Cape Cod. It also appears that the Beaufort–Cape Hatteras region of North Carolina is the northern limit of species in the warm-temperate or Carolinian province, with the area between there and Cape Cod a transitional one holding a mixture of warm-temperate and cold-temperate species.

The distribution of polyclads in the tropical waters of the western Atlantic indicates that there is a faunistic area extending from the Gulf of Mexico to southern Brazil, including the Antilles and the Bermudas. This area may constitute a West Indian Province in polyclad zoogeography.

With the exception of *Notoplana atomata*, the only known amphiboreal species, the 20 or so European boreal species have not been found in the North American Atlantic boreal. Some species occurring in the Mediterranean have been collected in the English Channel and among the Cape Verde Islands, but, with the exception of the widely-distributed *Thysanozoon brocchii*, none has been found on the American coasts. It would seem, therefore, that there are barriers in the North Atlantic preventing an interchange of European and American temperate polyclads.

A study of the distribution of the polyclads of the Pacific coast of North America from Lower California to Alaska shows the existence of two or three faunal provinces: an Aleutian Province extending from the Alaskan Peninsula to Puget Sound, and corresponding to the boreal of the North Atlantic; the Californian Province extending roughly between Oregon and Lower California; and the Cortez Province embracing the Gulf of California.

From available information, it appears that the most widely-distributed families are the Stylochidae, the Leptoplanidae, the Pseudocerotidae and the Prosthiostomidae, although none has been found in Antarctic waters. The Stylochidae seem to be restricted to waters between 50°N and 50°S. About 90 species have been assigned to this family and two-thirds of them occur in the Indo-west Pacific region. The Pseudocerotidae and the Prosthiostomidae are essentially inhabitants of warm-temperate and especially tropical seas. The Leptoplanidae appear to be evenly distributed in the Atlantic, Indian and Pacific Oceans, with the genus *Notoplana* ranging from the Arctic through the tropics to about 45°S.

The family Euryleptidae is also interesting zoogeographically, for it constitutes eight or nine genera containing over 70 species, of which about 90 per cent are known from the Atlantic and eastern Pacific regions. One species of *Eurylepta*, either a subspecies of *E. cornuta* of the north-east Atlantic and Mediterranean, or a distinct species, has been recorded from Antarctica. The euryleptid subfamily Laidlawiinae consists almost entirely of species endemic to the Antarctic.

Genera containing several species show varying patterns of distribution. The tropical and subtropical cotylean genus *Pseudoceros*, for instance, has had about 120 species assigned to it, and of these 7 have been recorded from the Pacific coast of North America, 14 from the Atlantic and nearly 100 from the Indo-west Pacific. In contrast, the acotylean genus *Notoplana* includes about 75 species, and these are, as mentioned above, widely but evenly distributed in all oceans except the Antarctic. Another interesting fact concerning the genus *Pseudoceros* is that although a very large number of its species occur in the Indo-west Pacific region none seems to have found its way into the Atlantic, except *P. spendidus*. This species has been recorded from Vietnam, the Mediterranean, the Bermudas and the Galapagos, a wide longitudinal distribution difficult to explain. It would seem that the cold waters of the West-Wind or Antarctic Drift and the Benguela Current off South Africa inhibit the migration of Indo-west Pacific species and their pelagic larvae round the Cape of Good Hope into the eastern Atlantic. Likewise, the Humboldt or Peruvian Current along the west coast of South America inhibits the tropical and subtropical species of the eastern Pacific from entering the Atlantic round Cape Horn or through the Magellan Straits.

From a review of the literature, it seems that the most wide-spread species of polyclad is the pseudocerotid *Thysanozoon brocchii.* It is an inhabitant of warm-temperate and tropical waters, ranging in the Atlantic from the Mediterranean to Florida, through the Caribbean to Brazil, and in the Indo-Pacific from Japan westwards through the Indo-Malaysian region to False Bay, South Africa and through the western Pacific Islands to New Zealand. It is not yet possible to understand how this pattern of distribution has arisen, but pelagic larvae of the genus *Thysanozoon* have been found in the open sea, thus suggesting that *T. brocchii* is distributed by ocean currents. Nevertheless, the distribution of the adult of this species indicates an aversion to cool waters, and in view of the cold-water barriers in the South Atlantic, it again seems unlikely that there is an interchange of the larvae or adults of this species between the Atlantic and the Indian or Pacific Oceans by means of water currents. In view of the apparent dislike of cool waters, and if existing records are correct, it is difficult to explain satisfactorily the presence of *T. brocchii* in the Indo-Pacific on the one hand and in the Atlantic on the other. Speciation in the genus *Thysanozoon* is considerable in the Indo-west Pacific region, which hints at this region being the centre of its dispersal. Possibly the genus entered the Atlantic from the Pacific at a time when South America was isolated from the North, for, with the exception of *T. Brocchii,* all known species of *Thysanozoon* in the Atlantic seem to occur only in the region between Florida and northern Brazil. *T. brocchii,* like *Pseudoceros splendidus* mentioned above, is common in the Mediterranean, and this perhaps indicates that both species are relicts of the Tethys Sea and spread westwards with the formation of the Atlantic Ocean during Tertiary times.

Evidence of the free movement of tropical polyclads between the Pacific and the Atlantic, prior to the formation of the Panamanian isthmus some time in the late Tertiary, is shown by the fact that of the 48 genera known to occur in the tropical western Atlantic, 22 of these occur in the eastern Pacific. It should, however, be noted that the polyclad fauna of the Pacific coast of the isthmus has been little explored.

Widely-separated geographical regions sometimes show a general similarity in ecological conditions, and yet they foster widely-differing species of polyclads, indicating that there are natural obstacles to free migration. Nevertheless, in recent years a few species have shown vast gaps in the continuity of their geographical distribution. For instance, *Euplana gracilis* is very common along the Atlantic seaboard of North America, from Prince Edward Is. to Florida, but is now known from the Victorian coast of southern Australia, and *Echinoplana celerrima,* a common species on the coasts of New South Wales and South Australia, has recently been found in the Mediterranean. It seems probable that species may be artificially transported as adults, juveniles or egg masses among encrustations on the hull of ships from the endemic areas to new locations. Plehn (1896*a*) records *Stylochus pilidium,* an inhabitant of the Mediterranean, having been found in Chilean waters on the bottom of a ship from Italy. There is also the likelihood of worms as eggs or otherwise being transported with oysters for cultivation in other parts of the world, and Hyman (1955*a*) cites *Pseudostylochus ostreophagus* being imported with seed-oysters into Puget Sound, U.S.A., from Japan. The worms soon became a successful addition to the fauna of Puget Sound, and it was estimated that in one locality there were 600 000 worms per acre.

Probably the most interesting of polyclads showing discontinuous distribution is *Stylostomum ellipse,* for it is common along the European coast from Spitzbergen to the Mediterranean, and Bock (1913) has recorded it from the lighthouse at Cape Town, South Africa, and from Tierra del Fuego, the Falkland Isles and South Georgia. This intermittent distribution is puzzling, but may have arisen through egg masses being transported by some means or other. The genus *Stylostomum,* however, is a cotylean and supposedly has an indirect life cycle involving a planktonic Müller's larva, which might be swept by cold deep-water currents from one subpolar region to the other. There is also the possibility that the boreal and subantarctic forms of *S. ellipse* are descendents of two species, which lived in very similar ecological conditions and evolved by convergence into two forms with identical morphological features. Clearly, the question of discontinuous distribution, such as that of *S. ellipse*, is open to considerable speculation.

Concerning species of the polar regions, so far only two species, *Plehnia arctica* and *Notoplana kuekenthali*, both acotyleans, may be regarded as endemics of the Arctic, whereas five species, all cotyleans, are endemics of the Antarctic and are possibly relict forms. The scarcity of species in the polar regions does suggest that the limit of distribution of each species of polyclad is regulated by its physiological response to the physical elements of the environment.

About 20 species have been found floating in the sea, but whether they should all be regarded as entirely pelagic is questionable. Some ten or twelve species are certainly so, and of these *Planocera pellucida*, *P. simrothi*, *Gnesioceros sargassicola* and *Hoploplana grubei* appear to be common and widespread in the central and western regions of the North Atlantic, where they frequently occur among floating *Sargassum* weed. *P. pellucida* has also been found in the South Atlantic, Indian and Pacific Oceans, while of the other species only *G. sargassicola* has been recorded as occurring beyond the North Atlantic, for a single specimen was found on driftwood off the New Guinea coast (Graff, 1892*a*), and an invasion of the species in the Gulf of California has been recorded by Cheng & Lewin (1975). *G. sargassicola*, *Acerotisa notulata* and one or two other species frequently found among floating *Sargassum* also occur in the intertidal zone of the Caribbean region.

It remains to be mentioned that the Indo-west Pacific is usually considered to be exceedingly rich in marine littoral and sublittoral invertebrates, but comparatively little has been done to elucidate its polyclad fauna, and only a few scattered accounts have so far been published.

To sum up, the available information on the vertical and horizontal distribution of polyclads is so scanty that it is difficult to interpret what effects physical and other conditions of the seas and oceans have on these planarians, and at the moment one is left to assume or speculate on the influences behind the differing patterns of distribution.

CLASSIFICATION *Phylogenetic notes.* It is generally accepted that polyclad turbellarians have an ancient lineage, but their origins cannot be determined by the study of fossil polyclads since only one, *Platyendron ovale* Simonetta & Delle Cave, has been described. Since, however, this shows only a vague indication of what may be dendriform intestinal branches, the evidence for it being a polyclad fossil is extremely superficial. The lack of palaeontological evidence offers room for speculation and so theories of their origin have from time to time been developed from investigations into comparative anatomy and development. The first of these theories to appear was the 'ctenophore-polyclad theory' expressed by Kowalewsky (1880), discussed by Selenka (1881*c*) and Chun (1882), and refined by Lang (1884). The roots of this theory lie in the similarities in the structure and embryology of polyclads and two aberrant genera of platycteneid ctenophores, *Coeloplana* and *Ctenoplana*, both of which are dorso-ventrally flattened and modified for creeping. Lang pointed to similarities in the presence of a pharynx, two dorsal tentacles, aboral sense organs and epidermal gland cells, together with certain features of the nervous system and some resemblance between the eight ciliated plates in ctenophores and the eight ciliated lobes of the free-swimming Müller's larva of cotylean polyclads. There is also a degree of resemblance in development, both ctenophores and polyclads having eggs of determinate cleavage with the production of micromeres and macromeres. The implication of Lang's reasoning is that all other groups of Turbellaria have sprung from the Polycladida. Hadzi

(1944) transposed the basis of the theory that polyclads were derived from ctenophores by contending that the Ctenophora had evolved from the Polycladida, principally through the neotenous retention of certain structural elements of Müller's larva. There are also certain fixed features in the cleavage of eggs of both groups, but the ctenophore egg undergoes a biradial type of cleavage, as found in sponges, coelenterates and echinoderms, whereas the polyclad egg undergoes spiral cleavage, as found in nemertines, polychaetes and some molluscs. All things considered, the early development of a ctenophore does not closely approach that of the polyclad, as it might do if there were a close relationship. There is also the fact that the neotenic Müller's larvae of the cotyleans *Graffizoon lobatum* and *Cycloporus gabriellae* bear no resemblance to any stage in the later development of ctenophores.

It is now accepted by authorities on Turbellaria that ctenophore similarities with the polyclads are merely due to convergence and have no evolutionary significance, and Hadzi (1958) himself discards any theory connecting the Ctenophora with the Polycladida in favour of one in which the Turbellaria, through the order Acoela, derive directly from multinucleate ciliate Protozoa.

Another theory on the origin of the Polycladida was offered by Graff (1882), who considered that the order Acoela was the most primitive of the Turbellaria and gave rise directly to the Polycladida. He presented the so-called planuloid-acoeloid theory, and pointed out that acoel turbellarians resemble the planula stage of coelenterates in many respects, and that the acoels originated from an acoeloid planula-like ancestor. Hyman (1951) details the supposed evidence of a phylogenetic relationship between the Coelenterata and the Acoela by way of the planula larva. There is, however, one important embryological difference, because the cleavage of the egg is determinate, equal and radial in planulae, whereas it is determinate, unequal and spiral in the Acoela. However, the classification of cleavage-types might have little significance in the systematic grouping of animals, as exemplified by the difference in cleavage-types of the Cephalopoda and the remainder of the Mollusca.

It is, perhaps, worth noting that Bock (1922) points out that there are certain histological features in the body wall of chromoplanids which are unusual in polyclads and remind one of the proseriate (alloeocoele) turbellarians.

Recent investigators into the phylogenetic relationships within the Turbellaria accept neither the Acoela nor the Polycladida as the stem form of the genealogical tree of the Turbellaria, and indicate that it is not possible at present to trace one turbellarian order directly from another. It is, however, thought that the polyclads should be placed in the lower half of such a tree, perhaps forming a branch of a stem arising from a turbellarian archetype (Ax, 1963).

Since it appears that the origin of the Polycladida has not yet been elucidated, phylogeny within the order is likewise obscure. There have, however, been suggestions made over the years that certain structural features are primitive, the most prominent of these being the central position of the mouth and pharynx. Bresslau (1928–33) appears to accept this and states that the family Anonymidae represents the primitive form of the Cotylea and the family Planoceridae the primitive form of the Acotylea. It is interesting to note that the larvae of those forms with an indirect life cycle and the newly-hatched worm of those with a direct life cycle, both have the mouth placed centrally or in the posterior third of the body, thus suggesting that the more or less centrally-placed mouth is a primitive condition.

It is also worthy of note that in *Discocelides langi* the intestinal trunk passes into an unpaired anterior, median, intestinal branch and its hinder gives off two powerful posteriorly-directed intestinal branches. This reminds one of the three main intestinal branches found in proseriate and seriate (triclad) turbellarians.

The longitudinal nerve cords and their connections, and the branching intestinal system, have been quoted as relicts of radial symmetry, but it is more likely that these features are associated with the extreme dorso-ventrally flattened body of polyclads and have no phylogenetic importance.

It has also been suggested that the presence of eyes on the margins of the body is a primitive feature among polyclads, having arisen from a radiate ancestor.

In the opinion of Laidlaw (1902), 'tentacles of the Polyclads, whether marginal or nuchal, are a structure peculiar to the group' and that 'they originated as anterior [marginal] organs in connection with eye-spots', retained in the Cotylea, and that 'they have shifted back from the margin and come to lie dorsally' in the region of the cerebral organ, as in the Acotylea. The position of the cerebral organ posterior to the anterior margin of the body is constant throughout the Polycladida.

It seems possible that earlier polyclads had a highly complex male copulatory system and that relicts of this system, especially the prostatoids, have persisted in certain recent forms, such as those found in the acotylean families Discocelididae and Polyposthiidae, and the cotylean families Anonymidae and Boniniidae.

Clearly, to develop a satisfactory system for a 'natural classification' of the Polycladida from existing information is exceedingly difficult, if not impossible. Available information on functional morphology in polyclads is scanty and little is known of the various aspects of their biology.

Systematics. Early workers on the systematics of the Turbellaria tended to lump the polyclads with the triclads to form a hotch-potch of genera to which Ehrenberg (1831) gave the name Dendrocoela, a group later variously regarded as an order, a suborder or a tribe of the Turbellaria. This uncertainty persisted until Lang (1884) separated the polyclads from the remainder of the Turbellaria and erected for them the suborder Polycladidea in the order Dendrocoela. He proposed an entirely new classification for the species he had placed in the Polycladidea and grouped them into two tribes, the Cotylea and the Acotylea. These tribes were established chiefly on the presence of a muscular ventral sucker, placed in the middle region of the body and posteriorly to the female copulatory complex, and a pair of tentacles on the anterior margin of the body, features characteristic of the Cotylea, whereas these features are absent in the Acotylea. Hallez (1893) emended the name Polycladidea to Polyclada and regarded it as a class of the order Turbellaria, whereas Gamble (1893*a*) emended it to Polycladida–a name now universally accepted–and regarded it as a suborder of the Turbellaria. Laidlaw (1903*a*) lists the Acotylea and the Cotylea as suborders, thus implying the Polycladida as an order. Shortly after Laidlaw (1903*e*) reviewed the systematic relationships of acotylean genera known at that time, and, although disagreeing with Lang in some respects, retained the Acotylea and Cotylea without comment. He also gave a diagnostic table for the acotylean genera and emphasized the importance of the reproductive system in classifying these genera. Meixner (1907*b*) generally accepts Laidlaw's acotylean families and disagrees only in the constitution of the family Planoceridae. Bock (1913) reviewed work done on the systematics of the Polycladida since Lang and paid attention to Laidlaw's classification of the Acotylea. He added new families and genera, and divided the suborder into three sections, the Craspedommata, the Schematommata and the Emprosthommata, solely on differences in the distribution of the eyes. These sections were raised to the rank of superfamilies by Poche (1926) as Stylochides, Planocerides and Cestoplanides, respectively. These latter names have been emended to Stylochoidea, Planoceroidea and Cestoplanoidea in accordance with Recommendation 29A of the International Code of Zoological Nomenclature (1961).

Bock also briefly reviewed the Cotylea and pointed out that several genera placed in the suborder following Lang's work might well be included in either suborder. In fact, the family Enantiidae appears to be a link between the two suborders, and Bock (1923*b*) implies that the family Boniniidae is similarly linked. Nevertheless, the presence of a distinct ventral sucker centrally situated and posterior to the female pore, or a ventral adhesive depression placed in the posterior region of the body, still appear to be basic features supporting the retention of the Cotylea.

It should be mentioned, however, that two genera *Nymphozoon* and *Simpliciplana*, both said to be without any form of ventral adhesive organ, have been placed in the cotylean family Pseudocerotidae, of which they are typical in their gross morphology. Hyman (1959*a*) described *Nymphozoon bayeri* and said 'A sucker is definitely wanting', but it does seem possible that individual openings in a median row of several female copulatory complexes might also function as suckers. Kaburaki (1923*b*) in describing *Simpliciplana marginata* says that it 'appears to be devoid of any ventral sucking disk', but he accepts that his specimen 'is too incomplete to admit of a satisfactory diagnosis'. Superficially, *Simpliciplana* undoubtedly appears to be a pseudocerid, and this suggests that a ventral sucker might have been lost accidentally. Similarly, *Amakusaplana ohshimai* is said to be without a ventral sucker, but the gross morphology of this species is undoubtedly that of the cotylean family Prosthiostomidae. Two species of the genus *Diposthus* apparently lack a ventral sucker, but both have a pair of marginal tentacles with eyes in and around them, typical of the Cotylea. Occasionally, the ventral sucker is small and may be withdrawn into the body to such a degree that all outward signs of its existence disappear, particularly at fixation. Moreover, it sometimes happens that when a collector prizes a cotylean off a surface to which it is attached, the sucker is torn off and the wound is often difficult to detect.

The anatomical features upon which modern classifications of the families and genera of polyclads have been proposed are mainly the arrangement of the eyes and the structure of the copulatory complexes. There are, however, instances of where the recognition of genera is difficult, because in some families, particularly those of the Cotylea, the copulatory complexes are practically identical in structure and the genera and species are recognized largely by superficial features. Among the Acotylea, however, the recognition of such taxa is rarely possible by superficial features alone, and at present a study of the structure of the copulatory organs is essential.

Unfortunately, our knowledge of morphological variation among individual species is almost non-existent. The descriptions of so many of them have been based solely upon the examination of just one specimen, sometimes not fully developed and sometimes poorly preserved.

Concerning classification, Wharton (1938) holds the opinion that 'a thorough knowledge of life histories is important in the formation of a natural classification'. He further suggests that three types of larvae are found among polyclads, 'and it is therefore probable that there should be three main groups of the order Polycladida'. These worms have a direct and an indirect type of development, the latter with two kinds of larvae—Götte's larva (the primitive larva according to Wharton) and Müller's larva, both of which undergo metamorphosis. The direct type of development and Götte's larva, are so far known only among the Acotylea, but the type of life cycle does not allow for two distinct groups within the suborder, for each of the genera *Stylochus* and *Notoplana* embrace both forms of life cycles among their morphological species. The free-swimming Müller's larva appears at present to have some systematic importance, since it has so far been found only among cotyleans. But, in the acotylean *Planocera reticulata*, the embryo develops into Müller's larva, which metamorphoses within the egg capsule and hatches in a form resembling a miniature adult, thus showing a type of life cycle intermediate between the direct and the indirect. It would seem, therefore, that in the systematics of polyclads the type of life cycle has apparently only specific importance among acotyleans.

Apart from the presence or absence of organs of attachment and the disposition of eyes, the classification of the Polycladida is based on certain features of the copulatory complexes, and it is perhaps worth considering the value of certain superficial features along with other organ-systems in the systematics of these worms.

There are indications that the outline of the dorso-ventrally flattened body has importance in the recognition of some families. The body may be elongate or ribbon-like (Latocestidae; Prosthiostomidae), broadly oval or almost discoid (Planoceridae), oval, tapering somewhat at both ends (Polyposthiidae), oval, frequently widened anteriorly and somewhat tapering posteriorly (Leptoplanidae). At fixation, however, the normal shape of the body may be entirely lost through excessive contraction. Moreover, the external appearance might differ according to the sexual stage reached by the worm. For instance, Marcus & Marcus (1951) find that in the male phase of *Latocestus ocellatus* the body is thin and narrow (1·5 mm wide), but in the female phase it is thick and broad (4 mm wide).

Size is a feature which might be thought to have some taxonomic weight, but this feature should be considered with reserve. Among some families mature specimens of a species measure from 9 mm to 40 mm in length, and Yeri & Kaburaki (1918*b*) found specimens of *Planocera reticulata* measuring between 10 mm long and 6 mm wide and 80 mm long and 45 mm wide, all having copulatory organs. Adult specimens of *Notoplana australis* in a preserved condition have been recorded as measuring 15–37 mm long, but living worms have been found to reach a length of about 76 mm.

The colour and markings of the dorsal surface of the body appear at present to be the only means by which many cotylean species have been determined, and this is especially so among species of the genus *Pseudoceros*. But even these features have their limitations, because little is known of variation in colour and markings. It is worthy of note that Lang (1884) found a specimen of *Pseudoceros splendidus* with an orange-yellow band around the margin of the body, but when the specimen was killed with hot sublimate the band suddenly became blood-red and a similarly-coloured fluid issued from the body. An instance of colour variation may be found in the acotylean *Callioplana marginata* for its dorsal surface is reddish brown to jet-black with a margin of tawny brown, orange, pink or white. It is known to the present writer that variation in colour and markings will be found in *Pseudoceros zebra* and *P. fuscopunctatus* and that preservation may either alter the depth of colour or completely destroy it.

The presence or absence of nuchal tentacles may be regarded as taxonomic features, but these are likely to occur only when the cerebral and tentacular eye-clusters are distinctly separated from each other. Sometimes these tentacles are withdrawn into the body and may only be detected by translucent areas dorsal to the tentacular eyes.

The arrangement of the eyes is generally accepted as systematically important, but again caution must be exercised in the use of this feature. In the superfamily Cestoplanoidea and in the family Prosthiostomidae, eyes are confined to anterior region of the body, and on close examination they leave the peripheral zone completely free. However, if the body contracts at fixation, the eyes will often appear to be marginal instead of submarginal. Generally, the arrangement of the eyes is important for generic and specific determination, especially among some cotylean families. Again, caution must be employed in the use of the distribution of marginal eyes, for Bock (1913) found in *Discocelides langi* that the young have marginal eyes all round the body, but in older specimens such eyes are confined to the anterior region. A similar variation was recorded by the present writer (1977) for a stylochid species. On the other hand, in some members of stylochoid families eyes are minute and might be overlooked.

The disposition of the cerebral and tentacular groups of eyes is also valuable taxonomically, but the number of eyes in each group often tends to increase with body growth.

There is very little difference overall in the structure of the alimentary system. The mouth and pharynx are placed at various points along the median line, and their positions have varying degrees of systematic importance. In the genus *Stylochus* both almost always occur in the hind region of the body, and in the genus *Emprosthopharynx* they occur anteriorly. The position of these structures has some significance at family level in the Cotylea. The ruffled type of pharynx predominates, but a tubular or bell-shaped pharynx occurs mainly among the cotylean families. In the description of some species, attention has been given specifically to the number of pairs of lateral folds of the ruffled pharynx when enclosed in the pharyngeal chamber. In some species the number is fairly constant, but in others the number increases with the growth of the body. The intestinal trunk gives rise to several pairs of ramifying lateral branches, which may or may not anastomose, the systematic importance of which has not yet been determined. It does, however, seem that in some families the branches do not anastomose, whereas in others they do, and in others both conditions occur. The fact is that this feature of the alimentary system requires a much closer study than it has so far received. According to the literature, the intestinal branches among species of the genus *Notoplana* may or may not anastomose, but the present writer has often found in this genus that the branches anastomose in the vicinity of the intestinal trunk, but do not in the submarginal zones of the body.

The nervous system is little known, but it does seem that differences in the pattern of the nerve net might have some taxonomic importance, for there is a noticeable difference in the number of nerve cords found in the Leptoplanidae and that found in the Planoceridae.

Lang (1884) grouped polyclad spermatozoa into three types: filiform, flagellate and flanged. Kato (1940) found that the spermatozoa of eight species representing four families of acotyleans were all of the filiform type. A point of possible systematic interest is that Kato found the spermatozoa of *Stylochus aomori* have a distinguishable head, middle piece and tail, whereas Thomas (1970) found no such divisions in the spermatozoa of *Stylochus zebra.*

The classification of genera and species has, since Lang's monograph, been based mainly on the morphology of the copulatory organs, which may be satisfactorily determined only when examined in longitudinal (sagittal) serial-sections. However, the copulatory complexes are so similar in some families of the Cotylea that generic and specific determination based on these complexes alone is difficult.

Unfortunately, the present methods of fixation almost invariably cause contraction of the body. This usually alters to positional relationship of the organs of these complexes, which may, therefore, appear to vary in specimens of the same species. One must consequently try to visualize the normal relationship of various organs to one another.

There are also problems arising from functional morphology, because some taxonomically important features of the copulatory complexes are not always apparent until full sexual maturity has been reached. As an item of caution, it may be mentioned that a feature sometimes used diagnostically is whether the vasa deferentia open independently into the seminal vesicle, or whether they unite to form a simple duct opening into the vesicle. In some forms the vasa deferentia always open independently, irrespective of the degree of maturity attained, whereas in others the vasa deferentia in the early stage of the male phase unite to open into the seminal vesicle by a common duct, but as the vesicle becomes swollen with sperm, so the duct disappears, and the vasa deferentia appear to open into the vesicle separately.

A further point worth considering is the course of the vasa deferentia, which might be of much diagnostic importance, but often it is seen in its fullest extent only when the male phase is predominant.

Several species have been described as being without any form of prostatic organ. There must be doubt as to the truth of this in many instances, for in some leptoplanids a mass of extracapsular gland cells invests a portion of the ejaculatory duct and pass their eosinophilic secretions into the lumen of the duct, which functions as a tubular prostatic organ. The present writer has found that in some instances these gland cells do not become noticeable until the male complex is fully developed, thus a leptoplanid with a distinct, but not fully functional male complex might be said to be without a definite or implied prostatic organ. Another instance of a species having a diagnostic feature which might not appear until a height of sexual development has been reached is found in the cotylean *Oligoclado floridanus.* In this species, each uterine vesicle gives off a uterine duct which opens to the exterior through a common mid-ventral pore, but it is present only in the most mature specimens, sometimes not even in specimens with eggs in the uterine canals and sperm in the vasa deferentia. Again, the cotylean *Diposthus corallicola* possesses a pair of spermiducal bulbs which may disappear during the female phase of the worm.

The structure of the female copulatory complex has limited value in the classification of families of the Acotylea, but has more importance among the Cotylea.

The karyological aspect of polyclad systematics has not been seriously investigated, but what little has been done suggests that the chromosome complement in some families might be fairly uniform, whereas in others the complement seems to be variable.

Lang (1884) included in his monograph brief descriptions taken from the original accounts of many species published prior to his work. Early descriptions of polyclads, except those of the more colourful forms, are most unsatisfactory for determination by modern investigators, as details of internal structures were very rarely noted. Consequently generic determination of the species is often a matter of considerable uncertainty. The existing nomenclature within the Polycladida would be in disarray if the Articles laid out in the International Code of Zoological Nomeclature were applied. For instance, Grube (1840) described a new genus and species under the name *Orthostomum rubrocinctum,* but for no apparent reason Lang erected the genus *Cestoplana* for it and it is now known as *Cestoplana rubrocincta* (Grube, 1840) Lang, 1884, but according to the Code *Cestoplana* is a synonym of *Orthostomum.* In another instance, Lang accepted the genus *Discocelis,* erected by Ehrenberg (1836) for *Planaria lichenoides* Mertens, 1832, an unrecognizable species. He clearly based his definition of *Discocelis* on the morphology of the species *Polycelis tirgina* Blanchard, 1847, which is certainly not a member of the genus *Polycelis* Ehrenberg, 1831 (a triclad), nor is it congeneric with *D. lichenoides* (Mertens). Yet, his conception of *Discocelis* has been accepted for several decades, thus usage is the dominant factor in this problem, and the genus is here presented as: *Discocelis* Ehrenberg, 1836, *sensu* Lang, 1884, implying the problem involved.

There are many nomenclatural difficulties of this kind to be found among the Polycladida, and strictly such cases should be submitted to the International Commission of Zoological Nomenclature for resolution, but it is very likely that usage would be the sole factor in all decisions. Usage is therefore applied to such problems met with in the present synopsis.

In conclusion, since it does not seem possible at present to determine the evolution of the families and genera of polyclads, the classification adopted herein follows that accepted by all authorities in placing the Acotylea before the Cotylea. It should, however, be realized that there is a degree of artificiality in recognizing these two suborders.

PRACTICAL METHODS The difficulty in dealing technically with polyclads is probably the reason why they have been largely ignored by taxonomists.

Collection. Polyclads may be found between the tide marks under stones, in crevices in rock pools and among seaweeds and their holdfasts. In fact, anywhere that shields them from direct light. Many of the forms frequenting the intertidal zone are translucent and may easily be overlooked by the inexperienced eye, unless careful searching is employed. They may also be obtained by collecting seaweeds, scrapings from encrustations of colonial animals and from piles of wharfs, etc., and by dredging in sublittoral waters and placing this material in a container of sea water, leaving it undisturbed for a time to allow any planarians present to swim to the surface of the water or crawl to the sides of the container.

Pelagic polyclads must be carefully searched for among *Sargassum* weed, on driftwood and other flotsam and in tow-nettings.

These animals require gentle handling, as they are fragile and tend to break-up if treated roughly. The smaller forms must be lifted off the surface over which they are moving with a camel-hair brush and immediately placed in a dish of sea water. Larger forms should have a spatula, a section-lifter or a thin-bladed knife placed in front of them, and when the worm moves over the blade, it is then lifted gently into a suitable container of sea water.

Living worms should be examined with a hand-lens, the larger the better, and notes made of size, body shape, coloration, the form and position of tentacles, if present, and other noticeable features. Appearance in life can seldom be preserved and some features observed in the living animal may be completely lost at fixation, and at that time the body may become distorted. The best means of recording the coloration and markings of the living worm is photographically, using colour transparencies. Notes on habitat, temperature of water and its degree of salinity, and date of collection are important. The mode of locomotion should be observed and noted.

Polyclads are difficult animals to keep in an aquarium for long, for they require well-oxygenated water with no trace of pollution. If such conditions are available they may be kept for a time in marine aquaria containing algae and fed fragments of polychaetes, molluscs and crustaceans.

Fixation and preservation. The fixation of polyclads is difficult, because immediately they are disturbed a thick coating of mucus covers the body, thus forming a barrier against the action of fixatives. Anaesthetizing fluids are useless against the mucus-covered body, which will eventually appear to dissolve. A method often recommended for killing these animals is to cover them with a minimum amount of sea water, and as the worms extend hot sea water saturated with mercuric chloride is quickly poured over them. This procedure reduces the amount of contraction of the body, but varying degrees of contraction invariably occur, sometimes reducing a worm to half of its original size. With large specimens even a hot solution is unsatisfactory, and it is often better to drop them into a tall jar containing a 10 per cent solution of formalin in sea water neutralized with borax. The worms will sink to the bottom of the jar and often appear to be fixed in a relaxed state. For histological examination, however, fixation with mercuric chloride is better than with formalin. After two or three minutes this fixative must be washed out by several changes of tap water. Small, more delicate forms may be fixed satisfactorily by dropping them into hot 4 per cent formalin or 80 per cent alcohol. Specimens may be stored in these preservatives. Hot tap water poured over the worm as it extends its body is also an effective killing agent, but the body must be placed in a preservative immediately after death. Highly-coloured specimens will retain their coloration for some time when fixed and stored in formalin and not exposed to daylight for long periods. The writer has examined specimens that were frozen in sea water and then fixed in Bouin's fluid. Histologically, the specimens were in good condition, but picric acid in the fixative destroyed all coloration and markings in the worms, which became yellowish. Fixing the worms in 10 per cent formalin after freezing would probably have been more successful in retaining natural colour and markings.

Examination. Specimens may become curled or otherwise distorted during fixation, and in this condition they are exceedingly difficult to examine satisfactorily. Such a specimen is flattened between two slides held tightly together by a strong rubber band and passed up through the alcohols. Sometimes the body of a distorted specimen needs to be soaked in a 0·5 per cent solution of trisodium phosphate or in a mixture of equal parts of glycerine, 70 per cent alcohol and distilled water. After a period in either of these solutions the specimen having softened is flattened and compressed, held between two slides, and passed through the alcohols. After about 30 minutes in 80 per cent alcohol, the worm is freed. If mercuric chloride has been used as a fixative two or three drops of iodine are added to the alcohol for the extraction of the fixative. Fading of the brownish colour of the iodine indicates the continued presence of the fixative in the tissues, and small amounts of iodine are added until the colour ceases to fade. The specimen is again placed between two slides, which are bound together, and passed into absolute or isopropyl alcohol. After about 45 minutes the worms will have stiffened and pressure can be released.

To study the arrangement of eyes and other internal characters after preservation and dehydration, a specimen must be cleared by a clearing agent, preferably methyl salicylate. After clearing the specimen is then examined using a low-powered microscope with transmitted light.

The alimentary system and the copulatory organs of delicate specimens may be satisfactorily studied as stained whole-mounts if placed in Mayer's paracarmine for about 15 minutes. Excess stain is removed by washing in 80 per cent alcohol and destained overnight in a solution of 0·5 per cent hydrochloric acid in 50 per cent alcohol. For permanent preparation fix the stain by placing the specimen in distilled water for a few minutes, then dehydrate, clear normally and mount in Canada balsam. Other kinds of carmine solutions are acceptable for delicate specimens.

Heath (1907) clearly made out the nervous network in a specimen of *Planocera hawaiiensis* examined in an equal mixture of glycerine and formaldehyde. Some genera contain species that are, at present, separated only on differences in the structure of the copulatory organs. Generally, to observe these differences it is necessary for the portion of the body containing the organs to be cut out, longitudinally serially sectioned and stained by the usual haematoxylin-eosin method. Histological details of the various organ systems are, however, better understood in transverse section.

Classification of the Order Polycladida

KEY TO SUBORDERS

A. Forms with muscular sucker, adhesive pad or depression on ventral surface of body posterior to female genital pore and, or, with pair of anterior marginal tentacles or tentacular eye-clusters . COTYLEA

B. Forms without ventral sucker or other adhesive structures posterior to female copulatory complex; tentacles, when present, nuchal . . . . ACOTYLEA

Suborder **ACOTYLEA** Lang, 1884*
(=ACOTYLINA Pearse, 1936)

Forms without ventral sucker or adhesive structure posterior to female genital pore. Tentacles, when present, nuchal. Tentacular and cerebral eye-clusters; sometimes with marginal or submarginal band of eyes. Pharynx usually ruffled. Copulatory complexes situated posteriorly to pharynx. Male complex tends to be directed anteriorly from male pore. Uterine canals invariably extend anteriorly from vagina.

KEY TO SUPERFAMILIES OF ACOTYLEA

A. Eyes in tentacular, cerebral and frontal clusters and in marginal zones of body; eyes rarely absent STYLOCHOIDEA

B. Eyes distributed anteriorly from cerebral organ to submarginal zones of body . . . . CESTOPLANOIDEA

C. Eyes only in region of cerebral organ . . . . PLANOCEROIDEA

Superfamily **STYLOCHOIDEA** Poche, 1926, emended Nicoll, 1936
(=CRASPEDOMMATA Bock, 1913; CRASPEDOMMATIDEA Marcus & Marcus, 1966)

ACOTYLEA with rows of marginal eyes of variable extent; sometimes with frontal eyes in addition to cerebral and tentacular eye-clusters, or anterior region of body strewn with small eyes not differentiated into particular groups. Exceptionally without eyes.

*Since this work was completed an entirely new classification of the Acotylea, based on the structure of the prostatic organ, has been proposed by Dr Anno Faubel (1983) (*Mitt. hamb. zool. Mus. Inst.* **80:** 17–121).

KEY TO FAMILIES OF STYLOCHOIDEA

1. Male copulatory complex with prostatoids 2
   Without prostatoids 3
2. Prostatoids open into male antrum; Lang's vesicle usually U-shaped . . DISCOCELIDIDAE
   Prostatoids open on ventral surface of body; Lang's vesicle sacciform . POLYPOSTHIIDAE
3. Prostatic organ interpolated or not apparent . . . . . . CRYPTOCELIDIDAE
   Prostatic organ more or less independent 4
4. Vasa deferentia enter distal region of prostatic organ . . . . . . PLEHNIIDAE
   Prostatic organ independent of vasa deferentia 5
5. Ejaculatory duct not connected with prostatic organ or penis-papilla and opens ventrally between male and female genital pores . . . . *Nephtheaplana* gen. nov. (?PLEHNIIDAE)
   Ejaculatory duct unites with prostatic duct and penis-papilla 6
6. Small eyes scattered fanwise over cephalic region; no tentacular or cerebral eye-clusters . . LATOCESTIDAE
   Eyes disposed marginally and in cerebral and tentacular clusters . . .STYLOCHIDAE

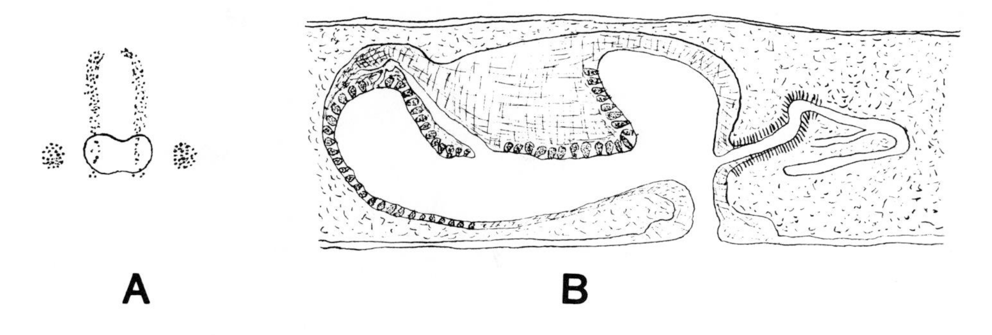

Fig. 3 *Discocelis tigrina*. **A**, tentacular and cerebral eye-clusters; **B**, sagittal section of copulatory organs (after Lang).

Family **DISCOCELIDIDAE** Laidlaw, 1903, emended Poche, 1926

STYLOCHOIDEA Fleshy, oval or broadly-oval forms, sometimes cuneate in outline; without tentacles. Marginal eyes in band of variable extent; additional eyes disposed in two elongate cerebral clusters anterior to, or between, a pair of rounded tentacular clusters, or both sets of eyes on each side merge to form an elongate cluster; no frontal eyes. Pharynx in middle third of body; intestinal branches not anastomosing. Male copulatory complex immediately posterior to pharynx. Vasa deferentia confluent posteriorly to female copulatory complex; at hind level of pharynx they turn medially to lead into male complex. Seminal vesicle feebly developed or absent. Prostatic organ, when present, interpolated. Penis-papilla variously developed, suspended from dorsal wall of male antrum. Numerous unarmed prostatoids embedded in thick musculature investing male antrum into which they open. Vagina long. Lang's vesicle sacciform or horseshoe-shaped, limbs directed anteriorly.

KEY TO DISCOCELIDID GENERA

1.	With interpolated prostatic organ; no penis-papilla	***Adenoplana***
	Without prostatic organ; penis-papilla distinct	2
2.	Muscular prostatoids embedded in penis-papilla and male antrum	. ***Discocelis***
	Similar prostatoids in semi-circle round male antrum	***Coronadena***

Genus ***Discocelis*** Ehrenberg, 1836, *sensu* Lang, 1884 Fig. 3

SYNONYMY *Thalamoplana* Laidlaw, 1904.

DISCOCELIDIDAE Body oval, somewhat broader anteriorly. Marginal eyes confined to anterior third of body. Intestinal trunk as long as pharynx. At posterior level of pharynx, vasa deferentia converge medially to unite and form ejaculatory duct, proximal region of which may be modified as a small muscular seminal vesicle. Ejaculatory duct leads into broad truncate penis-papilla formed from a protuberance of the dorsal wall of the muscular male antrum, which also bears smaller protuberances. Without typical prostatic organ; numerous prostatoids embedded in wall of male antrum and its protuberances. Vagina runs dorso-posteriorly from its pore; Lang's vesicle horseshoe-shaped.

TYPE-SPECIES *D.* [*Planaria*] **lichenoides* (Mertens, 1832) Ehrenberg, 1836. Type by original designation.

*Throughout this work the generic name in square brackets is that originally given to the type-species.

TYPE-LOCALITY Attached to stones at high-water mark, 'Sitcha' [Sitka], Baranof I., Alaska.

NOTE The genus *Discocelis* has been generally accepted on the basis of Lang's (1884) generic definition, which appears to have been based on features found in *Polycelis tigrinus* Blanchard, transferred by Lang to *Discocelis*. Ehrenberg (1836), however, erected the genus for *Planaria lichenoides* Mertens,

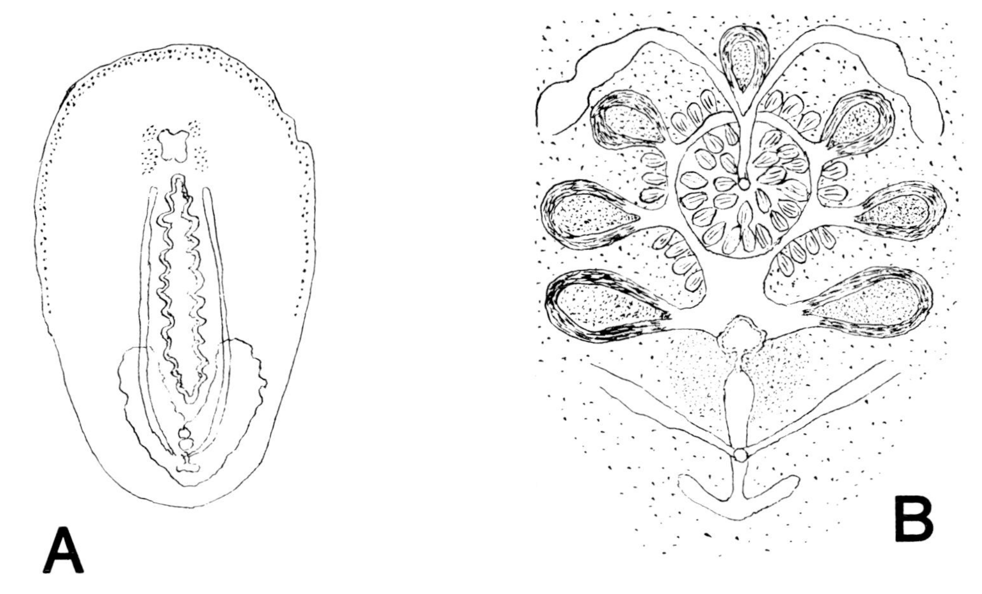

Fig. 4 *Coronadena mutabilis.* **A**, complete worm (cleared); **B**, copulatory complexes (ventral view) (after Hyman).

but Mertens' description is useless for taxonomic analysis. *P. lichenoides* and the genus *Discocelis* are, therefore, unrecognizable taxa. On the other hand, since the genus *Discocelis sensu* Lang has been generally acknowledged during the past hundred years, it seems practical on the grounds of usage to accept Lang's conception of the genus, as indicated above.

The genus *Thalamoplana* was erected by Laidlaw (1904*b*) for the reception of *T. herdmani* Laidlaw collected in the Gulf of Manaar, Sri Lanka. Laidlaw distinguished this genus from *Discocelis* by the presence of muscular projections of the wall of the male antrum carrying pyriform glandular organs and by the presence of a common gonopore. Bock (1913) rightly considered the projections bearing glandular organs to be comparable with those found in *Discocelis,* and he also regarded the union or separation of the genital pores not to be important generically. Nevertheless, Marcus (1950) lists the two genera as distinct solely on the relationship of the male and female genital pores to one another in a differential key to the genera of the Discocelididae, and de Beauchamp (1961) and Marcus & Marcus (1966) accept the separation. Hyman (1955*c*) described *Discocelis insularis* in which the genital pores are separated, but very near each other and might be regarded as a condition intermediate between that found in *Discocelis tigrina* and that in *Thalamoplana herdmani.* There is, however, a further difference which might have some taxonomic importance. In *D. insularis* and *T. herdmani,* both with separated genital pores, the tentacular eyes merge with the cerebral eyes, whereas in *D. tigrina* and species with a single genital pore the two sets of eyes are distinctly separated. These differences are rather superficial and here considered to warrant *Thalamoplana* subgeneric status only.

Genus ***Coronadena*** Hyman, 1940 — Fig. 4

DISCOCELIDIDAE Oval body streaked and speckled with yellow-brown or tan dorsally. Marginal eyes extend posteriorly to about middle level of body; tentacular and cerebral eyes in distinct paired groups. Male and female genital canals open by a common gonopore. Course of vasa deferentia similar to that of *Discocelis.* No spermiducal bulbs, seminal vesicle or typical prostatic organ. Short truncate penis-papilla in roof of spacious male antrum. In addition to many small prostatoids, male antrum contains in its anterior and lateral walls a semicircle of seven to eleven pockets, each of which holds from one to three pyriform glandular organs. Female system as in *Discocelis.*

TYPE-SPECIES *C. [Polycelis] mutabilis* (Verrill, 1873) Hyman, 1940. Type by original designation.

TYPE-LOCALITY Among red algae in 3 to 5 metres, off Thimble I., Connecticut, U.S.A.

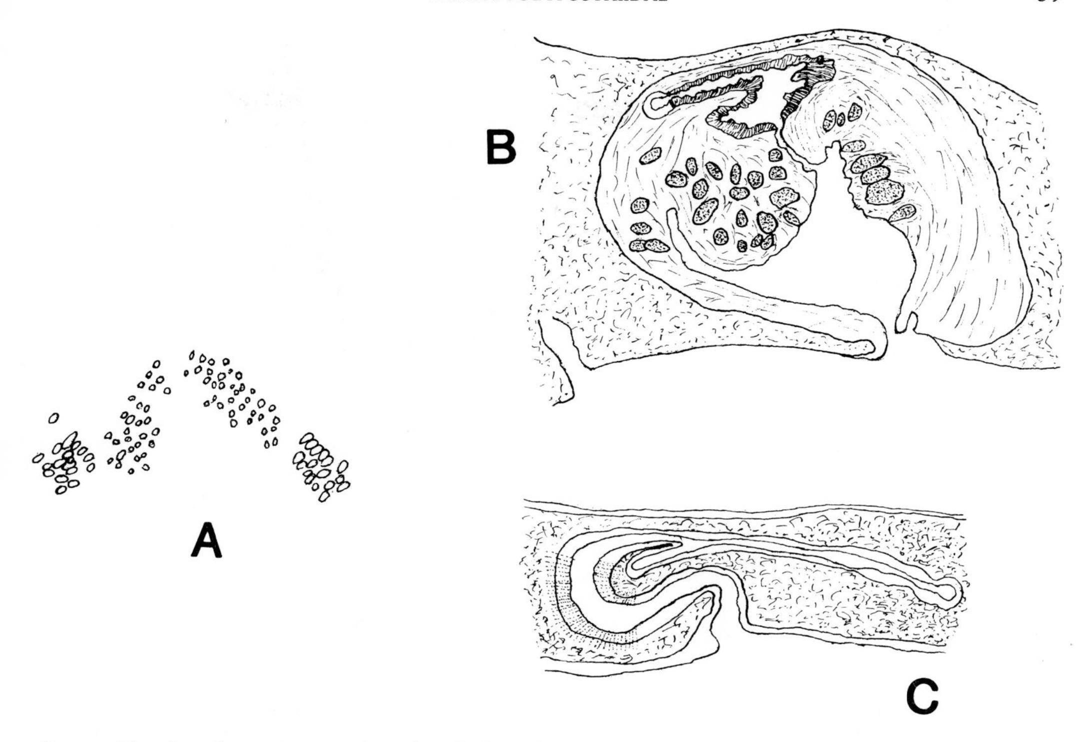

Fig. 5 *Adenoplana obovata.* **A**, tentacular and cerebral eye-clusters; **B**, male copulatory organs; **C**, female copulatory organs (after Stummer-Traunfels).

Genus ***Adenoplana*** Stummer-Traunfels, 1933 Fig. 5

DISCOCELIDIDAE Body elongate-oval to obovate. Marginal eyes may extend round body. Genital pores well separated. Pair of spermiducal bulbs open separately into interpolated fusiform prostatic organ lying in muscular dorsal wall of male antrum. No penis-papilla. Many prostatoids with thin muscular walls and of similar size lying in wall of male antrum, which forms a trilobed structure, as seen dorso-ventrally. Lang's vesicle horseshoe-shaped or sacciform.

TYPE-SPECIES *A. [Polycelis] obovata* (Schmarda, 1859). Type by subsequent designation.

TYPE-LOCALITY Jamaica, West Indies.

Family **POLYPOSTHIIDAE** Bergendal, 1893, emended Bock, 1913

(Includes CRYPTOCELIDIDAE Bergendal, 1893; POLYPOSTIADAE Bergendal, 1893)

STYLOCHOIDEA Generally rather fleshy oval forms, extremities of which tend to taper, but often more rounded posteriorly. Nuchal tentacles ill defined or absent. Eyes absent or very small and scattered over cephalic region, or arranged in two elongate cerebral clusters, posterior to which may lie two rounded tentacular clusters; frontal eyes few, widely distributed; marginal eyes confined to anterior half of body. Short pharynx in middle third of body. Male copulatory complex variable in structure; without seminal vesicle or spermiducal bulbs. Typical prostatic organ may be present. Prostatoids arranged in one or several rosettes along median line of body, each rosette with its own ventral aperture. Many additional prostatoids may be distributed posteriorly or throughout the body. Vasa

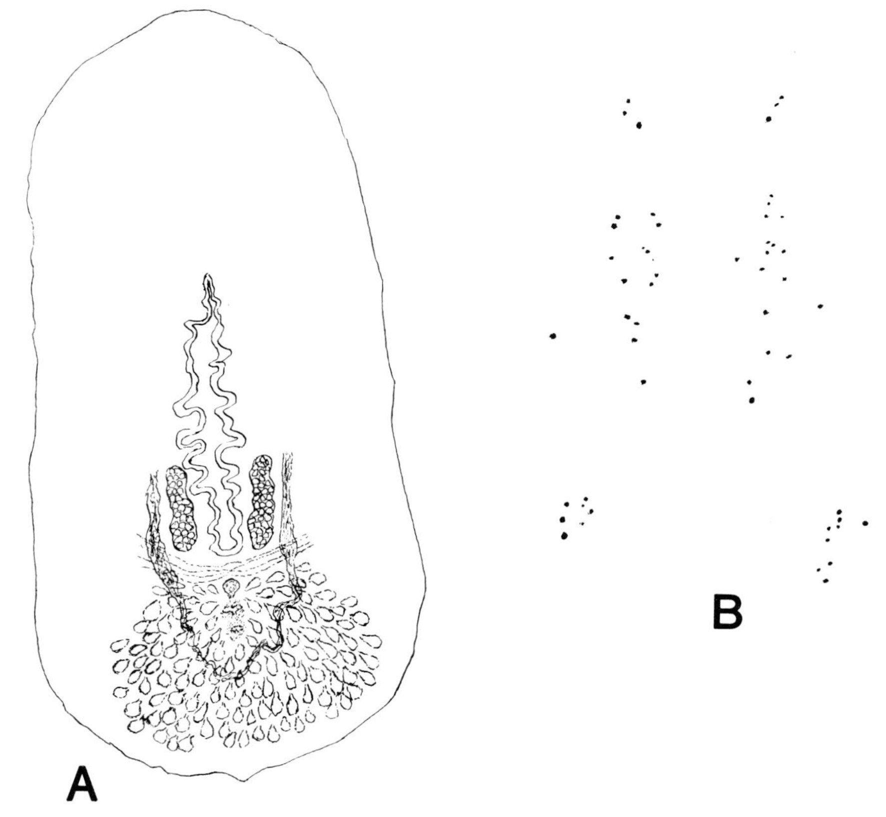

Fig. 6 *Polyposthia similis.* **A**, complete worm (ventral view); **B**, tentacular and cerebral eye-clusters (after Bock).

deferentia well developed, lateral to uterine canals. Female copulatory complex simple, opening in median line, posteriorly to pharynx and anteriorly or posteriorly to male genital pore. Lang's vesicle may or may not occur.

KEY TO POLYPOSTHIID GENERA

1. Eyes absent; with ventral epidermal cushion posterior to female genital pore . . . ***Polyphalloplana***
   Eyes present; without epidermal cushion 2
2. Prostatoids scattered throughout body; some arranged in rosettes along median line . . . . ***Polyposthides***
   Prostatoids confined to posterior half of body 3
3. Male pore anterior to female; prostatic organ typical . . . . . . ***Polyposthia***
   Male pore posterior to female; prostatic organ absent 4
4. 2 to 8 prostatoids in rosette around male antrum . . . . ***Cryptocelides***
   Numerous prostatoids in several rings around genital pores . . . . . ***Metaposthia***

Genus ***Polyposthia*** Bergendal, 1893, emended Bock, 1913 Fig. 6

POLYPOSTHIIDAE Body tapering towards both ends. Nuchal tentacles ill defined or not apparent. Marginal eyes inconspicuous; frontal eyes few; indistinct tentacular eye-clusters well posterior to elongate cerebral eye-clusters. Intestinal branches not anastomosing. Male genital pore immediately anterior to female pore, but both may form an antrum. From pharyngeal region, a pair of vasa deferentia extend posteriorly and near male copulatory complex each vas deferens bifurcates, the inner limbs

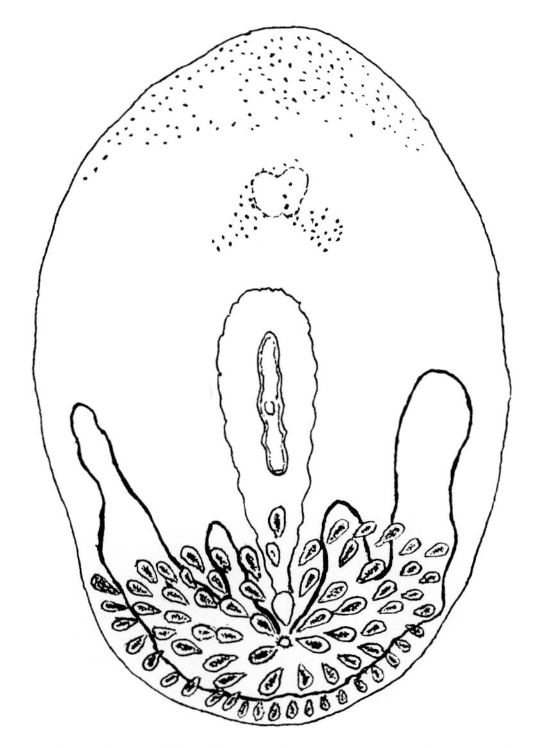

Fig. 7 *Metaposthia norfolkensis.* Complete worm (ventral view) (after Palombi).

running medially to open into large pyriform prostatic organ, whereas the outer limbs continue posteriorly and unite posteriorly to female complex. Typical prostatic organ invested with extracapsular gland cells and opens into conical penis-papilla lying in spacious male antrum. Many prostatoids scattered throughout posterior region of body; several of those disposed around female complex are larger than remainder, which open independently on ventral surface of body. Vasa deferentia open into those organs lying near female complex. Vagina forming an anteriorly-directed loop; vagina externa very muscular; with Lang's vesicle. Uterine canals short and wide.

TYPE-SPECIES *P. similis* Bergendal, 1893. Type by original designation.

TYPE-LOCALITY On mud, sublittoral, Gullmarfjord, Sweden.

Genus ***Metaposthia*** Palombi, 1924 Fig. 7

POLYPOSTHIIDAE Body oval, of delicate consistency, without tentacles. Frontal eyes distributed over cephalic region and merge with narrow anterior band of marginal eyes; cerebral eyes in two elongate clusters; large tentacular eyes in two groups. Intestinal branches anastomosing. Two principal genital pores situated posteriorly, male posterior to female. Prostatic organ absent. Vasa deferentia unite posteriorly, extend into sublateral fields of body to middle level, where they turn inwardly to open into several large pyriform prostatoids encircling the male antrum and female copulatory complex. Many smaller prostatoids lateral and posterior to larger organs, the latter having a thicker musculature than the former, but both types of organs have their narrow ends directed towards the male antrum. No indication of prostatoids forming rosettes in median line. Uterine canals confluent anteriorly to pharynx. Large vagina bulbosa. No Lang's vesicle.

TYPE-SPECIES *M. norfolkensis* Palombi, 1924. Type by original designation.

TYPE-LOCALITY 28°20′S, 170°5′E, near Norfolk I., Pacific Ocean (pelagic).

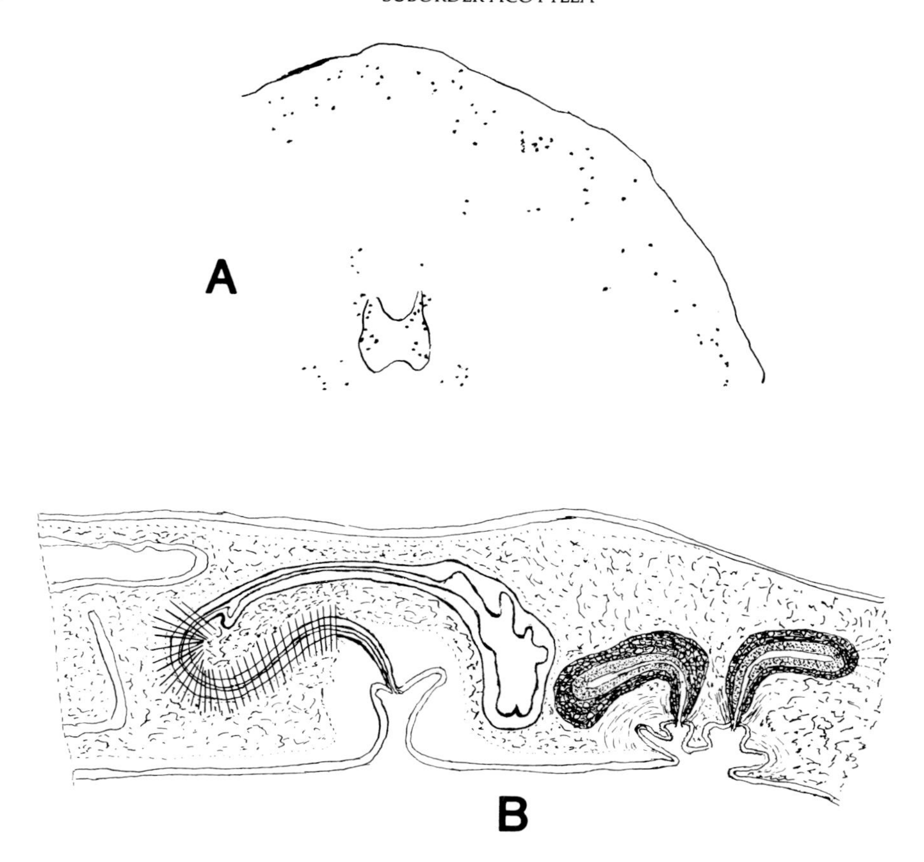

Fig. 8 *Cryptocelides loveni.* **A,** anterior arrangement of eyes; **B,** sagittal section of copulatory complexes (after Bock).

Genus *Cryptocelides* Bergendal, 1890

Fig. 8

POLYPOSTHIIDAE Body usually tapering anteriorly, more rounded posteriorly. Without tentacles. Numerous small eyes along anterior margin of body distinct from isolated frontal eyes; paired elongate group of cerebral eyes alongside cerebral organ, posteriorly and laterally to which is a pair of tentacular eye-clusters. Intestinal branches not anastomosing. Male genital pore posterior to female pore. No prostatic organ. Two to eight pyriform prostatoids form a rosette around spacious male antrum, into which they open, their narrow ends forming penial papillae. Two vasa deferentia, lateral to uterine canals, send narrow ducts opening into anteriorly-placed prostatoids. Many of the smaller prostatoids lying posteriorly to the male pore are without such connections and open independently on ventral surface of body. Vagina narrow, forming an anteriorly-directed loop; vagina externa modified as a vagina bulbosa; 'shell'-chamber long; Lang's vesicle sacciform. Uterine canals not confluent anteriorly.

TYPE-SPECIES *C. loveni* Bergendal, 1890. Type by original designation.

TYPE-LOCALITY On mud in sublittoral, Bohuslan, Gullmarfjord, Sweden.

NOTE This genus has been found in Scandinavian and northern British waters and at depth off the Moroccan coast and pelagic off Samoa. *Cryptocelides loveni* is an inhabitant of the European boreal and has been recorded from southern Australia, but Laidlaw (1904*a*) doubts the authenticity of this record, which Prudhoe (1982*b*) points out might be genuine.

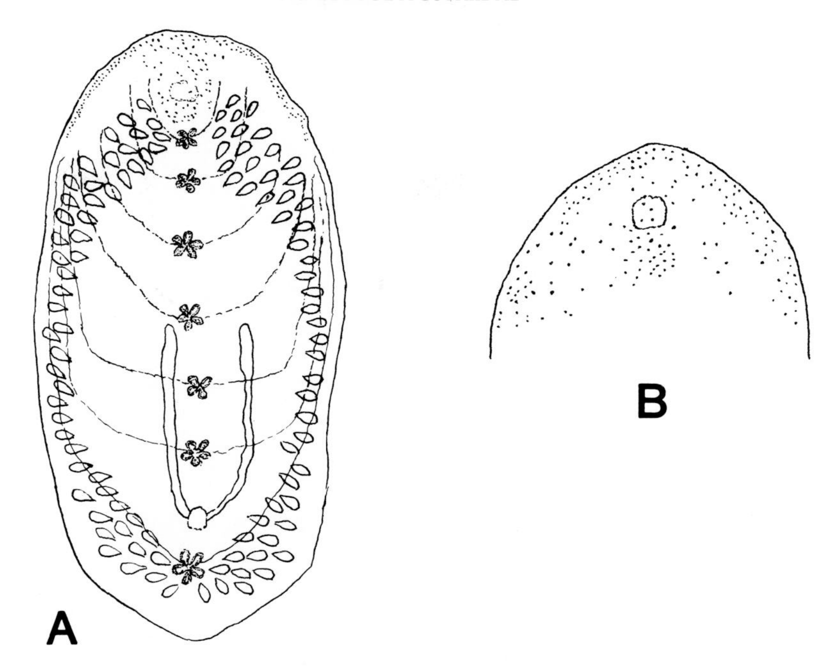

Fig. 9 *Polyposthides karimatensis.* **A,** complete worm (ventral view); **B,** arrangement of eyes (after Palombi).

Genus ***Polyposthides*** Palombi, 1924 Fig. 9

POLYPOSTHIIDAE Body tapering anteriorly, rounded posteriorly. Without tentacles. Marginal eyes in anterior half of body; cerebral eyes in two elongate clusters; tentacular eye-clusters may occur. Intestinal branches not anastomosing. No typical prostatic organ. Numerous prostatoids distributed throughout body, many of which are arranged in several rosettes, each of 4 or 5 elements, disposed along median line, extending from closely posterior to cerebral organ to near posterior end of body. Each rosette opens individually on ventral surface of body, thus several male pores occur intermittently along the median line. Vasa deferentia arise in sublateral fields and furnish each rosette of prostatoids with a pair of ducts. Female genital pore posterior to hinder, or penultimate, male pore, and well separated from posterior margin of body. Vagina bulbosa present. Uterine canals not confluent anteriorly. No Lang's vesicle.

TYPE-SPECIES *P. karimatensis* Palombi, 1924. Type by subsequent designation.

TYPE-LOCALITY Dredged at 90 metres, 1°20′S, 107°57′E, in Karimata Straits, between Borneo and Sumatra, East Indies.

NOTE This genus is so far known only from Indonesian and Caribbean waters.

Genus ***Polyphalloplana*** Bock, in de Beauchamp, 1951 Fig. 10

POLYPOSTHIIDAE Body lanceolate; without eyes or tentacles. Intestinal branches not anastomosing. Male complex consists of numerous prostatoids forming rounded cluster lying in ventral epidermal cushion situated posteriorly to female copulatory complex. Vasa efferentia forming network of canals opening into pair of vasa deferentia lying laterally to epidermal cushion and sending a narrow branch towards each prostatoid. Each of these and the vas deferens branch open together into ejaculatory duct passing through penis-papilla lying in antrum opening to exterior. Female copulatory organs between pharynx and male organs. Vagina forming anteriorly-directed loop. Lang's vesicle elongate; anterior extent of uterine canals not known.

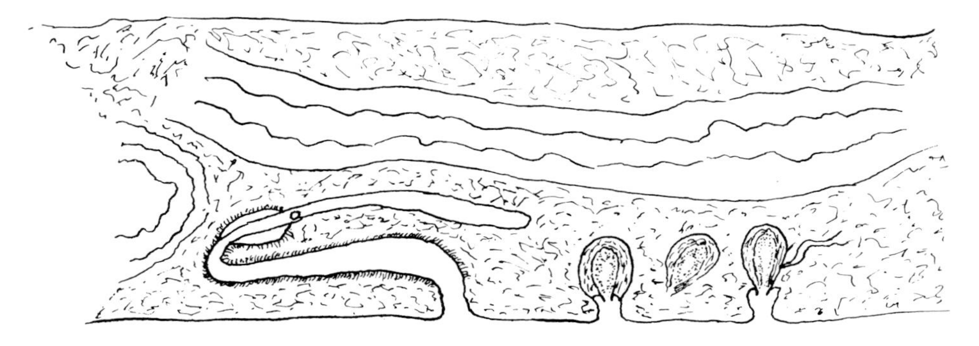

Fig. 10 *Polyphalloplana bocki.* Sagittal section of copulatory organs (after de Beauchamp).

TYPE-SPECIES *P. bocki* de Beauchamp, 1951 (monotypic)

TYPE-LOCALITY Dredged at 85 metres, 9°50′40″ to 9°51′10″W, 30°28′10″ to 30°25′30″N, off coast of Morocco.

Family **PLEHNIIDAE** Bock, 1913

STYLOCHOIDEA Elliptical or broadly-oval forms of variable size, without tentacles. Eyes absent or minute, when present arranged in paired tentacular and cerebral clusters and in marginal band of varying extent; frontal eyes few. Pharynx in middle third of body. Male and female genital pores closely approximate, well separated from hind margin of body. Male copulatory organs anterior to male pore. Vasa deferentia arise in hind region of body and extend to pharyngeal level, where they turn medially and posteriorly to lead into male complex. Spermiducal bulbs or seminal vesicle present. Efferent ducts of spermiducal bulbs pass separately into ventral wall of narrow distal region of partially independent prostatic organ; or ejaculatory duct from seminal vesicle opens to exterior independently of prostatic organ and penis-papilla. Latter short, broad, and without penis-sheath. Long vagina forming anteriorly-directed loop from female pore; vagina bulbosa or ductus vaginalis may be present; Lang's vesicle of variable size. Uterine canals separated anteriorly.

KEY TO PLEHNIID GENERA

1. With ductus vaginalis . . . . ***Discocelides***
   No ductus vaginalis 2
2. With seminal vesicle; prostatic organ elongate . . . . . ***Nephtheaplana*** gen. nov.
   With spermiducal bulbs; prostatic organ pyriform 3
3. Prostatic organ with thin muscular wall enclosing large glandular chamber . . . ***Plehnia***
   Prostatic organ exceedingly muscular; its glandular chamber confined to small distal area . . ***Paraplehnia***

Genus ***Plehnia*** Bosk, 1913

Fig. 11

PLEHNIIDAE Elliptical forms, broadly acuminate anteriorly. Eyes absent, or in paired cerebral and tentacular clusters and in marginal band. Elongate spermiducal bulbs ventral to prostatic organ into which they open separately. No seminal vesicle. Prostatic organ large, with relatively thin muscular wall and lined with a tall glandular epithelium thrown into radial folds. Small penis-papilla unarmed. Lang's vesicle small and bulbous. Uterine canals short.

TYPE-SPECIES *P. [Acelis] arctica* (Plehn, 1896). Type by original designation.

TYPE-LOCALITY East coast of Spitzbergen, 79°N, 22°E.

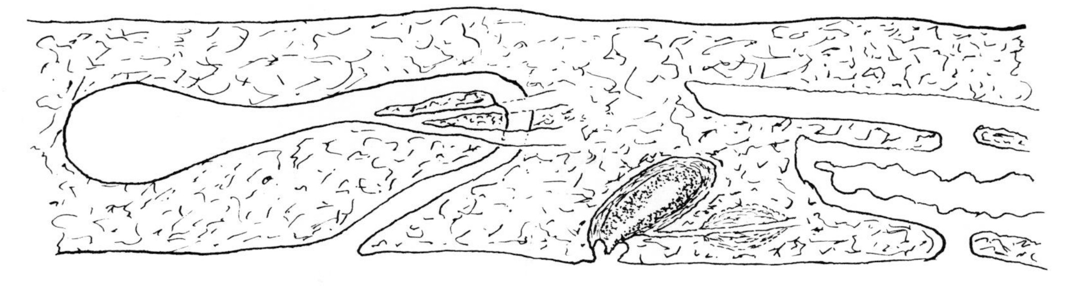

Fig. 11 *Plehnia arctica*. Sagittal section of copulatory organs (modified after Plehn).

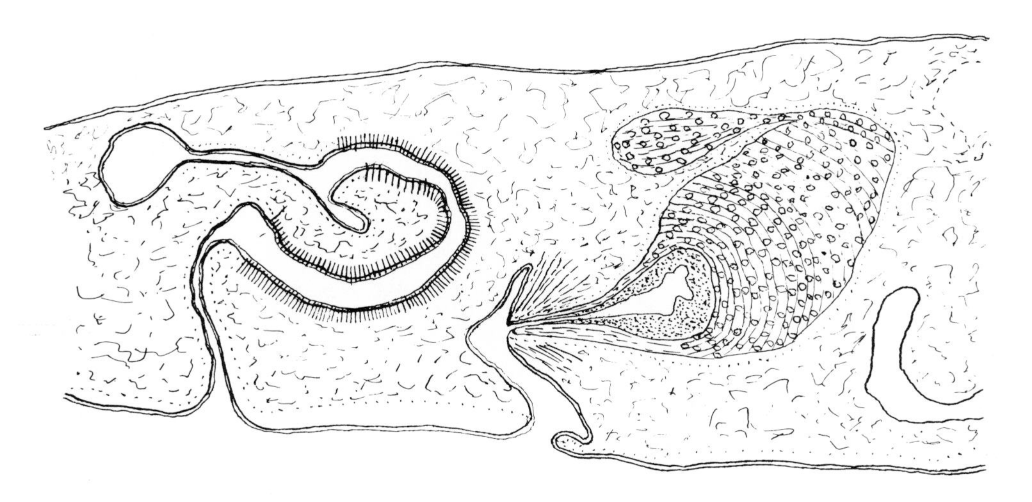

Fig. 12 *Paraplehnia japonica*. Sagittal section of copulatory organs (after Bock).

Genus *Paraplehnia* Hyman, 1953 — Fig. 12

PLEHNIIDAE Body oval, somewhat broader posteriorly. Eyes minute, in marginal, frontal, cerebral and tentacular groups. Intestinal caeca ramifying without anastomosing. Vasa deferentia lateral to uterine canals. Elongate spermiducal bulbs well developed. No seminal vesicle. Pyriform prostatic organ highly muscular, with a glandular chamber confined to a small portion of the narrow distal region, and invested with extracapsular gland-cells. Penis-papilla unarmed, short and broad. Vagina bulbosa or strongly muscular vagina externa. Lang's vesicle small. Uterine canals short.

TYPE-SPECIES *P. [Plehnia] japonica* (Bock, 1923). Type by original designation.

TYPE-LOCALITY On mud in 12–15 metres, Kobe Bay, Inland Sea of Japan.

Genus *Discocelides* Bergendal, 1893 — Fig. 13

PLEHNIIDAE Elliptical or broadly-oval forms. Eyes in paired cerebral and tentacular groups; frontal eyes present; marginal eyes in several rows anteriorly, becoming fewer laterally, with individual eyes posteriorly in young specimens. Intestinal branches anastomosing in anterior and posterior regions of body. No true seminal vesicle; spermiducal bulbs well developed, opening separately into prostatic organ. Large, muscular, prostatic organ pyriform, with spacious lumen lined with a low epithelium

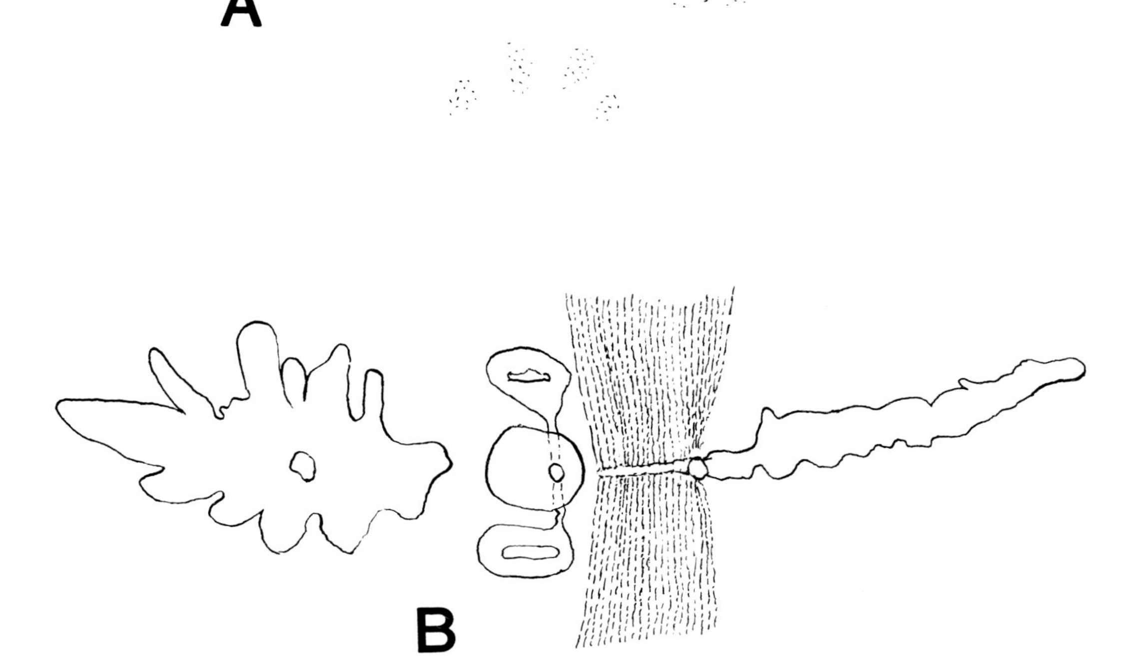

Fig. 13 *Discocelides langi.* **A**, arrangement of eyes; **B**, pharynx and copulatory organs (ventral view) (after Bock).

without gland-cells and heavily invested with extracapsular glands, the efferent ducts of which pass through the thick musculature to open into the lumen of the prostatic organ. Penis-papilla unarmed. Lang's vesicle sacciform; muscular ductus vaginalis connects 'stalk' of Lang's vesicle with vagina externa. Uterine canals long and convoluted, lateral to vasa deferentia.

TYPE-SPECIES *D. langi* Bergendal, 1893. Type by original designation.

TYPE-LOCALITY Gullmarfjord, Sweden.

Genus ***Nephtheaplana*** gen. nov. Fig. 14

PLEHNIIDAE Small, broadly-oval forms without eyes. Small seminal vesicle anterior to prostatic organ. Long narrow ejaculatory duct runs along dorsal wall of prostatic organ and opens into a shallow ventral depression situated closely anterior to female genital pore. Large elongate-oval prostatic organ with tall, smooth, epithelial lining opens into conical penis-papilla with a cuticularized tip. Lang's vesicle small. Uterine canals short, lateral to vasa deferentia.

TYPE-SPECIES *N. [Plehnia] tropica* (Hyman, 1959) nov. comb.

TYPE-LOCALITY Associating with an alcyonarian (*Nephthea*) on reef south of Ngaremediu, east end of Urukthapel, Palau Isles.

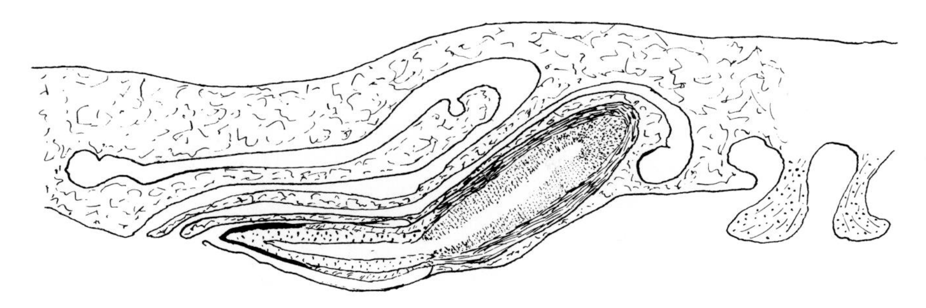

Fig. 14 *Nephtheaplana tropica.* Sagittal section of copulatory organs (after Hyman).

SYSTEMATIC NOTE Hyman (1959*a*) originally placed her species in the genus *Plehnia,* presumably because of its lack of eyes, but this feature alone does not warrant this generic assignment, for the male copulatory complex of this species is quite different from that of the type-species of *Plehnia.* For this reason, a new genus, *Nephtheaplana,* is here proposed for the reception of *Plehnia tropica* Hyman, 1959. It must, however, be stated that since the lack of a connection between the ejaculatory duct and the prostatic organ in this species appears to be unique among polyclads this feature requires confirmation.

Family **LATOCESTIDAE** Laidlaw, 1903

STYLOCHOIDEA Elongate-oval or ribbon-like forms without tentacles. Numerous small eyes scattered more or less fanwise, often in streaks directed obliquely forward, over anterior region of body. Marginal eyes invariably present; cerebral and tentacular eye-clusters not apparent. Mouth and pharynx centrally or posteriorly situated. Genital pores separated, near hind end of pharynx. Male copulatory complex directed more or less anteriorly from male pore. Vasa deferentia arise in pharyngeal region and run posteriorly to unite and form an ejaculatory duct; with or without a seminal vesicle. Large muscular prostatic organ pyriform and independent, usually dorsal to ejaculatory duct, and provided with tall epithelial lining. Prostatic canal and ejaculatory duct unite before entering base of thick unarmed penis-papilla situated in penis-pocket or shallow male antrum. Vagina narrow, forming a more or less anteriorly-directed loop. Lang's vesicle or ductus vaginalis present. Uterine canals open into vagina through a common duct.

NOTE The region in which the genital pores are closely approximate is sometimes withdrawn into the ventral wall of the body and gives a false impression of a genital atrium.

KEY TO LATOCESTID SUBFAMILIES

Lang's vesicle present . . . . LATOCESTINAE subfam. nov.
Without Lang's vesicle . . . . TRIGONOPORINAE Laidlaw, 1903

KEY TO LATOCESTINE GENERA

1. Lang's vesicle opens on ventral surface of body . . . . ***Nonatona***
   Lang's vesicle without external opening 2
2. With distinct seminal vesicle . . . . ***Aprostatum***
   Without seminal vesicle . . . . ***Latocestus***

Genus ***Latocestus*** Plehn, 1896

Fig. 15

SYNONYMY *Oculoplana* Pearse, 1938; *Alleena* Marcus 1947.

LATOCESTIDAE: LATOCESTINAE Body narrowing towards extremities. Marginal eyes (said to be absent in type-species) in band of variable extent; further eyes commencing posteriorly to cerebral organ and

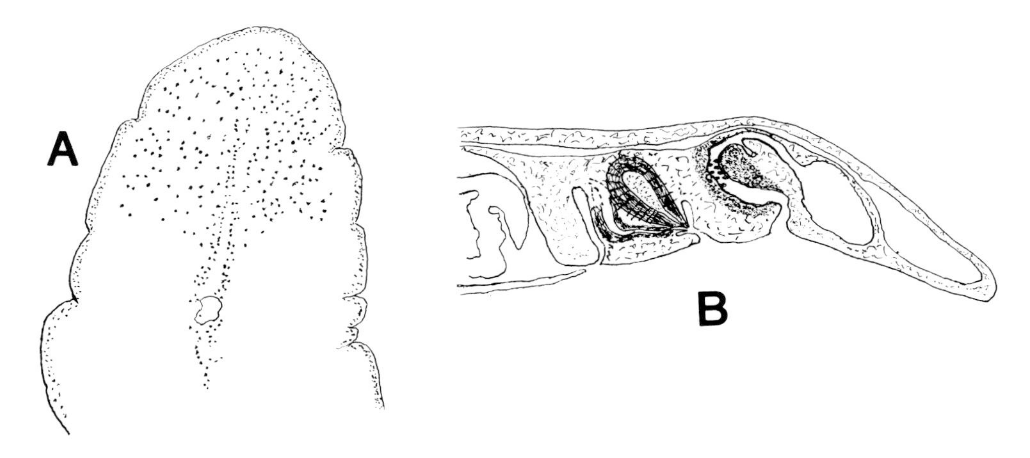

Fig. 15 *Latocestus*. **A**, arrangement of eyes in *L. caribbeanus* (after Prudhoe); **B**, sagittal section of copulatory organs of *L. argus* (after Prudhoe).

spreading fanwise over cephalic region. Pharynx in hind third of body; anterior branch of intestinal trunk long; intestinal branches ramifying but not anastomosing. Vasa deferentia may be simple or possess elongate spermiducal bulbs before uniting to form an ejaculatory duct, which may be distinctly muscular. Well-developed prostatic organ pyriform, heavily invested with extracapsular gland-cells, and directed anteriorly or vertically from male pore. No penis-pocket. Vagina narrow; lining of 'shell'-chamber thrown into a helical fold. Lang's vesicle bulbous or sacciform. Uterine canals not anteriorly confluent.

Widely distributed in tropical and subtropical regions.

TYPE-SPECIES *L. atlanticus* Plehn, 1896. Type by original designation.

TYPE-LOCALITY Cape Verde I., Atlantic Ocean.

SYSTEMATIC NOTE *Alleena* is said to differ from *Latocestus* only in the vertical disposition of the prostatic organ above the male genital pore, whereas in *Latocestus* the organ is directed anteriorly from the genital pore, especially in *L. whartoni* (Pearse). These differences are here considered to be of no generic importance, because intergradations between the two conditions are to be found among other species of *Latocestus*.

Genus ***Nonatona*** Marcus, 1952 — Fig. 16

LATOCESTIDAE: LATOCESTINAE Body elongate oval. Innumerable eyes range round entire body, forming an exceptionally wide band in submarginal and marginal zones of body, so that only a median area between cerebral organ and hind end of pharynx is free from eyes. Pharynx in posterior region of body; intestinal trunk long. Male and female copulatory complexes as in *Latocestus*, except for a duct from Lang's vesicle opening externally posterior to female genital pore.

TYPE-SPECIES *N. euscopa* Marcus, 1952. Type by original designation.

TYPE-LOCALITY In littoral zone, south of Paranaguá Bay, State of Parana, Brazil.

According to Marcus the haploid number of chromosomes in type-species is 2.

Genus ***Aprostatum*** Bock, 1913 — Fig. 17

LATOCESTIDAE: LATOCESTINAE Broadly-oval body of firm consistency. Marginal eyes in several irregular rows in complete series round body. Numerous small eyes distributed fanwise in streaks over cephalic region. Pharynx in hinder half of body. Pair of well-developed spermiducal bulbs open

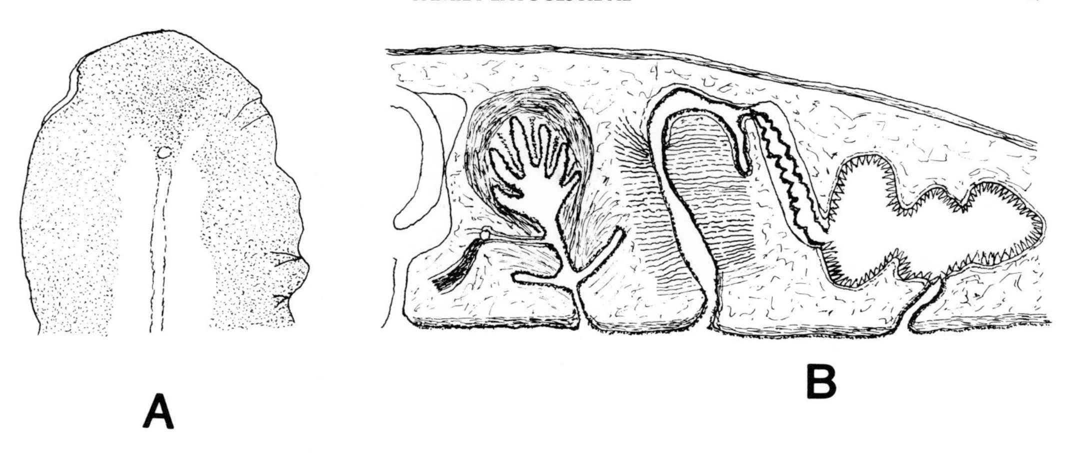

Fig. 16 *Nonatoma euscopa*. **A**, arrangement of eyes anteriorly; **B**, sagittal section of copulatory organs (after Marcus).

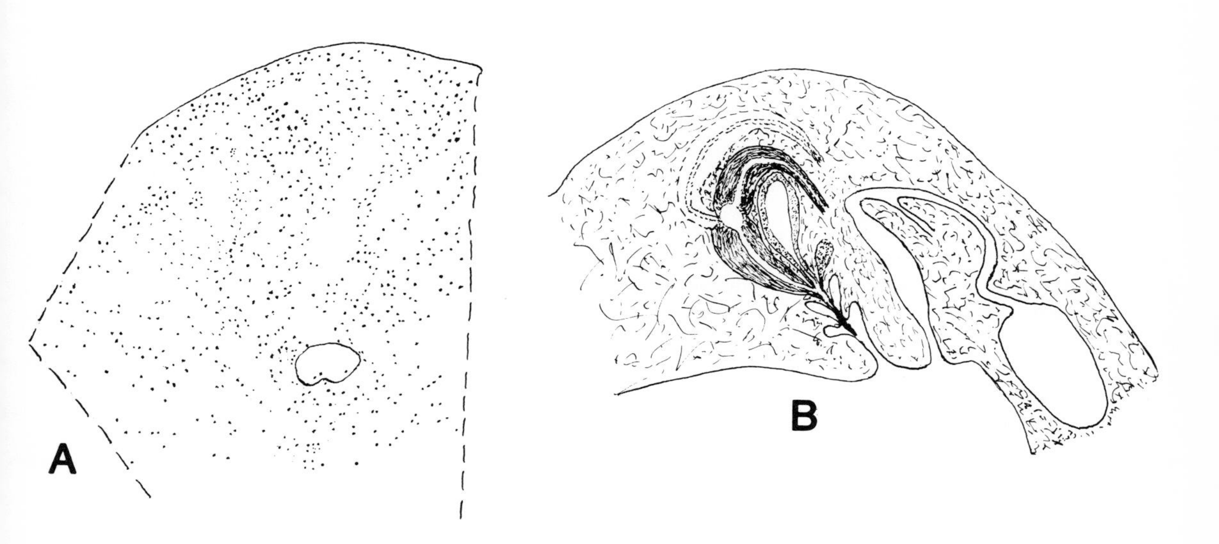

Fig. 17 *Aprostatum stiliferum*. **A**, arrangement of eyes anteriorly (after Bock); **B**, *sagittal* section of copulatory organs (modified, after Bock).

together into an elongate muscular seminal vesicle, lying ventrally to prostatic organ. Together the spermiducal bulbs and the seminal vesicle form a trilobed or anchor-shaped organ. Prostatic organ elongate, lined with low epithelium. After receiving prostatic duct, the ejaculatory duct is joined by a small pyriform accessory organ having no lumen and formed of a reticulum of muscle-fibres enclosing an accumulation of nuclei. Penis-papilla with stylet and enclosed in penis-pocket. Vagina relatively long and muscular. Lang's vesicle large.

TYPE-SPECIES *A. stiliferum* Bock, 1913. Type by original designation.

TYPE-LOCALITY On sand and gravel in 20 metres, off Cape Alman, Golfo de Corcovado, Chile.

SYSTEMATIC NOTE Bock (1913) observed neither a seminal vesicle nor a prostatic organ and placed the genus in the family Cryptocelididae, Marcus (1954*a*) examined new material of *A. stiliferum* from

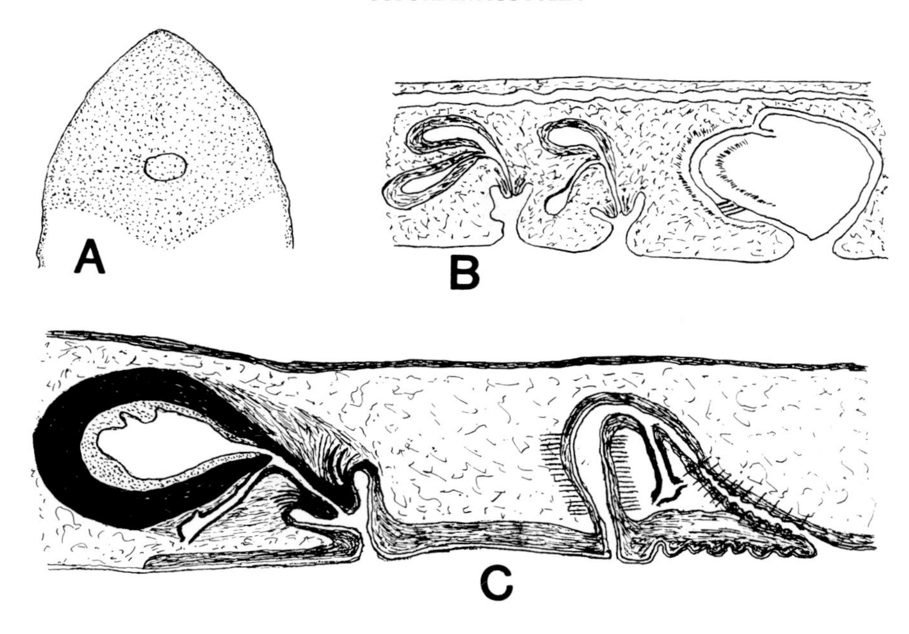

Fig. 18 *Bergendalia diversa.* **A,** arrangement of eyes (after Yeri & Kaburaki); **B,** sagittal section of copulatory organs (after Y. & K.); **C,** *Trigonoporus cephalophthalmus* sagittal section of copulatory organs (after Lang).

Chile and found a seminal vesicle and an independent prostatic organ to be present, and he transferred the genus to the family Stylochidae. *Aprostatum,* however, does not possess nuchal tentacles, nor distinct cerebral and tentacular eye-clusters, and in this respect resembles members of the family Latocestidae, to which *Aprostatum* is here assigned.

KEY TO TRIGONOPORINE GENERA

1. Vagina interna prolonged into ductus vaginalis . . . . ***Trigonoporus***
   Without ductus vaginalis . . . . ***Pentaplana***

Genus ***Trigonoporus*** Lang, 1884 — Fig. 18

SYNONYMY *Bergendalia* Laidlaw, 1903

LATOCESTIDAE: TRIGONOPORINAE Elongate body tapering towards extremities. Marginal eyes in band of varying extent; further eyes distributed fanwise over cephalic region. Pharynx centrally situated; intestinal trunk long; intestinal branches anastomosing. Genital pores in hind third of body. Male copulatory complex directed anteriorly or vertically from male pore. No seminal vesicle; two moderately-developed spermiducal bulbs may occur. Prostatic organ with smooth epithelial lining. Penis-papilla short and broad; no penis-pocket. Duplicate prostatic organ with independent external aperture sometimes present. Vagina externa narrow; dilated 'shell'-chamber with epithelium often thrown into a helical fold. Ductus vaginalis opening into female antrum or near female genital pore. Epidermis around female pore may appear corrugated to form adhesive pad.

TYPE-SPECIES *T. cephalophthalmus* Lang, 1884. Type by original designation.

TYPE-LOCALITY *Melobesia* zone of Secca di Gajola, Bay of Naples, Italy.

SYSTEMATIC NOTE Without discussion, Kato (1944) sinks *Bergendalia* Laidlaw as a synonym of *Trigonoporus* Lang, but Marcus & Marcus (1966) appear not to accept this action, although they do concede that *Bergendalia mirabilis* Kato, 1938, belongs to *Trigonoporus.* The only difference between

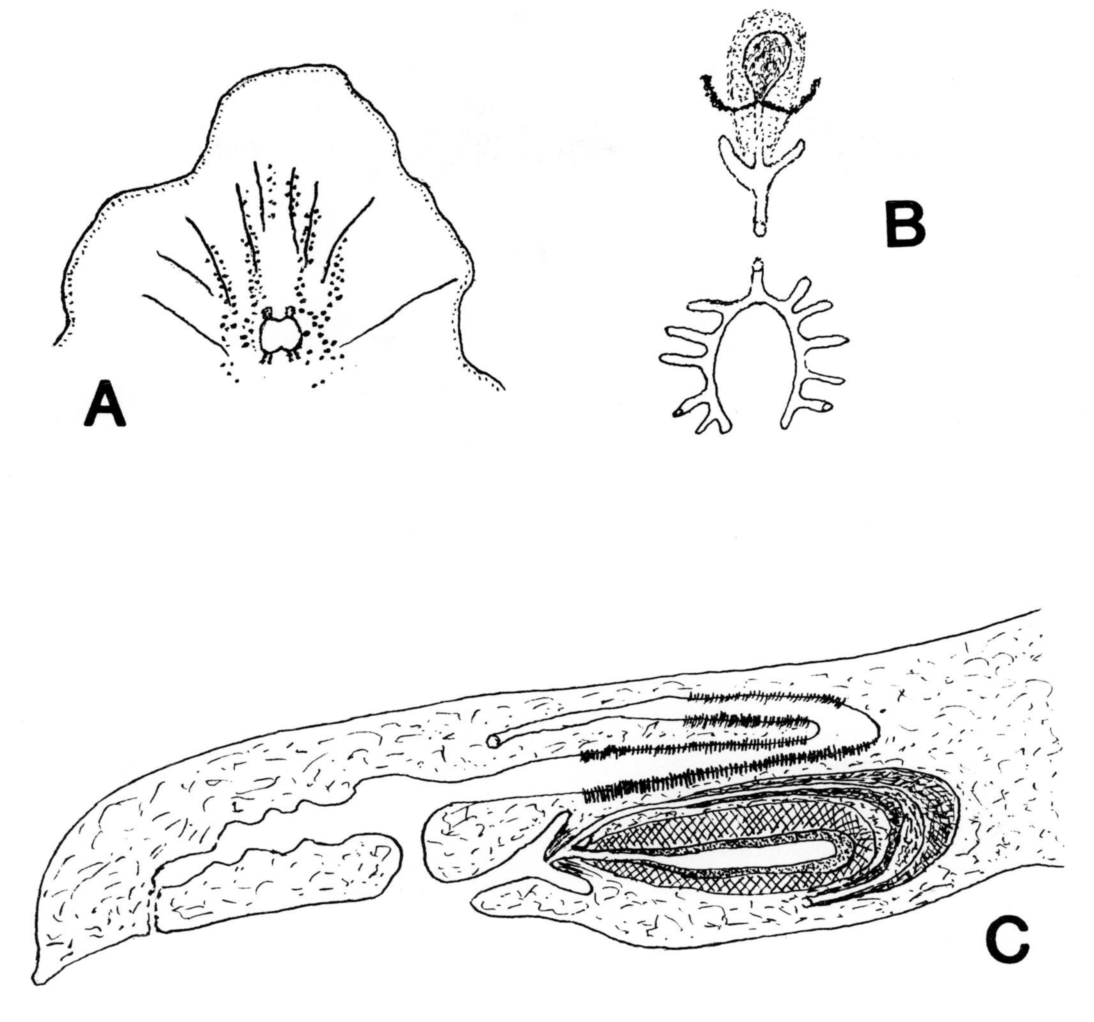

Fig. 19 *Pentaplana divae*. **A**, arrangement of eyes anteriorly; **B**, copulatory organs (ventral view); **C**, sagittal section of copulatory organs (after Marcus).

the two genera is that *Bergendalia anomala* Laidlaw, 1903, and *B. diversa* Yeri & Kaburaki, 1918, each possess a duplicate prostatic organ independent of the functional male copulatory complex. Otherwise, the diagnostic features of both genera are so similar as to indicate that they are probably congeneric.

Genus ***Pentaplana*** Marcus, 1949 Fig. 19

LATOCESTIDAE: TRIGONOPORINAE Small, elongate-oval forms narrowing anteriorly and rounded posteriorly. Marginal eyes in anterior rows; further eyes disposed in three chevron-rows extending anteriorly from cerebral region. Pharynx in hinder half of body; intestinal trunk reaching to cerebral organ; intestinal branches not anastomosing. Narrow seminal vesicle and large pyriform prostatic organ enclosed in muscular sheath. Ejaculatory duct dorsal to prostatic organ. Penis-papilla stoutly conical; no penis-pocket. Female antrum leads into an ampulla, from which a pair of ducts, each with lateral diverticula, extend posteriorly to open separately near posterior end of body. Vagina extends anteriorly from the ampulla, above the male complex, to the pharynx, where it turns acutely to run posteriorly and unite with the uterine canals. No Lang's vesicle. Uterine canals not anteriorly confluent.

TYPE-SPECIES *P. divae* Marcus, 1949. Type by original designation.

TYPE-LOCALITY Among algae in Bay of Santos, Ilha Porchet, Brazil.

Family **STYLOCHIDAE** Stimpson, 1857

STYLOCHOIDEA Generally thick freshy forms varying in shape and size. Nuchal tentacles in varying degrees of development often present. Marginal eyes small. Numerous additional eyes distributed over cephalic region, arranged in tentacular and cerebral clusters and sometimes frontal or cerebro-frontal groups. Mouth rarely in anterior half of body; intestinal branches seldom anastomosing. Male and female genital pores usually separated in posterior third of body. Male copulatory complex anterior, rarely dorsal, to male pore. Vasa deferentia originate laterally to pharynx and extend posteriorly to open into male complex. Seminal vesicle and, or, pair of spermiducal bulbs usually placed more or less ventrally to prostatic organ. Variably developed prostatic organ independent. Penis-papilla elongate or bluntly conical. Vagina of variable length, usually forming an anteriorly-directed loop; its proximal region may continue as Lang's vesicle, a genito-intestinal canal, a ductus vaginalis, or it may unite with the uterine canals, which are usually not anteriorly confluent.

KEY TO SUBFAMILIES OF STYLOCHIDAE

1.	Uterine canals open together into inner end of vagina	STYLOCHINAE
	Vagina extending posteriorly beyond entrance of uterine canals	2
2.	Lang's vesicle present	IDIOPLANINAE
	Lang's vesicle absent	3
3.	Ductus vaginalis present	CRYPTOPHALLINAE
	Genito-intestinal canal present	ENTEROGONIINAE

Subfamily **STYLOCHINAE** Laidlaw, 1903

Female copulatory complex without Lang's vesicle; uterine canals open into proximal end of vagina.

KEY TO STYLOCHINE GENERA

1.	Seminal vesicle present	2
	Seminal vesicle absent	3
2.	Prostatic organ ventral to ejaculatory duct	***Indistylochus***
	Prostatic organ dorsal to ejaculatory duct	***Stylochus***
3.	With penis-sheath	***Meixneria***
	Without penis-sheath	4
4.	Prostatic organ and ejaculatory duct unite before opening into penis-papilla	***Parastylochus***
	Prostatic organ and ejaculatory duct opening independently into male antrum through protuberance	***Ilyplana***

Genus ***Stylochus*** Ehrenberg, 1831

Fig. 20

SYNONYMY *Dicelis* Schmarda, 1859; *Diopsis* Diesing, 1862; *Imogine* Girard, 1853; *Eustylochus* Verrill, 1893.

STYLOCHIDAE: STYLOCHINAE Oval or somewhat discoid forms with retractile nuchal tentacles. Marginal eyes in band of variable extent; cerebral eyes in single mass or two distinct clusters; tentacular eyes within, or at bases of, tentacles; frontal eyes often present. Pharynx long, more or less in middle of body; intestinal branches not anastomosing. Male and female genital pores closely approximate in posterior quarter of body. Seminal vesicle globular or elongate, or merged with

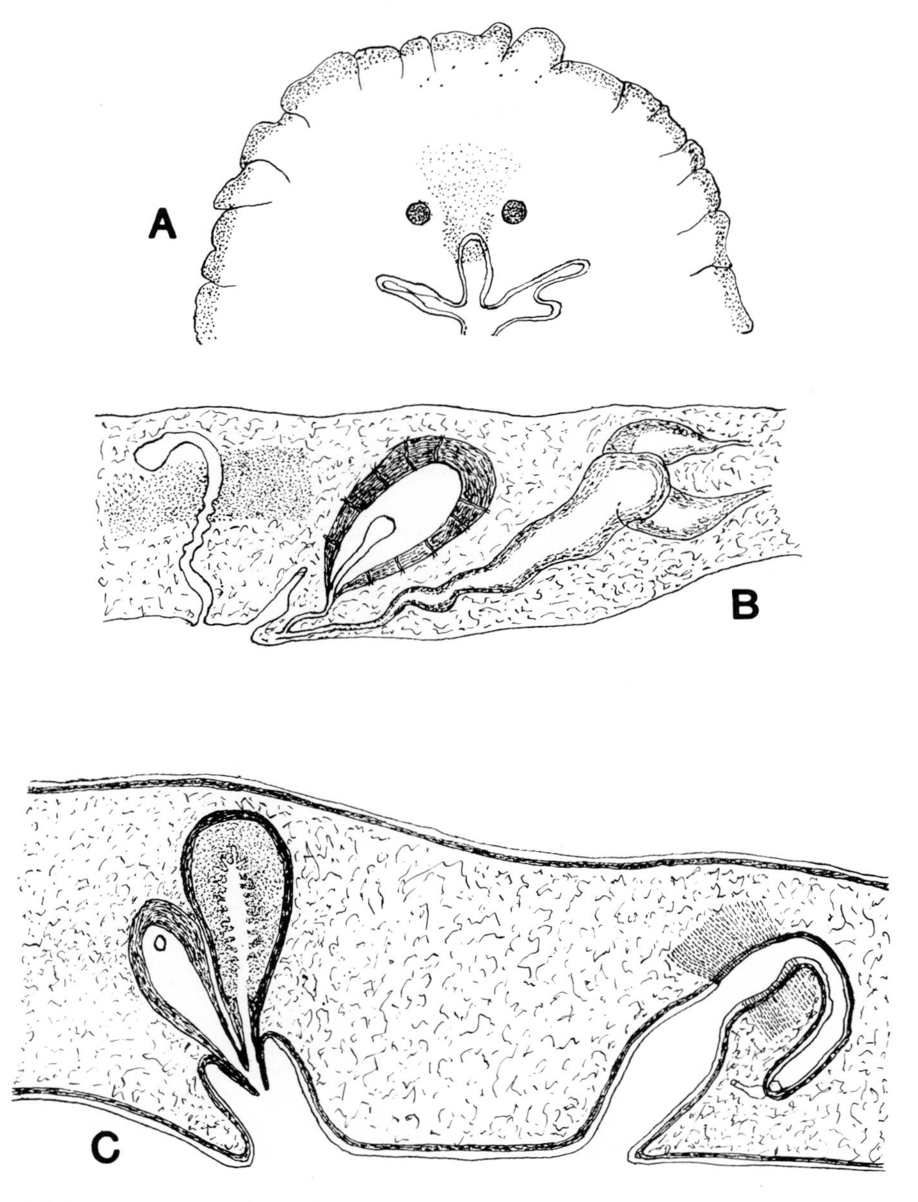

Fig. 20 *Stylochus.* **A,** arrangement of eyes of *S. (Imogine) oculiferus* (after Hyman); **B,** sagittal section of copulatory organs of *S. (I.) oculiferus* (after Hyman); **C,** sagittal section of copulatory organs of *S. (Stylochus) suesensis* (after Palombi).

muscular spermiducal bulbs to form tripartite anchor-shaped structure. Large prostatic organ horizontal and dorsal to ejaculatory duct, or vertical to penis-papilla; epithelial lining thrown into radial folds. Generally, penis-papilla stout, unarmed, lying in male antrum. Vagina short, crozier-like in lateral view; epithelial lining of 'shell'-chamber often thrown into a helical fold. Life-history direct, or indirect through Götte's larva.

In a list, prepared by Galleni & Puccinelli (1975), of the chromosomes found among polyclads, the haploid number found among species of *Stylochus* varies between 6 and 10.

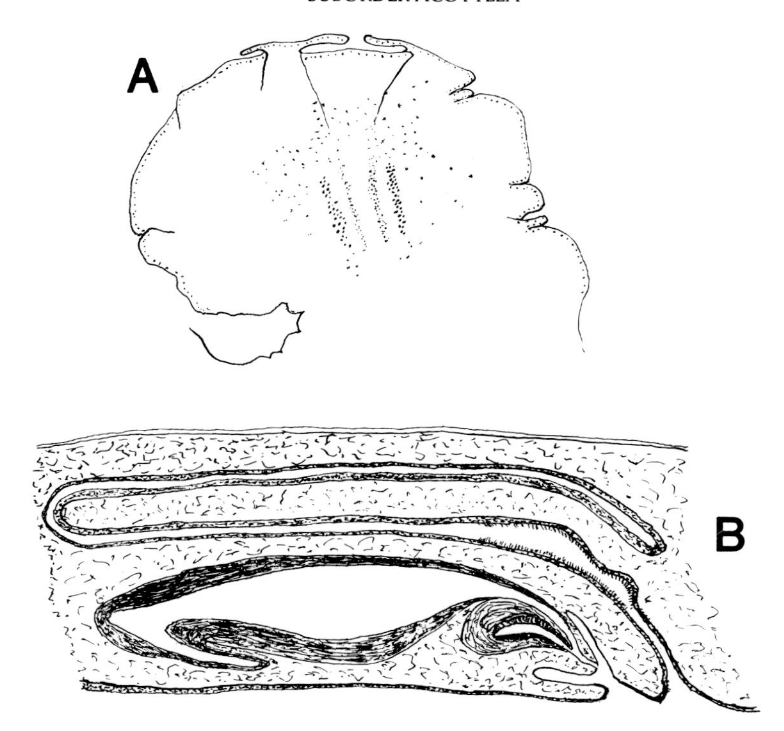

Fig. 21 *Indistylochus hewatti*. **A**, arrangement of eyes; **B**, sagittal section of copulatory organs (after Hyman).

TYPE-SPECIES *S. (Stylochus) suesensis* Ehrenberg, 1831, *sensu* Palombi, 1928. Type by monotypy.

TYPE-LOCALITY Tor, Red Sea.

SYSTEMATIC NOTE du Bois–Reymond Marcus & Marcus (1968) point out that the fifty or so species so far assigned to the genus *Stylochus* may be separated into two distinct groups on the basis of the structure of the seminal vesicle in mature specimens. These authors point out that in the type-species of *Stylochus* the seminal vesicle is a simple structure receiving thin-walled vasa deferentia, which may or may not unite before entering the vesicle. Species possessing this feature the authors assign to the subgenus (*Stylochus*). In the second group, the distal regions of the vasa deferentia form muscular spermiducal bulbs, which fused with the inner end of the seminal vesicle to form a trilobed or anchor-shaped organ, and for species with this feature the authors lower the genus *Imogine* Girard, 1853, to the rank of subgenus, with *Stylochus (Imogine) oculiferus* (Girard, 1853) as its type-species.

MISCELLANEOUS NOTES (i) instances have been recorded among species of *Stylochus* in which mature worms lack marginal eyes posteriorly, while young specimens have them; (ii) *S. ellipticus* is said to possess a penis-stylet; (iii) in *Stylochus ticus* the penis-papilla lies in a penis-pocket.

Genus ***Indistylochus*** Hyman, 1955 Fig. 21

STYLOCHIDAE: STYLOCHINAE Elongate oval forms without tentacles. Marginal eyes small, confined to a single irregular row anteriorly; cerebral eyes in two elongate clusters lying between two similar clusters of tentacular eyes and merging with scattered frontal eyes. Much-folded pharynx mainly in

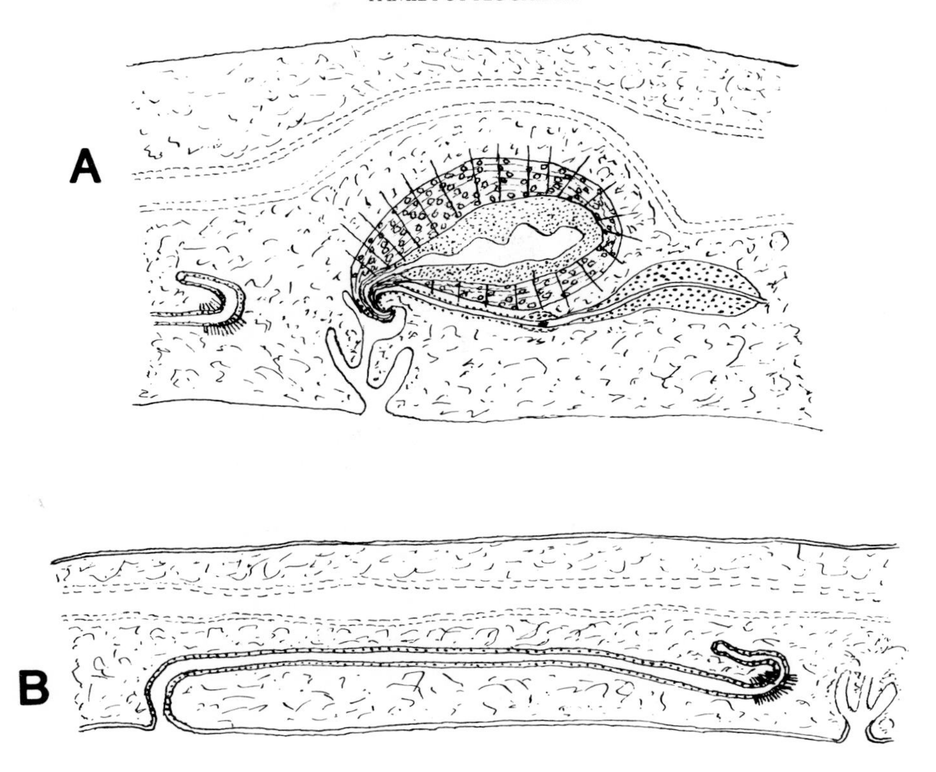

Fig. 22 *Meixneria furva.* **A,** sagittal section of ♂ copulatory organs; **B,** sagittal section of ♀ copulatory organs (after Bock).

hinder half of body. Genital pores approximate in posterior region. Vasa deferentia enter separately into large elongate-oval seminal vesicle. Prostatic organ relatively small, ventral to seminal vesicle, and provided with a thick muscular wall and a smooth epithelial lining. Penis-papilla occupying much of male antrum. Very long vagina forming an anteriorly-directed loop dorsal to male complex and terminating at entrance of uterine canals, dorsal to female pore.

TYPE-SPECIES *I. hewatti* Hyman, 1955. Type by original designation.

TYPE-LOCALITY Under beds of *Mytilus* between tides at East Point Beach, Boquerón, Puerto Rico.

Genus *Meixneria* Bock, 1913

Fig. 22

STYLOCHIDAE: STYLOCHINAE Elongate-oval forms with tentacles containing many eyes. Marginal eyes anterior; cerebral eyes smaller than tentacular eyes, between which they are arranged in two clusters; frontal eyes few. Pharynx about half as long as body, centrally situated. Genital pores widely separated; the female near hind margin of body. Male copulatory complex immediately posterior to pharynx. Spermiducal bulbs, but no seminal vesicle; ejaculatory duct long. Prostatic organ large, horizontally disposed dorsally to ejaculatory duct, with tall epithelial lining thrown into radial folds. Small, unarmed penis-papilla enclosed in penis-pocket with well-developed penis-sheath. Vagina exceptionally long, forming an anteriorly-directed loop reaching to near male complex. Vagina externa long; 'shell'-chamber short.

TYPE-SPECIES *M. furva* Bock, 1913. Type by original designation.

TYPE-LOCALITY Under tree trunk on shore at Lem Ngob, Gulf of Siam.

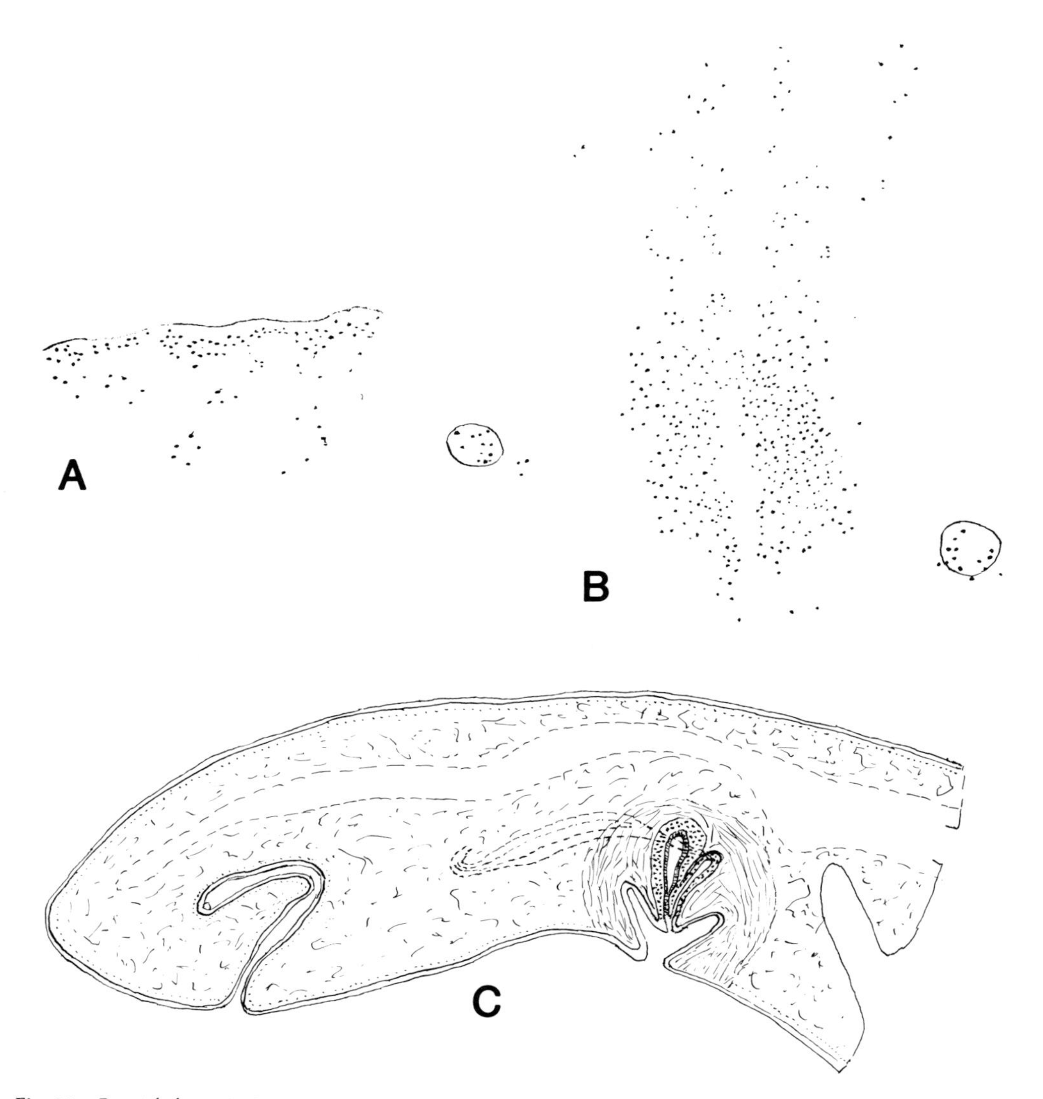

Fig. 23 *Parastylochus astis.* **A,** anterior marginal eyes; **B,** cerebral and tentacular eyes; **C,** sagittal section of copulatory organs (after Bock).

Genus *Parastylochus* Bock, 1913

Fig. 23

STYLOCHIDAE: STYLOCHINAE Oval forms with tentacles containing eyes. Marginal eyes in complete series round body; cerebral eyes in two ill-defined elongate clusters between tentacles; tiny frontal eyes widely distributed. Pharynx long, much folded laterally. Genital pores widely separated; female pore near hind end of body. Male copulatory complex closely posterior to pharynx and dorsal to male pore. Spermiducal bulbs elongate, posterior to prostatic organ and male pore; no seminal vesicle. Ejaculatory duct short, unites with prostatic duct on apex of penis-papilla. Pyriform prostatic organ small, with smooth epithelial lining. Unarmed penis-papilla short and wide, lying in male antrum. Ejaculatory duct, prostatic organ and penis-papilla vertically-disposed in bulbous muscular sheath. Vagina short, crozier-shaped in lateral view.

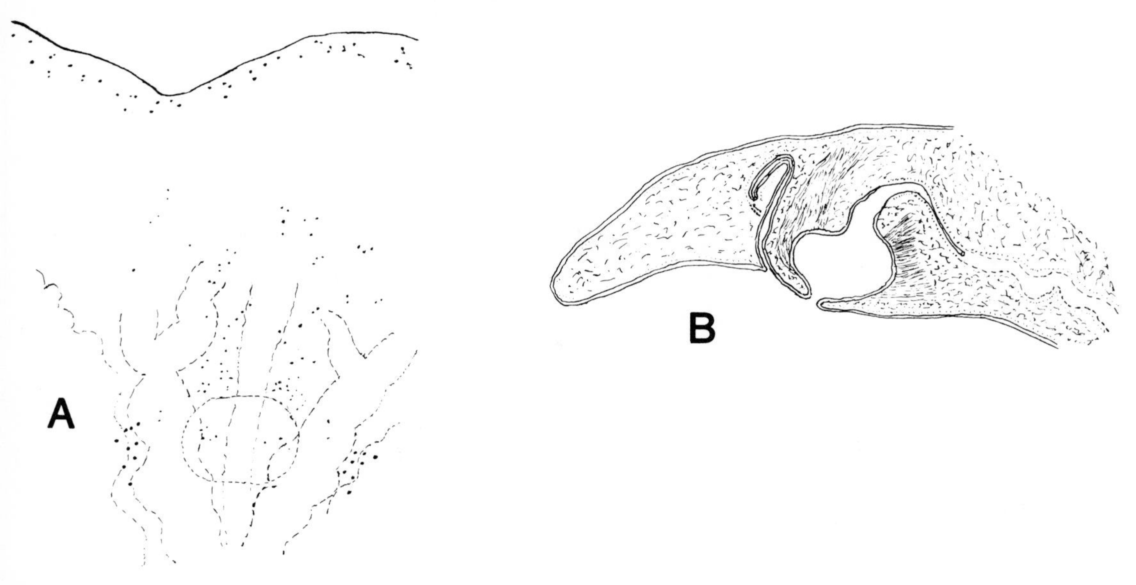

Fig. 24 *Ilyplana aberrans.* **A,** cerebral and tentacular eyes, cerebral organ and adjacent intestinal branches; **B,** sagittal section of copulatory organs (after Bock).

TYPE-SPECIES *P. astis* Bock, 1913. Type by original designation.

TYPE-LOCALITY Edam I., Java Sea.

Genus *Ilyplana* Bock, 1925 Fig. 24

STYLOCHIDAE: STYLOCHINAE Delicate, elongate-oval forms with tentacles containing few eyes. Marginal eyes in complete series round body. Cerebral eyes in two elongate clusters; frontal eyes apparently absent. Pharynx in middle third of body. Genital pores closely approximate posteriorly. No spermiducal bulbs or seminal vesicle. Prostatic organ vertically disposed posteriorly to ejaculatory duct, with feeble musculature and smooth epithelial lining and invested with extracapsular gland-cells. This organ and the ejaculatory duct lie within a large protuberance in wall of male antrum. Male antrum and protuberance hold efferent ducts of a mass of cyanophilic gland-cells enclosing the male copulatory complex. Vagina short, crozier-shaped in lateral view.

TYPE-SPECIES *I. aberrans* Bock, 1925. Type by original designation.

TYPE-LOCALITY On sand and mud in 60 metres, Colville Channel, New Zealand.

NOTE At present, this genus is based on the anatomy of one specimen, which appears to be not fully developed. Bock himself suspects that this specimen possesses an abnormal male copulatory complex, and he implied that further specimens needed to be examined before the genus could be firmly established.

Subfamily **IDIOPLANINAE** Bresslau, 1933

STYLOCHIDAE Penis-papilla may be armed with a stylet. Lang's vesicle bulbous, tubular or crescentic. Uterine canals open into vagina by a common canal

KEY TO IDIOPLANINE GENERA

1. Penis-papilla provided with stylet 2
   Penis-papilla without stylet (exc. *Limnostylochus borneensis*) 3

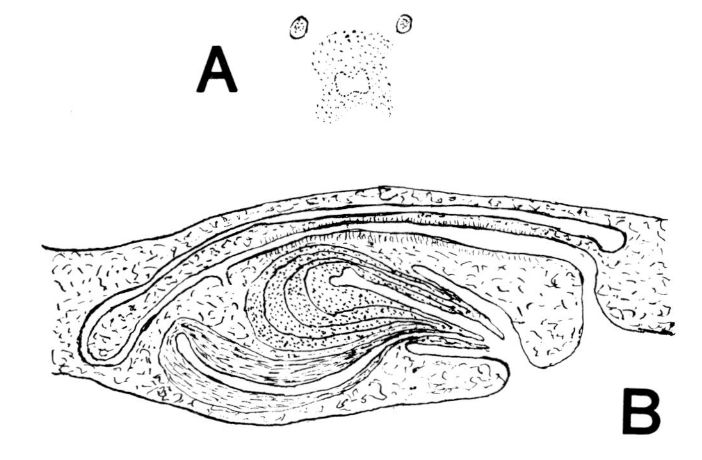

Fig. 25 *Idioplana australiensis.* **A**, cerebral and tentacular eyes; **B**, sagittal section of copulatory organs (modified after Woodworth).

2. Seminal vesicle elongate; vagina bulbosa present . . . . ***Neostylochus***
   Seminal vesicle trilobed; no vagina bulbosa . . . . ***Ancoratheca***
3. Lang's vesicle bulbous or tubular 4
   Lang's vesicle crescentic 6
4. Vagina short, directed posteriorly from genital pore . . . . . ***Leptostylochus***
   Vagina long, looping anteriorly beyond male complex 5
5. Seminal vesicle absent . . . . ***Idioplana***
   Seminal vesicle small and simple. . . . . ***Crassiplana***
6. Tentacles long; epithelial lining of prostatic organ with three longitudinal chambers . . . ***Idioplanoides***
   No tentacles; epithelial lining of prostatic organ smooth. . . . . ***Limnostylochus***

Genus ***Idioplana*** Woodworth, 1898 Fig. 25

STYLOCHIDAE: IDIOPLANINAE Oval forms, widened anteriorly. Tentacles distinctly anterior to cerebral organ and contain eyes. Marginal eyes variable in extent; cerebral eyes in dense cluster over cerebral organ. Pharynx with deep lateral folds in mid-third of body. Spermiducal bulbs well developed; no seminal vesicle. Prostatic organ horizontally disposed dorsally to spermiducal bulbs its epithelial lining shallow and smooth. Ejaculatory duct joins prostatic duct in apex of stoutly-conical penis-papilla lying in male antrum. Vagina long, forming anteriorly-directed loop. Lang's vesicle bulbous.

TYPE-SPECIES *I. australiensis* Woodworth, 1898. Type by original designation.

TYPE-LOCALITY On reef at Hope I., Great Barrier Reef, Australia.

SYSTEMATIC NOTE Palombi (1928) described a specimen from the Red Sea as *I. australiensis,* but a close examination of his description shows that his specimen bears a closer morphological resemblance to the genus *Idioplanoides* than to *Idioplana.* Palombi's description is, however, incomplete, and it is therefore difficult to assess the true relationship of the specimen within the family Stylochidae. On the other hand, there are in Palombi's specimen two or three features, such as the disposition of the prostatic organ ventral to a seminal vesicle, the radial folding of the epithelial lining of the prostatic organ and the possible presence of a penis-sheath. Moreover, Palombi's figure of the female organs suggests the later formation of a crescentic Lang's vesicle. If these features are genuine, they suggest that Palombi's specimen does not fit into any known stylochid genus.

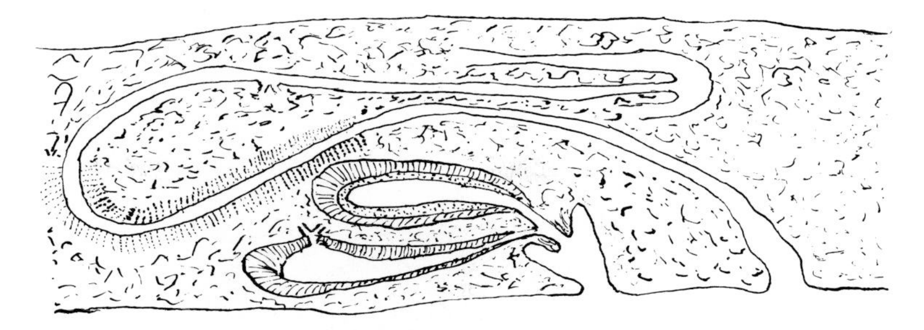

Fig. 26 *Idioplanoides insignis.* Sagittal section of copulatory organs (after Laidlaw).

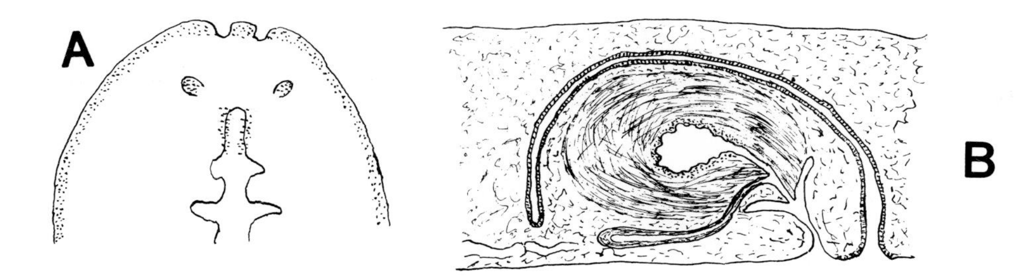

Fig. 27 *Crassiplana albatrossi.* **A**, arrangement of eyes anteriorly; **B**, sagittal section of copulatory organs (after Hyman).

Genus ***Idioplanoides*** Barbour, 1912 Fig. 26

SYNONYMY *Woodworthia* Laidlaw, 1904.

STYLOCHIDAE: IDIOPLANINAE Broadly-oval forms with tentacles containing eyes. Marginal eyes in complete series round body; cerebral eyes in simple cluster or in two indistinct groups; no frontal eyes. Pharynx in middle third of body. Genital pores closely approximate, posteriorly situated. Seminal vesicle elongate; no spermiducal bulbs. Prostatic organ well developed, horizontally disposed dorsally to seminal vesicle. Penis-papilla broadly conical, unarmed and lying in male antrum. Vagina long, looped anteriorly beyond male copulatory complex. Lang's vesicle crescentic or U-shaped, with anteriorly-directed limbs.

TYPE-SPECIES *I. [Woodworthia] insignis* (Laidlaw, 1904). Type by original designation.

TYPE-LOCALITY Pearl Banks, Gulf of Manaar, Sri Lanka.

Genus ***Crassiplana*** Hyman, 1955 Fig. 27

STYLOCHIDAE: IDIOPLANINAE Oval fleshy forms with nuchal tentacles near anterior margin of body and containing eyes. Marginal eyes in complete series round body; cerebral eyes in two elongate clusters lying posteriorly to tentacles; no frontal eyes. Pharynx mainly in anterior half of body. Genital pores closely separated in hind third of body. Seminal vesicle small; prostatic organ dorsal to ejaculatory duct and with exceptionally thick muscular wall, epithelial lining low. Ejaculatory duct penetrates wall of prostatic organ to join with prostatic duct. Unarmed penis-papilla stoutly conical, lying in male antrum. Vagina long, curving anteriorly round male complex. Lang's vesicle small and narrow (?)

TYPE-SPECIES *C. albatrossi* Hyman, 1955. Type by original designation.

TYPE-LOCALITY In 19 metres off mouth of Rio de la Plata, Uruguay.

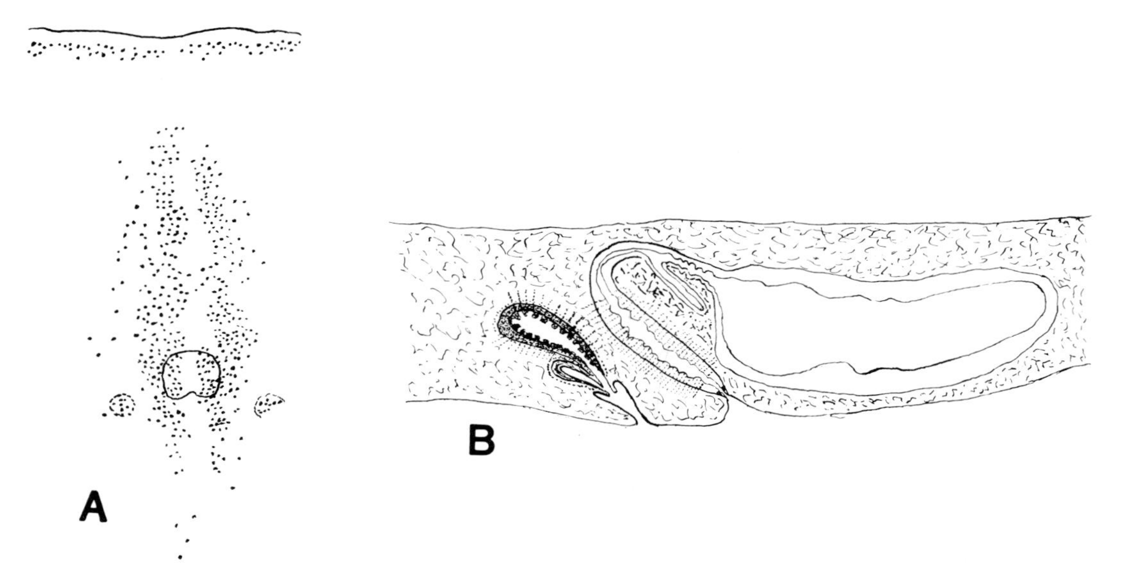

Fig. 28 *Leptostylochus elongatus.* **A,** arrangement of eyes anteriorly; **B,** sagittal section of copulatory organs (after Bock).

Genus ***Leptostylochus*** Bosk, 1925 Fig. 28

STYLOCHIDAE: IDIOPLANINAE Oval or elongate forms, often thin and delicate. Tentacles feebly developed or apparently absent, although tentacular groups of eyes occur, sometimes mingling with cerebral eyes. Marginal eyes confined to anterior half of body; cerebral eyes in two elongate clusters, which may merge into a single group; occasionally a few frontal eyes. Pharynx mainly in middle third of body, long, with several pairs of shallow lateral folds. Genital pores closely separated in posterior third of body. Spermiducal bulbs well developed; no definite seminal vesicle. Prostatic organ elongate or pyriform, epithelial lining usually thrown into radial folds, horizontally disposed dorsally to ejaculatory duct. Penis-papilla small, unarmed, sometimes lying in ill-defined penis-pocket. Vagina moderately long, forming anteriorly-directed loop. Female antrum may possess sphincter; 'shell'-chamber enlarged, usually dorso-ventrally compressed. Lang's vesicle bulbous.

TYPE-SPECIES *L. elongatus* Bock, 1925. Type by original designation.

TYPE-LOCALITY Under stones on shore of Ponui I., Auckland, New Zealand.

Genus ***Limnostylochus*** Bock, 1913 Fig. 29

SYNONYMY *Shelfordia* Stummer-Traunfels, 1902

STYLOCHIDAE: IDIOPLANINAE Elongate-oval forms without tentacles. Marginal eyes in band of variable extent; cerebral eyes in one or two clusters, which may blend with two distinct elongate clusters of tentacular eyes; frontal eyes few. Pharynx mainly in anterior half of body. Genital pores well separated, situated posteriorly. Seminal vesicle may be present. Prostatic organ dorsal to ejaculatory duct, very elongate and undulating, with smooth epithelial lining. Penis-papilla small, stoutly conical, sometimes with stylet. Vagina very long, looped anteriorly beyond male complex. Lang's vesicle crescentic or U-shaped, limbs directed anteriorly.

TYPE-SPECIES *L. [Shelfordia] borneensis* (Stummer-Traunfels, 1902). Type by monotypy.

TYPE-LOCALITY In stagnant fresh-water, Kuching, Sarawate, Borneo.

NOTE The three known species of this genus are known only from stagnant fresh or brackish water.

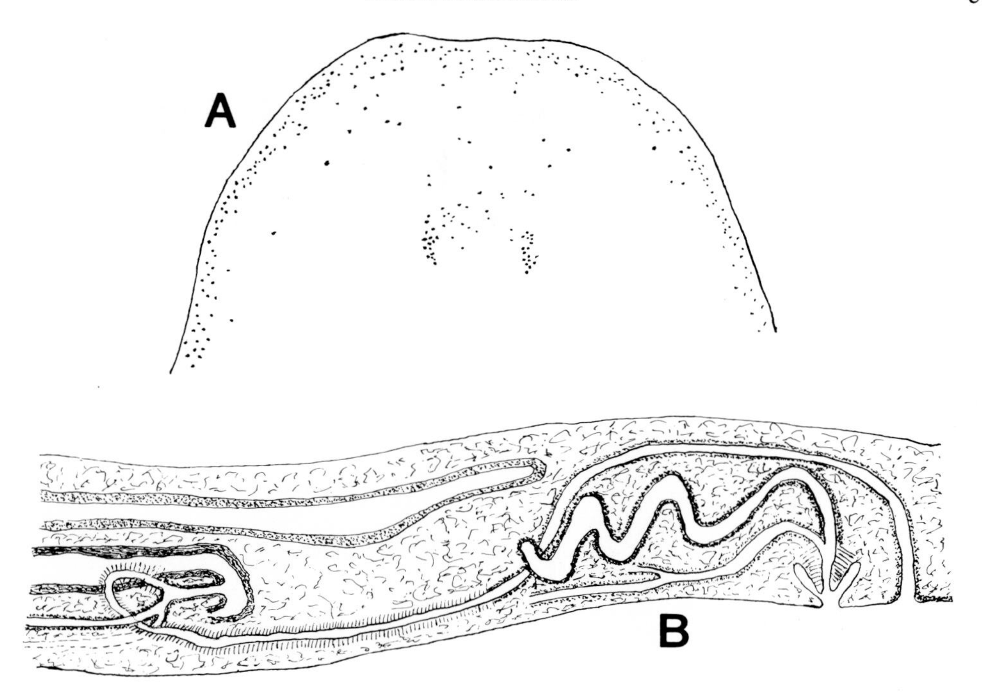

Fig. 29 *Limnostylochus annandalei.* **A,** arrangement of eyes anteriorly; **B,** sagittal section of copulatory organs (after Kaburaki).

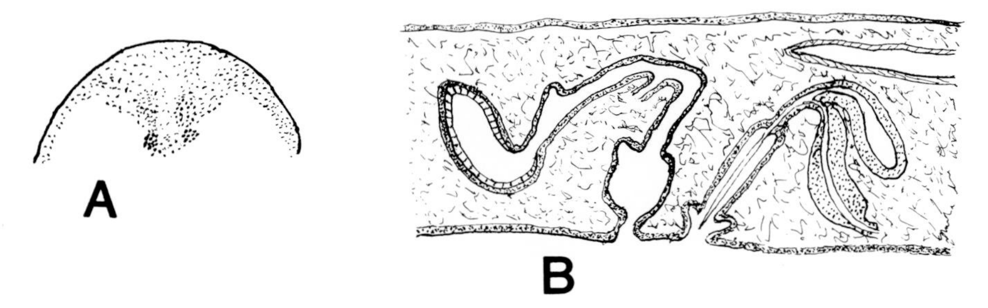

Fig. 30 *Neostylochus fulvopunctatus.* **A,** arrangement of eyes anteriorly; **B,** sagittal section of copulatory organs (after Yeri & Kaburaki).

SYSTEMATIC NOTE The type-species apparently possesses a seminal vesicle and a penis-papilla with a stylet, whereas the other known species do not. It therefore seems possible that the two groups may need to be separated subgenerically, or even generically, but further material of all three species needs to be examined.

Genus *Neostylochus* Yeri & Kaburaki, 1920

Fig. 30

STYLOCHIDAE: IDIOPLANINAE Delicate and translucent oval forms without tentacles. Marginal eyes in anterior band; tentacular eyes in two rounded clusters; cerebral eyes merge with tentacular and widely-distributed frontal eyes. Pharynx in middle third of body. Genital pores closely approximate,

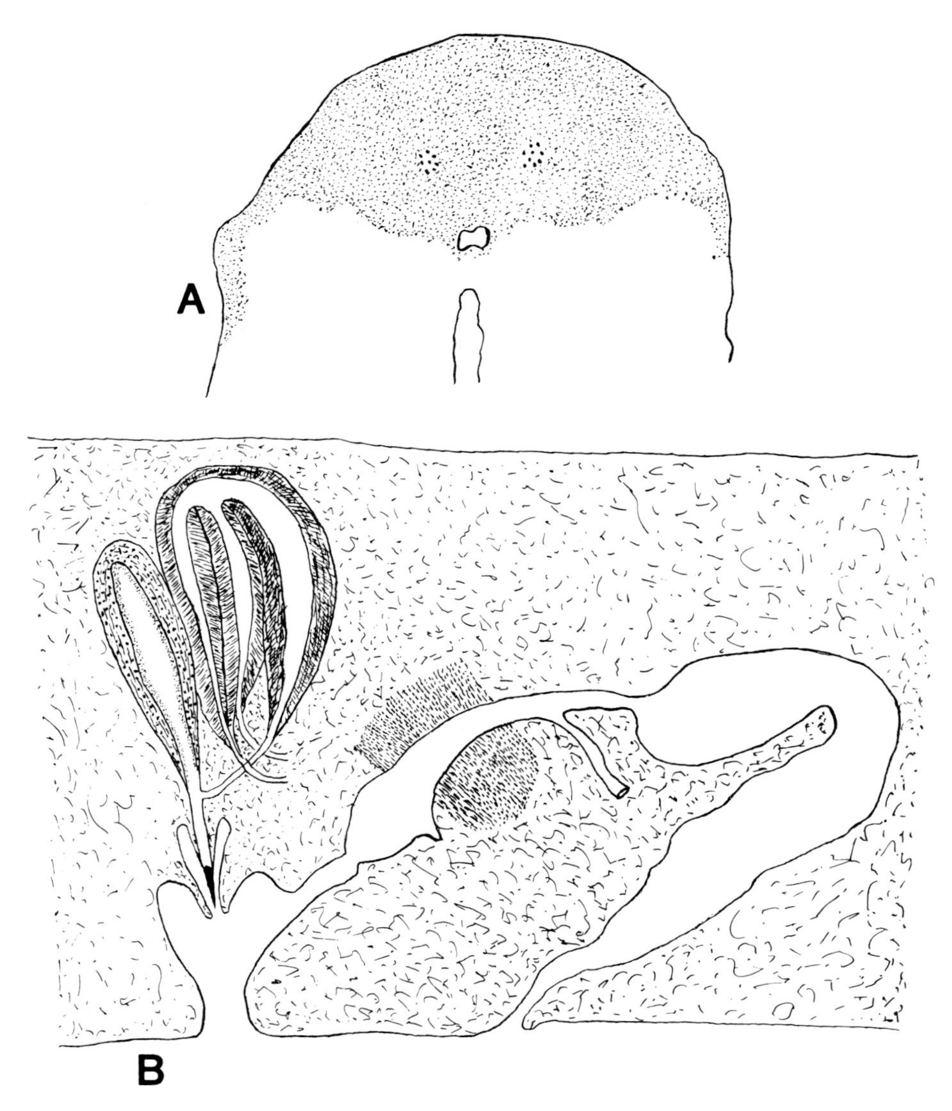

Fig. 31 *Ancoratheca australiensis.* **A**, arrangement of eyes anteriorly; **B**, sagittal section of copulatory organs (after Prudhoe).

well separated from hind margin of body. Vasa deferentia open side by side into narrow muscular seminal vesicle. Prostatic organ elongate, relatively small, with smooth epithelial lining; horizontally disposed dorsally to seminal vesicle. Slender penis-papilla in penis-pocket. Vagina bulbosa with thick plicate wall. Lang's vesicle bulbous. Uterine canals anteriorly confluent.

TYPE-SPECIES *N. fulvopunctatus* Yeri & Kaburaki, 1918. Type by original designation.

TYPE-LOCALITY Between tide marks at Misaki, Japan.

Genus ***Ancoratheca*** Prudhoe, 1982 Fig. 31

STYLOCHIDAE: IDIOPLANINAE Body oval. Marginal eyes in band of variable extent; from pharynx, cerebro-frontal eyes spread fanwise over cephalic region; tentacular eyes in two small clusters among

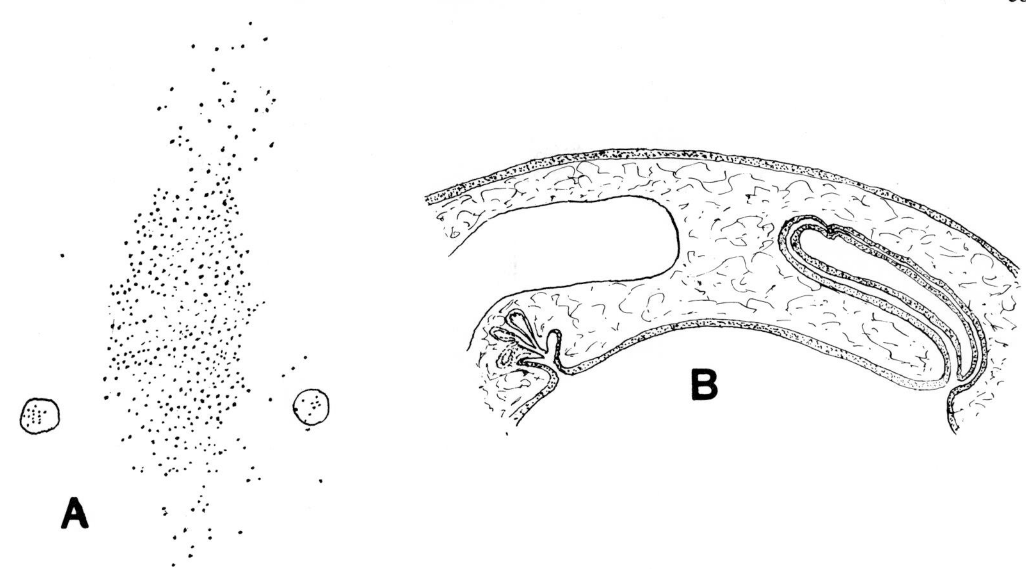

Fig. 32 *Cryptophallus wahlbergi*. **A**, cerebral and tentacular eye-clusters; **B**, sagittal section of copulatory organs (after Bock).

cerebro-frontal eyes. Pharynx in middle third of body or more posteriorly placed. Genital pores adjacent in posterior region of body. Spermiducal bulbs and seminal vesicle merge to form a trilobed or anchor-shaped structure. Prostatic organ well developed, elongate, with smooth shallow epithelial lining, disposed more or less dorsally to seminal vesicle. Penis-papilla bearing short stylet, lying in penis-pocket with shallow penis-sheath. Vagina in a more or less posteriorly-directed loop, terminating in Lang's vesicle, which may open on ventral surface of body.

TYPE-SPECIES *A. australiensis* Prudhoe, 1982. Type by original designation.

TYPE-LOCALITY In 15 metres, Upper Spencer Gulf, South Australia.

Subfamily **CRYPTOPHALLINAE** Bresslau, 1933

STYLOCHIDAE Female complex without Lang's vesicle; with ductus vaginalis opening into female antrum, or to exterior ventrally through an independent aperture.

KEY TO CRYPTOPHALLINE GENERA

1.	Dorsal surface of body with numerous tubercles or whitish spots	***Mexistylochus***
	Dorsal surface without tubercles or spots	2
2.	Mouth and genital pores in 'genito-buccal' antrum	***Mirostylochus***
	Mouth well separated from genital pores	3
3.	Prostatic organ small, vertically disposed	***Cryptophallus***
	Prostatic organ large, longitudinally disposed	***Kaburakia***

Genus ***Cryptophallus*** Bock, 1913

Fig. 32

STYLOCHINAE: CRYPTOPHALLINAE Fleshy, somewhat broadly-oval forms with shallow tentacles containing eyes. Marginal eyes in series round body; cerebral eyes numerous, arranged in elongate cluster or two indistinct clusters lying between tentacles and merging with frontal eyes. Pharynx long, with posterior appendix, more or less centrally situated. Male copulatory complex relatively small, ventral to pharyngeal appendix. No seminal vesicle. Spermiducal bulbs pyriform or elongate. Small

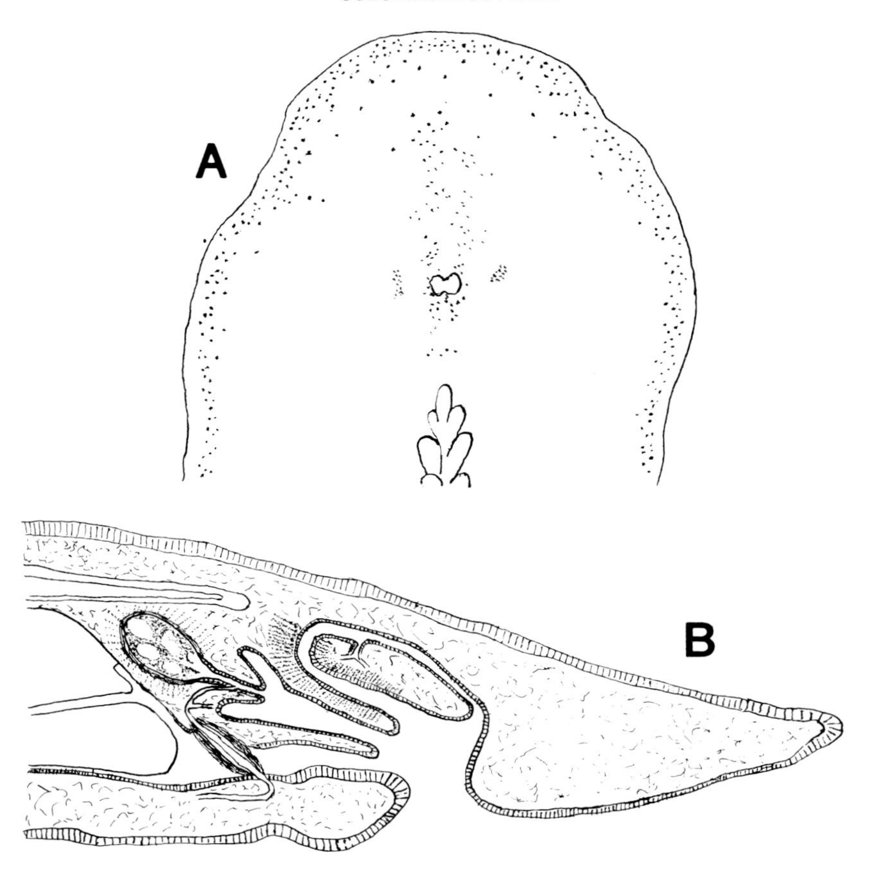

Fig. 33 *Mirostylochus akkashiensis.* **A**, arrangement of eyes anteriorly; **B**, sagittal section of copulatory organs (after Kato).

pyriform prostatic organ disposed vertically to male pore and posteriorly to spermiducal bulbs, with shallow epithelial lining slightly folded. Conical penis-papilla short and thick, without penis-pocket. Ductus vaginalis opening into female antrum.

TYPE-SPECIES *C. wahlbergi* Bock, 1913. Type by original designation.

TYPE-LOCALITY Port Natal, Durban, South Africa.

Genus ***Mirostylochus*** Kato, 1937 Fig. 33

STYLOCHIDAE: CRYPTOPHALLINAE Fleshy, elongate-oval forms without tentacles. Marginal eyes in complete band around body; two tentacular eye-clusters; cerebral eyes in elongate cluster lying between tentacular clusters and widening anteriorly; frontal eyes few. Pharynx more than half as long as body. Digestive system and both copulatory complexes open into 'genito-buccal atrium' placed near hind end of body. No seminal vesicle; spermiducal bulbs elongate. Prostatic organ pyriform, dorsal to ejaculatory duct and lined with tall epithelial lining containing chambers. Penis-papilla small; no penis-pocket. Male antrum long and narrow. Vagina and ductus vaginalis open separately into 'genito-buccal atrium'.

TYPE-SPECIES *N. akkeshiensis* Kato, 1937. Type by monotypy.

TYPE-LOCALITY Dredged at about 8 metres, mouth of Lake Akkeshi, Hokkaido, Japan.

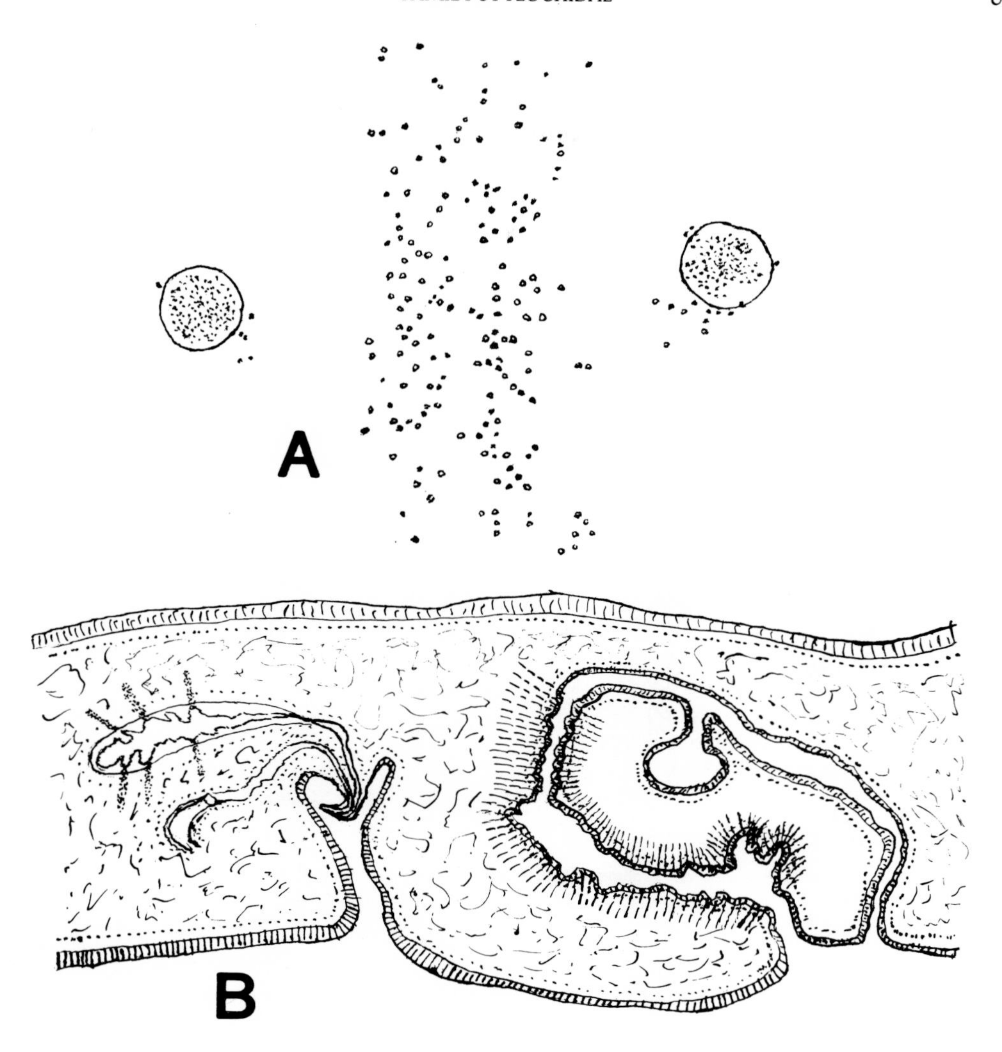

Fig. 34 *Kaburakia excelsa.* **A,** arrangement of eyes anteriorly; **B,** sagittal section of copulatory organs (after Bock).

Genus ***Kaburakia*** Bock, 1925

Fig. 34

STYLOCHIDAE: CRYPTOPHALLINAE Broadly-oval, fleshy forms with long retractile tentacles containing many eyes. Marginal eyes in several rows forming a band around body: cerebral eyes in one or two elongate clusters between tentacles; no frontal eyes. Long pharynx in middle third of body. Genital pores separated in hind region of body. Male copulatory complex well separated from pharynx. No seminal vesicle. Spermiducal bulbs elongate. Large claviform prostatic organ horizontally disposed dorsally to ejaculatory duct; epithelial lining thrown into radial folds. Penis-papilla small; no penis-pocket. Ductus vaginalis opening into female antrum or on ventral surface of body closely posterior to female genital pore.

TYPE-SPECIES *K. excelsa* Bock, 1925. Type by original designation.

TYPE-LOCALITY Under stones at low water, False Narrows, Nanaimo, Vancouver I., British Columbia, Canada.

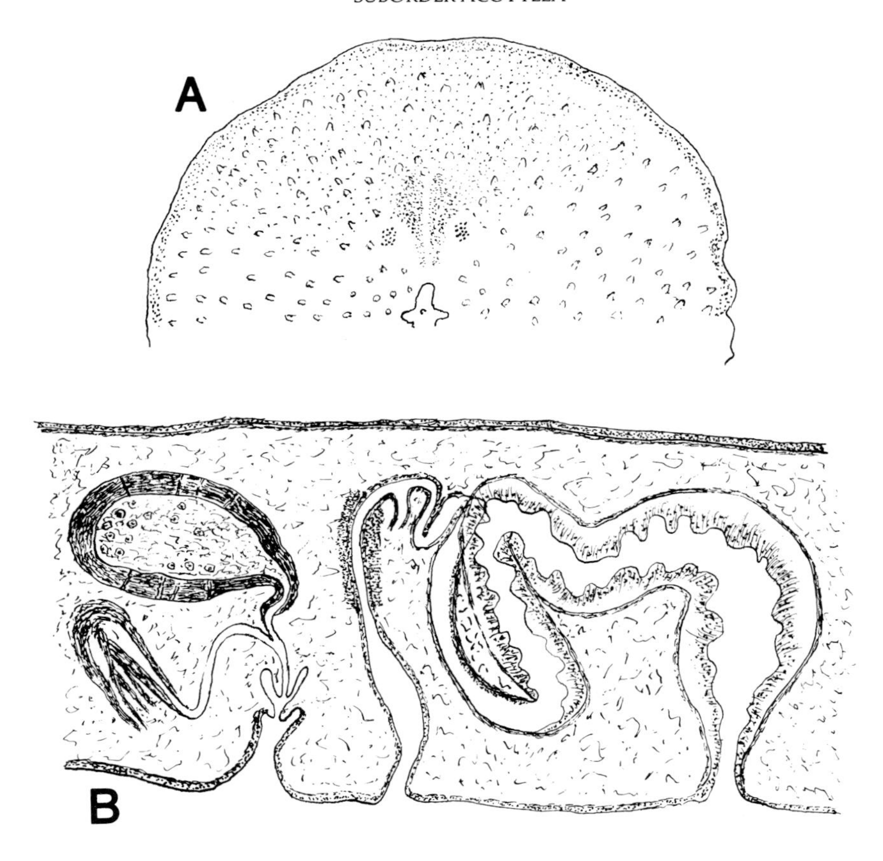

Fig. 35 *Mexistylochus tuberculatus.* **A,** arrangement of eyes anteriorly; **B,** sagittal section of copulatory organs (after Hyman).

Genus ***Mexistylochus*** Hyman, 1953 Fig. 35

STYLOCHIDAE: CRYPTOPHALLINAE Broadly-oval forms covered with low blunt tubercles or whitish spots dorsally. Boss-like nuchal tentacles may be present. Marginal eyes in series round body; tentacular eye-clusters distinctive among scattered frontal eyes; cerebral eyes between tentacular eye-clusters and merging with frontal eyes which may extend to posterior level of cerebral organ. Pharynx long; intestinal branches anastomosing. Genital pores closely separated posteriorly. Spermiducal bulbs and seminal vesicle form muscular tripartite or anchor-shaped structure. Prostatic organ well developed, lined with tall, chambered or folded epithelium, and horizontally or vertically disposed dorsally to ejaculatory duct. Penis-papilla small; penis-sheath present. Ductus vaginalis with tall epithelial lining thrown into radial folds and opening to exterior posteriorly to female genital pore.

TYPE-SPECIES *M. tuberculatus* Hyman, 1953. Type by original designation.

TYPE-LOCALITY On rocky shore, Miramar Beach, Guaymas, Sonora, Mexico.

SYSTEMATIC NOTE Hyman (1955*a* & *c*) regards *Mexistylochus* as a synonym of *Ommatoplana* Laidlaw, 1903, but this synonymy is here not accepted. There is, in fact, considerable uncertainty as to the systematic position of the genus *Ommatoplana,* because of the incomplete and confused description of *O. tuberculata* given by Laidlaw (1903*d*), but the description and later information given by Laidlaw (1903*e*) do suggest that the species is not a stylochid. *Ommatoplana* is here placed in the family Cryptocelididae in agreement with Bock (1913).

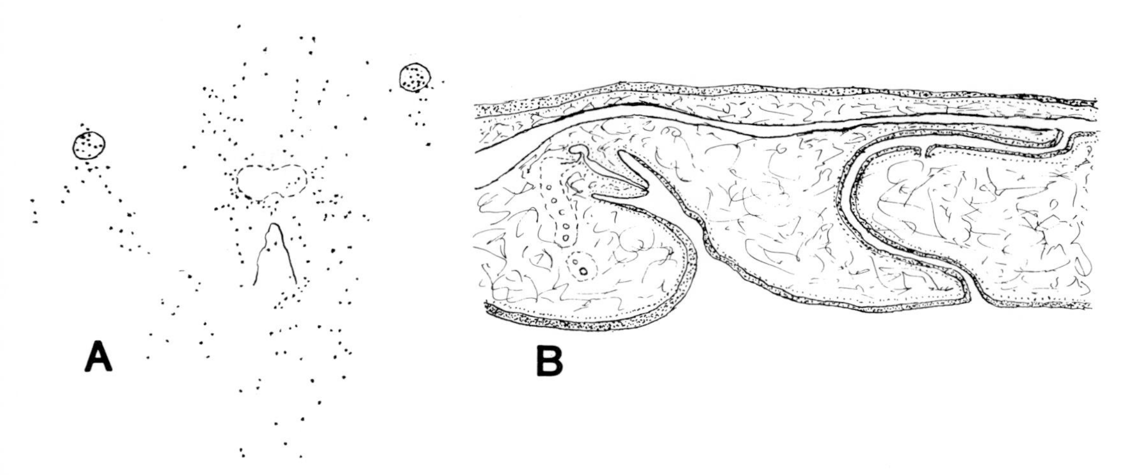

Fig. 36 *Enterogonia orbicularia.* **A**, arrangement of cerebral and tentacular eyes; **B**, sagittal section of copulatory organs (modified after Bock).

Subfamily **ENTEROGONIINAE** Bresslau, 1933

STYLOCHIDAE Female genital complex with genito-intestinal canal.

KEY TO ENTEROGONIINE GENERA

1. With spermiducal bulbs . . . . ***Discostylochus***
   Without spermiducal bulbs . . . . ***Enterogonia***

Genus ***Enterogonia*** Haswell, 1907 — Fig. 36

STYLOCHIDAE: ENTEROGONIINAE Elliptical forms with rudimentary tentacles containing eyes, anterior to level of cerebral organ. Marginal eyes in complete band around body; cerebral eyes in broad elongate cluster lying between tentacles and merging with numerous frontal eyes to spread fanwise over cephalic region. Pharynx long, in middle third of body. Vasa deferentia open into relatively long ejaculatory duct. No spermiducal bulbs or seminal vesicle. Prostatic organ ill-defined, consisting merely of a small swelling of the ejaculatory duct and lined with a glandular epithelium. Penis-papilla as a thick projection in roof of male antrum; no penis-pocket. Vagina long, forming anteriorly-directed loop; 'shell'-chamber with epithelium thrown into a helical fold. No Lang's vesicle; vagina terminating in genito-intestinal canal.

TYPE-SPECIES *E. [Polycelis] orbicularis* (Schmarda, 1859). Type by subsequent designation.

TYPE-LOCALITY Chile; but according to Stummer-Traunfels (1933), the species was originally found in New Zealand.

Genus ***Discostylochus*** Bock, 1925 — Fig. 37

STYLOCHIDAE: ENTEROGONIINAE Broadly- or elongate-oval forms with inconspicuous tentacles, each bearing several eyes. Marginal eyes confined to anterior half of body; cerebral eyes in single cluster, extending anteriorly from well behind cerebral organ to merge with frontal eyes diffusely scattered over cephalic region. Pharynx in middle third of body, of variable length and deeply folded. Genital pores closely approximate. Muscular spermiducal bulbs open into ejaculatory duct. Prostatic organ large, ovoid or elongate, horizontally disposed dorsally to ejaculatory duct and provided with low epithelial lining. Penis-papilla short and broad; no penis-pocket. Vagina of moderate length, forming anteriorly-directed loop and ending in genito-intestinal canal.

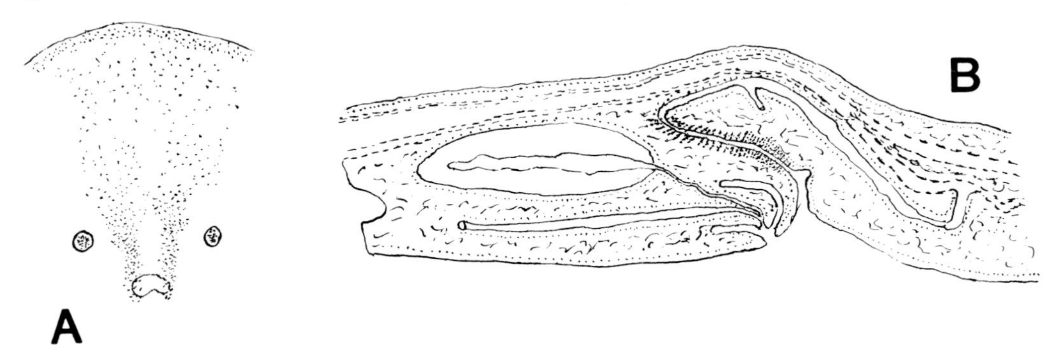

Fig. 37 *Discostylochus.* **A,** arrangement of eyes anteriorly in *D. yatsui* (after Kato); **B,** sagittal section of copulatory organs in *D. parcus* (after Bock).

TYPE-SPECIES *D. parcus* Bock, 1925. Type by original designation.

TYPE-LOCALITY Beneath sea-urchin, Hilo, Hawaiian Is.

SYSTEMATIC NOTE Two species, *D. parcus* Bock and *D. yatsui* Kato, have so far been assigned to this genus, but there is some doubt as to whether or not they are congeneric. In *D. parcus* the outline of the body is broadly oval, the spermiducal bulbs open by narrow canals into a thin-walled ejaculatory duct, and the uterine canals open into the vagina interna by a common canal, whereas in *D. yatsui* the body is elongate oval, the spermiducal bulbs unite with a very muscular ejaculatory duct to form, superficially, a trilobed seminal vesicle, and the uterine canals open individually into the vagina interna. It therefore seems that further specimens of both species need to be examined to establish their relationship with one another.

Family **CRYPTOCELIDIDAE** Laidlaw, 1903, emended Poche, 1926

(Includes PHAENOCELIDAE Stummer-Traunfels, 1933.)

STYLOCHOIDEA Elongate to rounded-oval forms, in which nuchal tentacles are absent or indistinct. Marginal eyes small; additional eyes distributed fanwise anteriorly or disposed in cerebral, tentacular and frontal groups. Mouth in middle third of body; intestinal trunk relatively short. Genital pores situated centrally or in posterior half of body and may unite to form a common antrum. Male copulatory complex closely posterior to pharynx and anterior to male pore. Prostatic organ variably developed, interpolated. Vagina often short, sometimes without Lang's vesicle.

KEY TO CRYPTOCELIDID SUBFAMILIES.

1.	Prostatic organ well developed	CRYPTOCELIDINAE subfam. nov.
	Prostatic organ not apparent or scarcely differentiated	TAENIOPLANINAE

Subfamily **CRYPTOCELIDINAE** subfam. nov.

CRYPTOCELIDIDAE Prostatic organ highly muscular, bulbous or ovoid.

KEY TO CRYPTOCELIDINE GENERA

1.	Body with tubercles on dorsal surface	***Ommatoplana***
	Body with smooth dorsal surface	2
2.	Penis-papilla with penis-pocket	3
	Penis-papilla without penis-pocket	4

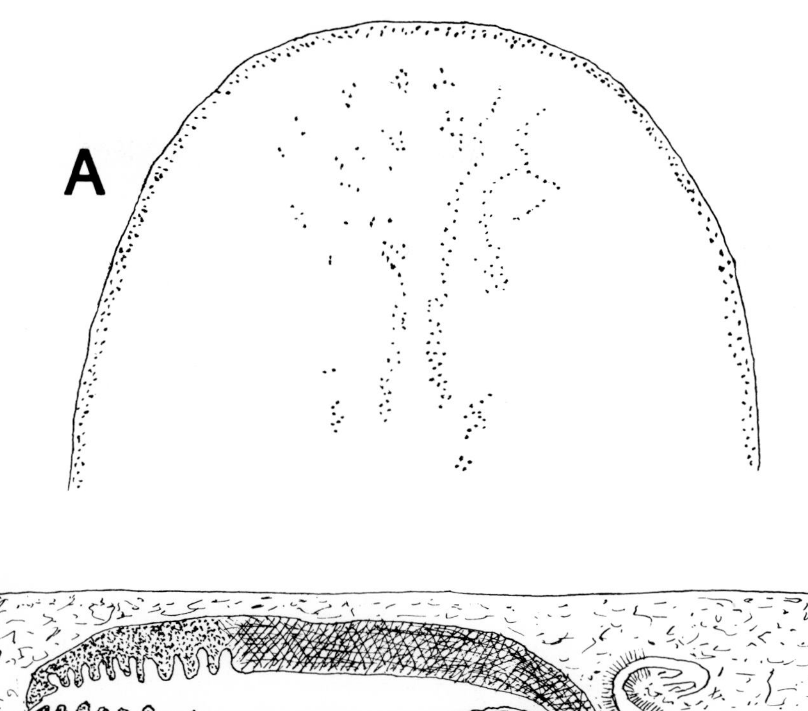

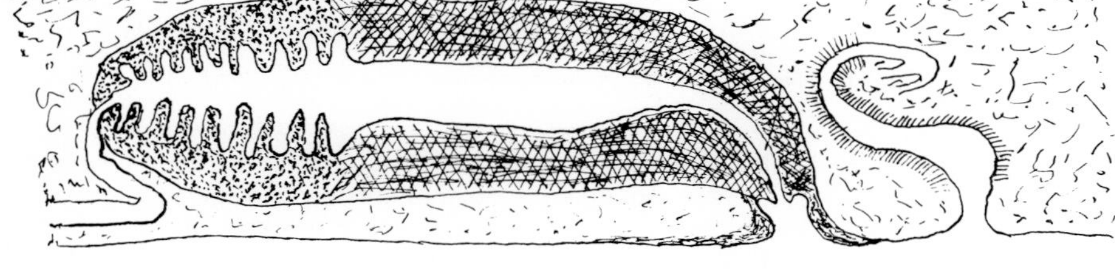

Fig. 38 *Cryptocelis alba.* **A,** arrangement of eyes anteriorly; **B,** sagittal section of copulatory organs (after Lang).

3. Lang's vesicle present
   (*a*) Spermiducal bulbs muscular; no seminal vesicle . . . . ***Igluta***
   (*b*) No spermiducal bulbs; seminal vesicle muscular . . . . ***Amemiyaia***
   Lang's vesicle absent . . . . ***Triadomma***
4. Without Lang's vesicle . . . .***Cryptocelis***
   Lang's vesicle large 5
5. Numerous cephalic eyes not in cerebral and tentacular clusters . . . . . ***Microcelis***
   Cerebral and tentacular clusters distinct 6
6. Without spermiducal bulbs or seminal vesicle . . . . ***Longiprostatum***
   With seminal vesicle. 7
7. Pharynx in middle third of body. . . . . ***Phaenocelis***
   Pharynx in posterior third of body . . . . ***Mesocela***

Genus ***Cryptocelis*** Lang, 1884 Fig. 38

CRYPTOCELIDIDAE: CRYPTOCELIDINAE Elongate-oval to broadly-oval forms of firm consistency and without tentacles. Marginal eyes usually in complete band round body; frontal eyes sometimes absent; cerebral and frontal eyes may merge into a single wide-spread mass; tentacular eyes may be well differentiated, or scarcely distinguishable, from cerebral eye clusters. Pharynx in middle third of body; intestinal branches many, may anastomose. Genital pores separated in middle or hind third of body.

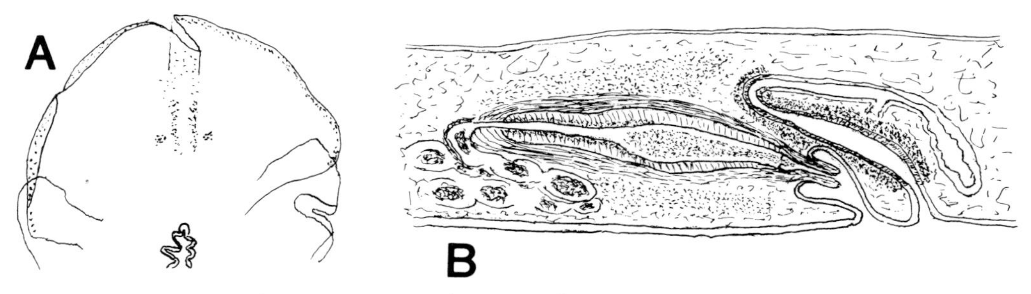

Fig. 39 *Longiprostatum rickettsi.* **A,** arrangement of eyes anteriorly; **B,** sagittal section of copulatory organs (after Hyman).

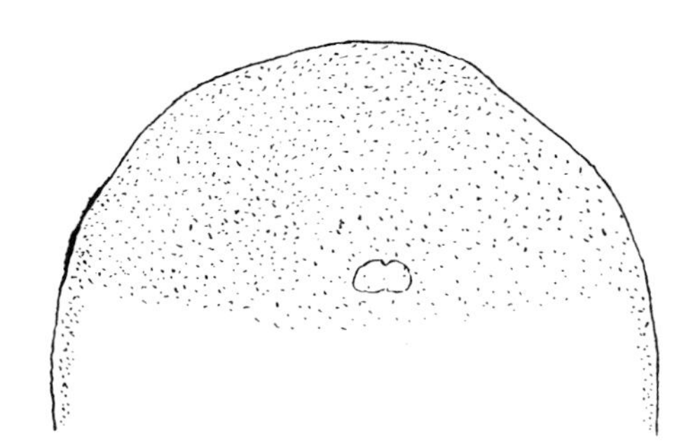

Fig. 40 *Microcelis schauinslandi.* Arrangement of eyes anteriorly (after Plehn).

Vasa deferentia or spermiducal bulbs open independently or by a common duct into a large prostatic organ, which consists of two chambers: a glandular proximal or anterior chamber, the epithelial lining of which is often thrown into deep radial folds; and a sinous distal or posterior chamber, the lining of which is smooth or slightly folded. Penis-papilla feebly developed or not apparent. Vagina simple, directed anteriorly from female pore; without Lang's vesicle; 'shell'-chamber spacious.

TYPE-SPECIES *C. alba* Lang, 1884. Type by subsequent designation.

TYPE-LOCALITY On mud at 4–10 metres, off Posillipo, Bay of Naples, Italy.

Genus ***Longiprostatum*** Hyman, 1953 Fig. 39

CRYPTOCELIDIDAE: CRYPTOCELIDINAE Somewhat elongate-oval forms, without tentacles. Marginal eyes limited to anterior half of body; cerebral eyes in two elongate clusters merging with frontal eyes; tentacular eye clusters distinct. Long, much-folded pharynx in middle third of body; intestinal branches anastomosing peripherally. Genital pores separated. No seminal vesicle. Vasa deferentia pass separately into slender, muscular prostatic organ, consisting of a narrow proximal portion and a distal portion having a wide lumen and lined with a smooth glandular epithelium. Distal portion opens directly into small conical penis-papilla, the efferent duct of which is lined with cuticle. Vagina simple, forming an anterior loop and ending in an elongate Lang's vesicle.

TYPE-SPECIES *L. rickettsi* Hyman, 1953. Type by original designation.

TYPE-LOCALITY Angeles Bay, Gulf of California, Mexico.

Genus ***Microcelis*** Plehn, 1899 Fig. 40

CRYPTOCELIDIDAE: CRYPTOCELIDINAE Oval body of firm consistency. Small marginal eyes in band extending to hind quarter of body. Numerous small eyes distributed fanwise anteriorly; cerebral and tentacular clusters not differentiated. Pharynx long, in middle third of body. Genital pores separated,

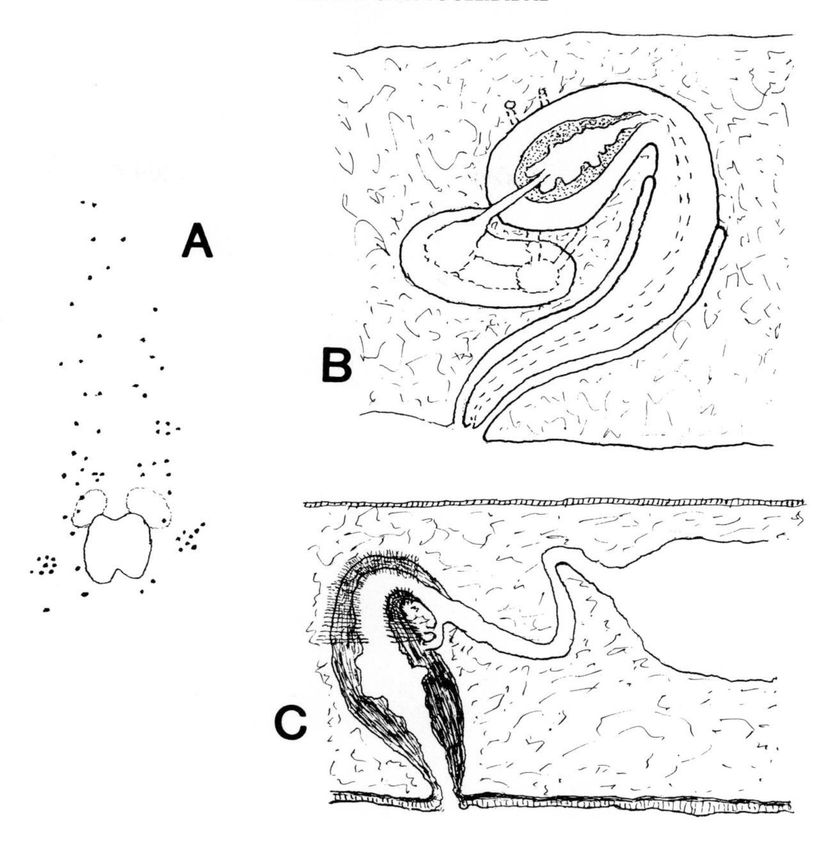

Fig. 41 *Phaenocelis purpurea.* **A**, cerebral and tentacular eye-clusters (after Hyman); **B**, sagittal section of ♂ copulatory organs; **C**, sagittal section of ♀ copulatory organs (after Stummer-Traunfels).

posterior. Vasa deferentia open into muscular ejaculatory duct. No seminal vesicle. Prostatic organ small and rounded. Penis-papilla unarmed. Lang's vesicle elongate and swollen.

TYPE-SPECIES *M. schauinslandi* Plehn, 1899. Type by original designation.

TYPE-LOCALITY. Tasmania.

NOTE This type-species is poorly described by modern standards and no figures of the copulatory organs in section are available.

Genus ***Phaenocelis*** Stummer-Traunfels, 1933 — Fig. 41

CRYPTOCELIDIDAE: CRYPTOCELIDINAE Elongate-oval forms of rather delicate consistency. Nuchal tentacles may be present. Very small marginal eyes in band of variable extent; cerebral and tentacular eye clusters differentiated; frontal eyes, when present, few. Pharynx mainly in anterior half of body, but well separated from cerebral organ. Genital pores separated, posterior to middle of body. Seminal vesicle well developed, dorso-ventrally arcuate. From seminal vesicle, ejaculatory duct forms a papilla projecting into lumen of muscular prostatic organ lined with a tall glandular epithelium. Penis-papilla long, unarmed; no penis-pocket. Vagina long, may have vagina bulbosa. Lang's vesicle elongate.

TYPE-SPECIES *P. [Leptoplana] purpurea* (Schmarda, 1859). Type by subsequent designation.

TYPE-LOCALITY Jamaica, Caribbean.

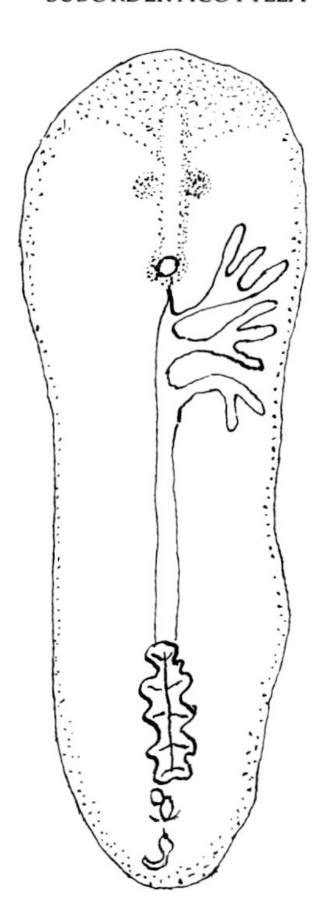

Fig. 42 *Mesocela caledonica.* Complete worm (after Jacubowa).

Genus ***Mesocela*** Jacubowa, 1906 Fig. 42

CRYPTOCELIDIDAE: CRYPTOCELIDINAE Elongate fleshy forms without tentacles. Marginal eyes in band around body. Two bands of cerebral eyes extend anteriorly from cerebral organ, passing between two rounded clusters of tentacular eyes and spreading fanwise in streaks to the anterior margin of body, where they merge with the marginal eyes. Pharynx relatively short, in hind third of body; intestinal trunk reaching anteriorly beyond cerebral organ; intestinal branches not anastomosing. Separate genital pores in hind region of body. Vasa deferentia open into small muscular seminal vesicle. Prostatic organ 'ungesonderter' (interpolated (?)). Conical penis-papilla unarmed, lying in male antrum. Vagina short; female copulatory complex said to be as in the genus *Cestoplana.*

TYPE-SPECIES *M. caledonica* Jacubowa, 1906. Type by original designation.

TYPE-LOCALITY Isle of Pines, New Caledonia, Pacific Ocean.

NOTE This genus is based on an incompletely developed worm, but Jacubowa placed it in the family Cestoplanidae, whereas Bock (1913) provisionally referred it to the family Cryptocelididae, which seems, at present, to be acceptable.

Genus ***Igluta*** du Bois Reymond Marcus & Marcus, 1968 Fig. 43

CRYPTOCELIDIDAE: CRYPTOCELIDINAE Elongate forms without tentacles. Marginal eyes in a band around body. Cerebral and tentacular eyes merge to form two elongate rows; no frontal eyes. Pharynx in anterior half of body. Genital pores approximate in middle third of body. Pair of spermiducal bulbs unite to form ejaculatory duct, which passes through the thick musculature of the prostatic organ, into the lumen of which it opens through a distinct papilla. Penis-papilla embedded in prostatic musculature

Fig. 43 *Igluta tipica.* **A**, arrangement of eyes anteriorly; **B**, sagittal section of copulatory organs (after du B.-R. Marcus & Marcus).

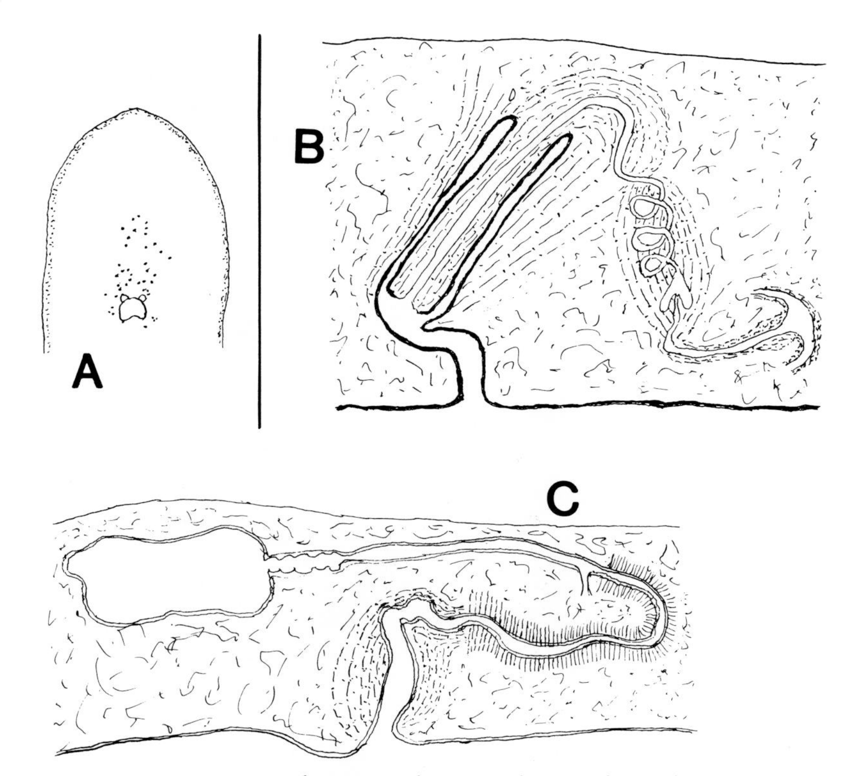

Fig. 44 *Amemiyaia pacifica.* **A**, arrangement of eyes anteriorly; **B**, sagittal section of ♂ copulatory organs; **C**, sagittal section of ♀ copulatory organs (after Kato).

and leads into a long stylet lying in penis-pocket with thick muscular walls. Male antrum short and narrow. Vagina rather muscular. Lang's vesicle large.

TYPE-SPECIES *I. tipica* du Bois Reymond Marcus & Marcus, 1968. Type by original designation.

TYPE-LOCALITY Piscadera Baai, Curaçao, Caribbean.

Genus ***Amemiyaia*** Kato, 1944 Fig. 44

CRYPTOCELIDIDAE: CRYPTOCELIDINAE Slender body of delicate consistency, without tentacles, but with tentacular and cerebral eye-clusters; marginal eyes in band around body. Pharynx short, in anterior third of body; intestinal branches few, not anastomosing. Genital pores widely separated; male pore more or less centrally situated. Vasa deferentia enter seminal vesicle separately. Ejaculatory duct

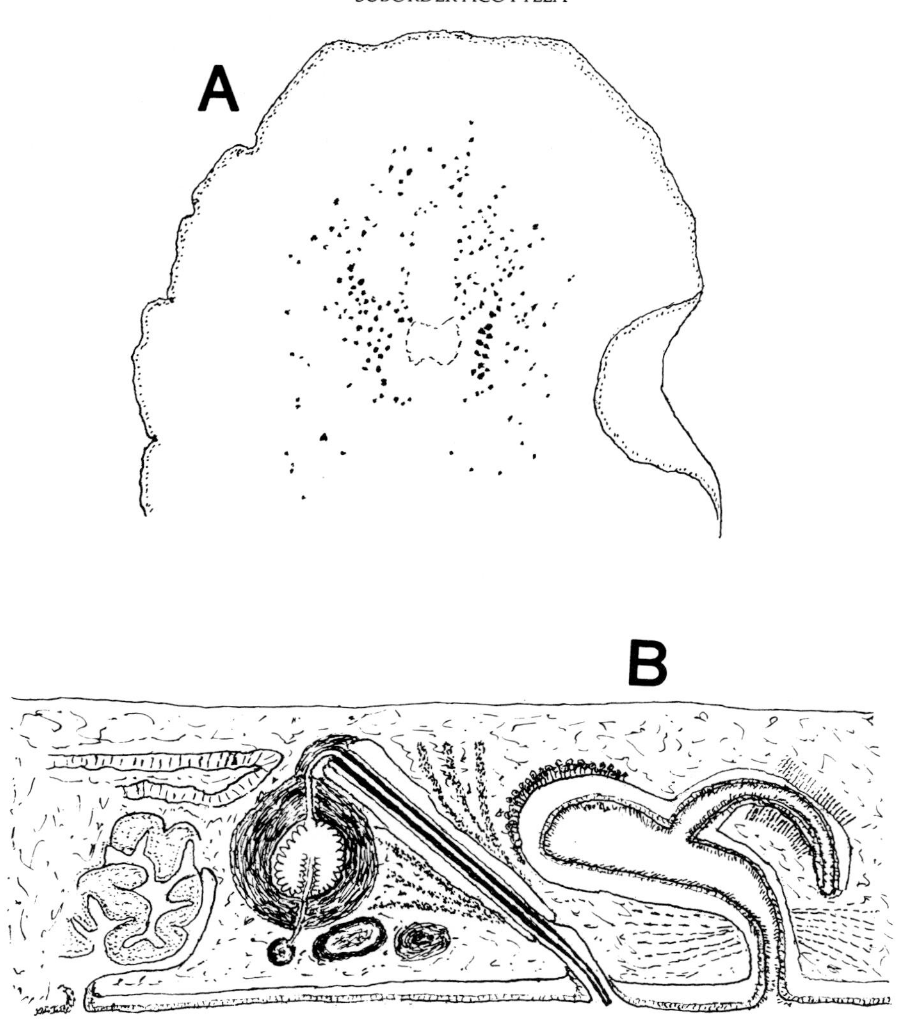

Fig. 45 *Triadomma evelinae.* **A,** arrangement of eyes anteriorly; **B,** sagittal section of copulatory organs (after Marcus).

passes into prostatic organ through a distinct papilla. Prostatic organ a coiled canal surrounded by a thick musculature, through which the ducts of extracapsular gland-cells pass to open into its lumen. Penis-papilla muscular and coated with a cuticular membrane along its entire length; penis-pocket muscular. Vagina bulbosa well developed. Lang's vesicle large.

TYPE-SPECIES *A. pacifica* Kato, 1944. Type by monotypy.

TYPE-LOCALITY Susaki, Amakusa, Japan.

Genus ***Triadomma*** Marcus, 1947

Fig. 45

CRYPTOCELIDIDAE: CRYPTOCELIDINAE Oval forms in which marginal eyes are confined to anterior region. Tentacular eyes in two elongate clusters; cerebral eyes irregularly distributed anteriorly

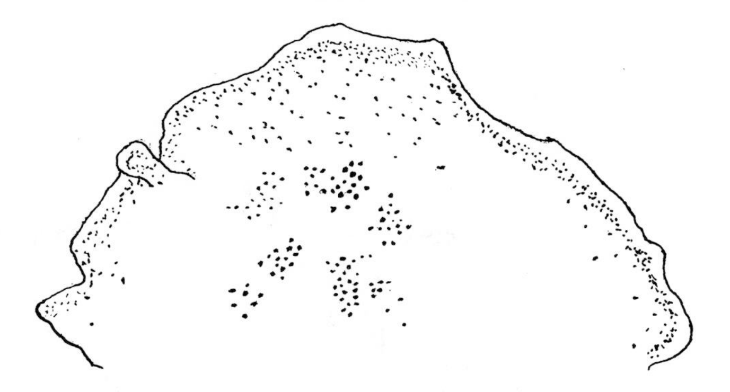

Fig. 46 *Ommatoplana tuberculata.* Arrangement of eyes anteriorly (after Laidlaw).

to cerebral organ. Pharynx centrally situated; intestinal branches anastomosing anteriorly and posteriorly. Genital pores near each other in posterior region of body. Vasa deferentia arise laterally to pharynx and extend posteriorly to open into seminal vesicle, *via* spermiducal bulbs. Prostatic organ bulbous. Penis-papilla with long stylet in penis-pocket. Female antrum modified as a vagina bulbosa. No Lang's vesicle. Uterine canals not anteriorly confluent.

TYPE-SPECIES *T. evelinae* Marcus, 1947. Type by original designation.

TYPE-LOCALITY Palmas I., Bay of Santos, Brazil.

Genus ***Ommatoplana*** Laidlaw, 1903 Fig. 46

CRYPTOCELIDIDAE: CRYPTOCELIDINAE Broadly-oval fleshy forms provided with numerous dorsal tubercles or papillae. Marginal eyes confined to anterior third of body; numerous small eyes scattered in irregular groups over cephalic region. Mouth centrally placed. Genital ducts open into genital atrium closely posterior to mouth. Seminal vesicle not described. Prostatic organ with thick muscular wall, a tall epithelial lining containing longitudinal canals and is assumed to be interpolated. Penis-papilla unarmed, enclosed in penis-pocket. Lang's vesicle present.

TYPE-SPECIES *O. tuberculata* Laidlaw, 1903. Type by original designation.

TYPE-LOCALITY Prison I., East Africa.

SYSTEMATIC NOTE The systematic relationship of the genus *Ommatoplana* is most uncertain, although Laidlaw (1903*d*) originally placed it in the family Cestoplanidae. Laidlaw's description of *O. tuberculata* is unsatisfactory and confusing, because he does not say whether the prostatic organ is interpolated or independent, and he also says that, after receiving the common duct of the uterine canals, the vagina 'runs forwards and downwards towards the atrium (possibly opening into it?)', which might suggest the existence of a ductus vaginalis. On the other hand, in his definition of the genus he says 'Vagina provided with an accessory vesicle' (presumably Lang's vesicle), thus suggesting that he was earlier referring to the course of the outer region of the vagina. In a later paper (1903*e*), Laidlaw again classified *Ommatoplana* in the family Cestoplanidae, with which it is said to agree in having no tentacular eye clusters, in the occurrence of the 'pharynx behind centre of body' and in the possession of a small penis-papilla and a penis-pocket. Moreover, in a key to acotylean genera Laidlaw places this genus among those forms with an interpolated prostatic organ. Hyman (1955*a*) regards *Mexistylochus* Hyman, 1953, as a synonym of *Ommatoplana,* but, as stated earlier, this is not so for Hyman's genus is a stylochid.

Bock (1913) considers that *Ommatoplana* is not a cestoplanid and redefines the genus under the family Cryptocelididae and on the evidence available this action appears to be justified. Nevertheless, the true systematic relationship of this genus will remain uncertain until specimens resembling *O. tuberculata* are collected from east African waters and adequately described.

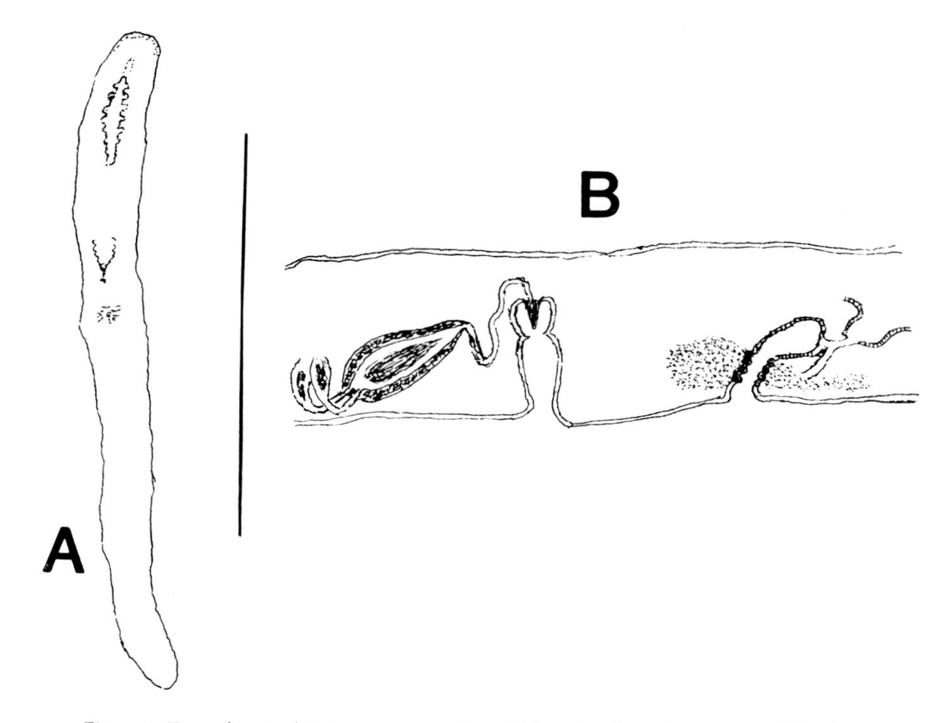

Fig. 47 *Taenioplana teredini.* **A,** entire worm; **B,** sagittal section of copulatory organs (after Hyman).

Subfamily **TAENIOPLANINAE** Marcus & Marcus, 1966, *sensu nov.*

(=TAENIOPLANIDAE Marcus & Marcus, 1966)

CRYPTOCELIDIDAE Prostatic organ not apparent or scarcely differentiated.

KEY TO TAENIOPLANINE GENERA

1. With Lang's vesicle . . . . **Laidlawiana** gen. nov
   Without Lang's vesicle 2
2. Body very elongate . . . . ***Taenioplana***
   Body more or less oval 3
3. Distinct tentacular and cerebral eye-clusters 4
   No tentacular eye-clusters 5
4. Male antrum with thick muscular walls and glandular pocket . . . . . ***Anandroplana***
   Male antrum without such walls or pocket . . . . ***Ilyplanoides***
5. Seminal vesicle well developed . . . . ***Semonia***
   No seminal vesicle . . . . ***Paranandroplana*** gen. nov.

Genus ***Taenioplana*** Hyman, 1944 Fig. 47

CRYPTOCELIDIDAE: TAENIOPLANINAE Body elongate or ribbon-like. Marginal eyes limited to band across anterior margin. Cerebral eyes in two elongate clusters; frontal eyes few. Much folded pharynx in anterior quarter of body. Intestinal trunk extending through length of body and giving off numerous lateral branches. Genital pores separated, more or less centrally situated. Copulatory organs centrally placed, some distance posterior to pharynx. Pair of spermiducal bulbs open separately into large muscular seminal vesicle. No distinct prostatic organ, Penis-papilla small and conical, lying in

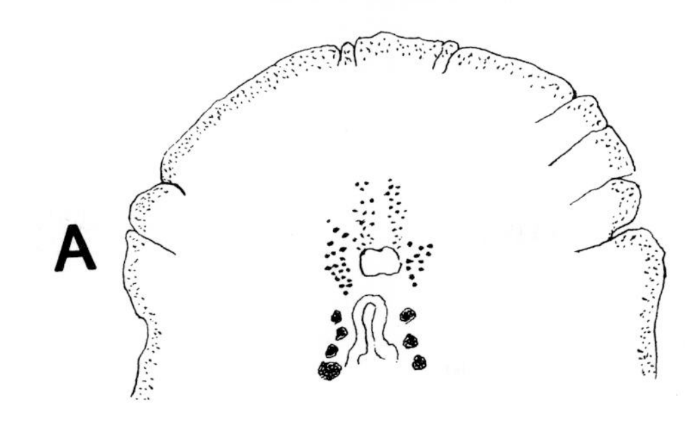

Fig. 48 *Anandroplana muscularis* **A,** arrangement of eyes anteriorly; **B,** sagittal section of copulatory organs (after Hyman).

penis-pocket. Vagina short, directed posteriorly from female pore. No Lang's vesicle, but short genito-intestinal canal.

TYPE-SPECIES *T. teredini* Hyman, 1944. Type by original designation

TYPE-LOCALITY In empty burrows of *Teredo,* Oahu, Hawaiian Is, Pacific Ocean.

SYSTEMATIC NOTE Hyman originally placed this genus in the family Latocestidae, but Marcus & Marcus (1966) erected the family Taenioplanidae for its reception; it is here considered that the genus fits well into the family Cryptocelididae.

Genus ***Anandroplana*** Hyman, 1955 Fig. 48

CRYPTOCELIDIDAE: TAENIOPLANINAE Elongate-oval forms with marginal eyes in band round body; cerebral and tentacular eyes in distinct paired clusters. Pharynx long with many pairs of lateral folds. Genital pores well separated. Male copulatory complex closely posterior to pharynx. No seminal vesicle, no distinct prostatic organ and no penis-papilla. Large male antrum with exceedingly thick muscular walls. Anterior ventral diverticulum of this antrum with distinct gland cells, probably of a prostatic nature. It has been suggested that the vasa deferentia lead into the inner end of the diverticulum. Female antrum expands into a 'shell'-chamber, from which a narrow vagina interna extends posteriorly to terminate on receiving the uterine canals, which are not anteriorly confluent.

TYPE-SPECIES *A. muscularis* Hyman, 1955. Type by original designation.

TYPE-LOCALITY In about 7 metres off Puntilla Point, Puerto Rico, Caribbean.

Genus ***Paranandroplana*** gen. nov. Fig. 49

CRYPTOCELIDIDAE: TAENIOPLANINAE Elongate-oval forms with marginal eyes in band around body. Cerebral eyes in triangular mass over cerebral organ and anteriorly to this mass there is a paired group of loosely arranged frontal eyes remaining separated from marginal eyes. No tentacular eyes. Pharynx mainly in anterior half of body. Genital pores separated. Male pore opens into a narrow antrum, which

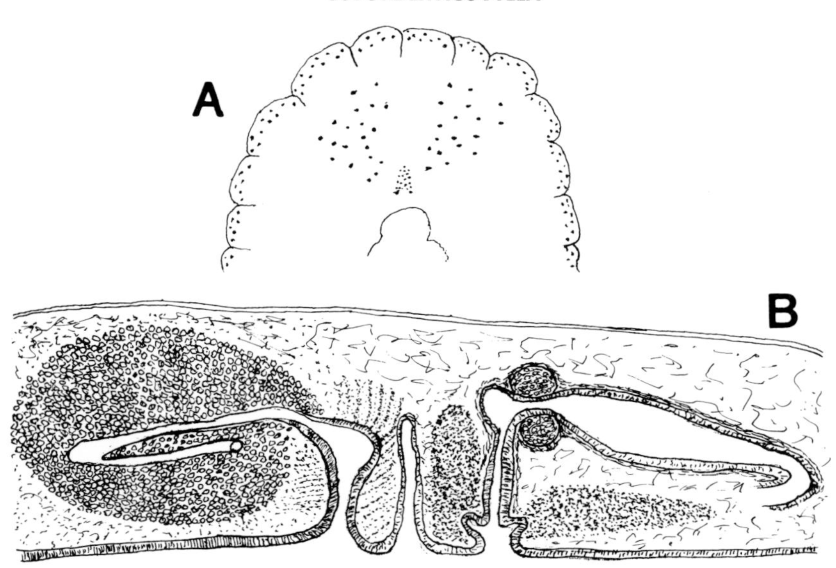

Fig. 49 *Paranandroplana portoricensis.* **A**, arrangement of eyes anteriorly; **B**, sagittal section of copulatory organs (after Hyman).

leads into a sacciform swelling lined with a glandular epithelium and invested with eosinophilic gland cells. This swelling is thought to represent a prostatic organ. From this structure, an ejaculatory duct forms an anteriorly-directed loop to merge with a pair of vasa deferentia. No penis-papilla; no seminal vesicle. Large mass of circular muscle-fibres envelops the ejaculatory duct. Female antrum expands to form a dorso-ventrally compressed 'shell'-chamber. Vagina interna ascends dorsally and dilates to form a small muscular chamber, then continues at right angles posteriorly as a short narrow canal initially enclosed by a mass of muscle fibres to furnish a sphincter. Vagina terminates on receiving uterine canals.

TYPE-SPECIES *P. [Anandroplana] portoricensis* (Hyman, 1955) nov. comb.

TYPE-LOCALITY Under flat rocks in surf-zone at Rincón Playa, Puerto Rico, Caribbean.

SYSTEMATIC NOTE The original description of this species was based on a damaged specimen, and it requires the examination of further material to understand the copulatory complexes more fully. Nevertheless, *Anandroplana portoricensis* appears to differ from *Anandroplana muscularis* Hyman, 1955, in its eye-pattern and in certain features of the male copulatory complex, and these differences appears to warrant generic distinction for each species. Therefore, the genus *Paranandroplana* is erected for the reception of the former species.

Genus **Ilyplanoides** Kato, 1944

Fig. 50

CRYPTOCELIDIDAE: TAENIOPLANINAE Elongate forms, bluntly pointed anteriorly and posteriorly, and without tentacles. Marginal eyes disposed around body. Few additional eyes arranged in paired cerebral and tentacular clusters. Pharynx centrally situated. Genital pores closely separated in posterior region of body. Vasa deferentia open into ejaculatory duct, the walls of which are not modified to form a seminal vesicle or a vesicular prostatic organ. Ejaculatory duct opens directly into a wide male antrum; no penis-papilla. Distal half of ejaculatory duct and male antrum invested with a mass of cyanophilic gland cells, the efferent ducts of which open into the antrum. Vagina dorso-posteriorly directed. No Lang's vesicle.

TYPE-SPECIES *I. mitsuii* Kato, 1944. Type by monotypy.

TYPE-LOCALITY Japan.

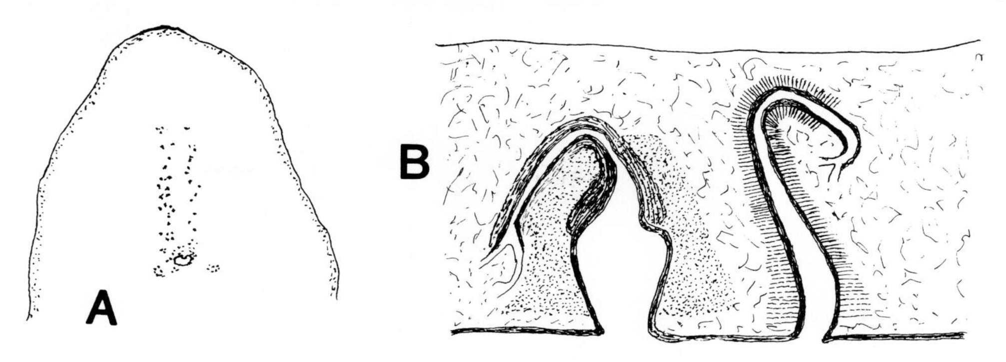

Fig. 50 *Ilyplanoides mitsuii.* **A,** arrangement of eyes anteriorly; **B,** sagittal section of copulatory organs (after Kato).

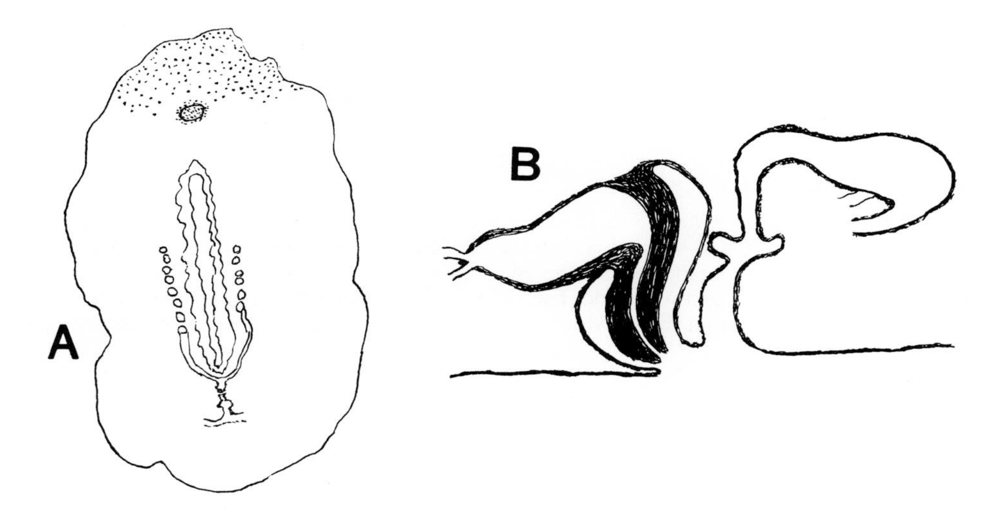

Fig. 51 *Semonia maculata.* **A,** complete worm; **B,** sagittal section of copulatory organs (after Plehn).

Genus ***Semonia*** Plehn, 1896

Fig. 51

CRYPTOCELIDIDAE: TAENIOPLANINAE Oval forms of moderate consistency; without tentacles. Marginal eyes confined to anterior region of body; cerebral eyes in a rounded cluster over cerebral organ; frontal eyes numerous; no tentacular eye-clusters. Pharynx long with many pairs of lateral folds. Common gonopore in hind region of body. Vasa deferentia arise laterally to pharynx and extend posteriorly to lead into a large seminal vesicle; proximal regions of vasa deferentia said to be modified into chains of six to eight glandular vesicles. No prostatic organ. Large conical penis-papilla unarmed; no penis-pocket. Vagina dorso-posteriorly directed from female pore; 'shell'-chamber dilated. Lang's vesicle absent.

TYPE-SPECIES *S. maculata* Plehn, 1896. Type by original designation.

TYPE-LOCALITY Java.

NOTE The presence of a row of glandular vesicles said to be formed by each vas deferens is so far unique among polyclads, Plehn may, however, have misinterpreted what she saw and figured, for when a vas deferens is filled with sperm it often becomes tightly convoluted and gives the appearance

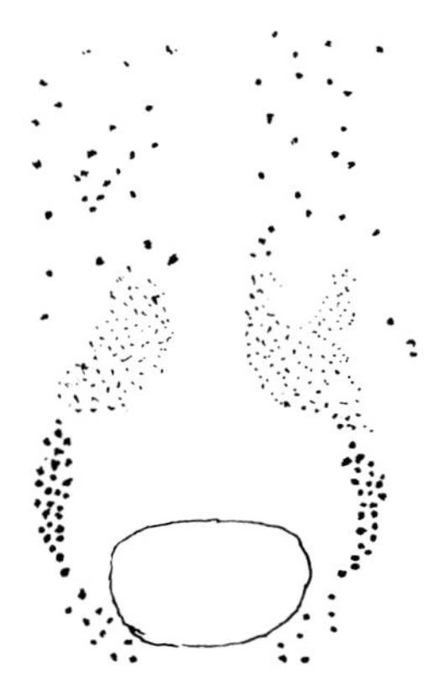

Fig. 52 *Laidlawiana penangensis.* Arrangement of eyes anteriorly (after Laidlaw).

of being moniliform or like a string of beads, as her illustration suggests. It therefore seems likely that Plehn mistook each 'bead' for a glandular vesicle.

Genus ***Laidlawiana*** gen. nov

Fig. 52

CRYPTOCELIDIDAE: TAENIOPLANINAE Oval forms of moderate consistency. Marginal eyes confined to anterior region of body. Small cerebral and larger tentacular eyes merge into an elongate group on either side of the cerebral organ; frontal eyes few. Pharynx in middle third of body. Genital pores separated. Copulatory complexes in hinder half of body. Vasa deferentia simple. Seminal vesicle well developed. No prostatic organ. Blunt, stout penis-papilla unarmed. 'Shell'-chamber narrow. Lang's vesicle very small.

TYPE-SPECIES *L. [Semonia] penangensis* (Laidlaw, 1903) nov. comb.

TYPE-LOCALITY Islet of Pulau Bidan, Straits of Malacca, Malaysia.

NOTE Though Laidlaw's description of *Semonia penangensis* is, by modern standards, inadequate, it does contain sufficient information to raise doubt as to the species being congeneric with *Semonia maculata* Plehn. The presence of glandular vesicles on the vasa deferentia in the latter species may not be considered generically important, especially as such structures seem to be unique among polyclads and their existence awaits confirmation. On the other hand, the disposition of the eyes and the morphology of the female copulatory complex seem sufficiently dissimilar in the two forms to warrant generic distinction. Thus the genus *Laidlawiana* is erected for Laidlaw's species.

Superfamily **PLANOCEROIDEA** Poche, 1926, emended Nicoll, 1935
(=SCHEMATOMMATA Bock, 1913; SCHEMATOMMATIDEA Marcus & Marcus, 1966)

ACOTYLEA without marginal eyes; generally with paired cerebral and tentacular groups of eyes lateral to cerebral organ and well separated from anterior margin of body. On either side of cerebral organ cerebral and tentacular groups may be separated or merge to form an elongate cluster.

KEY TO FAMILIES OF PLANOCEROIDEA

1.	Pharynx campanulate or tubular	2
	Pharynx ruffled	3
2.	Body with cuticular spines dorsally; no distinct prostatic organ	ENANTIIDAE
	Body without spines; prostatic organ interpolated	APIDIOPLANIDAE
3.	Male copulatory complex with a cirrus	4
	Male complex without a cirrus	5

4. Prostatic organ independent . . . . PLANOCERIDAE
   Prostatic organ interpolated . . . . GNESIOCEROTIDAE
5. Prostatic organ independent 6
   Prostatic organ interpolated or not apparent 7
6. Prostatic organ dorsal to ejaculatory duct or seminal vesicle . . . CALLIOPLANIDAE
   Prostatic organ ventral to seminal vesicle . . . . . . STYLOCHOCESTIDAE
7. Testes anterior, ovaries posterior, to copulatory organs . . . . . THEAMATIDAE
   Gonads irregularly scattered or in submarginal chains 8
8. Penial stylet attached to distal end of prostatic organ . . . . . HOPLOPLANIDAE
   Penial stylet, when present, attached to penis-papilla . . . . . LEPTOPLANIDAE

Family **LEPTOPLANIDAE** Stimpson, 1857

PLANOCEROIDEA Oval or somewhat elongate forms, often widened anteriorly and tapering posteriorly. Frequently with nuchal tentacles. Eyes alongside cerebral organ, arranged in two elongate groups or in two tentacular groups, between which lie two elongate cerebral groups. Occasionally, eyes may collect into a single mass over cerebral organ. Mouth central or subcentral; pharynx ruffled. Genital pores, which sometimes unite to form a common antrum, occur almost invariably in posterior half of body, usually some distance from hind margin. Gonads irregularly scattered or in submarginal chains. Male copulatory complex usually anterior to male pore. Generally with a muscular seminal vesicle; sometimes with two spermiducal bulbs. Prostatic organ variably developed, interpolated; it may be vesicular, with thick muscular walls, or narrow and cylindrical, represented merely by a portion of ejaculatory duct whose epithelial lining assumes a glandular character; sometimes the latter form may not be apparent. Penis-papilla often bears stylet. Vagina of variable length, often thrown into an anterior loop and sometimes ending in a ductus vaginalis. Lang's vesicle, when present, varies in size and shape.

KEY TO LEPTOPLANID SUBFAMILIES

1. Prostatic organ tubular and sometimes indistinct, scarcely differentiated from ejaculatory duct . . . . LEPTOPLANINAE
   Prostatic organ vesicular and muscular . . . . . . STYLOCHOPLANINAE

Subfamily **LEPTOPLANINAE** Marcus, 1947
(Includes EUPLANINAE Marcus, 1947; MUCROPLANINAE Sopott-Ehlers & Schmidt, 1975)

LEPTOPLANIDAE Prostatic organ cylindrical, often indistinct, or portion of ejaculatory duct or antrum masculinum invested with extracapsular prostatic gland cells.

KEY TO LEPTOPLANINE GENERA

1. Prostatic organ tubular, with ventral or lateral appendix proximally . . . . ***Leptoplana***
   Prostatic organ, if present, without appendix 2
2. Seminal vesicle not developed 3
   Seminal vesicle present 5
3. Lang's vesicle opens on ventral surface of body . . . . ***Euplanina***
   Lang's vesicle with no external aperture 4
4. Uterus as pair of canals . . . . ***Haploplana***
   Uterus as a median sac . . . . ***Mucroplana***
5. Seminal vesicle tripartite or anchor-shaped 6
   Seminal vesicle bulbous or elongate 7
6. Penis-papilla absent; prostatic granular mass investing ejaculatory duct . . ***Crassandros***
   Penis-papilla distinct; no prostatic granular mass . . . . . . ***Phylloplana***
7. Vaginal bursa present . . . . ***Pulchriplana***
   Vaginal bursa not developed 8
8. With ductus vaginalis . . . . ***Tripylocelis***
   Without ductus vaginalis 9

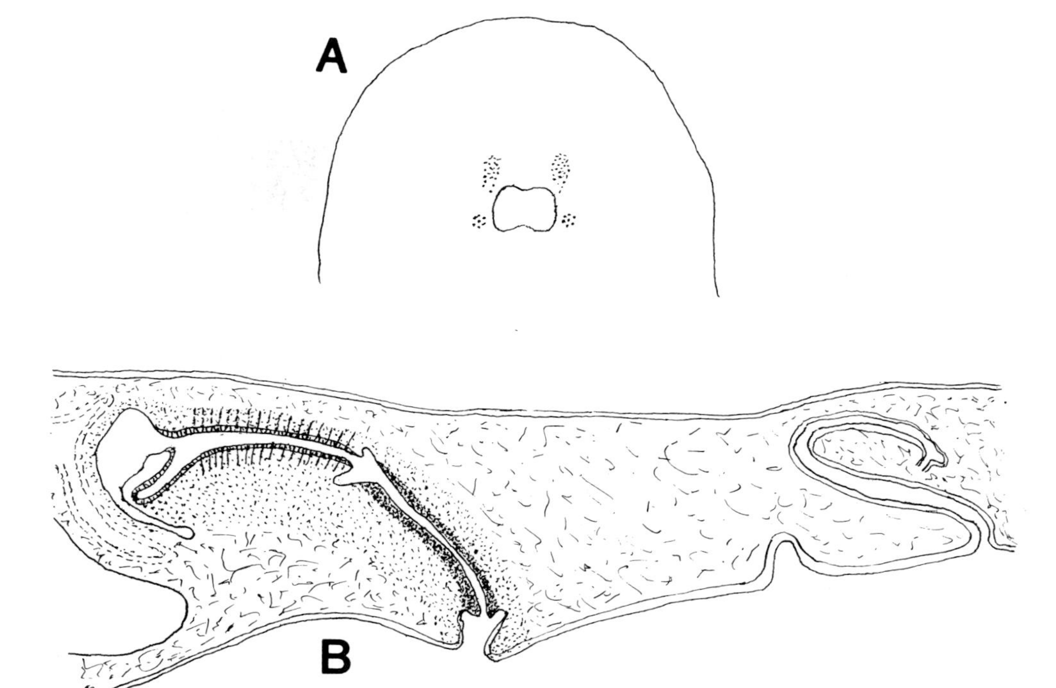

Fig. 53 *Leptoplana tremellaris.* **A,** arrangement of eyes (original); **B,** sagittal section of copulatory organs (after Bock).

9. Lang's vesicle absent . . . . ***Euplana***
   Lang's vesicle bulbous or crescentic . . . . ***Discoplana***

Genus ***Leptoplana*** Ehrenberg, 1831, *sensu* Lang, 1884 Fig. 53

LEPTOPLANIDAE: LEPTOPLANINAE Body more or less oval, sometimes wider anteriorly than posteriorly. Without tentacles. Eyes in paired tentacular and cerebral clusters. Pharynx much folded, centrally situated, and about half as long as body. Genital pores separated, sometimes with adhesive pad between them. Vasa deferentia lead into muscular seminal vesicle. Prostatic organ tubular, usually with a diverticulum proximally. Penis-papilla inconspicuous, unarmed, in a long narrow penis-pocket. Antrum masculinum slender. Thick musculature invests male copulatory complex. Vagina long, forming anteriorly-directed loop. Lang's vesicle of variable development. When fully developed, uterine canals anteriorly confluent between cerebral organ and pharynx.

TYPE-SPECIES *L. hyalina* Ehrenberg, 1831. Type by monotypy.

TYPE-LOCALITY Tor, Red Sea.

SYSTEMATIC NOTE Although the type-species of *Leptoplana* is unrecognizable, Lang (1884) considered it to be synonymous with *Planaria tremellaris* Müller, 1774, and it is on this basis that *Leptoplana* has been accepted as a recognizable genus. However, a study of the problem shows that *L. tremellaris* is a North Atlantic form ranging from the White Sea into the western Mediterranean, though a variety *taurica* has been described from the Black Sea, and is unknown in the Indian Ocean. The migration of *L. tremellaris,* in the form of *Leptoplana hyalina* as described by Ehrenberg (1831), from the Mediterranean through the Suez Canal into the Red Sea is probably out of the question, since the construction of the Suez Canal was not started until 1859. Nevertheless, Hyman (1953*a*) appears to accept the synonymy suggested by Lang and designates *tremellaris* as the type-species of *Leptoplana.*

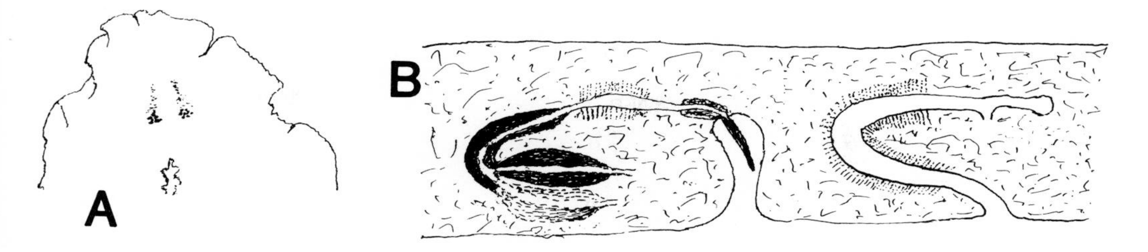

Fig. 54 *Phylloplana lactea.* **A,** arrangement of eyes (after Laidlaw); **B,** sagittal section of copulatory organs (diagrammatic).

The only other species of *Leptoplana* so far recorded from the Red Sea is *L. nadiae* Melouk, 1941, but this is actually a species of *Stylochoplana.*

Since its erection, over 120 species have been assigned to *Leptoplana,* but many have, over the years, been transferred to other genera, while others are unrecognizable, and only about 3 species now appear to be congeneric with *L. tremellaris.*

Genus ***Phylloplana*** Laidlaw, 1903 — Fig. 54

SYNONYMY *Indiplana* Stummer-Traunfels, 1933.

LEPTOPLANIDAE: LEPTOPLANINAE Elongate-oval to broadly-oval forms without tentacles. Eyes in paired tentacular and cerebral clusters or in two elongate groups. Pharynx mainly in anterior half of body; intestinal branches not anastomosing. Genital pores separated, in middle third of body. Spermiducal bulbs elongate and muscular, fusing with very muscular proximal portion of ejaculatory duct to form a somewhat trilobed or anchor-shaped seminal vesicle. Prostatic glands invest a narrow muscular portion of ejaculatory duct to function as a cylindrical prostatic organ. Penis-papilla distinct, may bear stylet. Vagina narrow. Lang's vesicle small.

TYPE-SPECIES *P. lactea* Laidlaw, 1903. Type by original designation.

TYPE-LOCALITY Shore, 'British East Africa.'.

SYSTEMATIC NOTE In distinguishing the genus *Indiplana* from *Phylloplana,* Stummer-Traunfels (1933) indicates that a prostatic organ is not apparent in the latter, but is in the former. According to Laidlaw (1903*d*), however, the epithelium of the proximal portion of the ejaculatory duct after leaving the seminal vesicle is, in *Phylloplana,* of a prostatic nature. Later, in a diagnostic table of the acotylean genera, Laidlaw (1903*e*) says 'prostate reduced' in this genus. It is, therefore, reasonably certain that Laidlaw recognized a prostatic organ of some kind in *Phylloplana,* although there seems to be little superficial difference between it and the ejaculatory duct. Such appears to be the case in *Indiplana,* notwithstanding the fact that the prostatic nature of its ejaculatory duct is more marked and actually represents a very narrow and elongate prostatic organ. A comparison of the descriptions of the type-species of *Phylloplana* and of *Indiplana* shows a few minor differences, the value of which appear, at the moment, to be no more than specific. Therefore, the two genera are here regarded as synonymous.

Genus ***Crassandros*** Hyman, 1955 — Fig. 55

LEPTOPLANIDAE: LEPTOPLANINAE Oval forms without tentacles. Each cluster of tentacular eyes enveloped by cerebral eyes. Pharynx long, centrally situated. Spermiducal vesicles merge with true seminal vesicle and together form a tripartite or anchor-shaped structure. Large granular prostatic mass invests ejaculatory duct immediately it leaves the seminal vesicle, and between this mass and the male antrum the duct is enclosed in a thick wall of circular muscle fibres. Penis-papilla not observed. Male antrum broad and shallow. Female pore well separated from male pore, leads into short vagina with radially-folded walls. Vagina runs anteriorly-arcuate course and terminates in Lang's vesicle.

TYPE-SPECIES *C. dominicanus* Hyman, 1955. Type by original designation

TYPE-LOCALITY In rock-pools, Marigot, Dominica I., Caribbean.

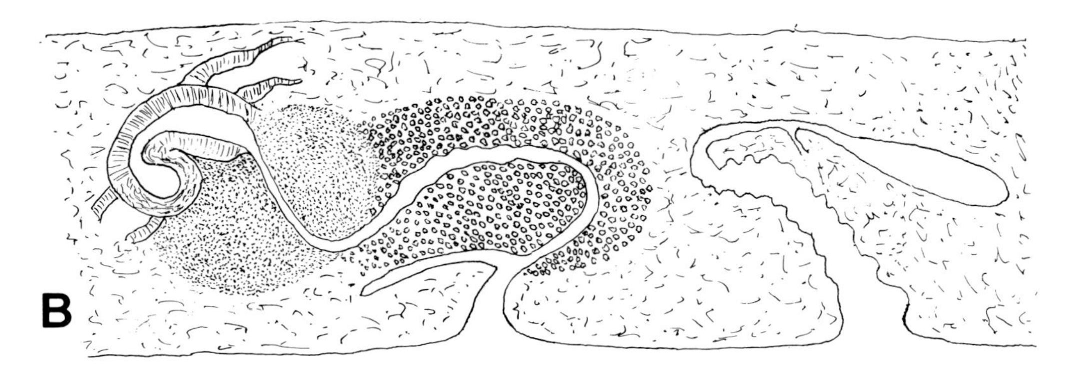

Fig. 55 *Crassandros dominicans*. **A**, arrangement of eyes; **B**, sagittal section of copulatory organs (after Hyman).

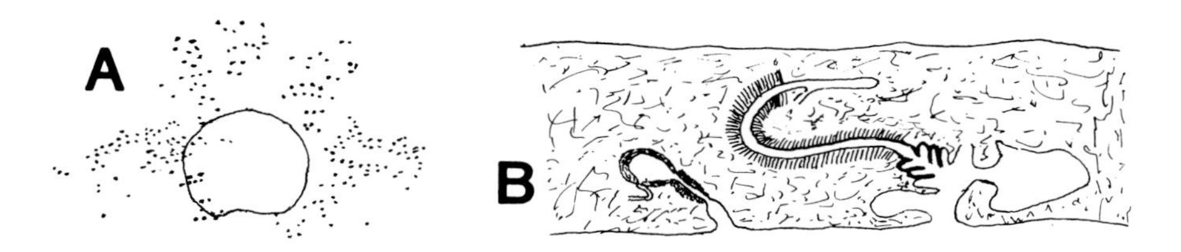

Fig. 56 *Pulchriplana insignis*. **A**, arrangement of eyes; **B**, sagittal section of copulatory organs (after Palombi).

Genus ***Pulchriplana*** Palombi, 1938

Fig. 56

LEPTOPLANIDAE: LEPTOPLANINAE Oval forms with tentacles. Eyes at base of each tentacle, between which occur two small groups of cerebral eyes. Pharynx in middle third of body; intestinal branches not anastomosing. Vasa deferentia originate in posterior region of body and extend anteriorly to open separately into small muscular seminal vesicle. Prostatic organ said not to be developed, but extracapsular eosinophilic glands invest distal region of ejaculatory duct, which therefore represents a tubular non-muscular prostatic organ. Penis-papilla inconspicuous, without stylet. Antrum femininum spacious; large thin-walled vesicle opens into hind region of antrum. Vagina narrow, bearing small bursa between antrum and 'shell'-chamber. Lang's vesicle slender.

TYPE-SPECIES *P. insignis* Palombi, 1938. Type by monotypy.

TYPE-LOCALITY Reef Bay, Port Elizabeth, South Africa.

Genus ***Tripylocelis*** Haswell, 1907

Fig. 57

LEPTOPLANIDAE: LEPTOPLANINAE Large elongate forms with well-developed tentacles containing eyes. Cerebral eyes in two or four clusters between tentacles. Pharynx in middle third of body; intestinal branches not anastomosing. Genital pores separated. Vasa deferentia open separately into an elongate seminal vesicle. Prostatic vesicle long and narrow, extracapsular gland-cells opening into its lumen. Penis-papilla small, glandular, without stylet. Antrum masculinum muscular, long and lined

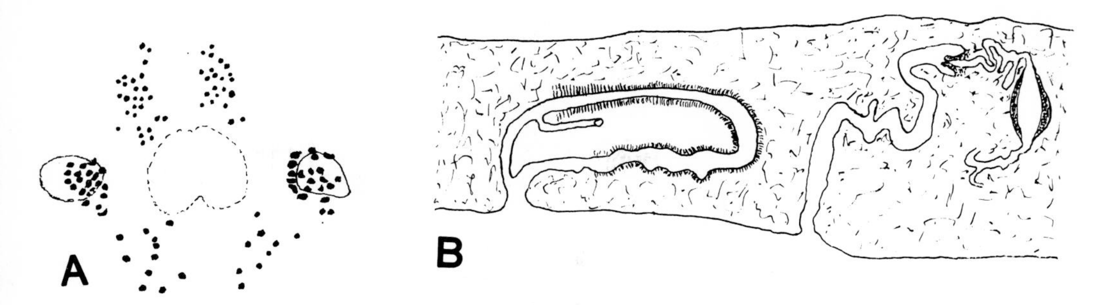

Fig. 57 *Tripylocelis typica.* **A**, arrangement of eyes; **B**, sagittal section of copulatory organs (after Prudhoe).

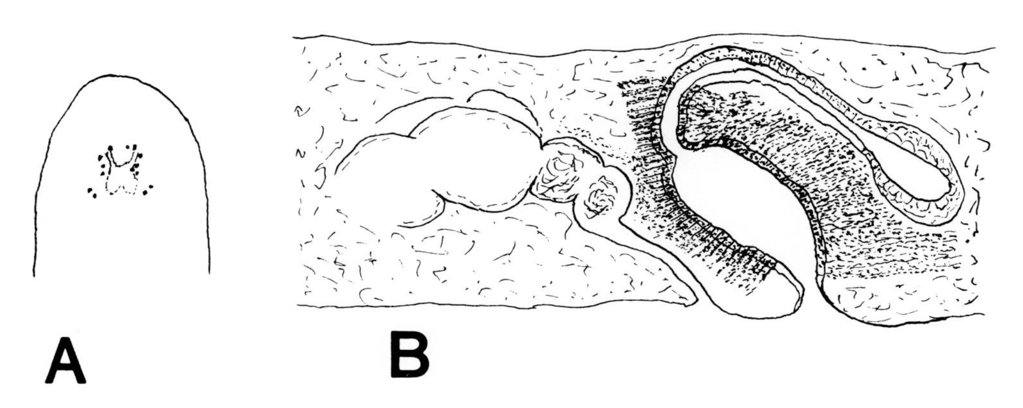

Fig. 58 *Euplana gracilis.* **A**, arrangement of eyes; **B**, sagittal section of copulatory organs (after Hyman).

with long cilia. Vagina narrow, looped anteriorly; 'shell'-chamber long; no Lang's vesicle. Ductus vaginalis opening into female antrum or to exterior behind female pore. Uterine canals not anteriorly confluent.

TYPE-SPECIES *T. typica* Haswell, 1907. Type by original designation.

TYPE-LOCALITY In algae between tidal limits, Port Jackson, New South Wales, Australia.

Genus ***Euplana*** Girard, 1893 Fig. 58

SYNONYMY *Conjuguterus* Pearse, 1938.

LEPTOPLANIDAE: LEPTOPLANINAE Relatively small, elongate forms, tapering somewhat posteriorly. Without tentacles. Eyes few, disposed around cerebral organ in two rounded tentacular clusters and two elongate cerebral clusters. Pharynx in anterior half of body; intestinal ramifications anastomosing initially, but not marginally. Genital pores close together and centrally placed. Thin-walled vasa deferentia open into weakly-developed seminal vesicle. No distinct prostatic organ. When present, penis-papilla insignificant. Vagina thrown into anterior loop. No Lang's vesicle. Uterine canals anteriorly confluent.

TYPE-SPECIES *E. [Prosthiostomum] gracilis* (Girard, 1850). Type by original designation.

TYPE-LOCALITY Boston Harbour, Massachusetts, U.S.A.

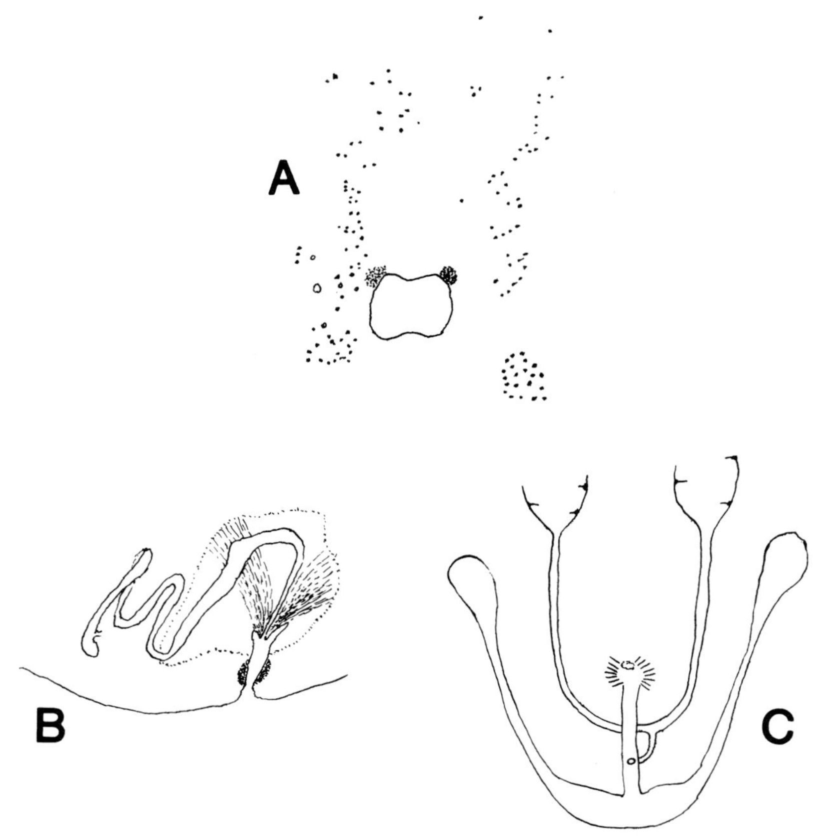

Fig. 59 *Discoplana gigas.* **A,** arrangement of eyes; **B,** sagittal section of ♂ copulatory organs (after Stummer-Traunfels); **C,** ventral view of ♀ copulatory organs (after Laidlaw).

SYSTEMATIC NOTES Hyman (1939*b*) regards the genus *Discoplana* Bock as a synonym of *Euplana*, and this opinion is accepted by Marcus (1947). This synonymy is here not accepted, because *Discoplana* may be separated by differences in its structure of the male complex and by the presence of Lang's vesicle. This leaves *Euplana* with 4 species not assignable to *Discoplana: E. gracilis, E. clippertoni* Hyman, 1939, *E. carolinensis* Hyman, 1940 and *E. hymanae* Marcus, 1947. However, the generic designation of *E. hymanae* is questionable, for this species possesses some important features which are absent in *E. gracilis*, the type-species. It has vasa deferentia running laterally to the uterine canals and, according to Marcus' figure of the male complex, there appears to be an elongate prostatic organ, into which the ejaculatory duct projects as a distinct papilla, a feature of the genus *Notoplana*. Moreover, the species possesses a long penis-papilla lying in a slender penis-pocket. Further, Hyman (1939*c*) in describing *Euplana clippertoni* states that a prostatic organ is absent, but it is labelled in a figure she gives of the male apparatus of this species. In fact, there is a suggestion that the male organs shown in the figured specimen may not be fully mature.

Genus *Discoplana* Bock, 1913

Fig. 59

SYNONYMY *Susakia* Kato, 1934

LEPTOPLANIDAE: LEPTOPLANINAE Large oval forms without tentacles. Species appear to be protogynous. Eyes in two tentacular groups, between which, or anterior to, lie two cerebral groups. Length

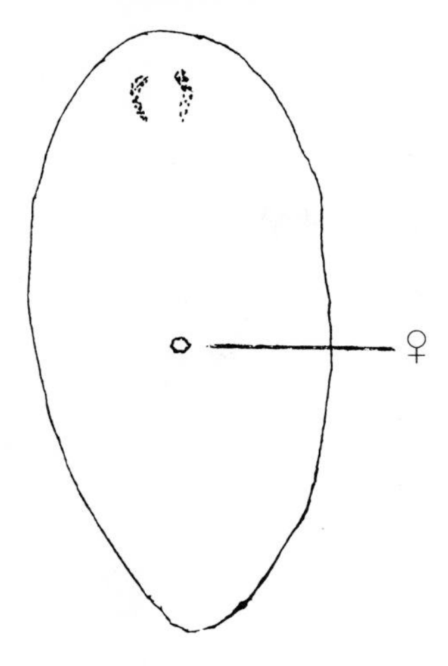

Fig. 60 *Haploplana elioti.* Entire worm, ventral view (after Laidlaw).

of pharynx one-quarter to one-third of body-length and centrally situated. Genital pores separated, some distance anteriorly to hind end of body. Vasa deferentia lead into pyriform or elongate seminal vesicle. Ejaculatory duct extends from seminal vesicle to penis-papilla without forming a prostatic organ, but at the height of the male phase the seminal vesicle, the ejaculatory duct and the male antrum are invested with a thick coat of extracapsular gland-cells possibly with a prostatic function. Penis-papilla distinct and muscular. Vagina narrow. Lang's vesicle bulbous or crescentic. Uterine canals not anteriorly confluent.

TYPE-SPECIES *D. [Leptoplana] gigas* (Schmarda, 1859) Stummer-Traunfels, 1933. Type by subsequent designation.

TYPE-LOCALITY Trincomalee and Bellingham, Sri Lanka.

Genus ***Haploplana*** Laidlaw, 1903 Fig. 60

LEPTOPLANIDAE: LEPTOPLANINAE Small, oval, rather stout forms without tentacles. Eyes in two elongate clusters adjacent to cerebral organ. Pharynx in anterior half of body; intestinal branches anastomosing. Genital pores more or less centrally situated. Apparently no seminal vesicle or prostatic organ. Vasa deferentia lead into a vesicle (?penis-papilla) lying in small antrum masculinum. Vagina short; Lang's vesicle small; uterine canals not anteriorly confluent.

TYPE-SPECIES *H. elioti* Laidlaw, 1903. Type by original designation.

TYPE-LOCALITY 'British East Africa'.

NOTE This genus is so little known that its systematic position within the family Leptoplanidae is uncertain. In the original description of the type-specimen, Laidlaw (1903*d*) states that the vasa deferentia 'open into a very small median vesicle, which latter appears to open directly by a minute pore in the [male] antrum'. Bock (1913) in his definition of the genus *Haploplana,* mentions a 'Echte Samenblase' as a character and is presumably referring to the 'small median vesicle'. Bock's interpretation of this 'vesicle' appears doubtful, because in another paper Laidlaw (1903*e*) says 'no vesicula seminalis present' in *Haploplana,* and he further implies that the 'median vesicle' is a penis-papilla. Clearly, the question of the presence of a seminal vesicle in this genus will not be settled until the type-specimen of *H. elioti* is re-examined, which seems most unlikely, or further material is obtained from East Africa.

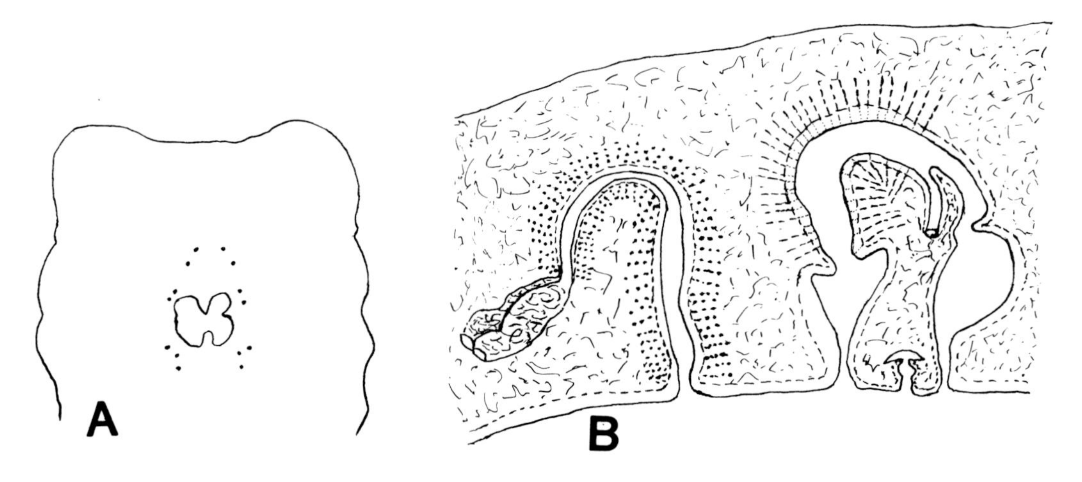

Fig. 61 *Euplanina horrida.* **A**, arrangement of eyes; **B**, sagittal section of copulatory organs (after Sopott-Ehlers & Schmidt).

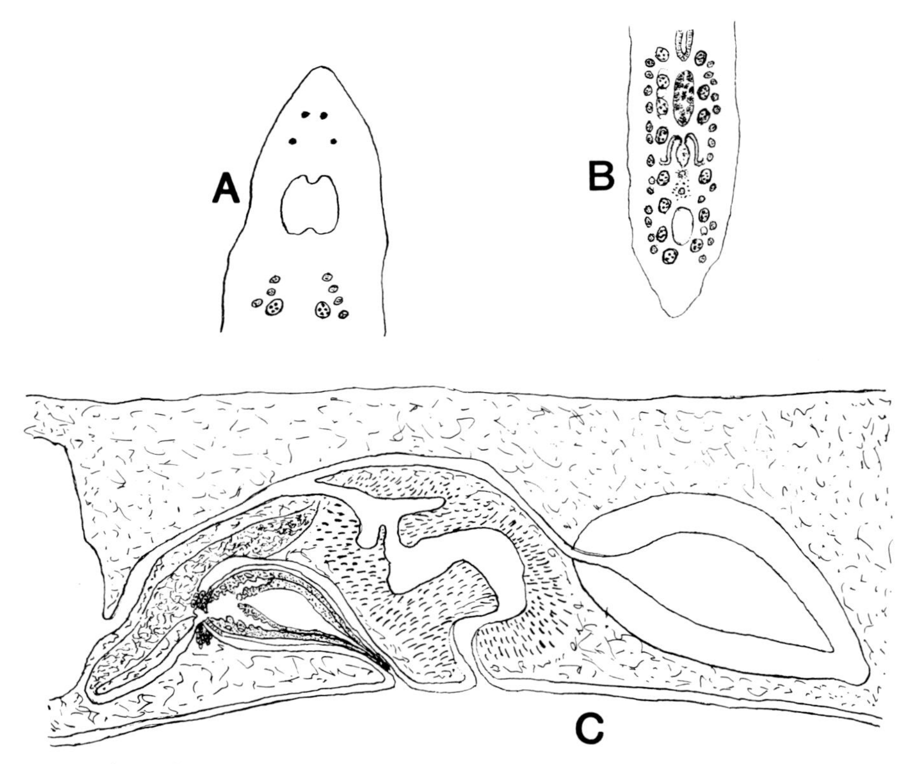

Fig. 62 *Mucroplana caelata.* **A**, anterior region of body showing eyes, cerebral organ and gonads; **B**, posterior region showing gonads and copulatory complexes; **C**, sagittal section of copulatory organs (after Sopott-Ehlers & Schmidt).

Genus ***Euplanina*** Sopott-Ehlers & Schmidt, 1975 Fig. 61

LEPTOPLANIDAE: LEPTOPLANINAE Small slender forms without tentacles. Few cerebral and tentacular eyes in two elongate clusters. Pharynx posterior to middle of body. Posteriorly-situated genital pores separated. Testes in two chains lateral to median line and extending from about anterior third of body to copulatory organs. No seminal vesicle or distinct prostatic organ. Paired spermiducal bulbs lead into long ejaculatory duct. Epithelium of ejaculatory duct and surrounding tissue contain innumerable gland-cells, but prostatic gland-cells have not been detected. No penis-papilla. Ovaries in two chains lateral to testes. Vagina forming a dorso-posterior loop. Lang's vesicle with ventral aperture, between which and female pore lies a deep depression. Uterine canals anteriorly confluent.

TYPE-SPECIES *E. horrida* Sopott-Ehlers & Schmidt, 1975. Type by original designation.

TYPE-LOCALITY 'Mesopsammon' Santa Cruz, Galapagos Is, Pacific Ocean.

Genus ***Mucroplana*** Sopott-Ehlers & Schmidt, 1975 Fig. 62

LEPTOPLANIDAE: LEPTOPLANINAE Small slender forms without tentacles. Eyes few, precerebral. Pharynx in middle third of body. Genital pores separated. Testes and ovaries in two chains on each side of median line and extending from post-cerebral region to near posterior end of body; testes lateral to ovaries. Spermiducal bulbs lead into ejaculatory duct, distal region of which is enclosed by prostatic gland-cells. No seminal vesicle or distinct prostatic organ. Ejaculatory duct opens into a two-chambered pyriform vesicular intromittent organ invested with a two-layered musculature. Vagina forming anteriorly-directed loop. 'Shell'-chamber long and wide; Lang's vesicle well developed. Unpaired uterine sac between pharynx and spermiducal bulbs.

TYPE-SPECIES *M. caelata* Sopott-Ehlers & Schmidt, 1975. Type by original designation.

TYPE-LOCALITY 'Mesopsammon', Santa Cruz, Galapagos Is, Pacific Ocean.

Subfamily **STYLOCHOPLANINAE** Meixner, 1907
(Includes NOTOPLANINAE Marcus, 1947)

LEPTOPLANIDAE Prostatic organ vesicular and muscular.

KEY TO STYLOCHOPLANINE GENERA

1.	Elongate pharynx transversely disposed	***Plagiotata***
	Pharynx longitudinally disposed	2
2.	With ductus vaginalis	3
	Without ductus vaginalis	5
3.	Longitudinal ducts in epithelial lining of prostate	***Copidoplana***
	Without such ducts	4
4.	Penial stylet; ductus vaginalis opens into female antrum	***Ceratoplana***
	No penial stylet; ductus vaginalis opens behind female pore	***Digynopora***
5.	Genito-intestinal canal present	***Macginitiella***
	Without genito-intestinal canal	6
6.	Male copulatory complex double	***Diplandros***
	Male complex single	7
7.	Paired spermiducal bulbs; no true seminal vesicle	***Freemania***
	True seminal vesicle; no spermiducal bulbs	8
8.	Longitudinal or radial canals in epithelial lining of prostate	9
	Without such ducts	13
9.	Lang's vesicle crescentic	***Leptocera***
	Lang's vesicle bulbous or absent	10
10.	Antrum femininum bulbous and muscular	***Parviplana***
	Antrum femininum neither bulbous nor muscular	11
11.	Pharynx posterior; uterine canals separated anteriorly	***Pucelis***
	Pharynx central; uterine canals confluent anteriorly	12

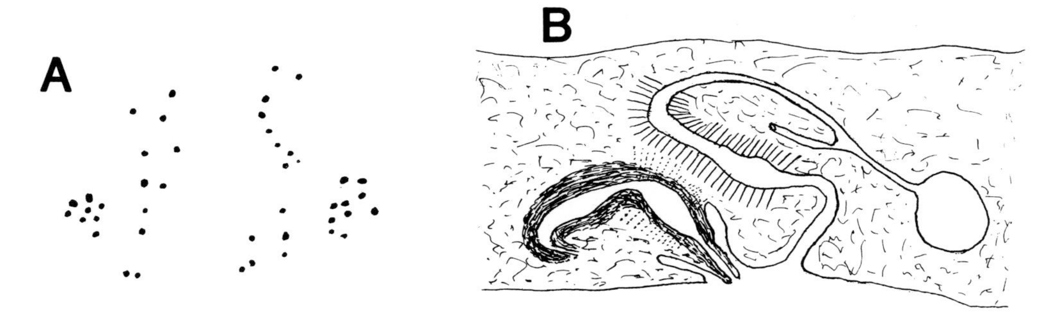

Fig. 63 *Stylochoplana maculata.* **A**, arrangement of eyes; **B**, sagittal section of copulatory organs (after Bock).

12.	Epithelial lining of prostate in deep radial folds	***Notoplanella***
	Epithelial lining with longitudinal chambers	. ***Notoplana***
13.	Copulatory complexes near hind end of body	. ***Zygantroplana***
	Copulatory complexes well separated from hind end	. 14
14.	With bursa copulatrix	. ***Candimboides***
	Without bursa copulatrix	. 15
15.	Uterine canals short, posterior to pharynx	. ***Candimba***
	Uterine canals long, confluent anteriorly to pharynx	. ***Stylochoplana***

Genus ***Stylochoplana*** Stimpson, 1857 Fig. 63

SYNONYMY *Alloioplana* Plehn, 1896; *Notoplanides* Palombi, 1928.

LEPTOPLANIDAE: STYLOCHOPLANINAE Generally rather pellucid or translucent forms, somewhat cuneate, broadly rounded anteriorly and tapering posteriorly. Nuchal tentacles may be present. Eyes in paired cerebral and tentacular clusters which may merge and form an elongate group on either side of cerebral organ. Seminal vesicle well developed. Bulbous prostatic organ with shallow and smooth epithelial lining. Ejaculatory duct uniting seminal vesicle and prostatic organ, but not projecting into lumen of latter. Penis-papilla variably developed and may bear a stylet, which lies in a penis-pocket. Vagina simple, with or without Lang's vesicle. Uterine canals may be anteriorly confluent.

TYPE-SPECIES *S. [Stylochus] maculata* (Quatrefages, 1845). Type by subsequent designation.

TYPE-LOCALITY St Malo, NW. France.

SYSTEMATIC NOTES The genus *Alloioplana*, type *A. delicata* Plehn, 1896, from Peru has been defined by Hyman (1953*a*) as a member of the family Planoceridae solely on the basis of two Californian species, *Planocera californicus* Heath & McGregor and *Planocera sandiegensis* Boone. These two latter species are certainly not congeneric with *A. delicata* and are members of the family Gnesiocerotidae, whereas *A. delicata* undoubtedly belongs to the leptoplanid genus *Stylochoplana*. In fact, Plehn herself regarded *Alloioplana* as a leptoplanid.

Palombi (1928) erected the genus *Notoplanides* for a form which he described as *N. opisthopharynx* and a young immature specimen which Bock (1924) merely regarded as an immature leptoplanid. This genus has been accepted in a classification of the Polycladida presented by Marcus & Marcus (1966), and du Bois-Reymond Marcus & Marcus (1968) erected a further species *N. alcha* and pointed out that *Notoplanides* differs from *Stylochoplana* merely in the posterior position of its pharynx. Hyman (1953*a*) questioned the distinctiveness of *Notoplanides* from *Stylochoplana*, but Palombi appears to differentiate his genus from other leptoplanid genera solely by reason of the pharynx lying in the posterior half of the body. The description of *N. opisthopharynx* is based on a single contracted specimen, the posterior end of which is missing. Judging from Palombi's figure of the incomplete worm, and also bearing in mind its contracted condition, it may be assumed that the pharynx actually occupies a more central position in the body of the complete worm than Palombi believes. Regarding Bock's immature form, in very young leptoplanids, as in many of the acotyleans, the pharynx constantly occupies a posterior position in the body, but with the development of the copulatory organs, the body lengthens posteriorly and the position of the pharynx becomes more central. Moreover, a study of the species so far assigned to *Stylochoplana* shows a gradation in the final position of the pharynx from a central to a posterior one. It seems, therefore, that *Notoplanides* is indistinguishable from *Stylochoplana. Stylochoplana* is a genus widely distributed geographically and to which over 40 species have been assigned. Bock (1913) arranged the then known species into three groups, A, B and C, and later (1924) erected the subgenus (*Stylochoplana*) for group A. These groupings appear to have been based on body shape, the presence or absence of tentacles, the penis-papilla with or without a stylet and the presence or absence of a penis-pocket. With the acquisition of many new species following Bock's work, *Stylochoplana* was divided into 9 groups of species by du Bois-Reymond Marcus & Marcus (1968). As so often happens when a large number of species are divided into several groups, the groups are confused and, in fact, may not be satisfactorily separated, and some species get placed erroneously. In the present writer's experience, for the species known at present, Bock's groupings are acceptable, with slight modifications, and with the addition of a fourth group, a further arrangement may be given as follows:

Group A (= subgenus (*Stylochoplana*). Tentacles variably developed or absent; cerebral and tentacular eye-clusters separated; penis-papilla without stylet. Includes, *aberrans* Kato, *agilis* Lang, *alcha* (du B.-R. Marcus & Marcus), *amica* Kato, *challengeri* (Graff), *conoceraea* (Schmarda), *gracilis* (H. & McG.), *graffii* (Laidlaw), *lynca* du B.-R. Marcus & Marcus, *maculata* (Quatrefages), *nadiae* (Melouk), *palmula* (Quatrefages), *parasitica* Kato, *pusilla* Bock, *selenopsis* Marcus, *tarda* (Graff), *utunomii* Kato.

Group B. Tentacles absent; eyes in two elongate groups; penis-papilla without stylet. Includes, *bayeri* du B.-R. Marcus & Marcus, *chilensis* (Schmarda), *genicotyla* Palombi, *inquilina* Hyman, *longipenis* Hyman, *minuta* Hyman, *opisthopharynx* Palombi, *pallida* (Quatrefages), *parva* Palombi, *suesensis* Palombi, *taiwanica* Kato, *walsergia* du B.-R. Marcus & Marcus.

Group C. Tentacles absent; eyes in two elongate or rounded groups, penis-papilla with stylet. Includes, *affinis* Palombi, *leptalea* Marcus, *panamensis* (Plehn), *plehni* Bock, *reishi* Hyman, *robusta* Palombi, *snadda* du B.-R. Marcus & Marcus, *suoensis* Kato, *tenuis* Palombi, *wyona* du B.-R. Marcus & Marcus.

Group D. Tentacles variably developed; tentacular and cerebral eye-clusters well separated; with penis-stylet or penis-papilla covered with cuticle. Includes, *aulica* Marcus, *clara* Kato, *delicata* (Plehn), *divae* Marcus, *hancocki* Hyman, *taurica* Jacubowa, *vesiculata* Palombi.

Bock (1924) also erected the subgenus *Stylochoplana* (*Stylochoplanoides*) for *S. pusilla* Bock, 1924, but under the above-mentioned arrangement it is synonymous with the nominate subgenus (*Stylochoplana*). It seems that (*Stylochoplanoides*) was based on the shape of the body, being widened posteriorly and tapering anteriorly, and on the posterior position of the copulatory organs in *S. pusilla*. Furthermore, Kato (1934*b*) found that each uterine canal bifurcates in this species. Whether these features are sufficiently important to sustain a subgenus is, at present, questionable. Finally, it remains to be pointed out that du Bois-Reymond Marcus & Marcus (1968) transfer *Leptoplana diaphana* Stummer-Traunfels and *L. limnoriae* Hyman to *Stylochoplana*, but this change is not here accepted.

Genus ***Digynopora*** Hyman, 1940 Fig. 64

LEPTOPLANIDAE: STYLOCHOPLANINAE Body elongate and slender. Eyes in two elongate clusters lateral to cerebral organ. Pharynx in middle third of body; intestinal branches anastomosing. Common genital atrium well posterior to pharynx and some distance from hind end of body. Vasa deferentia enter muscular wall of seminal vesicle separately. Prostatic organ very muscular, with simple low epithelial lining and shallow lumen. Penis-papilla conical and relatively stout. Vagina narrow, forming a ductus vaginalis opening to exterior closely posterior to genital atrium. Uterine canals anteriorly confluent.

TYPE-SPECIES *D. americana* Hyman, 1940. Type by original designation.

TYPE-LOCALITY Among ascidians on piles in St Joseph's Bay, Florida, U.S.A.

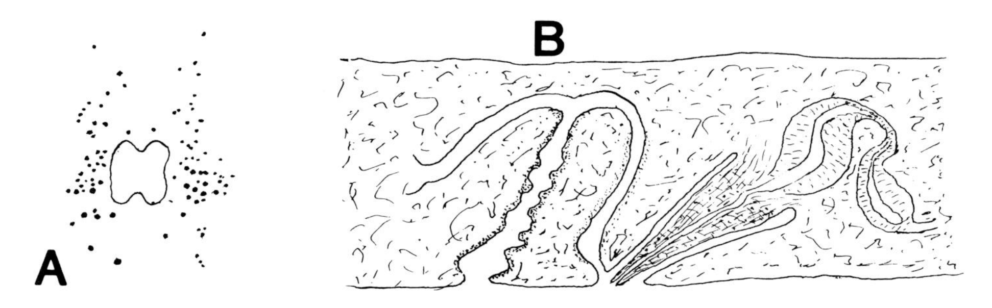

Fig. 64 *Digynopora americana.* **A**, arrangement of eyes; **B**, sagittal section of copulatory organs (after Hyman).

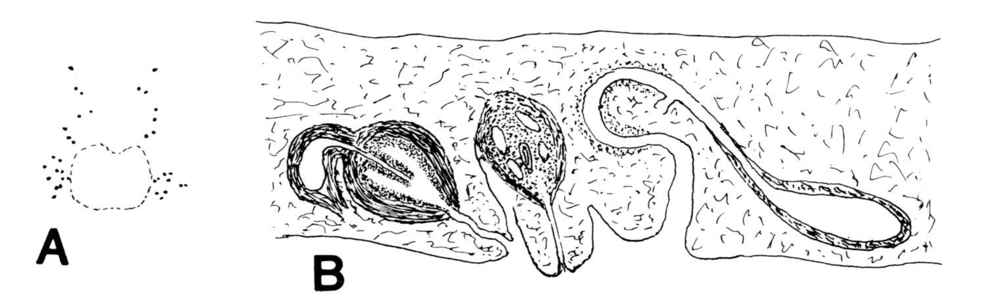

Fig. 65 *Diplandros singularis.* **A**, arrangement of eyes; **B**, sagittal section of copulatory organs (after Hyman).

Genus *Diplandros* Hyman, 1953 — Fig. 65

LEPTOPLANIDAE: STYLOCHOPLANINAE Elongate forms without tentacles. Eyes arranged in distinctly separated cerebral and tentacular clusters; former anterior to latter. Pharynx in middle third of body. Two male copulatory complexes in tandem, each with its own pore. Each complex consists of seminal vesicle and a bulbous prostatic organ, into the lumen of which projects the ejaculatory duct invested with a prostatic epithelium containing several longitudinal chambers lying parallel to the duct. No definite penis-papilla, although the male antrum may be everted to simulate a papilla. Vagina narrow, terminating in Lang's vesicle.

TYPE-SPECIES *D. singularis* Hyman, 1953. Type by original designation.

TYPE-LOCALITY La Jolla, California, U.S.A.

Genus *Plagiotata* Plehn, 1896 — Fig. 66

LEPTOPLANIDAE: STYLOCHOPLANINAE Broadly-oval fleshy forms, with pre-cerebral tentacles filled with eyes. Cerebral eyes widely distributed between tentacles and anteriorly to cerebral organ. Mouth anterior to pharyngeal chamber, into which it opens through a narrow 'oesophagus'. Pharynx transversely elongate, centrally placed. Genital pores separated. Vasa deferentia open separately into middle of elongate sinuous seminal vesicle. Prostatic organ pyriform, relatively small, with tall epithelial lining containing 5 or 6 narrow chambers lying parallel to ejaculatory duct, which projects well into the organ. Penis-papilla large and muscular, with short stylet. Male antrum spacious. Vagina wide, muscular, without Lang's vesicle.

TYPE-SPECIES *P. promiscua* Plehn, 1896. Type by original designation.

TYPE-LOCALITY West coast of Hong Kong, China Sea.

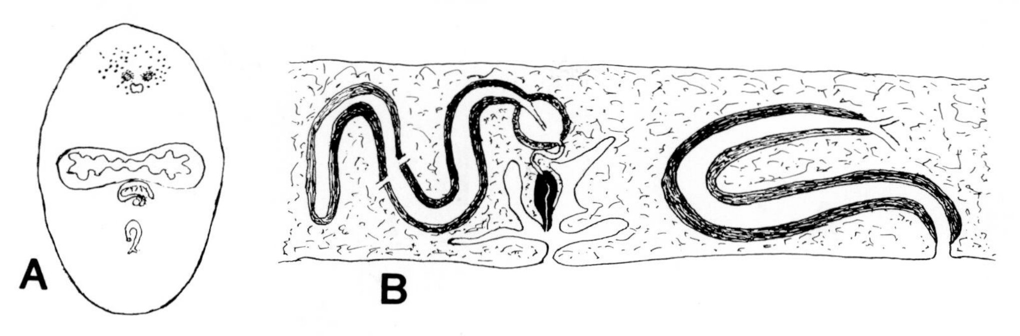

Fig. 66 *Plagiotata promiscua.* **A**, entire worm (cleared); **B**, sagittal section of copulatory organs (after Plehn).

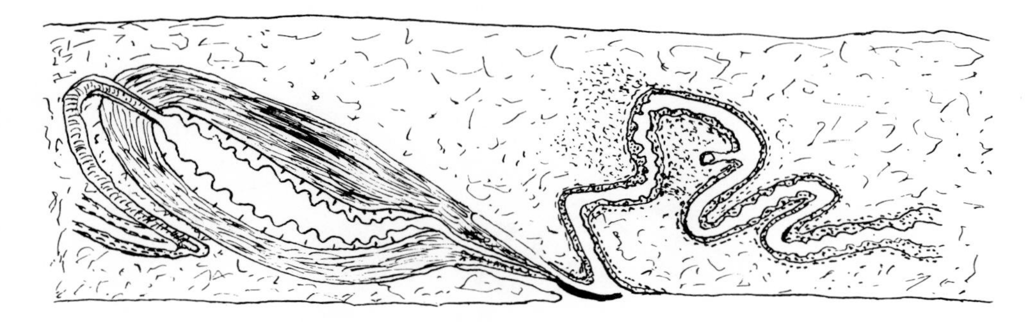

Fig. 67 *Macginitiella delmaris.* Sagittal section of copulatory organs (after Hyman).

SYSTEMATIC NOTE The copulatory complexes of *Plagiotata promiscua* bear a very strong resemblance to those of *Notoplana dubia* (Schmarda), except that Lang's vesicle is absent in the former and rudimentary in the latter. In fact, the relationship of *Plagiotata* with the genus *Notoplana* is so close that it seems doubtful whether the alleged transversely-elongate pharynx in the former, regarded as a generic feature by Plehn, is no more than an artifact produced perhaps by the method of fixing the only known specimen of *P. promiscua.*

Genus ***Macginitiella*** Hymen, 1953

Fig. 67

LEPTOPLANIDAE: STYLOCHOPLANINAE Elongate-oval forms without tentacles. Eyes few, limited to a pair of tentacular clusters. Pharynx long, mainly in middle third of body. Genital pores closely approximate in middle of hind half of body. No seminal vesicle. Spermiducal bulbs very large, muscular and elongate, directed anteriorly from vasa deferentia to unite and form a narrow ejaculatory duct opening into proximal region of prostatic organ. Latter an elongate structure, muscular, with epithelial lining thrown into several shallow radial folds. Conical penis-papilla elongate, with stylet, no penis-pocket. Vagina narrow, sinuous, mainly posterior to female pore, terminating in a genito-intestinal canal. Epithelium of 'shell'-chamber thrown into several radial folds. Uterine canals not anteriorly confluent.

TYPE-SPECIES *M. delmaris* Hyman, 1953. Type by original designation.

TYPE-LOCALITY Rocky shore at Corona del Mar, California, U.S.A.

Genus ***Freemania*** Hyman, 1953

Fig. 68

LEPTOPLANIDAE: STYLOCHOPLANINAE Elongate-oval, somewhat cuneate forms, without tentacles. Cerebral eyes in two linear clusters; tentacular eyes in two clusters lateral to cerebral organ and

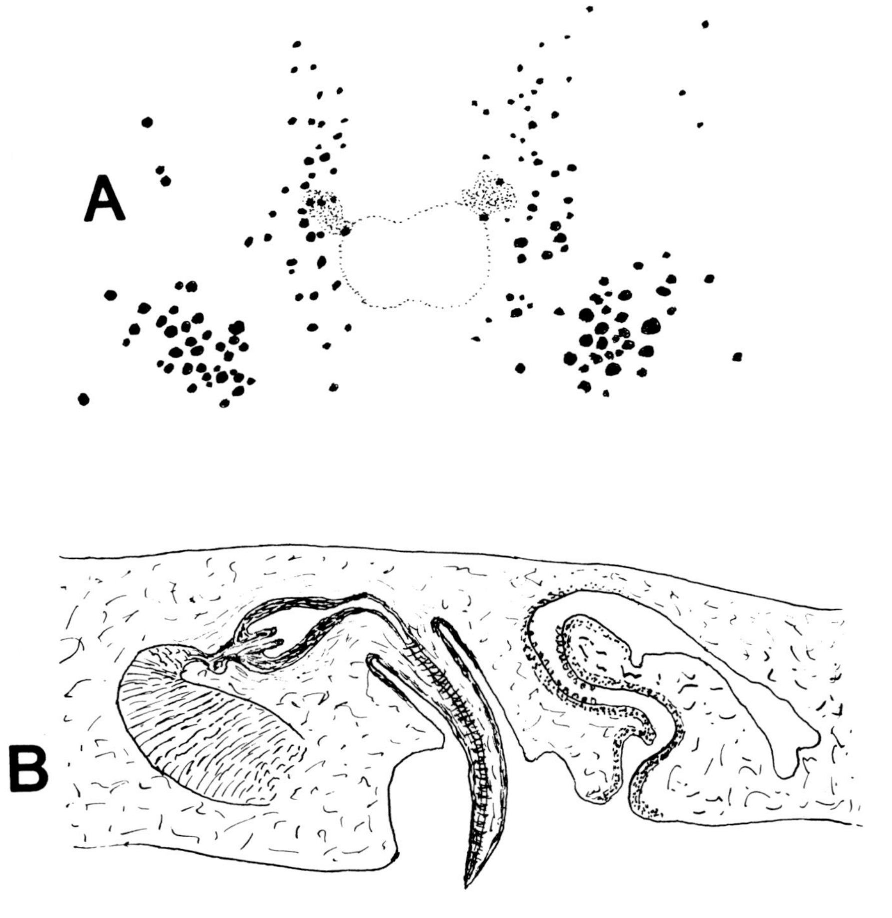

Fig. 68 *Freemania litoricola.* **A,** arrangement of eyes (after Heath & McGregor); **B,** sagittal section of copulatory organs (after Hyman).

posterior to cerebral clusters. Pharynx in middle third of body. Genital pores approximate. No seminal vesicle. Large spermiducal bulbs lead into an ejaculatory duct which enters and projects into lumen of well-differentiated, bulbous prostatic organ. Surrounding the ejaculatory duct inside the prostatic organ there is a tall epithelium containing several longitudinal chambers lying parallel to the duct. Penis-papilla elongate, muscular and unarmed. Vagina narrow, relatively short; Lang's vesicle small.

TYPE-SPECIES *F. [Phylloplana] litoricola* (Heath & McGregor, 1912). Type by original designation.

TYPE-LOCALITY Under stones below medium tide-mark along southern shore of Monterey Bay, California, U.S.A.

Genus *Copidoplana* Bock, 1913 — Fig. 69

LEPTOPLANIDAE: STYLOCHOPLANINAE Rather elongate forms without tentacles. Eyes in two elongate groups or in separate tentacular and cerebral clusters. Pharynx more or less in middle third of body. Genital pores closely separated or forming a common gonopore some distance anterior to posterior margin of body. Vasa deferentia open separately into an arcuate seminal vesicle lying postero-ventrally to prostatic organ. Latter well developed, globular, with tall epithelial lining containing several longitudinal chambers lying parallel to ejaculatory duct, which projects well into lumen of

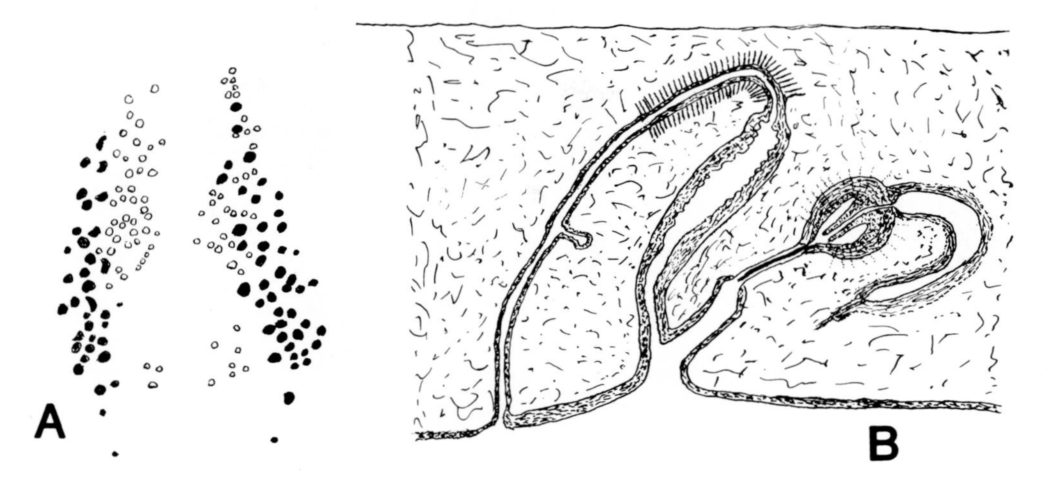

Fig. 69 *Copidoplana paradoxa.* **A,** arrangement of eyes; **B,** sagittal section of copulatory organs (after Bock).

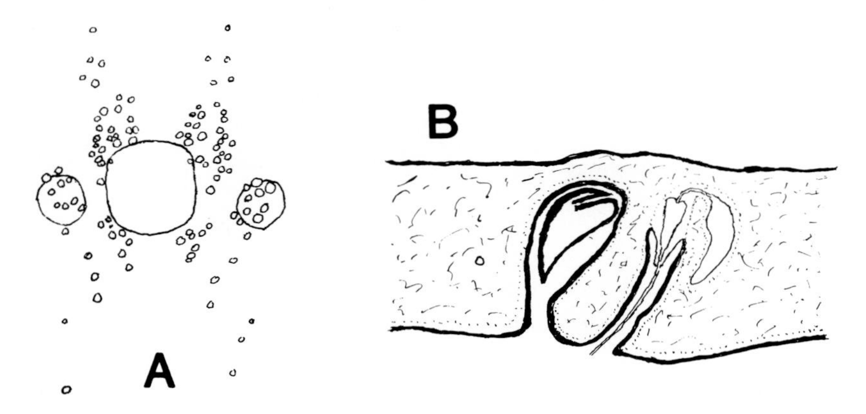

Fig. 70 *Ceratoplana colobocentroti.* **A,** arrangement of eyes; **B,** sagittal section of copulatory organs (after Bock).

prostatic organ. Penis-papilla small, with long stylet lying in long narrow penis-pocket. Antrum masculinum narrow. Vagina externa may have thick musculature; without Lang's vesicle. Ductus vaginalis opens to exterior close behind female pore. Uterine canals not anteriorly confluent.

TYPE-SPECIES *C. paradoxa* Bock, 1913. Type by original designation.

TYPE-LOCALITY From a depth of 2 metres, off coral reef on north point of Koh Chang, Gulf of Siam.

Genus *Ceratoplana* Bock, 1925

Fig. 70

LEPTOPLANIDAE: STYLOCHOPLANINAE Body somewhat cuneate, with long tentacles containing eyes. Cerebral eyes in two elongate clusters lying between tentacles. Pharynx centrally situated, more than one-third as long as body. Genital pores approximate. Vasa deferentia unite to form short ejaculatory duct opening into elongate seminal vesicle through a small protuberance. Prostatic organ large and vesicular, lined with a low, ciliated, secretory epithelium and invested with numerous extracapsular gland-cells. Penis-papilla with slender stylet. Male antrum deep and tubular. Vagina short, without Lang's vesicle. Ductus vaginalis long, opening to antrum femininum.

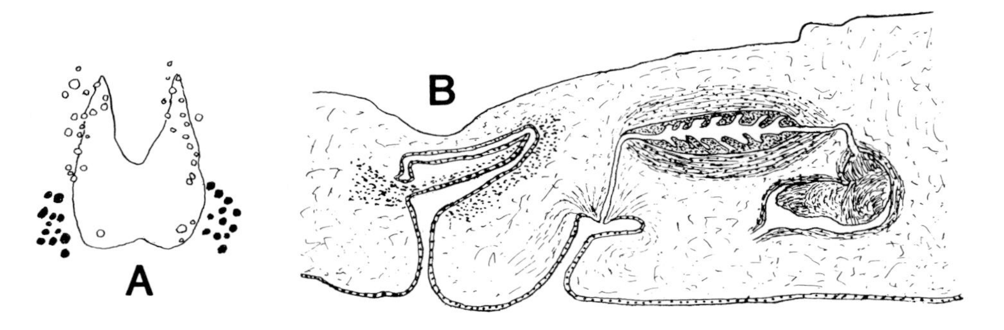

Fig. 71 *Notoplanella inarmata.* **A**, arrangement of eyes; **B**, sagittal section of copulatory organs (after Bock).

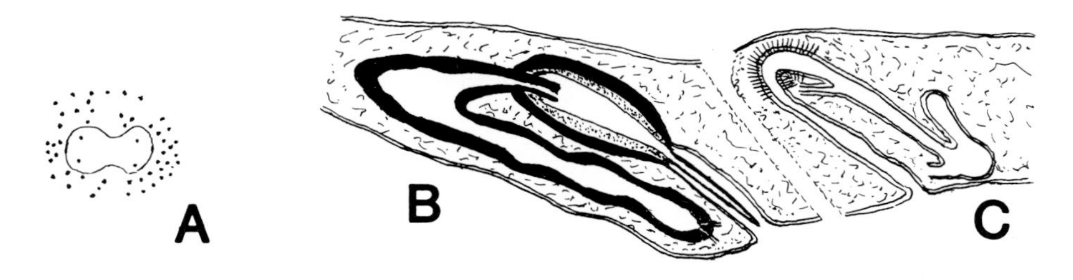

Fig. 72 *Leptocera delicata.* **A**, arrangement of eyes; **B**, sagittal section of ♂ copulatory organs; **C**, sagittal section of ♀ copulatory organs (after Plehn).

TYPE-SPECIES *C. colobocentri* Bock, 1925. Type by original designation.

TYPE-LOCALITY Beneath a sea-urchin (*Colobocentrotus atratus*), Krakatau, Sunda Strait, Java Sea.

Genus *Notoplanella* Bock, 1931 Fig. 71

LEPTOPLANIDAE: STYLOCHOPLANINAE Body elongate oval, without tentacles. Tentacular and cerebral eyes form a single elongate group on each lateral marginal of cerebral organ. Pharynx in middle third of body. Elongate, arcuate seminal vesicle connected to prostatic organ by a short narrow ejaculatory duct. Prostatic organ large, elliptical, dorsal to seminal vesicle, and lined with a very deep epithelium thrown into radial folds. Penis-papilla shallow. Vagina poorly developed, without Lang's vesicle.

TYPE-SPECIES *N. inarmata* Bock, 1931. Type by original designation.

TYPE-LOCALITY Simon's Bay, Cape Town, South Africa.

Genus *Leptocera* Jacubowa, 1906 Fig. 72

LEPTOPLANIDAE: STYLOCHOPLANINAE Small, delicate, oval forms with tentacles. Eyes in rounded cluster over cerebral organ. Pharynx in middle region of body. Genital pores approximate in hind region of body. Vasa deferentia open separately into long arcuate seminal vesicle lying antero-ventrally to prostatic organ. Latter pyriform; epithelial lining smooth; ejaculatory duct projects into its lumen. Penis-papilla inconspicuous, with strong stylet lying in male antrum. Vagina narrow; Lang's vesicle crescentic. Uterine canals medial to vasa deferentia, not anteriorly confluent.

TYPE-SPECIES *L. delicata* Jacubowa, 1906. Type by original designation.

TYPE-LOCALITY New Britain, Pacific Ocean.

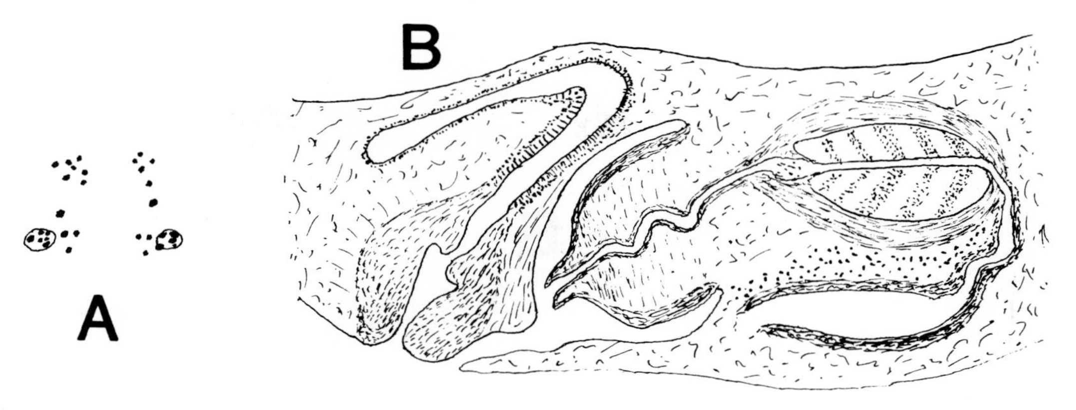

Fig. 73 *Parviplana californica.* **A,** arrangement of eyes (after Woodworth); **B,** sagittal section of copulatory organs (after Hyman).

Genus *Parviplana* Hyman, 1953

Fig. 73

LEPTOPLANIDAE: STYLOCHOPLANINAE Small elongate-oval forms. Eyes in paired cerebral and tentacular clusters, the former lying anteriorly to the latter. Pharynx in middle third of body. Genital pores approximate at about midway between pharynx and hind end of body. Vasa deferentia open by common canal into oval muscular seminal vesicle lying ventral to prostatic organ. From seminal vesicle, ejaculatory duct curves dorsally to open into oval, thick-walled prostatic organ surrounded by thick layer of extracapsular gland cells. Ejaculatory duct passes into prostatic organ to open into a small distal chamber. Penis-papilla large, in spacious male antrum. Vagina curves posteriorly from a vagina bulbosa. Lang's vesicle small.

TYPE-SPECIES *P. [Stylochoplana] californica* (Woodworth, 1894). Type by original designation.

TYPE-LOCALITY Gulf of California (surface, 26°48′0″N, 110°45′20″W)

NOTE In Woodworth's (1894) description of *Stylochoplana californica* nuchal tentacles are indicated, but Hyman (1953*a*) states that they are absent.

Genus *Notoplana* Laidlaw, 1903

Fig. 74

LEPTOPLANIDAE: STYLOCHOPLANINAE Elongate or elliptical body, sometimes broadly rounded anteriorly and narrowing somewhat posteriorly. Short, contractile tentacles may be present. Eyes arranged in paired cerebral and tentacular clusters, which may be distinctly separated or merged into two elongate clusters. Pharynx centrally situated. Male copulatory organ closely posterior to pharynx. Genital pores separated, but sometimes so close together as to form a common opening. Vasa deferentia often forming a loop posteriorly to the vagina. Seminal vesicle well developed. Ejaculatory duct passes some distance into well-developed prostatic organ, where it is enclosed in a very tall epithelium containing several longitudinal tubes lying parallel to the duct. Penis-papilla of variable development, sometimes covered with a cuticle or bearing a stylet of varying length. Vagina simple. Lang's vesicle of variable size, when present, and posterior to female genital pore.

TYPE-SPECIES *N. [Centrostomum] dubia* (Schmarda, 1895). Type by subsequent designation.

TYPE-LOCALITY Sri Lanka.

NOTES Some species of *Notoplana* show no indication of anastomosis of the secondary intestinal branches, but in others anastomosis of the branches occurs only in the pharyngeal or in the marginal regions.

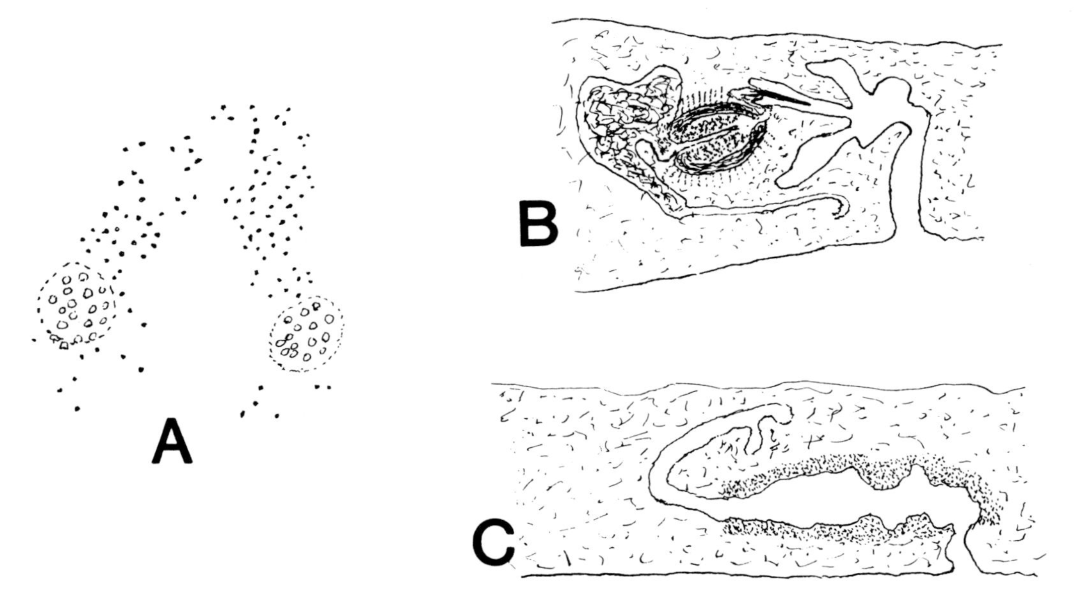

Fig. 74 *Notoplana dubia.* **A**, arrangement of eyes; **B**, sagittal section of ♂ copulatory organs; **C**, sagittal section of ♀ copulatory organs (after Stummer-Traunfels).

In the descriptions of some species of this genus, it appears that the vasa deferentia do not unite posteriorly to the vagina, but it seems that such a connection is not seen until the male phase is fully active.

Bock (1913) separated the then 17 recognizable species of *Notoplana* into 3 groups: **A** with *N. evansii* Laidlaw (s.o. *dubia*) as its type-species; **B** with *N. atomata* (Müller) as its type; and **C** with *N. alcinoi* (Schmidt) as its type. Since this arrangement was made, the number of species assigned to the genus has increased to over 60, and many of these cannot be placed satisfactorily into any one of Bock's groups. Consequently, du Bois-Reymond Marcus & Marcus (1968) divided the species into 9 groups, emphasizing the presence or absence of nuchal tentacles and features of the male copulatory complex. Unfortunately, such use of the tentacles is unreliable, because in *N. australis* (Schmarda) and several other species, tentacles may, in preserved specimens, be noticeable, ill-defined or not apparent. Therefore, a new division of species of *Notoplana* into 4 groups is here offered for assistance only in the determination of known species.

A. penis-papilla bearing stylet or covered with a cuticle; with penis-pocket. Includes, *annula* du B.-R. Marcus & Marcus, *atlantica* (Bock), *atomata* (Müller), *australis* (Schmarda), *cotylifera* Meixner, *delicata* Yeri & Kaburaki, *distincta* Prudhoe, *divae* Marcus, *dubia* (Schmarda), *fallax* (Quatrefages), *inquieta* (H. & McG.), *inquilina* Hyman, *kuekenthali* (Plehn), *lactoalba* (Verrill), *longastyletta* (Freeman), *longicrumena* Prudhoe, *megala* Marcus, *micheli* Marcus, *micronesiana* Hyman, *mortenseni* Bock, *nationalis* (Plehn), *parvula* Palombi, *plecta* Marcus, *puma* Marcus, *queruca* du Bois Reymond Marcus & Marcus, *sawayai* Kato, *stilifera* Bock, *vitrea* (Lang), *willeyi* Jacubowa,

B. penis-papilla bearing stylet or covered with cuticle, no penis-pocket or sheath. Includes, *alcinoi* (Schmidt), *igiliensis* Galleni, *insularis* Hyman, *sanpedrensis* Freeman, *serica* Kato.

C. penis-papilla without stylet or cuticular covering, with penis-pocket. Includes, *ferruginea* (Schmarda), *japonica* Kato, *koreana* Kato.

D. penis-papilla without stylet or cuticular covering, no penis-pocket. Includes, *acticola* (Boone), *celeris* Freeman, *chierchiae* (Plehn), *gardineri* (Laidlaw), *humilis* (Stimpson), *lapunda* du B.-R. Marcus & Marcus, *libera* Kato, longiducta Hyman, *longisaccata* Hyman, *martae* Marcus, *microsora* (Schmarda), *natans,* Freeman, *otophora* (Schmarda), *palaoensis* Kato, *palta* Marcus, *patellarum* (Stimpson), *rupicola* (Heath & McGregor), *sanguinea* Freeman, *sanjuania* Freeman, *saxicola* (Heath & McGregor), *sciophila* (Boone), *septentrionalis* Kato, *sophia* Kato, *syntoma* Marcus, *tavoyensis* Prudhoe.

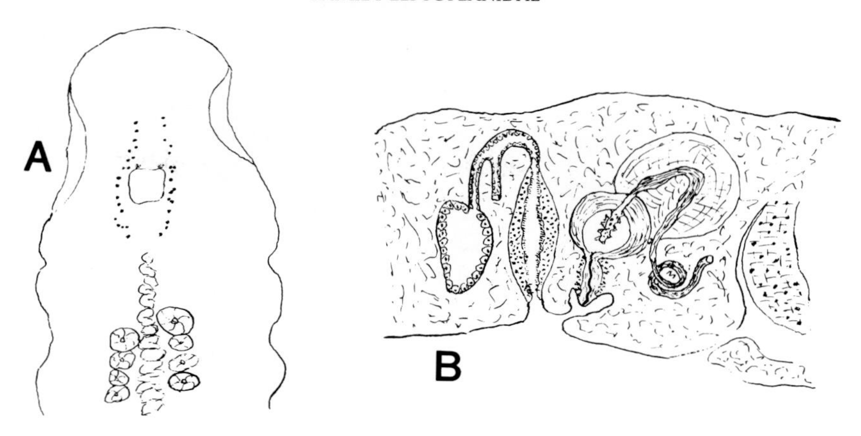

Fig. 75 *Pucelis evelinae.* **A**, arrangement of eyes; **B**, sagittal section of copulatory organs (after Marcus).

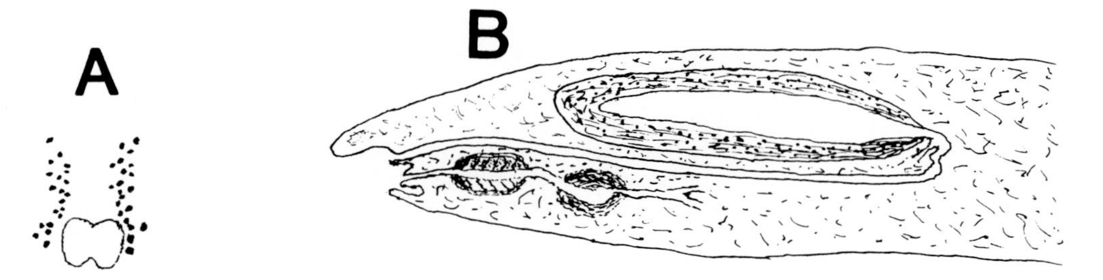

Fig. 76 *Zygantroplana verrilli.* **A**, arrangement of eyes; **B**, sagittal section of copulatory organs (after Laidlaw).

Genus ***Pucelis*** Marcus, 1947

Fig. 75

LEPTOPLANIDAE: STYLOCHOPLANINAE Small elongate forms, without tentacles. Eyes disposed in two narrow rows lateral to cerebral origin. Pharynx short, posterior to middle of body. Genital pores closely approximate, posterior to pharynx. Vasa deferentia unite before entering a large, arcuate, muscular seminal vesicle lying anteriorly to, and opening directly into, a globular, thick-walled, prostatic organ. Epithelial lining of this organ very tall and contains longitudinal chambers running parallel to and investing intruding ejaculatory duct. Penis-papilla without stylet. Vagina dorsally arcuate, with Lang's vesicle. Uterine canals not anteriorly confluent.

TYPE-SPECIES *P. evelinae* Marcus, 1947. Type by original designation.

TYPE-LOCALITY Santos Bay, Palmas Isle, Brazil.

NOTE Marcus distinguishes this genus from *Notoplana* on the grounds that its pharynx lies in the posterior half of the body, whereas in *Notoplana* it is centrally placed. It is doubtful whether this difference alone warrants generic separation.

Genus ***Zygantroplana*** Laidlaw, 1906

Fig. 76

LEPTOPLANIDAE: STYLOCHOPLANINAE Elongate-oval or somewhat oblong forms, without tentacles. Eyes in paired cerebral and tentacular clusters or in two elongate groups alongside cerebral organ. Pharynx more or less centrally situated. Copulatory complexes open into common atrium or by

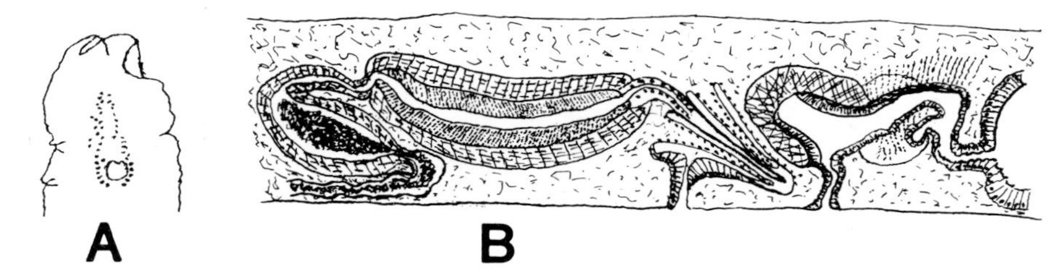

Fig. 77 *Candimboides rabita.* **A**, arrangement of eyes; **B**, sagittal section of copulatory organs (after du B.-R. Marcus & Marcus).

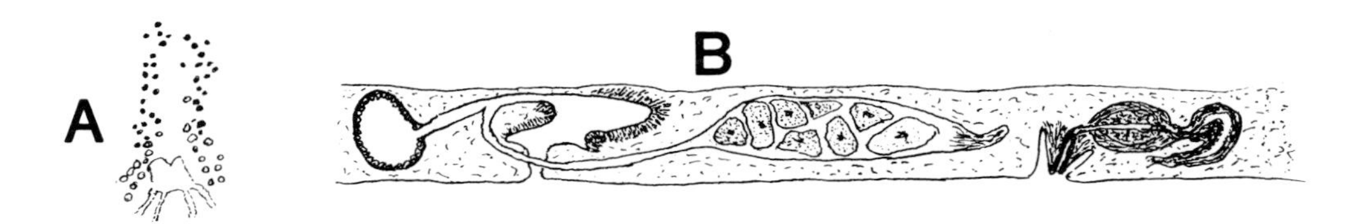

Fig. 78 *Candimba divae.* **A**, arrangement of eyes; **B**, sagittal section of copulatory organs (after Marcus).

independent pores closely associated with one another adjacent to posterior margin of body. Male and female complexes directed anteriorly from their respective openings, the former lying ventrally to the latter. Seminal vesicle moderately developed. Prostatic organ bulbous, or elongate, and lined with a smooth shallow epithelium. Penis-papilla, when present, small and pyriform, occasionally provided with a stylet lying in a long penis-pocket with a distinct penis-sheath. Vagina long, terminating in Lang's vesicle. Uterine canals extend anteriorly no further than the hind end of pharynx.

TYPE-SPECIES *Z. verrilli* Laidlaw, 1906. Type by original designation.

TYPE-LOCALITY Amongst weeds in St Vincent Harbour, Cape Verde Is, Atlantic Ocean.

NOTE The arrangement of the eyes described for *Zygantroplana clepeasta* Kato, 1944 is much like that found in the genus *Cestoplana.* In fact, the arrangement is so atypical of leptoplanid polyclads that, although Kato appears to have examined one complete and one damaged specimen, one must reasonably suspect whether the anterior and posterior portions of the worm described actually belonged to the same specimen.

Genus *Candimboides* Prudhoe, 1982 — Fig. 77

LEPTOPLANIDAE: STYLOCHOPLANINAE Body elongate, without tentacles. Eyes in two elongate clusters lateral to cerebral organ. Pharynx short, in middle third of body. Male and female genital pores approximate and well separated from hind margin of body. Male copulatory complex immediately posterior to pharynx. Vasa deferentia unite to form common duct opening into muscular seminal vesicle. Prostatic organ muscular and lined with tall, smooth, glandular epithelium. Penis-stylet in penis-pocket. Female copulatory complex with bursa copulatrix and Lang's vesicle.

TYPE-SPECIES *C. [Candimba] rabita* (du Bois-Reymond Marcus & Marcus, 1968). Type by original designation.

TYPE-LOCALITY Among algae, Piscadera Baai, Curaçao, Caribbean.

Genus *Candimba* Marcus, 1949 — Fig. 78

LEPTOPLANIDAE: STYLOCHOPLANINAE Body elongate, without tentacles. Tentacular and cerebral eyes form two elongate clusters laterally to median line, mainly anterior to cerebral organ. Short pharynx in anterior third of body. Genital pores widely separated in middle third of body. Seminal vesicle muscular, arcuate. Ovoid, muscular prostatic organ lined with tall, smooth, glandular epithelium. Penis-papilla stout, without stylet. Male antrum short. Vagina narrow, with dilated 'shell'-chamber and bulbous Lang's vesicle. Uterine canals very short, not anteriorly confluent.

TYPE-SPECIES *C. divae* Marcus, 1949. Type by original designation.

TYPE-LOCALITY Among algae in upper littoral of Ilha das Palma, Baia de Santos, São Paulo, Brazil.

Family **HOPLOPLANIDAE** Stummer-Traunfels, 1933

PLANOCEROIDEA Small, translucent, rounded or broadly-oval forms, sometimes narrower anteriorly than posteriorly; dorsal surface may be papillate. Conical nuchal tentacles long, each with ring of eyes at base; cerebral eyes in two linear clusters lateral to cerebral organ and between or anterior to tentacles. Pharynx with 4 to 7 pairs of deep lateral folds, centrally situated. Genital pores approximate. Male copulatory complex immediately posterior to pharynx. No seminal vesicle. Pair of widely-separated elongate spermiducal bulbs open into interpolated, bulbous, prostatic organ. No penis-papilla, but a strong penial stylet is attached to prostatic organ. Male antrum and prostatic organ may be enclosed in a thick muscular envelope. Vagina short, narrow and arcuate. No Lang's vesicle. Uterine canals short, not anteriorly confluent.

SYSTEMATIC NOTE There appears to be disagreement as to the validity of the family Hoploplanidae, since some authors have regarded the type-genus *Hoploplana* as a member of the family Leptoplanidae. In fact, Marcus & Marcus (1966) have relegated the status of Hoploplanidae to that of Hoploplaninae, a subfamily of the Leptoplanidae. There are, however, some features in the structure of *Hoploplana* which suggest that the genus is more closely related to the Planoceridae than to the Leptoplanidae. These are especially evident in the broadly-oval outline and texture of the body, the ring-like formation of eyes at the bases of the long tentacles and the very deep lateral folds to the pharynx. It would seem, therefore, that the Hoploplanidae is a link between the Leptoplanidae and the Planoceridae and warrants recognition of family status.

KEY TO HOPLOPLANID GENERA

1. Two ventral suckers lateral to female pore . . . . ***Itannia***
   No ventral suckers . . . . ***Hoploplana***

Genus ***Hoploplana*** Laidlaw, 1902
(= *Planocera* Group B of Lang, 1884)

Fig. 79

HOPLOPLANIDAE Dorsal surface sometimes raised into numerous slender processes. Elongate spermiducal bulbs highly muscular. Prostatic organ pyriform or spherical, muscular, with tall glandular epithelial lining. Male antrum long and narrow. Prostatic organ and antrum may be enclosed in a thick muscular envelope. 'Shell'-chamber long.

TYPE-SPECIES *H. [Planocera] insignis* (Lang, 1884). Type by indication.

TYPE-LOCALITY In shallow water, Castello dell'novo, Bay of Naples, Italy.

NOTE Development in this genus is either direct, as in *H. villosa,* or indirect, through Müller's larva, as in *H. inquilina.*

Genus ***Itannia*** Marcus, 1947

Fig. 80

HOPLOPLANIDAE Small, broadly-oval, non-papillate forms with widely-separated tentacles. Pharynx large, with six or seven pairs of deep lateral folds. Pair of small ventral suckers, lateral to female genital pore. Male and female copulatory complexes as in *Hoploplana.*

TYPE-SPECIES *I. ornata* Marcus, 1947. Type by original designation.

TYPE-LOCALITY Among algae, Ilha das Palmas, Baia de Santos, Brazil.

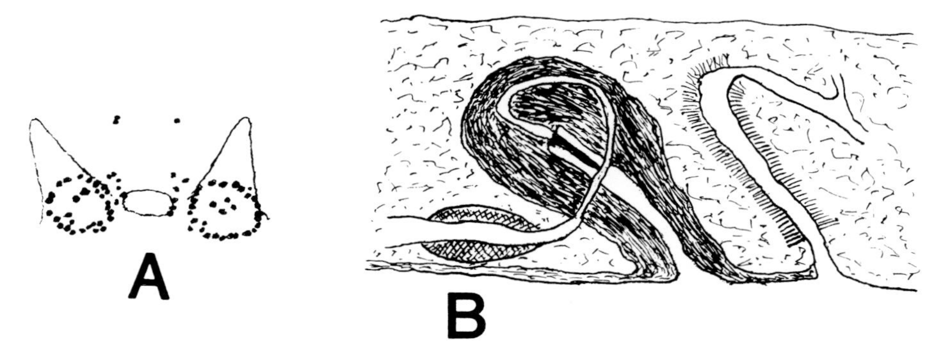

Fig. 79 *Hoploplana.* **A**, *H. ornata* arrangement of eyes (modified, after Yeri & Kaburaki); **B**, *H. insignis* sagittal section of copulatory organs (after Lang).

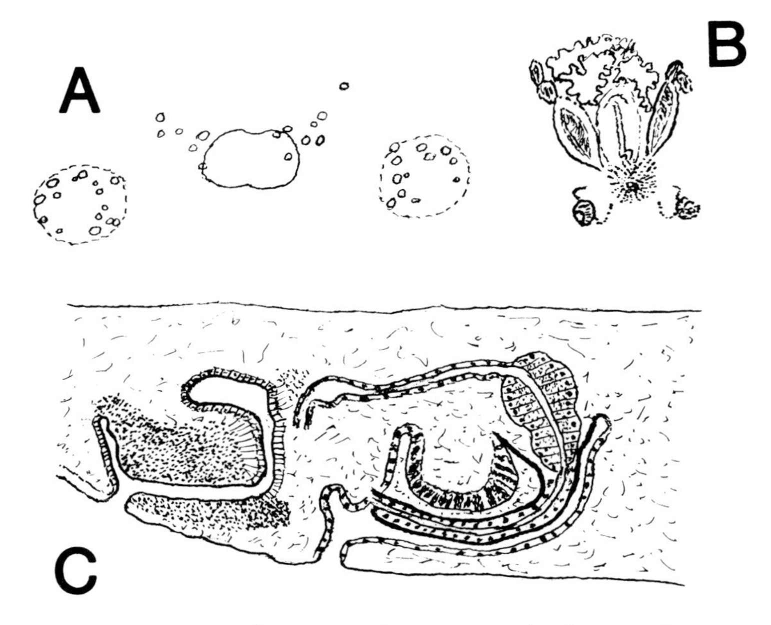

Fig. 80 *Itannia ornata.* **A**, arrangement of eyes; **B**, copulatory organs and suckers (ventral view); **C**, sagittal section of copulatory organs (after Marcus).

Family **GNESIOCEROTIDAE** Marcus & Marcus, 1966, emended Prudhoe, 1982 (= GNESIOCERINAE [*sic*] Marcus & Marcus, 1966)

PLANOCEROIDEA Elongate forms with or without tentacles. Eyes disposed in two elongate clusters alongside cerebral organ or in paired cerebral and tentacular clusters. Pharynx much folded and situated in middle third of body or somewhat anteriorly. Genital pores separated. Vasa deferentia may form a pair of spermiducal bulbs before opening into true seminal vesicle or the prostatic organ. Prostatic organ interpolated between sperm ducts or seminal vesicle and an eversible cirrus-sac. Epithelium of prostatic organ tall and thrown into several radial or longitudinal folds. Numerous extra-capsular gland cells may invest prostatic organ. Cirrus-sac may contain long cuticularized papilla or be

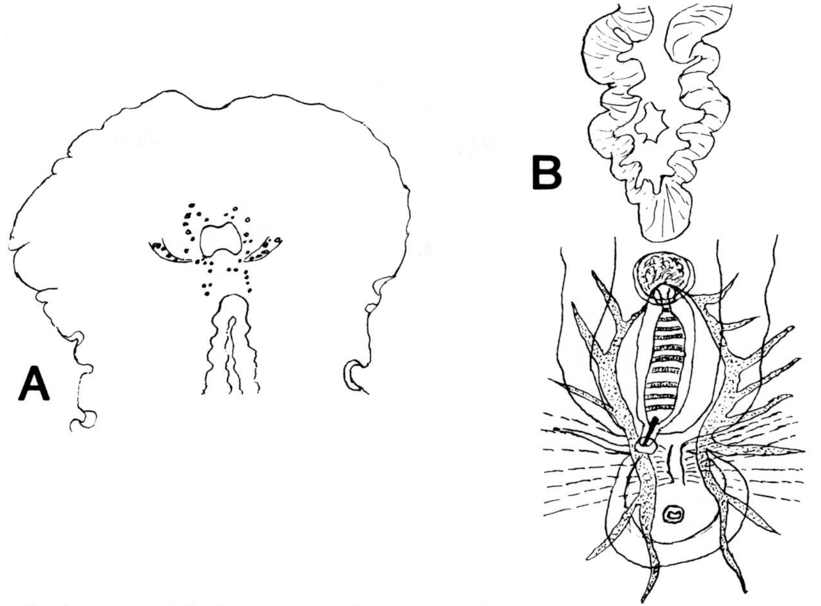

Fig. 81 *Gnesioceros sargassicola.* **A**, arrangement of eyes; **B**, ventral view of copulatory organs (semi-diagrammatic).

lined with cuticle thrown into ridges. Cuticular lining of cirrus-sac more often bears bristles, spines or hooks. It is also provided with a very thick musculature, which may also invest prostatic organ. Vagina simple, usually with Lang's vesicle.

KEY TO GNESIOCEROTID GENERA

1. With nuchal tentacles; eyes in paired cerebral and tentacular clusters 2
   Without nuchal tentacles; eyes in two elongate clusters 4
2. Row of tentacle-like tufts on body-margin . . . . *Styloplanocera*
   No special structures on body-margin 3
3. Conical cirrus covered with transverse serrated ridges . . . . . . *Gnesioceros*
   Cirrus-sac partially cuticularized, with strong tooth . . . *Pseudalloioplana* gen. nov.
4. Without spermiducal bulbs or seminal vesicle . . . . *Spinicirrus*
   With one or other type of sperm receptacles 5
5. With spermiducal bulbs; no seminal vesicle . . . . *Planctoplanella*
   With seminal vesicle; no spermiducal bulbs 6
6. Muscular pyriform organ in anterior wall of male antrum . . . . . *Neoplanocera*
   Male antrum without pyriform organ 7
7. Corrugated muscular pad between genital pores; cirrus spiny . . . . . *Echinoplana*
   No corrugated pad between genital pores; cirrus smooth . . . . . . *Amyris*

Genus ***Gnesioceros*** Diesing, 1861 Fig. 81

SYNONYMY *Pelagoplana* Bock, 1913.

GNESIOCEROTIDAE Pellucid, somewhat cuneate, forms expanded anteriorly and narrowed posteriorly. Conical tentacles containing eyes. Cerebral eyes in two scattered groups lateral to cerebral organ. Intestinal branches not anastomosing. Genital pores separated. Seminal vesicle well developed, opens directly into oval prostatic organ lined with tall epithelium thrown into several deep radial folds. Conical cirrus covered with several transversely serrated ridges of cuticle. Vagina externa with powerful glandulo-muscular fold invested with large gland cells. Lang's vesicle somewhat U-shaped or crescentic, limbs directed anteriorly.

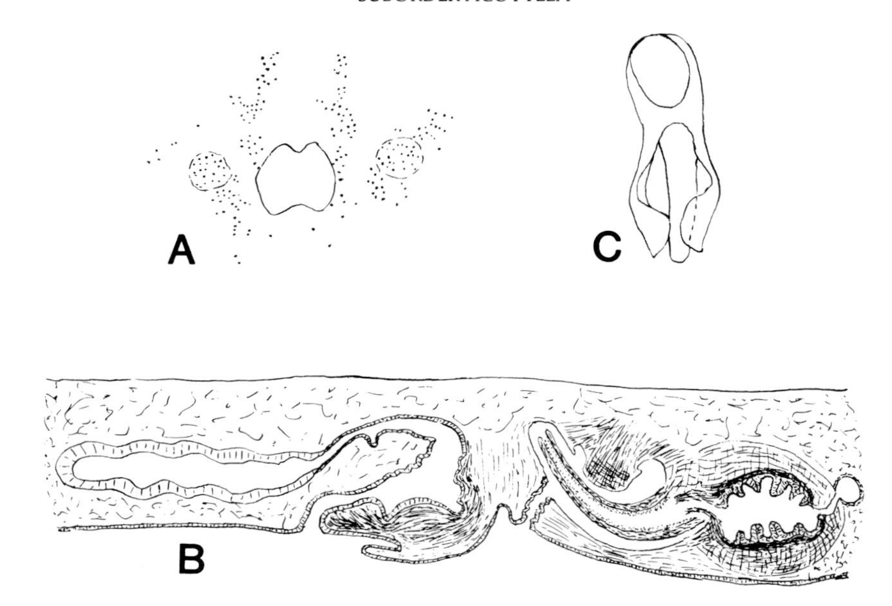

Fig. 82 *Pseudalloioplana californica*. **A**, arrangement of eyes (after Heath & McGregor); **B**, sagittal section of copulatory organs; **C**, cuticularized part of cirrus-sac (after Hyman).

TYPE-SPECIES *G. [Planaria] sargassicola* (Mertens, 1832). Type by subsequent designation.

TYPE-LOCALITY On *Sargassum* weed, Atlantic Ocean between 21° to 35°N and 36° to 38°W.

NOTE The species *sargassicola* has for the past ninety years or so been identified from Graff's (1892*b*) description and figures of *Stylochoplana sargassicola* (Mertens). The fact is, however, that Merten's excellent figure of *Planaria sargassicola* is identical with *Hoploplana [Planocera] grubei* (Graff, 1892). It is quite clear that Graff misidentified Mertens' *Planaria sargassicola*, but since Graff's concept of this species has been widely accepted since 1892, it seems reasonable, on the grounds of usage, to maintain this concept as *Gnesioceros sargassicola* (Mertens, 1832) *sensu* Graff, 1892.

Genus ***Pseudalloioplana*** gen. nov. Fig. 82

GNESIOCEROTIDAE Delicate elongate-oval forms with nuchal tentacles containing eyes. Cerebral eyes in two elongate clusters between tentacles. Intestinal branches not anastomosing. Genital pores separated near posterior end of body. Vasa deferentia open separately into seminal vesicle. Prostatic organ globular, provided with a thick muscular wall and lined with a tall folded epithelium. Cirrus-sac lined with cuticle and into lumen of its proximal region projects a papilla covered with a thick cuticle; lateral walls of cirrus-sac with pair of cuticularized lobes. Vagina narrow, forming anteriorly-directed loop, with vaginal bursa; Lang's vesicle trilobed. Uterine canals confluent anteriorly to pharynx.

TYPE-SPECIES *P. [Planocera] californica* (Heath & McGregor, 1913) nov. comb.

TYPE-LOCALITY Monterey Bay and vicinity, California, U.S.A.

NOTE Hyman (1953*a*) considered *Planocera californica* Heath & McGregor and *Planocera sandiegensis* Boone to be members of the genus *Alloioplana*, the type-species of which is *A. delicata* Plehn, 1896. The status of Plehn's genus has been discussed above and considered to be a synonym of the leptoplanid genus *Stylochoplana*. *P. californica* and *P. sandiegensis*, however, do not belong to *Stylochoplana* nor to the genus *Planocera*, therefore, a new genus, *Pseudalloioplana*, is erected for their reception.

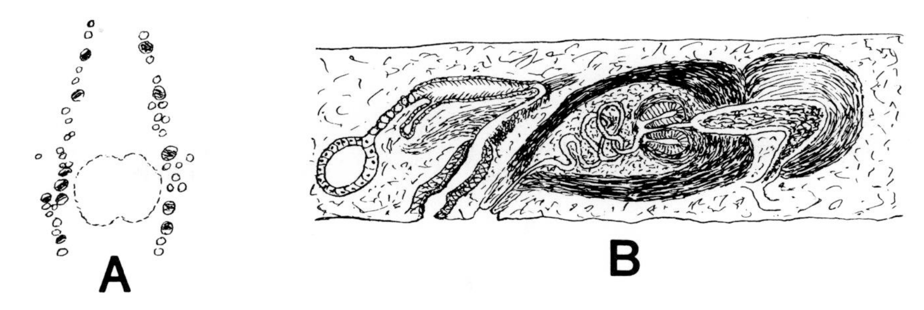

Fig. 83 *Amyris hummelincki.* **A,** arrangement of eyes; **B,** sagittal section of copulatory organs (after du B.-R. Marcus & Marcus).

Genus *Amyris* du Bois-Reymond Marcus & Marcus, 1968 — Fig. 83

GNESIOCEROTIDAE Small elongate forms without nuchal tentacles. Eyes in two elongate groups lateral to cerebral organ. Pharynx short, in posterior half of body; intestinal branches anastomosing. Genital pores approximate. Male copulatory complex immediately posterior to pharynx and directed anteriorly from male pore. Muscular seminal vesicle leads into globular prostatic organ lined with tall epithelium containing longitudinal chambers. From prostatic organ a coiled tubular and cuticularized, but not spinous, cirrus passes into male antrum. Loose connective tissue invests cirrus, and the whole is enclosed in a thick musculature arising from fibres of the muscular sheath around prostatic organ and seminal vesicle. Vagina simple, thrown into antero-dorsal loop, terminating in small Lang's vesicle. Uterine canals anteriorly confluent.

TYPE-SPECIES *A. hummelincki* du Bois-Reymond Marcus & Marcus, 1968. Type by original designation.

TYPE-LOCALITY From north side of cliff of coral rock, Boca Lagoon, Curaçao, Caribbean.

Genus *Planctoplanella* Hyman, 1940 — Fig. 84

GNESIOCEROTIDAE Oval forms with rounded extremities, broadest posteriorly. Nuchal tentacles absent. Eyes loosely in two elongate groups in cerebral region. Pharynx mainly in posterior half of body; intestinal branches anastomosing. Genital pores separated. No seminal vesicle. Proximal half of elongate prostatic organ with tall epithelium thrown into radial folds; distal half contains an indefinite tissue holding large cyanophilic masses. Cirrus-sac with very thick muscular wall enclosing narrow lumen containing a long slender papilla coated with cuticle. 'Shell'-chamber widened, otherwise vagina narrow and forming a short anterior loop. No Lang's vesicle. Uterine canals not anteriorly confluent.

TYPE-SPECIES *P. atlantica* Hyman, 1940. Type by original designation.

TYPE-LOCALITY In surface tow, Sheepshead Shoal, off Beaufort, North Carolina, U.S.A.

Genus *Echinoplana* Haswell, 1907 — Fig. 85

GNESIOCEROTIDAE Elongate forms without tentacles. Eyes in two elongate groups in region of cerebral organ. Pharynx mainly in anterior half of body; intestinal branches anastomosing. Male and female genital pores separated by genital sucker with corrugated surface. Seminal vesicle elongate. Prostatic organ muscular and elongate, connected with cirrus-sac by convoluted ejaculatory duct enclosed in mass of muscle-fibres. Cirrus-sac with thick musculature and lined with spines gradually increasing in size towards exterior. No male antrum. Vagina narrow, directed anteriorly, invested with 'shell'-glands over much of its length. Lang's vesicle rudimentary. Uterine canals anteriorly confluent.

TYPE-SPECIES *E. celerrima* Haswell, 1907. Type by subsequent designation.

TYPE-LOCALITY Port Jackson, New South Wales, Australia.

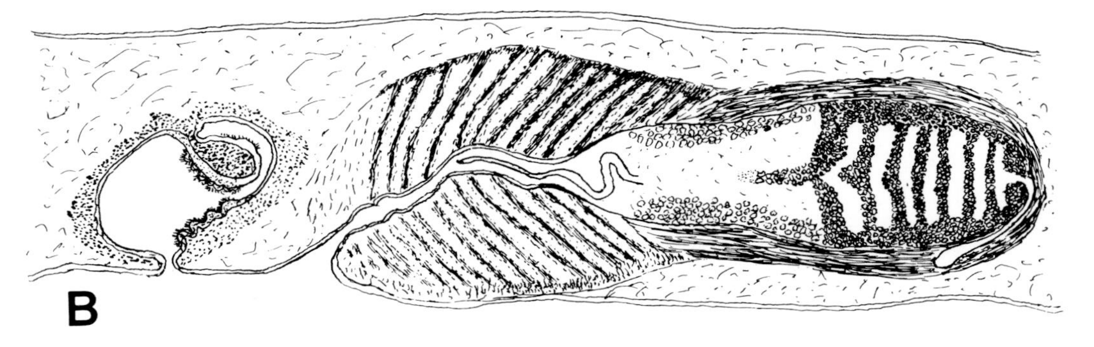

Fig. 84 *Planctoplanella atlantica.* **A,** arrangement of eyes; **B,** sagittal section of copulatory organs (after Hyman).

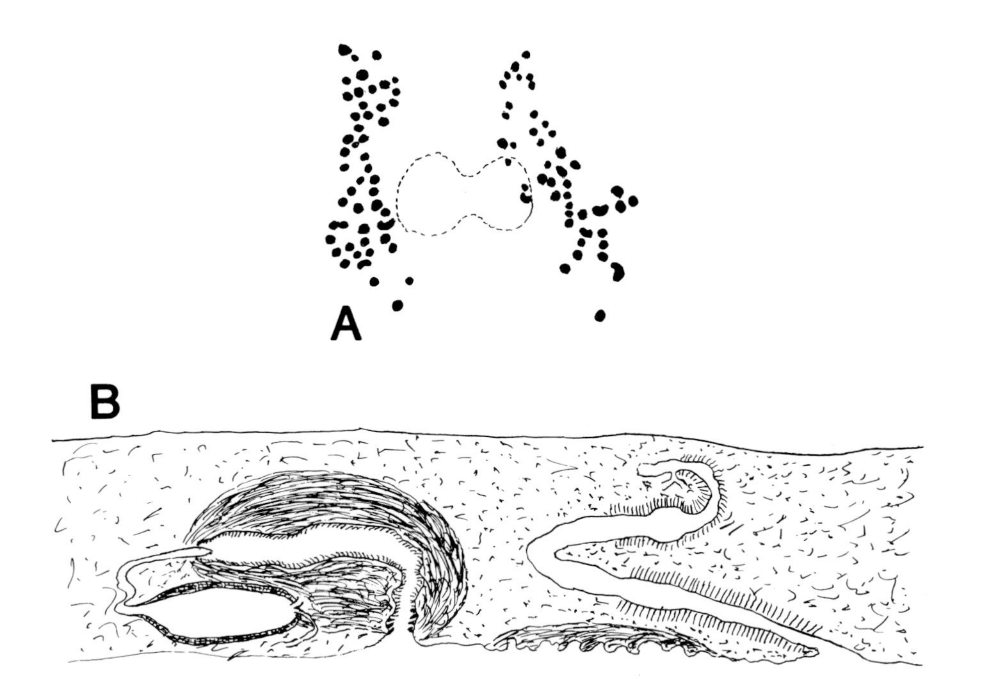

Fig. 85 *Echinoplana celerrima.* **A,** arrangement of eyes (after Haswell); **B,** sagittal section of copulatory organs (diagrammatic).

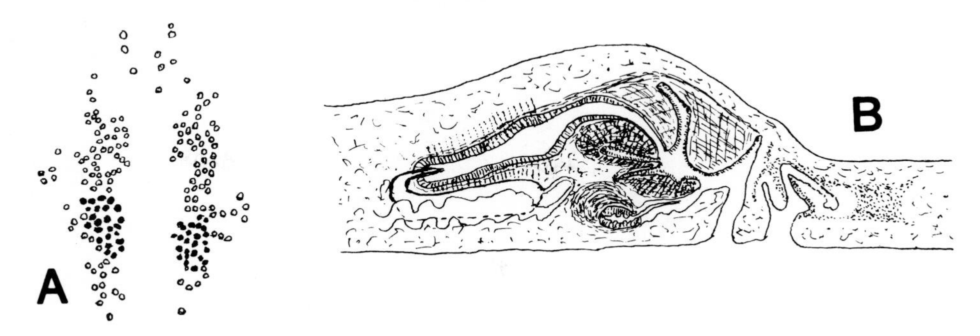

Fig. 86 *Neoplanocera elongata.* **A,** arrangement of eyes; **B,** sagittal section of copulatory organs (after Kato).

Genus *Neoplanocera* Yeri & Kaburaki, 1918 Fig. 86

SYNONYMY *Cirroposthia* Steinbock, 1937

GNESIOCEROTIDAE Elongate forms without tentacles. Eyes in two elongate groups lateral to cerebral organ. Pharynx in mid-third of body. Genital pores separated. Seminal vesicle elongate and thin walled. Elongate prostatic organ lined with tall smooth epithelium. Cirrus-sac lined with spines or stiff bristles. Prostatic organ and cirrus-sac not enclosed in muscular sheath. Pyriform glandular organ lying ventrally to cirrus-sac opens into male antrum. Vagina short and narrow; Lang's vesicle absent or rudimentary. Uterine canals anteriorly confluent.

TYPE-SPECIES *N. elongata* Yeri & Kaburaki, 1918. Type by original designation.

TYPE-LOCALITY Low-water mark, Awa, Japan.

Genus *Styloplanocera* Bock, 1913 Fig. 87

GNESIOCEROTIDAE Elongate forms with tentacles. Margins of body bearing row of contractile tentacle-like tufts. Two elongate clusters of cerebral eyes; tentacular eyes in and at base of each tentacle. Genital pores separated. Pharynx in anterior half of body. Seminal vesicle present. Prostatic organ lined with epithelium thrown into radial folds. Ejaculatory duct protrudes into lumen of prostatic organ. Cirrus-sac long and narrow, lined with strong spines. It opens into male antrum through a fairly-long cylindrical papilla with thick cuticular covering. Prostatic organ and cirrus-sac enclosed in muscular envelope. Well-developed vagina bulbosa. Lang's vesicle large, U-shaped. Uterine canals anteriorly separated.

TYPE-SPECIES *S. [Stylochus] fasciata* (Schmarda, 1859). Type by original designation.

TYPE-LOCALITY Jamaica, West Indies, Caribbean.

Genus *Spinicirrus* Hyman, 1953 Fig. 88

GNESIOCEROTIDAE Elongate forms without tentacles. Eyes in two bands lateral to cerebral organ. Pharynx in mid-third of body. Genital pores approximate. No spermiducal bulbs or seminal vesicle have been observed. Cirrus-sac anteriorly bearing two sets of large spines, posteriorly lined with small spines. Two pyriform glandular organs lie in thick muscular sheath around cirrus-sac, into the lumen of which one, lined with a basophilic epithelium, opens anteriorly, and the other, lined with an eosinophilic epithelium, opens posteriorly. The latter organ is larger than the former and is regarded as 'a prostatic vesicle of the free type'. No bursa copulatrix or Lang's vesicle. Uterine canals anteriorly confluent.

TYPE-SPECIES *S. inequalis* Hyman, 1953. Type by original designation.

TYPE-LOCALITY Off Pardita I., near La Paz Bay, Gulf of California, Mexico.

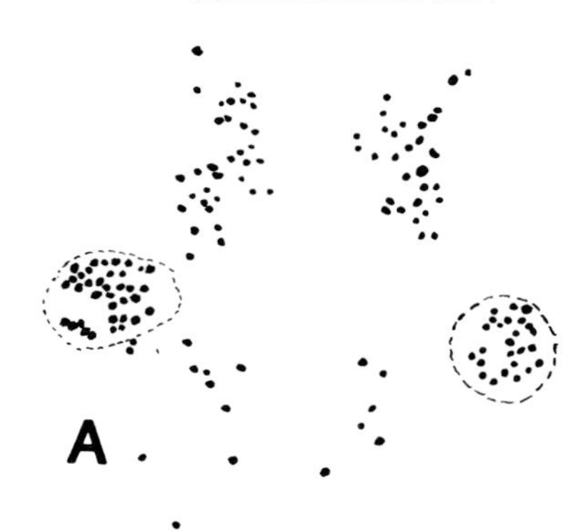

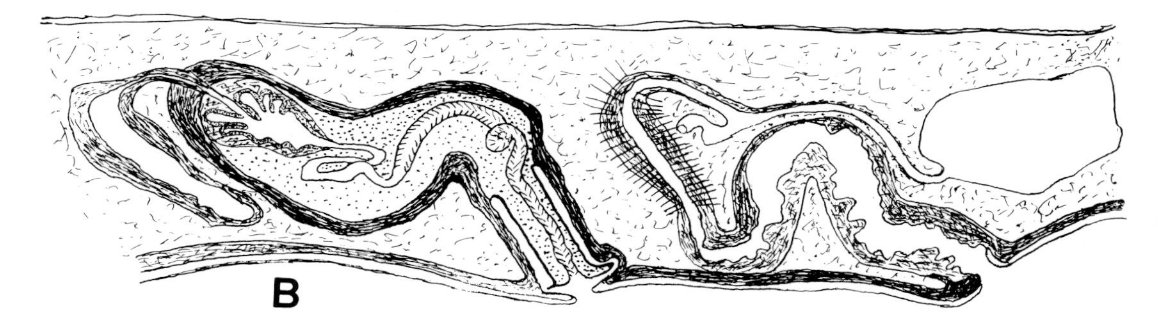

Fig. 87 *Styloplanocera fasciata.* **A**, arrangement of eyes; **B**, sagittal section of copulatory organs (after Bock).

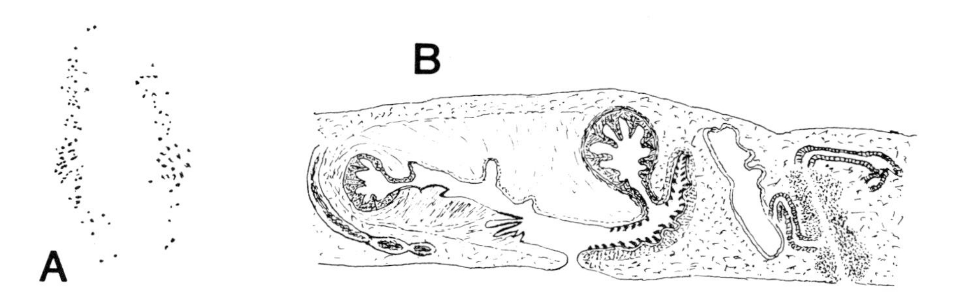

Fig. 88 *Spinicirrus inequalis.* **A**, arrangement of eyes; **B**, sagittal section of copulatory organs (after Hyman).

NOTE The only specimen of *S. inequalis* studied by Hyman was damaged, and it was not possible to give a complete account of the copulatory complexes. Miss Hyman, however, assumes that the 'prostatic vesicle' is not interpolated, and this has been accepted by Marcus & Marcus (1966), who placed *Spinicirrus* in the subfamily Planocerinae of the family Planoceridae (*sensu lato*). Nevertheless, it is here considered that because our knowledge of the copulatory organs of *Spinicirrus* is incomplete, the genus should be provisionally assigned to the family Gnesiocerotidae. It also bears a close resemblance to members of this family in some of its superficial features.

Family **PLANOCERIDAE** Lang, 1884

PLANOCEROIDEA Elongate to somewhat discoid forms of rather delicate consistency, often with slender nuchal tentacles. Eyes in paired tentacular and cerebral clusters, or in two elongate clusters. Pharynx in middle third of body. Genital pores separated at some distance from hinder margin of body. Male copulatory complex closely posterior to pharynx and may be enclosed in thick muscular sheath. It consists of an eversible cirrus-sac lined with spines, an independent prostatic organ and a seminal vesicle or a pair of spermiducal bulbs. Vagina variably developed, sometimes with muscular bursa, either as a vaginal bursa or a bursa copulatrix. Lang's vesicle variably developed.

NOTE The present writer (1982*b*) has pointed out the necessity of dividing Planoceridae (*sensu lato*) into two distinct families, the Planoceridae (*sensu stricto*), in which the prostatic organ is independent ('free'), and the Gnesiocerotidae, in which the prostatic organ is interpolated. The position of this organ in relation to the ejaculatory duct is of family importance among polyclads, differentiating, for instance, the Stylochidae from the Cryptocelididae and the Leptoplanidae from the Callioplanidae.

KEY TO PLANOCERID GENERA

1.	With nuchal tentacles	2
	Without nuchal tentacles	***Disparoplana***
2.	With seminal vesicle; no spermiducal bulbs	3
	With spermiducal bulbs; no seminal vesicle	4
3.	Eyes at bases of tentacles only	***Planocera***
	Eyes within and at bases of tentacles	***Planocerodes***
4.	Male antrum with two glandular pockets	***Paraplanocera***
	Male antrum without glandular pockets	***Aquaplana***

Genus ***Planocera*** Blainville, 1828, *sensu* Lang, 1884 — Fig. 89

PLANOCERIDAE Large, discoid or broadly-oval forms with slender nuchal tentacles well separated from anterior margin of body. Eyes in ring at base of each tentacle; cerebral eyes between tentacles. Main intestine with five to eight pairs of non-anastomosing lateral branches. Seminal vesicle highly muscular. Prostatic organ bulbous, muscular, with strongly-folded epithelium. Cirrus-sac lined with small spines, sometimes with one or more large spines or hooks at the opening of the sac into the male antrum. Ejaculatory duct and proximal region of cirrus-sac surrounded by an intermuscular space filled with loose connective tissue. Thick muscular sheath encloses prostatic organ and cirrus-sac. Vagina bulbosa present. Lang's vesicle variable. Uterine canals not anteriorly confluent.

TYPE-SPECIES *P. [Planaria] pellucida* (Mertens, 1832) (? = *Planocera gaimardi* Blainville, 1828.) Type by subsequent designation.

TYPE-LOCALITY Atlantic Ocean (7°48'N, 23°–56°W (*pellucida*)); not known (*gaimardi*).

NOTE Since the type-species of *Planocera*–*P. gaimardi* Blainville–is unrecognizable, it follows that Blainville's genus is unidentifiable. Lang (1884) defined the genus on the basis of *Planocera pellucida* and certain Mediterranean species, and since that time Lang's concept of the genus has been accepted. For a considerable number of years it has been generally thought that *P. gaimardi* and *Planaria pellucida* Mertens might be synonymous, and the present writer (1950*b*) suggested that if this synonymy were accepted the type-species of *Planocera* would then be a recognizable form, but it is now thought that on the grounds of usage *Planocera pellucida* (Mertens) is the more acceptable name for the type-species.

Genus ***Paraplanocera*** Laidlaw, 1903 — Fig. 90

PLANOCERIDAE Large, broadly-oval forms with slender nuchal tentacles at some distance from anterior margin of body. Eyes in paired tentacular and cerebral groups. Pharynx with four to seven pairs of deep lateral folds. Two spermiducal bulbs; no seminal vesicle. Prostatic organ bulbous, with epithelium thrown into deep folds. Cirrus-sac lined with strong cuticular spines. Short ejaculatory duct and much of cirrus-sac surrounded by an intermuscular space filled with a coarse connective tissue. Thick

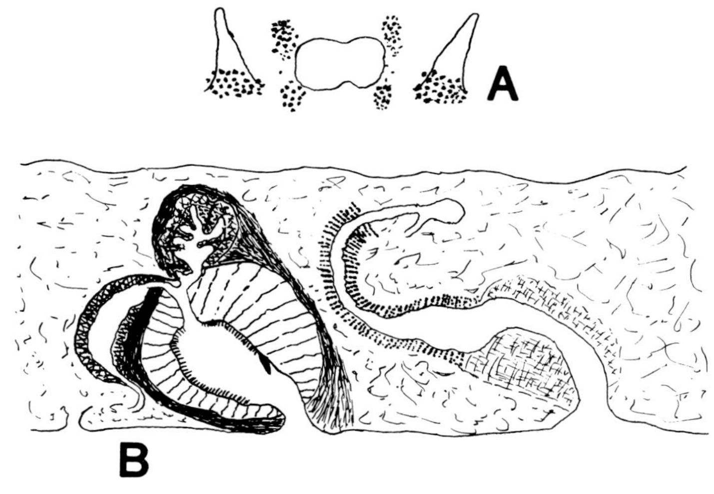

Fig. 89 *Planocera pellucida.* **A**, arrangement of eyes; **B**, sagittal section of copulatory organs (semi-diagrammatic).

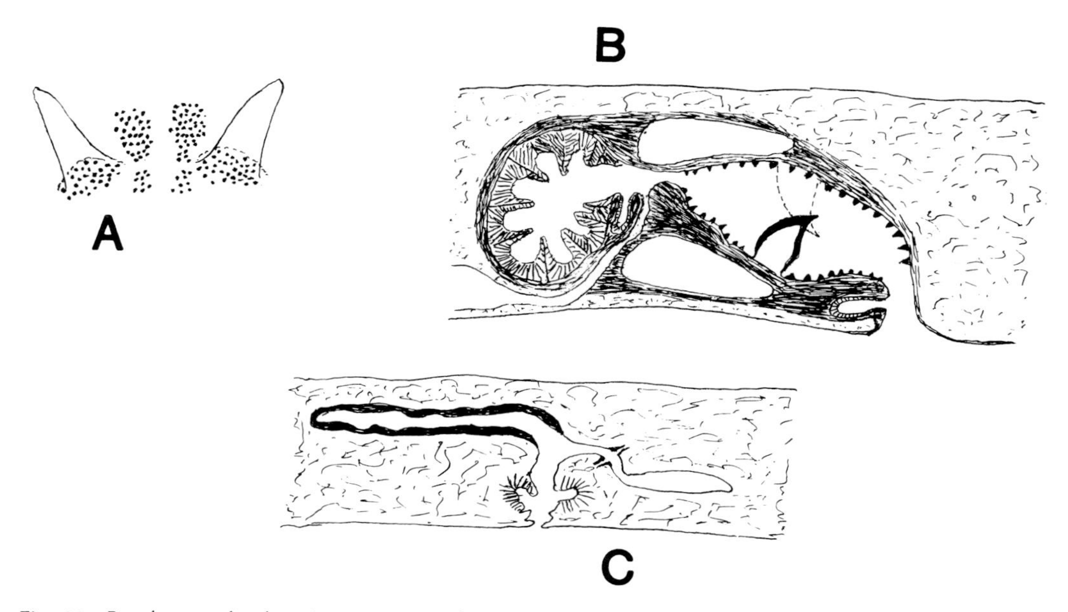

Fig. 90 *Paraplanocera oligoglena.* **A**, arrangement of eyes; **B**, sagittal section of ♂ copulatory organs (after Prudhoe); **C**, sagittal section of ♀ copulatory organs (after Laidlaw).

muscular sheath encloses prostatic organ and cirrus-sac. A pair of pockets lined with glandular epithelium open into male antrum. Large anteriorly-directed bursa copulatrix, innermost wall of which unites with thin-walled chamber, opens into vagina externa. Lang's vesicle large. Uterine canals not anteriorly confluent.

TYPE-SPECIES *P. [Stylochus] oligoglena* (Schmarda, 1859) (syn. *P. langii* Laidlaw, 1903). Type by subsequent designation.

TYPE-LOCALITY Sri Lanka, Indian Ocean.

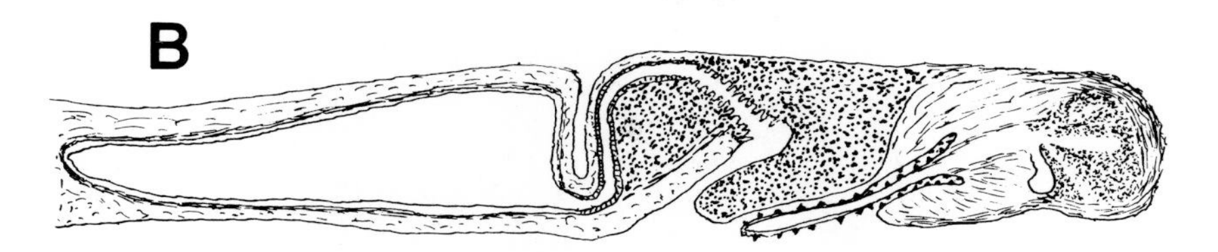

Fig. 91 *Aquaplana oceanica.* **A,** arrangement of eyes; **B,** sagittal section of copulatory organs (after Hyman).

Genus *Aquaplana* Hyman, 1953 — Fig. 91

PLANOCERIDAE Oval or somewhat discoid forms of delicate consistency, with paired nuchal tentacles, each with eyes at base. Cerebral eyes in two clusters more or less between and anterior to tentacles. Paired spermiducal bulbs; no seminal vesicle. Prostatic organ lined with glandular epithelium thrown into longitudinal folds. Cirrus-sac sparsely lined with spines, and partially everted as a spinous papilla. Prostatic organ and cirrus-sac enclosed in muscular sheath. Vagina narrow and muscular and may be provided with large antero-lateral bursa copulatrix. Lang's vesicle spacious. Uterine canals not anteriorly confluent.

TYPE-SPECIES *A. oceanica* Hyman, 1953. Type by original designation.

TYPE-LOCALITY On sandy bottom in 18–36 metres at Tagus Cove, Albermarle I., Galapagos, Pacific Ocean.

Genus *Planocerodes* Palombi, 1936 — Fig. 92

PLANOCERIDAE Large, oval forms with nuchal tentacles. Cerebral eyes in two irregular groups; tentacular eyes within and at base of each tentacle. Muscular seminal vesicle small. Prostatic organ similar to that of *Planocera.* Ductus communis prolonged into large conical papilla covered with series of long spines and lying in tubular cirrus-sac lined with spines. Vagina bulbosa unarmed. Lang's vesicle rudimentary.

TYPE-SPECIES *P. ceratommata* Palombi, 1936. Type by monotypy.

TYPE-LOCALITY Still Bay, South Africa.

NOTE Palombi says that his genus differs from *Planocera* in the occurrence of eyes within the tentacles and in the existence of a spinous penis-papilla. There must be some doubt as to the validity of *Planocerodes* when compared with *Planocera gilchristi* Jacubowa, 1908, also known from the Cape Town region of South Africa. In one of Jacubowa's original figures, the cirrus-sac is partially everted to form a papilla covered with spines, as in *Planocerodes.* The presence of eyes within the tentacles alone is not a generic feature, as shown in *Paraplanocera.*

Genus *Disparoplana* Laidlaw, 1903 — Fig. 93

PLANOCERIDAE Body elongate, without tentacles. Eyes in two elongate clusters lateral to cerebral organ. Seminal vesicle large. Small prostatic organ with folded epithelial lining. Cirrus-sac relatively small and narrow, lined with spines. Muscular sheath enclosing prostatic organ and cirrus-sac. Female copulatory complex simple, with rudimentary Lang's vesicle.

TYPE-SPECIES *D. dubia* Laidlaw, 1903. Type by original designation.

TYPE-LOCALITY East Africa.

Fig. 92 *Planocerodes ceratommata.* Arrangement of eyes (after Palombi).

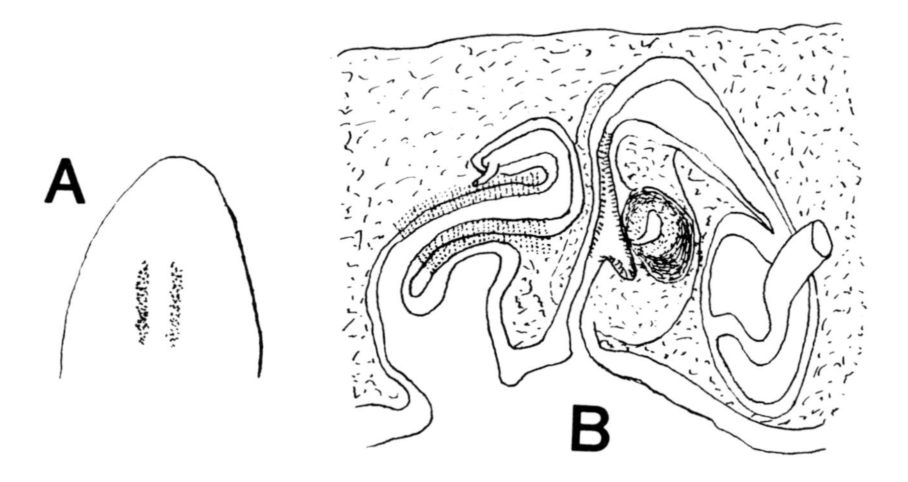

Fig. 93 *Disparoplana dubia.* **A**, anterior region of body (modified, after Laidlaw); **B**, sagittal section of copulatory organs (after Steinböck).

Family **CALLIOPLANIDAE** Hyman, 1953 (Includes DIPLOSOLENIIDAE Bock, 1913)

PLANOCEROIDEA Rather fleshy, discoid or oval form. Nuchal tentacles, when present, of variable development, each with eyes at its base. Two cerebral eye-clusters between tentacles. Pharynx mainly in middle third of body. One or two genital pores well separated from posterior margin of body; male aperture posterior to male copulatory complex. Seminal vesicle usually present, but may be replaced by two spermiducal bulbs. Prostatic organ vesicular, independent, lying dorsally to ejaculatory duct or seminal vesicle. Vagina narrow, thin walled, with or without Lang's vesicle.

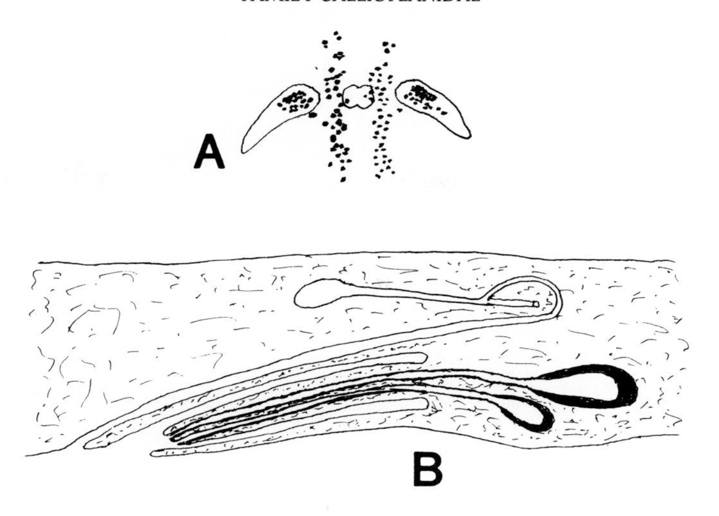

Fig. 94 *Callioplana marginata.* **A,** cerebral and tentacular eye-clusters; **B,** sagittal section of copulatory organs (after Yeri & Kaburaki).

KEY TO CALLIOPLANID GENERA

1.	Two elongate spermiducal bulbs; no seminal vesicle	***Discosolenia***
	Seminal vesicle; no spermiducal bulbs	2
2.	No tentacles; vagina short	. ***Asolenia***
	Tentacles distinct; vagina long	3
3.	Lang's vesicle crescentic	***Callioplana***
	Lang's vesicle bulbous or elongate	4
4.	Genital pores in tandem; uterine canals anteriorly confluent	***Pseudostylochus***
	Female pore to right of median line; uterine canals not anteriorly confluent .	***Monosolenia***

Genus ***Callioplana*** Stimpson, 1857, *sensu* Yeri & Kaburaki, 1918 Fig. 94

SYNONYMY *Diplosolenia* Haswell, 1907

CALLIOPLANIDAE Oval forms with prominent tentacles containing eyes. Cerebral eyes in two groups between tentacles. Intestinal branches occasionally anastomosing. Genital pores adjacent to one another. Seminal vesicle small, elongate and thin walled. Prostatic organ small, elongate, dorsal to seminal vesicle. Ejaculatory duct and prostatic duct pass through elongate, unarmed, penis-papilla and open independently at tip of papilla into male antrum. No penis-pocket. Seminal vesicle and prostatic organ often enclosed in thick muscular sheath. Vagina long, dorsal to male copulatory complex. Lang's vesicle crescentic, with anteriorly-directed limbs.

TYPE-SPECIES *C. marginata* Stimpson, 1857. Type by monotypy.

TYPE-LOCALITY 'Ousima', Japan.

Genus ***Discosolenia*** Freeman, 1933 Fig. 95

CALLIOPLANIDAE Large, rounded-oval forms with well-developed nuchal tentacles. Pharynx with 6 pairs of deep lateral folds; intestinal branches not anastomosing. Tentacular eyes in dense ring at base of each tentacle; cerebral eyes in four clusters, two on antero-lateral borders of cerebral organ, and two less dense clusters on postero-lateral borders. Genital pores closely separated. Pair of elongate spermiducal bulbs lateral to uterine canals. No seminal vesicle. Small, thick-walled prostatic organ lined

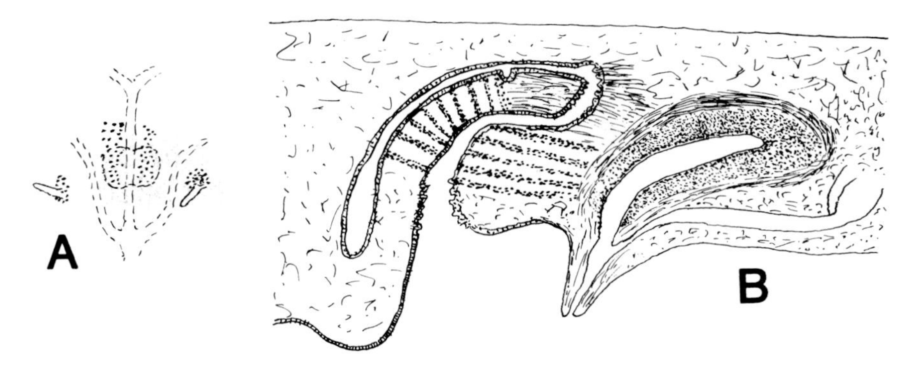

Fig. 95 *Discosolenia burchami.* **A,** cerebral and tentacular eye-clusters (after Freeman); **B,** sagittal section of copulatory organs (after Hyman).

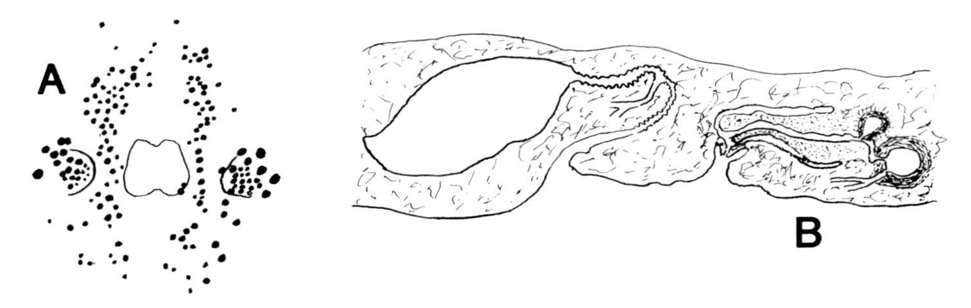

Fig. 96 *Pseudostylochus takeshitai.* **A,** cerebral and tentacular eye-clusters; **B,** sagittal section of copulatory organs (after Kato).

with tall smooth epithelium. Penis-papilla short, conical, lying in shallow male antrum. Vagina simple. Lang's vesicle large, oval or spherical. Uterine canals not anteriorly confluent.

TYPE-SPECIES *D. [Planocera] burchami* (Heath & McGregor, 1913) nov. comb. (syn. *D. washingtonensis* Freeman, 1933). Type by monotypy.

TYPE-LOCALITY At a depth of about 18 metres, Monterey Bay, California, U.S.A.

NOTE Hyman (1953*a*) regards *Discosolenia* as a synonym of *Pseudostylochus,* but the former possesses no seminal vesicle, but a pair of spermiducal bulbs, features not found in the latter.

Genus ***Pseudostylochus*** Yeri & Kaburaki, 1918 Fig. 96

CALLIOPLANIDAE Oval or elongate forms with distinct nuchal tentacles. Eyes at base of each tentacle; cerebral eyes in two elongate clusters between tentacles. Genital pores usually separated. Seminal vesicle well developed, elongate or spherical. Prostatic organ muscular, pyriform or elongate, provided with tall epithelial lining thrown into longitudinal or radial folds. Penis-papilla highly muscular, sometimes entering male antrum through a penis-sheath. Body wall around female aperture may be corrugated, suggesting an adhesive structure. Vagina of moderate length, forming an anteriorly-directed loop. Lang's vesicle bulbous or elongate. Uterine canals may be anteriorly confluent.

TYPE-SPECIES *P. takeshitai* Yeri & Kaburaki, 1918. Type by subsequent designation.

TYPE-LOCALITY Low-water mark, Matsuwa, near Misaki, Japan.

NOTE Kato (1939*b*) divides the species of *Pseudostylochus* into two groups on whether or not they have a corrugated adhesive pad around the female pore.

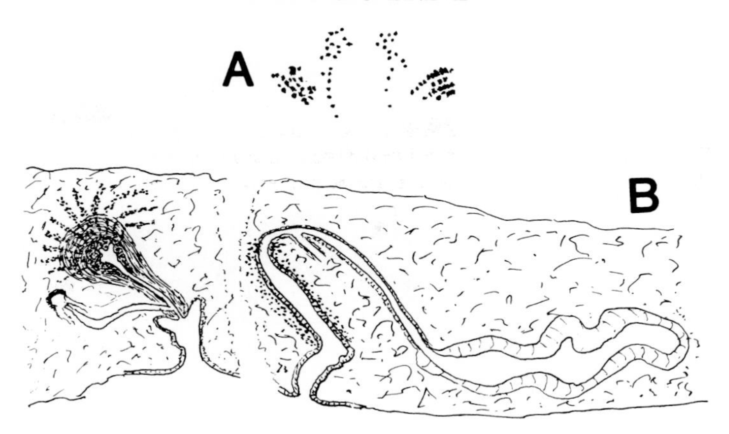

Fig. 97 *Monosolenia asymmetrica.* **A,** cerebral and tentacular eye-clusters; **B,** sagittal sections of copulatory organs (after Hyman).

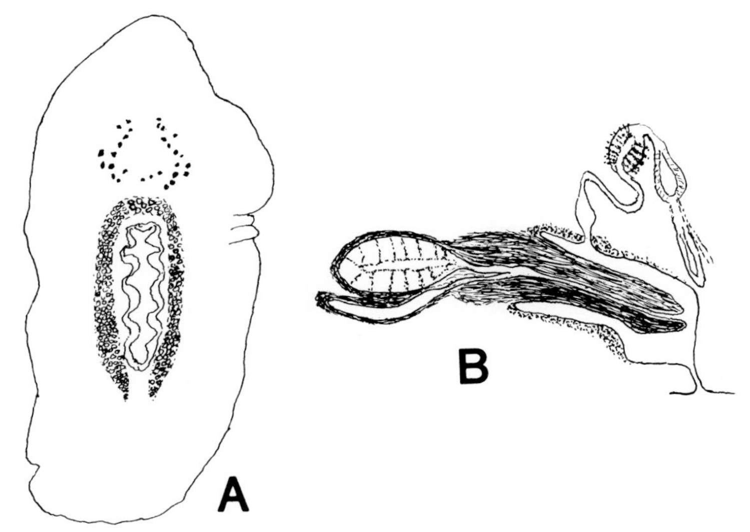

Fig. 98 *Asolenia deilogyna.* **A,** complete worm; **B,** sagittal section of copulatory organs (after Hyman).

Genus ***Monosolenia*** Hyman, 1953 Fig. 97

CALLIOPLANIDAE Oval or somewhat oblong forms without tentacles. Eyes in paired tentacular and cerebral clusters. Genital pores separated; female pore lying to one side of median line. Seminal vesicle small. Prostatic organ pyriform, with thick muscular wall, lined with tall glandular epithelium and invested with extracapsular gland cells. Conical penis-papilla short. Vagina narrow and simple, with elongate Lang's vesicle. Uterine canals not anteriorly confluent.

TYPE-SPECIES *M. asymmetrica* Hyman, 1953. Type by original designation.

TYPE-LOCALITY Under rock at Cape San Lucas, Lower California, Mexico.

Genus ***Asolenia*** Hyman, 1959 Fig. 98

CALLIOPLANIDAE Small elongate-oval forms without tentacles. Cerebral eyes in two rows; tentacular eyes few, in two clusters lateral to posterior cerebral eyes. Single gonopore. Seminal vesicle elongate

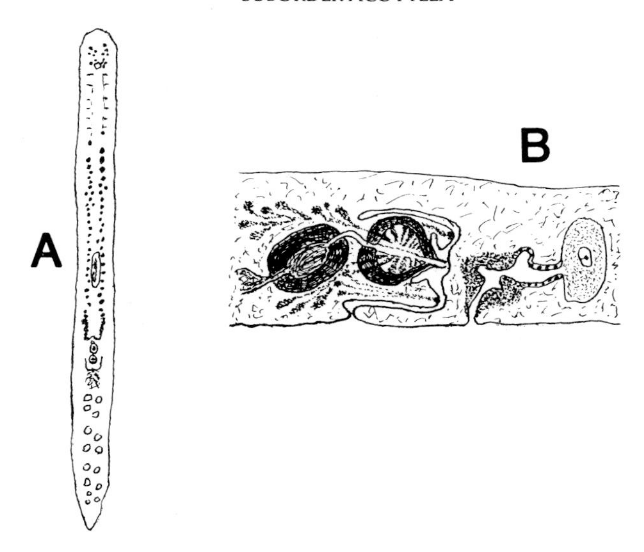

Fig. 99 *Theama evelinae.* **A**, entire worm; **B**, sagittal section of copulatory organs (after Marcus).

and thin walled. Prostatic organ dorsal to seminal vesicle, well developed and lined with tall epithelium thrown into radial folds. Penis-papilla elongate and muscular. Opening of female complex lies in roof of male antrum. Female antrum with sphincter. Vagina narrow; 'shell'-chamber dilated. No Lang's vesicle. Uterine canals anteriorly confluent.

TYPE-SPECIES *A. deilogyna* Hyman, 1959. Type by original designation.

TYPE-LOCALITY Among algae, Iwayana Bay, Auluptagel I., Palau Isles, Pacific Ocean.

Family **THEAMATIDAE** Marcus, 1949

PLANOCEROIDEA Elongate forms, without tentacles. Eyes small, few, disposed in paired tentacular and cerebral clusters anteriorly to cerebral organ. Short pharynx in middle region of body. Genital pores separated, midway between pharynx and posterior margin of body. Testes in two, lateral, longitudinal rows anterior to male copulatory organs, whereas fewer ovaries similarly disposed posteriorly to female complex. Seminal vesicle strongly muscular. Ejaculatory duct leads into interpolated prostatic organ with tall glandular epithelium thrown into radial folds. Penis-papilla, sometimes with stylet, in penis-pocket; special gland-cells in wall of penis-pocket which opens into male antrum through penis-sheath. Male antrum large. Vagina short, without Lang's vesicle, but its proximal or inner region forms a uterine sac.

Genus ***Theama*** Marcus, 1949 Fig. 99

THEAMATIDAE With characters of family.

TYPE-SPECIES *T. evelinae* Marcus, 1949. Type by original designation.

TYPE-LOCALITY In coarse sand, Ilha de São Sebastião, near Santos, Brazil.

Family **ENANTIIDAE** Graff, 1890, emended Gamble, 1896

PLANOCEROIDEA Small, delicate forms without tentacles. Row of well-developed cuticular spines on dorsal surface, near lateral and posterior margins. Eyes in two small tentacular groups and two elongate cerebral groups. Mouth in anterior half of body, immediately posterior to cerebral organ, and

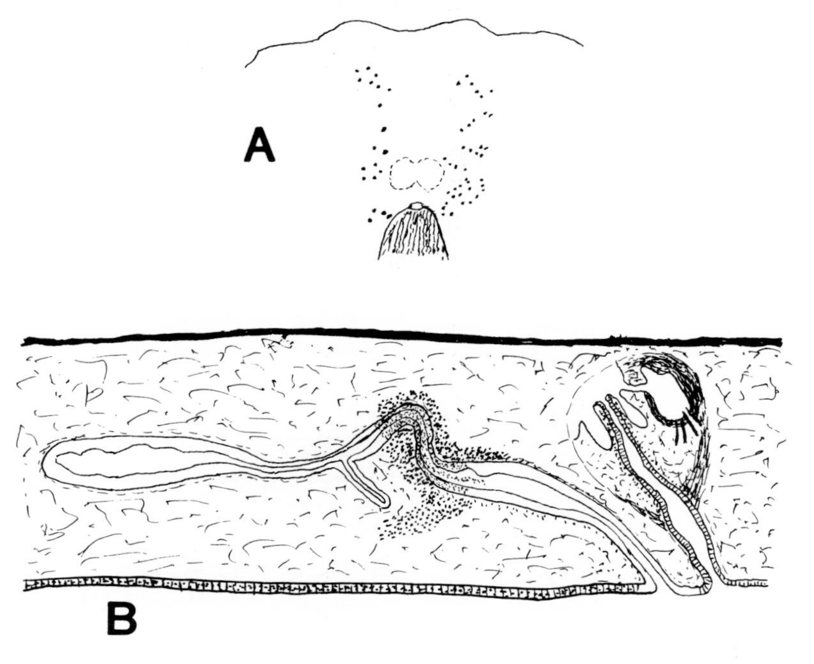

Fig. 100 *Enantia.* **A**, arrangement of eyes in *E. spinifera* (after Graff); **B**, sagittal section of copulatory organs of *E. pellucida* (after Hyman).

opens into anterior region of pharyngeal chamber containing bulbous tubular pharynx; intestinal trunk without anterior branch; its branches ramify and anastomose. Genital pores separated. Male copulatory complex small, near posterior end of pharynx, dorsally or dorso-posteriorly directed from male pore, and enclosed in muscular envelope. Prostatic organ absent. Short ejaculatory duct unites muscular seminal vesicle with spacious cavity which narrows and opens into male antrum through small, elongate, unarmed penis-papilla. Vagina relatively long, directed posteriorly from female pore. Lang's vesicle very elongate. One or two pairs of short uterine canals or vesicles extend laterally from vagina.

Genus ***Enantia*** Graff, 1890 Fig. 100

ENANTIIDAE With characters of family.

TYPE-SPECIES *E. spinifera* Graff, 1890. Type by original designation.

TYPE-LOCALITY Under a stone in about 3 metres, Trieste, Mediterranean.

Family **APIDIOPLANIDAE** Bock, 1926

PLANOCEROIDEA Elongate-oval forms with poorly-developed nuchal tentacles. Eyes not numerous, in small paired tentacular and cerebral clusters. Without marginal eyes. Mouth immediately posterior to cerebral organ, at anterior end of pharyngeal chamber enclosing tubular pharynx; intestinal branches not anastomosing. Genital pores separated, near hind end of body. Male copulatory complex anterior to male pore. Seminal vesicle well developed. Large elongate prostatic organ interpolated, with smooth epithelial lining. Penis-papilla large and covered with a hard cuticular material forming a 'cuff' or 'ruffle' at its apex. Several pyriform muscular organs irregularly distributed throughout body and opening independently on ventral surface. Female copulatory complex directed anteriorly from female pore. Vagina with bursa copulatrix. Lang's vesicle not developed. Uterine canals extend anteriorly to pharyngeal region.

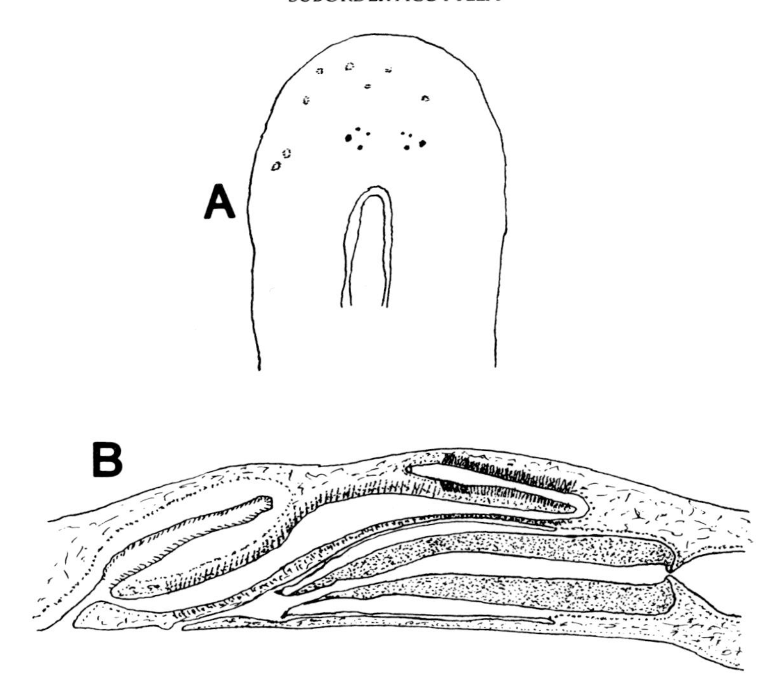

Fig. 101 *Apidioplana mira.* **A,** anterior region of body; **B,** sagittal section of copulatory organs (after Bock).

Genus ***Apidioplana*** Bock, 1926 Fig. 101

APIDIOPLANIDAE With characters of family

TYPE-SPECIES *A. mira* Bock, 1926. Type by original designation.

TYPE-LOCALITY *Melitodes* colonies in surf zone, Mbau, Fiji.

Family **STYLOCHOCESTIDAE** Bock, 1913

PLANOCEROIDEA Elongate forms without tentacles. Eyes in two elongate clusters alongside cerebral organ. Pharynx in middle third of body; intestinal branches not anastomosing. Genital pores separated, at some distance anteriorly to hind margin of body. Male copulatory complex anterior to male pore. Elongate seminal vesicle dorsal to large, independent, prostatic organ lined with smooth epithelium. Ejaculatory duct and prostatic duct open together at the apex of small penis-papilla bearing a short stylet. Vagina simple, forming an anteriorly-directed loop. Lang's vesicle not developed.

Genus ***Stylochocestus*** Laidlaw, 1904 Fig. 102

STYLOCHOCESTIDAE With characters of family.

TYPE-SPECIES *S. gracilis* Laidlaw, 1904. Type by original designation.

TYPE-LOCALITY Sri Lanka, Indian Ocean.

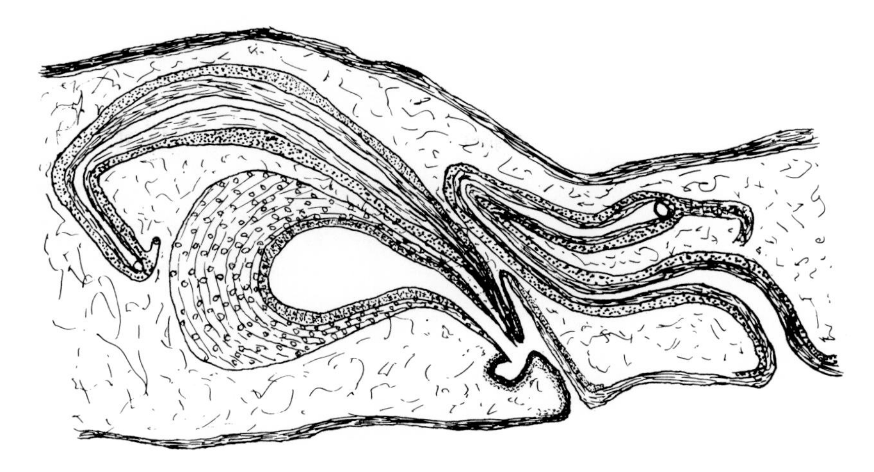

Fig. 102 *Stylochocestus gracilis.* Sagittal section of copulatory organs (after Meixner).

Superfamily **CESTOPLANOIDEA** Poche, 1926, emend. Prudhoe, 1983
(Includes, Emprosthommata Bock, 1913; Cestoplanides Poche, 1926, Emprosthommatidea Marcus & Marcus, 1966).

ACOTYLEA With eyes widely distributed over cephalic region, except marginal zone.

KEY TO CESTOPLANOID FAMILIES

1.	Pharynx in posterior half of body	CESTOPLANIDAE
	Pharynx in anterior half of body.	2
2.	Prostatic organ interpolated	EMPROSTHOPHARYNGIDAE
	No prostatic organ	DIPLOPHARYNGEATIDAE

Family **CESTOPLANIDAE** Lang, 1884

CESTOPLANOIDEA Elongate or ribbon-like forms in which a ventral adhesive area or pad sometimes occurs at posterior end of body. Tentacles absent. Eyes numerous, situated anteriorly and distributed somewhat fanwise from well posterior to cerebral organ to submarginal zones of cephalic region, leaving a clear narrow marginal field. Mouth and short pharynx posterior. Intestinal trunk long, with many pairs of non-anastomosing branches. Genital pores separated. Male copulatory complex more or less dorsal or posterior to its aperture. Seminal vesicle well developed. Prostatic organ pyriform, interpolated, and lined with smooth epithelium. Penis-papilla variably developed, may be slightly cuticularized and lie in a penis-pocket. Duplicate male complex may occur. Vagina short, forms dorso-posterior loop; 'shell'-chamber often spacious, but dorso-ventrally compressed. Lang's vesicle may occur. Uterine canals extend into anterior third of body and are not confluent. Testes and ovaries in longitudinal rows on either side of intestinal trunk.

Genus ***Cestoplana*** Lang, 1884 Fig. 103

SYNONYMY *Orthostomum* Grube, 1840; *Tricelis* Quatrefages, 1845.

CESTOPLANIDAE With characters of family.

TYPE-SPECIES *C. [Orthostomum] rubrocincta* (Grube, 1840). Type by page-precedence.

TYPE-LOCALITY Mediterranean.

NOTE In *C. raffaeli* Ranzi the pharynx is situated in the anterior quarter of the body length, and in *C. polypora* Meyer there is a row of five to thirty female pores.

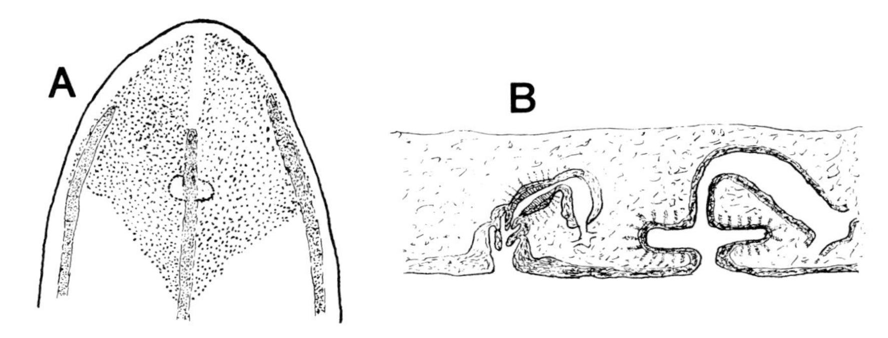

Fig. 103 *Cestoplana rubrocincta*. **A**, anterior region of body; **B**, sagittal section of copulatory organs (after Lang).

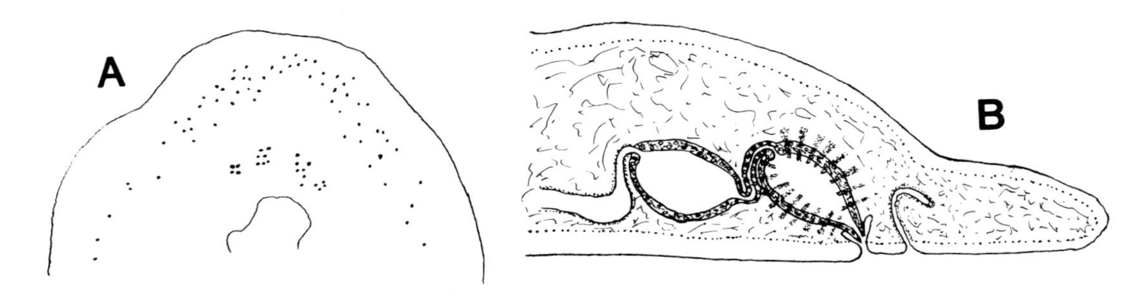

Fig. 104 *Emprosthopharynx opisthoporus*. **A**, anterior region of body; **B**, sagittal section of copulatory organs (after Bock).

Family **EMPROSTHOPHARYNGIDAE** Bock, 1913

CESTOPLANOIDEA Elongate-oval forms, rather fleshy and without tentacles. Submarginal eyes arranged anteriorly in an irregular band of varying width and extent; small frontal eyes may be scattered between submarginal eyes and cerebral eye-clusters. Pharynx closely posterior to cerebral organ. Intestinal trunk moniliform, reaching copulatory organs; intestinal branches not anastomosing. Genital pores separated, near posterior margin of body. Male copulatory complex anterior to its genital pore. Vasa deferentia unite to form long muscular canal opening into well-developed seminal vesicle. Pyriform prostatic organ interpolated, with smooth epithelial lining; its distal end forms base of blunt penis-papilla, which may be armed with a tiny stylet. Antrum masculinum shallow. Vagina short, forming anteriorly-directed loop. No Lang's vesicle. Uterine canals not anteriorly confluent.

Genus ***Emprosthopharynx*** Bock, 1913

Fig. 104

EMPROSTHOPHARYNGIDAE

With characters of family.

TYPE-SPECIES *E. opisthoporus* Bock, 1913. Type by original designation.

TYPE-LOCALITY Galapagos Is, Pacific Ocean.

NOTE Species of this genus tend to associate with hermit crabs.

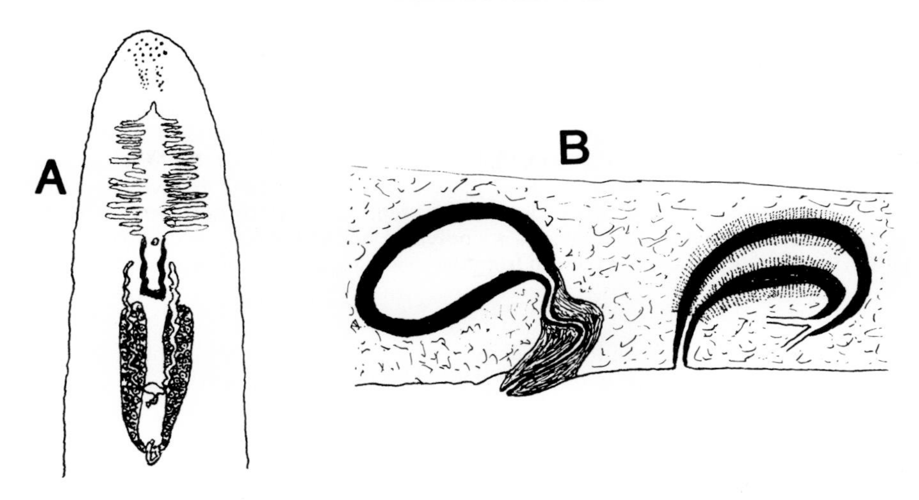

Fig. 105 *Diplopharyngeata filiformis.* **A**, anterior region of body; **B**, sagittal section of copulatory organs (after Plehn).

Family **DIPLOPHARYNGEATIDAE** Plehn, 1896

CESTOPLANOIDEA Very elongate forms without tentacles. Eyes in two elongate cerebral clusters, from which frontal eyes spread fanwise anteriorly, but leaving marginal zone of body free from eyes. Pharyngeal apparatus immediately posterior to cerebral organ; pharynx in two sections, anterior large and ruffled, posterior small and narrow; anterior section twice as long as posterior and with many pairs of deep narrow lateral pockets. Mouth at junction of two sections. Intestinal trunk extends from cerebral region to posterior end of body, its lateral branches not ramifying or anastomosing. Genital pores separated, at about first third of body. Testes and ovaries intermingle in dorsal parenchyma. Male copulatory complex anterior to its genital pore. Vasa deferentia lined with glandular epithelium. Seminal vesicle well developed. No prostatic organ. Penis-papilla muscular, without stylet. Vagina short and wide, directed posteriorly from female pore; without Lang's vesicle. Uterine canals short, extending anteriorly to pharynx, not anteriorly confluent.

Genus ***Diplopharyngeata*** Plehn, 1896 — Fig. 105

DIPLOPHARYNGEATIDAE With characters of family.

TYPE-SPECIES *D. filiformis* Plehn, 1896. Type by original designation.

TYPE-LOCALITY On a floating tree trunk off north coast of Sumatra.

NOTE The condition of the pharynx in this genus is most unusual and its diagnostic value needs confirmation. The present writer has met with leptoplanids in which one half of the pharynx is constricted longitudinally and the other half expanded and ruffled deeply to give the appearance of a bipartite structure. In these latter instances, the condition appears to have occurred during fixation.

Suborder **COTYLEA** Lang, 1884

Forms with well-developed ventral sucker, or depression on ventral surface of body, acting as an adhesive organ, occurs posteriorly to female genital pore. Tentacles, when present, antero-marginal (except Stylochoididae and Opisthogeniidae, in which nuchal tentacles occur). Marginal eyes variously distributed, more frequently in distinct groups at base of tentacles. Prostatic organ generally independent. 'Shell'-chamber enlarged, but dorso-ventrally compressed. Uterine canals extend posteriorly from vagina (except Boniniidae and Pericelididae), often possessing one or more pairs of accessory vesicles. Lang's vesicle generally absent.

NOTE Palombi (1928) described a maricolous triclad turbellarian from the Suez Canal under the name of *Ditremagenia macropharynx* and represented by Ditremageniidae among triclad families. Marcus & Marcus (1966), presumably following Bresslau's (1928–33) classification of the Turbellaria, regard this family as of cotylean polyclads. Their conclusion appears to have been based on the presence of multiple ovarian follicles, no vitelline follicles and possibly two uterine vesicles. Palombi, however, makes no mention of the absence of vitelline follicles, nor whether the ovaries produce endolecithal eggs, and he describes the two vesicles as 'shell'-glands. On the other hand, the only known specimen of *D. macropharynx* has a typical triclad digestive system, so it seems necessary to study further material to decide whether or not this species is a polyclad, but on the basis of its alimentary system it is a typical triclad and is, therefore, excluded from the following cotylean classification.

KEY TO COTYLEAN FAMILIES

1.	Mouth and pharynx in anterior third of body	2
	Mouth and pharynx in middle or posterior third of body	6
2.	Pharynx tubular or campanulate	3
	Pharynx ruffled or irregularly ring-like	5
3.	With tentacles	4
	Without tentacles	PROSTHIOSTOMIDAE
4.	Tentacles marginal	EURYLEPTIDAE
	Tentacles submarginal or nuchal	STYLOCHOIDIDAE
5.	Nuchal tentacles; male copulatory organs posterior to female	OPISTHOGENIIDAE
	Marginal tentacles; male copulatory organs anterior to female	PSEUDOCEROTIDAE
6.	Without tentacles	7
	Tentacles marginal	8
7.	Pyriform organs in two lateral longitudinal rows	ANONYMIDAE
	Without such organs	CHROMOPLANIDAE
8.	Male complex with prostatoids	BONINIIDAE
	Male complex without prostatoids	9
9.	Uterine canals directed anteriorly from female pore	PERICELIDIDAE
	Uterine canals run posteriorly from female pore	DIPOSTHIDAE

Family **BONINIIDAE** Bock, 1923

COTYLEA Body narrow and elongate, with two narrow outgrowths on antero-lateral margins. Ventral adhesive organ a shallow depression at posterior end of body. Eyes in anterior region, chiefly marginal or submarginal, with remaining few eyes in cerebral area. Pharynx ruffled in middle third of body. Main intestine long; secondary branches not anastomosing. Male copulatory complex with unarmed penis-papilla and includes a single prostatoid with stylet, or several such organs opening into male antrum or on ventral surface of body; these organs appear to radiate from male antrum. Vagina with dilated, dorso-ventrally compressed, 'shell'-chamber. Lang's vesicle developed. Uterine canals open separately into vagina, form H-shaped figure and bear several rounded vesicles.

KEY TO BONINIID GENERA

1.	Single prostatoid posterior to penis-papilla	***Traunfelsia***
	With several prostatoids	2
2.	Prostatoids one above other in pairs; opening into male antrum	***Boninia***
	Prostatoids not in pairs; opening independently into depression on ventral surface	***Paraboninia***

Genus ***Boninia*** Bock, 1923

Fig. 106

BONINIIDAE Seminal vesicle and interpolated prostatic organ present. Penis-papilla well developed. Several prostatoids radiating from male antrum; arranged one above other in pairs.

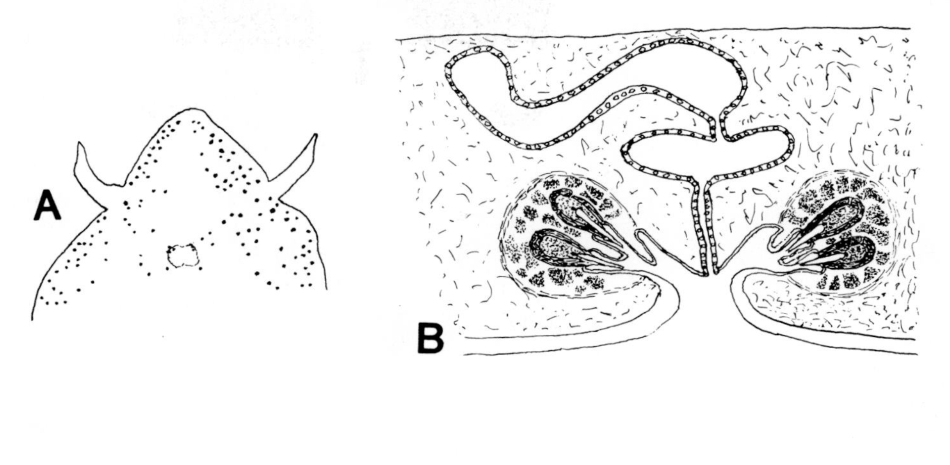

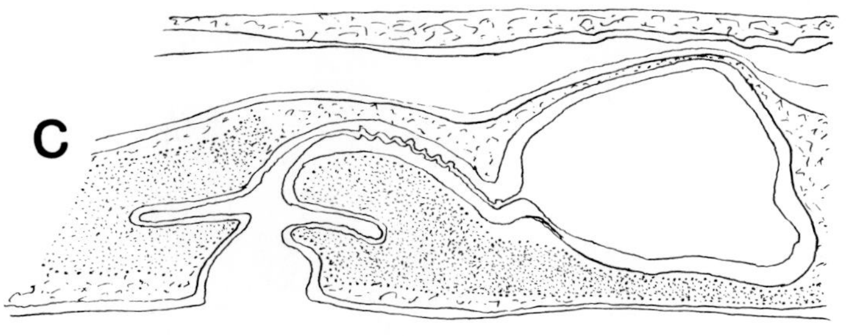

Fig. 106 *Boninia.* **A,** arrangement of eyes in *B. antillarum* (after du B.-R. Marcus & Marcus); **B,** sagittal section of ♂ copulatory organs of *B. mirabilis*; **C,** sagittal section of ♀ copulatory organs of *B. mirabilis* (after Bock).

TYPE-SPECIES *B. mirabilis* Bock, 1923. Type by orginal designation.

TYPE-LOCALITY On *Melitodes* in surf-zone, Bonin I., Pacific Ocean.

Genus ***Paraboninia*** Prudhoe, 1944 Fig. 107

BONINIIDAE Male copulatory complex with two-chambered seminal vesicle; no prostatic organ; penis-papilla small. Several prostatoids not in pairs but irregularly distributed round male complex and opening independently on ventral surface of body.

TYPE-SPECIES *P. caymanensis* Prudhoe, 1944. Type by monotypy.

TYPE-LOCALITY South Sound, Grand Cayman I., Caribbean.

Genus ***Traunfelsia*** Laidlaw, 1906 Fig. 108

BONINIIDAE Male complex without seminal vesicle or prostatic organ. Penis-papilla small. Single prostatoid, with stylet, posterior to penis-papilla and opens into depression on ventral surface of body. On each side of penis-papilla, a narrow canal opens into male antrum and terminates inwardly by branching into a number of small chambers invested with large gland cells.

TYPE-SPECIES *T. elongata* Laidlaw, 1906. Type by monotypy.

TYPE-LOCALITY Cape Verde Is, Atlantic Ocean.

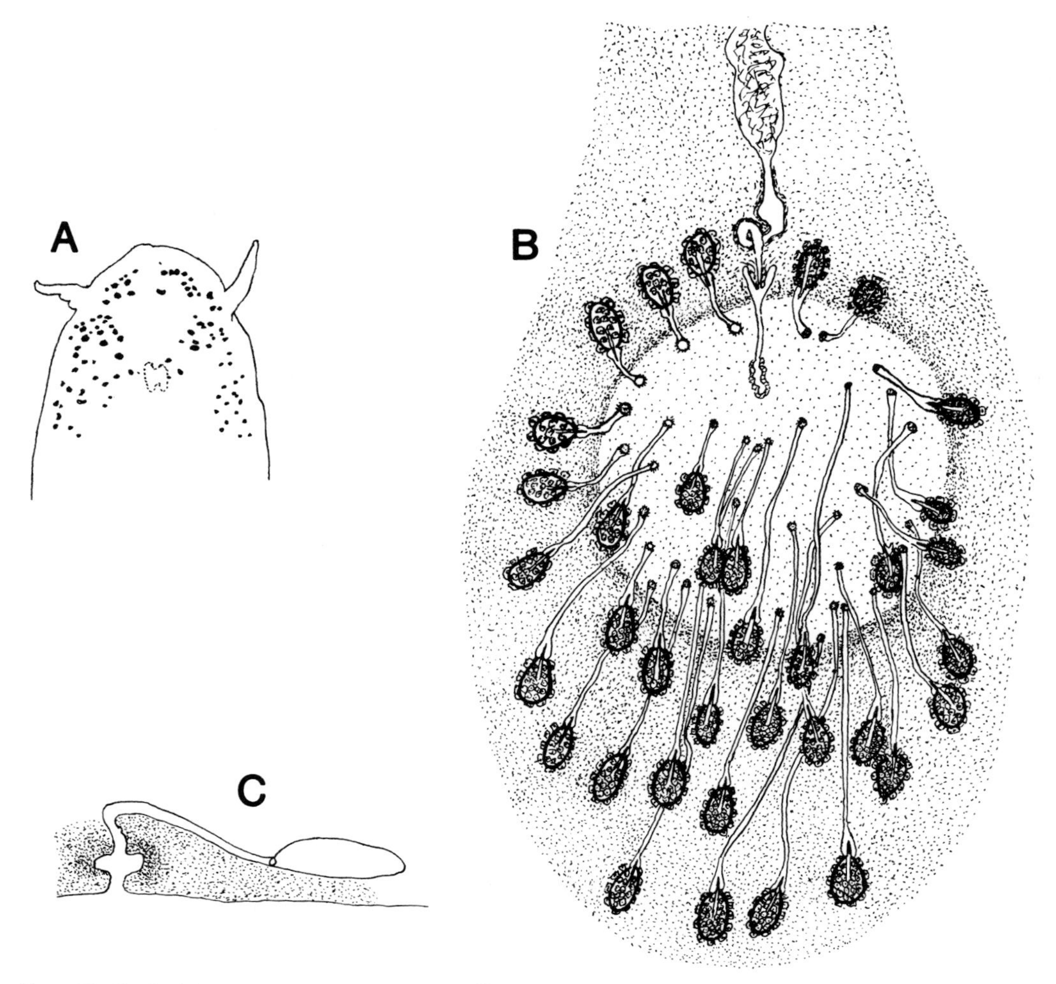

Fig. 107 *Paraboninia caymanensis.* **A,** anterior region of body; **B,** male copulatory complex (ventral view); **C,** sagittal section of female copulatory organs (after Prudhoe).

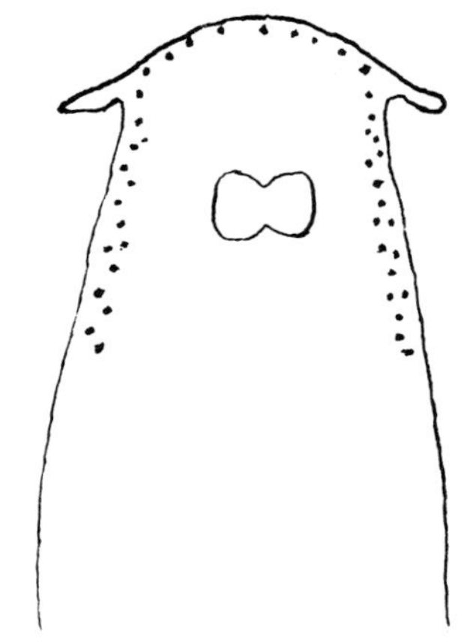

Fig. 108 *Traunfelsia elongata.* Anterior region of body (after Laidlaw).

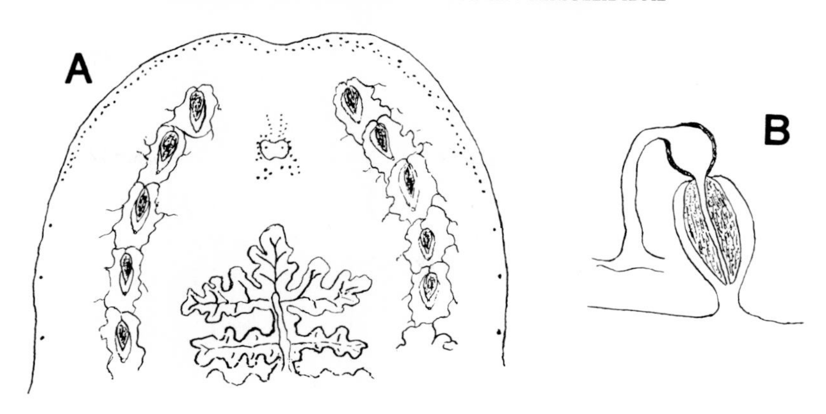

Fig. 109 *Anonymus virilis.* **A**, anterior region of body; **B**, sagittal section of ♂ copulatory complex (after Lang).

Family **ANONYMIDAE** Lang, 1884

COTYLEA Delicate, broadly-oval forms without tentacles. Eyes in two elongate clusters in cerebral region and in continuous series round margin of body. Ventral sucker posterior to centrally-situated mouth. Pharynx ruffled, with four pairs of large lateral folds in mid-third of body. Male copulatory complex includes several pyriform organs in two longitudinal rows in lateral fields of body. Each organ is supplied by a canal from the vasa deferentia, opens independently on ventral surface and consists of seminal vesicle and muscular penis-papilla without stylet. No prostatic organ. Female complex simple, between mouth and ventral sucker.

Genus ***Anonymus*** Lang, 1884 — Fig. 109

ANONYMIDAE With characters of family.

TYPE-SPECIES *A. virilis* Lang, 1884

TYPE-LOCALITY Bay of Naples, Mediterranean

Family **PERICELIDIDAE** Laidlaw, 1902, emend. Poche, 1926

COTYLEA Elongate- to broadly-oval forms, with pair of small, anterior, widely-separated marginal tentacles, each bearing eyes. Ventral sucker situated posteriorly. Marginal eyes in continuous series round body; cerebral eyes in two elongate clusters; frontal eyes present. Ruffled pharynx in middle third of body. Genital pores united or approximate, lying between pharynx and ventral sucker. Male copulatory complex anterior to male pore. Seminal vesicle distinct; no distinct prostatic organ, but proximal region of ejaculatory duct is lined with granular gland-cells when male phase of worm is fully active. Penis-papilla small, without stylet or penis-pocket. Vagina simple, 'shell'-chamber dilated and dorso-ventrally compressed. Uterine canals extend anteriorly from proximal end of vagina and possess several uterine vesicles; occasionally uteri form H-shaped configuration.

Genus ***Pericelis*** Laidlaw, 1902 — Fig. 110

SYNONYMY *Marcusia* Hyman, 1953.

PERICELIDIDAE With characters of family

TYPE-SPECIES *P. [Typhlolepta] byerleyana* (Collingwood, 1876). Type by monotypy.

TYPE-LOCALITY Under piece of coral on Pulo Barundum, off west coast of Borneo.

NOTE Bock (1925*a*) mentions a new pericelidid genus, *Thalattoplana*, but gives no description, except to imply that it has a common gonopore.

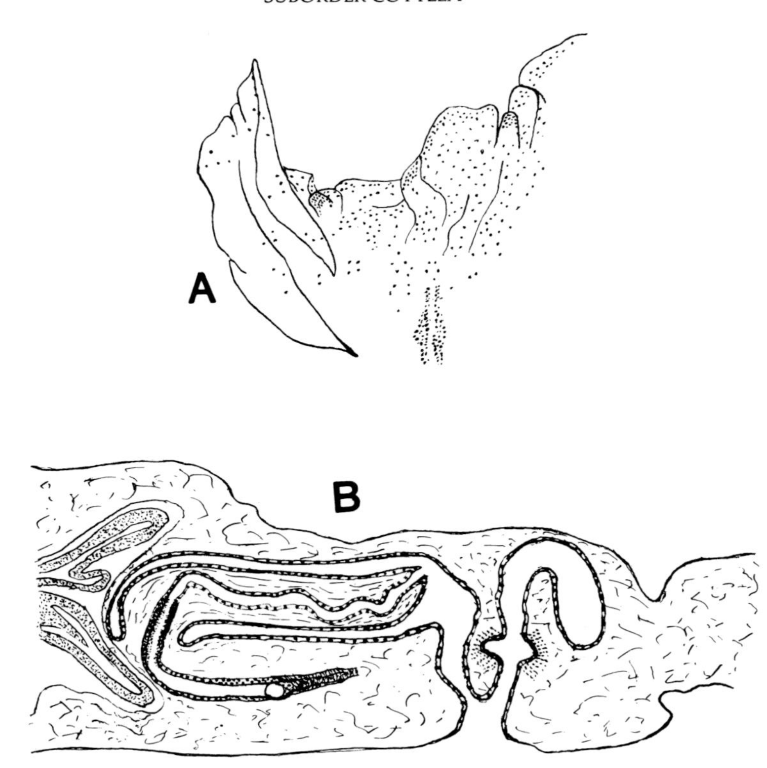

Fig. 110 *Pericelis byerleyana.* **A,** anterior region with eyes; **B,** sagittal section of copulatory organs (after Meixner).

Family **DIPOSTHIDAE** Woodworth, 1898

COTYLEA Discoid or oval forms, may have pair of conical marginal tentacles containing eyes. Ventral sucker, when present, centrally situated. Pharynx ruffled in mid-third of body. Genital pores distinctly separated. Pair of spermiducal bulbs or a seminal vesicle open directly into penis-papilla without stylet. Independent prostatic organ leads directly into male antrum. Short uterine canals posteriorly directed from proximal end of vagina and have two or three pairs of uterine vesicles.

KEY TO DIPOSTHID GENERA

1. Prostatic organ posterior to penis-papilla . . . . ***Diposthus***
   Prostatic organ anterior to penis-papilla . . . . ***Asthenoceros***

Genus ***Diposthus*** Woodworth, 1898 Fig. 111

DIPOSTHIDAE Oval forms with well-developed marginal tentacles containing eyes. Cerebral eyes in paired cluster. Pair of spermiducal bulbs lead into stout penis-papilla. Pyriform prostatic organ posterior to penis-papilla. Uterine canals with two or three pairs of vesicles.

TYPE-SPECIES *D. corallicola* Woodworth, 1898. Type by original designation.

TYPE-LOCALITY Under coral-rock on reef at Hope I., Great Barrier Reef, Australia.

NOTE The above definition is based entirely upon information available for the type-species, because the second species, *D. popae,* assigned to this genus by Hyman (1959*c*), does not appear to be congeneric with *D. corallicola.*

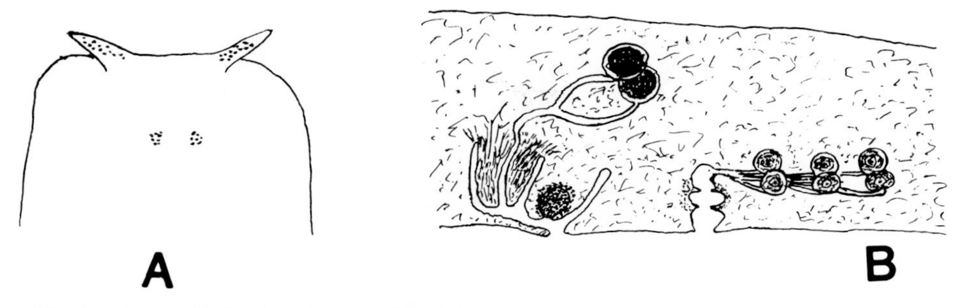

Fig. 111 *Diposthus corallicola.* **A**, anterior region of body; **B**, sagittal section of copulatory organs (modified, after Woodworth).

Genus *Asthenoceros* Laidlaw, 1903

DIPOSTHIDAE Broadly-oval forms apparently without definite marginal tentacles, although a small group of eyes occurs on either side of median line on anterior margin. No cerebral eyes observed. Seminal vesicle relatively large. Prostatic organ anterior to large penis-papilla. Uterine canals with two pairs of vesicles. Ventral sucker in middle of body.

TYPE-SPECIES *A. woodworthi* Laidlaw, 1903. Type by original designation.

TYPE-LOCALITY Pulau Bidan I., Straits of Malacca.

NOTE Laidlaw did not provide figures of the arrangement of the eyes nor of the structure of the copulatory complexes. In fact, it seems that in the specimen upon which he based his description the copulatory complexes were only partially developed.

Family **CHROMOPLANIDAE** Bock, 1922

COTYLEA Small, highly-coloured, elongate forms without tentacles. Ventral sucker when present, subcentral. Marginal eyes few, confined to frontal margin of body; cerebral eyes few, disposed in two groups. Mouth at about junction of anterior and middle thirds of body; pharynx ruffled; intestinal trunk long, with few non-anastomosing branches. Genital pores between mouth and ventral sucker, when present. Male copulatory complex directed anteriorly from its pore, with well-developed seminal vesicle and muscular, interpolated, prostatic organ. Penis-papilla without stylet. Vagina short, simple, directed posteriorly from female pore. Uterine canals have not been observed.

NOTE Members of this family bear a strong superficial resemblance to marine triclad turbellarians.

KEY TO CHROMOPLANID GENERA

1.	Genital pores unite with oral aperture	***Chromyella***
	Genital pores and mouth separated	2
2.	Male pore close behind mouth; female pore far behind mouth	***Chromoplana***
	Genital pores closely approximate, far behind mouth	***Amyella***

Genus *Chromoplana* Bock, 1922

Fig. 112

CHROMOPLANIDAE Cerebral eyes very few, including two pairs of large eyes; marginal eyes disposed anteriorly in a single row, divided into three groups, one median and two lateral. Genital pores well separated, male pore adjacent to mouth. Ventral sucker in posterior half of body. Male complex with spermiducal bulbs, a muscular seminal vesicle and a penis-papilla in a penis-pocket; pyriform prostatic organ well developed.

TYPE-SPECIES *C. bella* Bock, 1922. Type by original designation.

TYPE-LOCALITY Among *Corallina* weed in tidal pools and shallow water, Sagami Bay, Misaki, Japan.

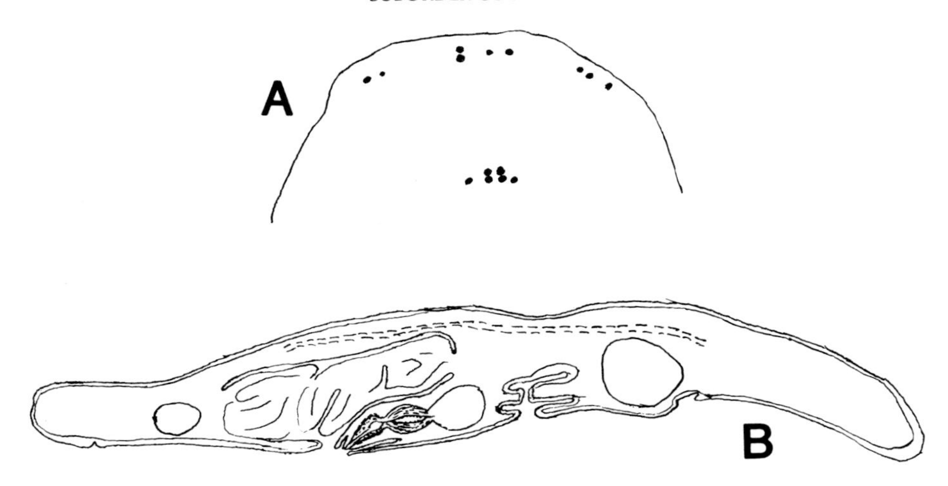

Fig. 112 *Chromoplana bella.* **A**, arrangement of eyes; **B**, sagittal section of body (after Bock).

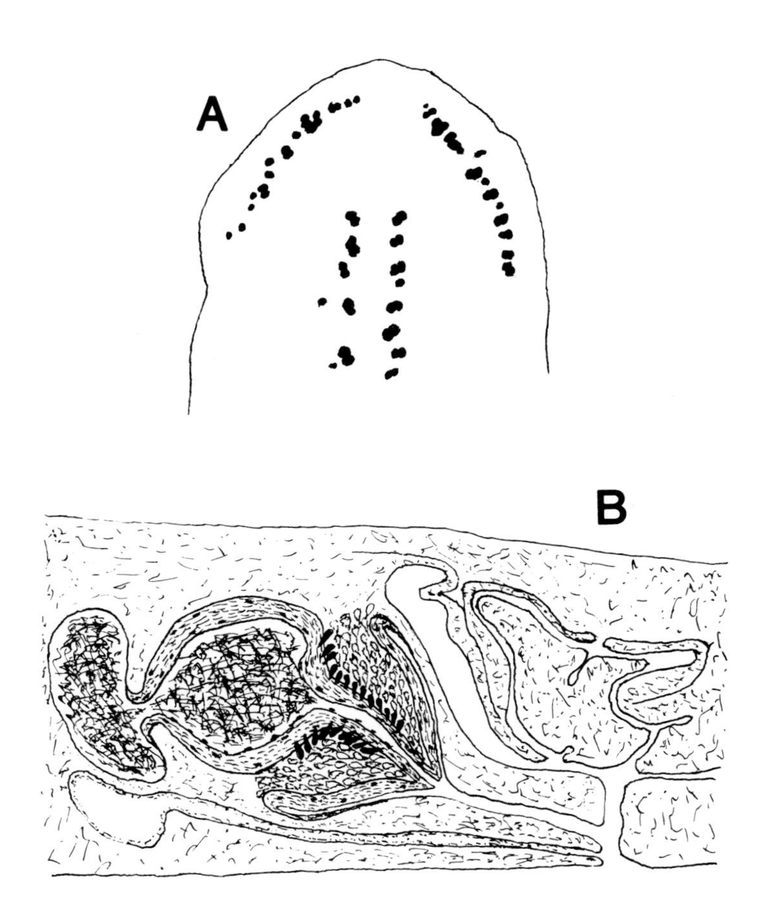

Fig. 113 *Chromyella saga.* **A**, arrangement of eyes; **B**, sagittal section of copulatory organs (after Corrêa).

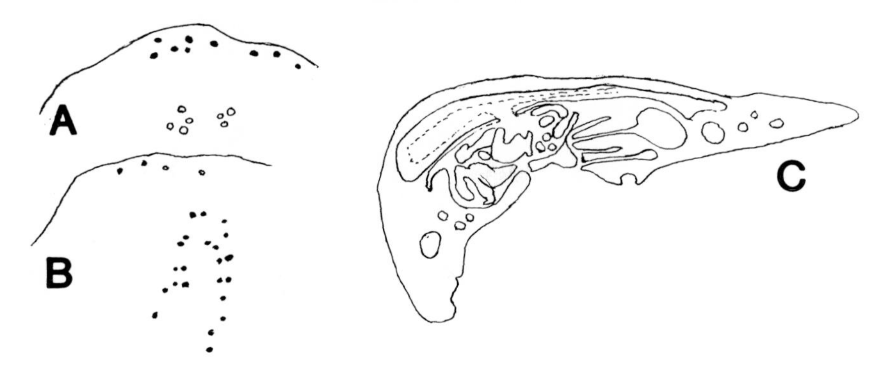

Fig. 114 *Amyella lineata.* **A,** arrangement of eyes ventrally; **B,** eyes dorsally; **C,** sagittal section of body (after Bock).

Genus ***Chromyella*** Corrêa, 1958 Fig. 113

CHROMOPLANIDAE Cerebral eyes in two elongate groups; marginal eyes in a row anteriorly, interrupted in median line. Genital pores united with mouth. An unpaired, thin-walled, spermiducal bulb opens into muscular seminal vesicle. No prostatic organ, but glands of two kinds invest distal region of seminal vesicle and ejaculatory duct within penis-papilla. Vagina terminates internally in a vesicle. Female copulatory complex lies ventrally to male. Ventral sucker not observed.

TYPE-SPECIES *C. saga* Corrêa, 1958. Type by original designation.

TYPE-LOCALITY In tide pool at Cabo Frio, Rio de Janeiro, Brazil.

NOTE In the original specimen of *C. saga* the female complex is not fully developed.

Genus ***Amyella*** Bock, 1922 Fig. 114

CHROMOPLANIDAE Several cerebral eyes in two elongate clusters; marginal eyes in band, divided into two groups by a median interruption. Genital pores closely approximate between mouth and ventral sucker. Male copulatory complex posterior to its pore and dorsal to vagina. No spermiducal bulbs. Prostatic organ reduced to a canal passing through penis-papilla. Vagina short and narrow, terminating internally in a small vesicle.

TYPE-SPECIES *A. lineata* Bock, 1922. Type by original designation.

TYPE-LOCALITY Among *Corallina* weed in tidal pools and shallow water, Sagami Bay, Misaki, Japan.

Family **OPISTHOGENIIDAE** Palombi, 1928, emended

COTYLEA Oval forms with a pair of nuchal tentacles containing eyes at their apices. Ventral sucker central. Marginal eyes confined to anterior region of body; cerebral eyes lie in oblong group between tentacular eyes. Mouth situated anteriorly; pharynx ruffled in anterior third of body; intestinal trunk very long, bifurcating posteriorly; intestinal branches anastomosing. Male genital pore ringed by large gland cells and situated between ventral sucker and posterior margin of body. Seminal vesicle anterior to male pore, independent prostatic organ posterior to pore. Penis-papilla small, without stylet. Female genital pore situated at about midway between pharynx and ventral sucker. Female copulatory complex anterior to its external opening. Vagina simple; uterus consists of two lateral groups of narrow interlacing canals opening into proximal end of vagina.

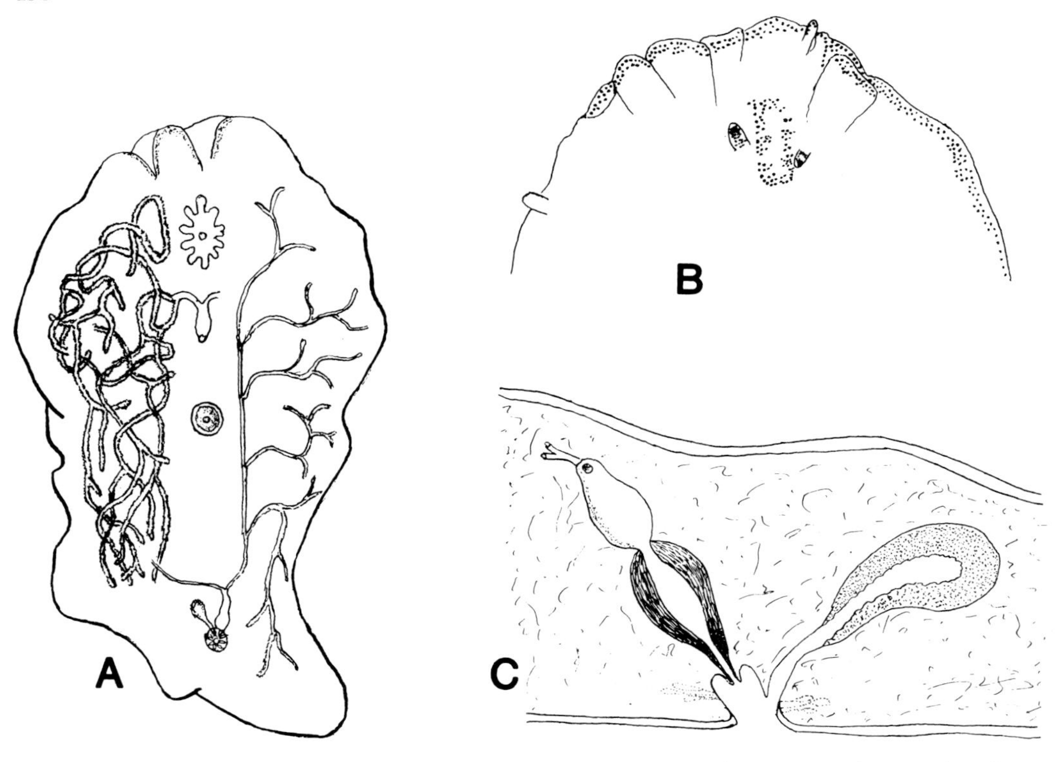

Fig. 115 *Opisthogenia tentaculata.* **A,** ventral view of entire worm; **B,** arrangement of eyes; **C,** sagittal section of copulatory organs (after Palombi).

Genus ***Opisthogenia*** Palombi, 1928 Fig. 115

OPISTHOGENIIDAE With characters of family.

TYPE-SPECIES *O. tentaculata* Palombi, 1928. Type by original designation.

TYPE-LOCALITY Dredged at Kubri, Suez Canal.

Family **PSEUDOCEROTIDAE** Lang, 1884, emended Poche, 1926

COTYLEA Broadly-oval or elongate forms varying considerably in size and often brilliantly coloured; dorsal surface sometimes papillate. Ventral sucker, when present, more or less central. Two tentacles formed by folds of anterior margin of body. Eyes within and at base of tentacles. Eyes also in a rounded or transversely arcuate cluster over cerebral organ. Pharynx ruffled or smooth margined, anteriorly situated. Intestinal trunk extends from pharynx to near posterior end of body and bears several pairs of anastomosing lateral branches. Genital pores distinctly separated between mouth and ventral sucker. One or two sets of male copulatory organs, when double disposed symmetrically. Male pore closely posterior or partially ventral to pharynx. Vasa deferentia extend on either side of intestinal trunk from posterior region of body to copulatory organs. Seminal vesicle well developed (absent in *Simpliciplana*). Prostatic organ (absent in *Dicteros* and *Simpliciplana*) independent. Penis-papilla small, generally provided with stylet enclosed in penis-pocket. Female copulatory complex closely posterior to male pore (*Nymphozoon* with several complexes). Vagina short, arcuate. 'Shell'-chamber dilated, but dorso-ventrally compressed. Uterine canals form H-shaped figure connected with anastomosing system of oviducts; uterine vesicles often present.

So far as is known, the development is indirect, and the larval stage is represented by Müller's larva.

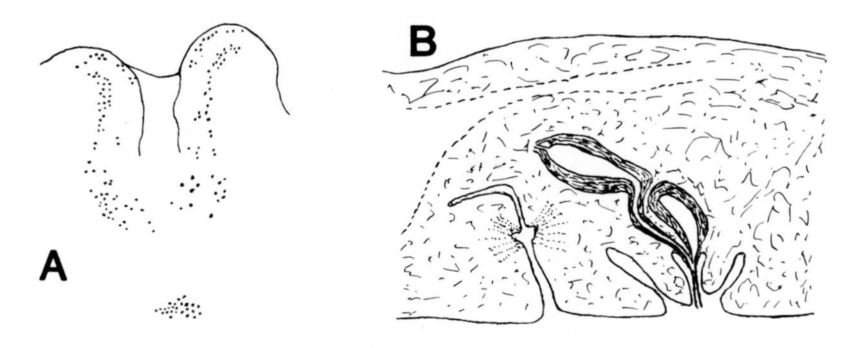

Fig. 116 *Pseudoceros litoralis* **A**, arrangement of eyes; **B**, sagittal section of copulatory organs (after Bock).

KEY TO PSEUDOCEROTID GENERA

1.	Prostatic organ present	2
	Prostatic organ lacking	6
2.	Body with dorsal papillae or tubercles	3
	Dorsal surface of body smooth	4
3.	Tentacles with eyes	***Thysanozoon***
	Tentacles without eyes	***Licheniplana***
4.	Without ventral sucker; with several female complexes	***Nymphozoon***
	With ventral sucker; one female complex	5
5.	Intestinal branches with anal pores	***Yungia***
	Intestinal branches without anal pores	***Pseudoceros***
6.	Without seminal vesicle	***Simpliciplana***
	With seminal vesicle	***Dicteros***

Genus ***Pseudoceros*** Lang, 1884 Fig. 116

SYNONYMY *Amblyceraeus* Plehn, 1897

PSEUDOCEROTIDAE Dorsal surface of body smooth, often highly coloured and variously patterned. Ventral sucker well developed. Intestinal branches without dorsal diverticula or anal pores. Cerebral eyes in a rounded cluster or in two short oval or elongate groups which may converge anteriorly or in two semicircles lying one behind other. Male copulatory complex may be duplicated. Prostatic organ smaller than seminal vesicle, dorsally to which it lies. One female copulatory complex.

TYPE-SPECIES *P. [Proceros] velutinus* (Blanchard, 1847). Type by subsequent designation.

TYPE LOCALITY Genoa, Italy.

NOTE The species of *Pseudoceros* are exceedingly difficult to determine satisfactorily, because the morphology of the copulatory complexes is fairly constant throughout, and Hyman (1954*a*) considers that only markings and coloration of the dorsal surface provide a basis for specific recognition. In the present writer's experience, markings and coloration may be reasonably constant in some species, but noticeably variable in others. It seems, therefore, that much more investigation is required to fully understand speciation in the genus *Pseudoceros.*

Genus ***Yungia*** Lang, 1884 Fig. 117

PSEUDOCEROTIDAE Forms with folded marginal tentacles and smooth dorsal surface. Many short canals leave intestinal branches to open on dorsal surface of body. Male copulatory complex as in *Pseudoceros.*

TYPE-SPECIES *Y. [Planaria] aurantiaca* (delle Chiaje, 1822). Type by subsequent designation.

TYPE-LOCALITY Sicily, Mediterranean.

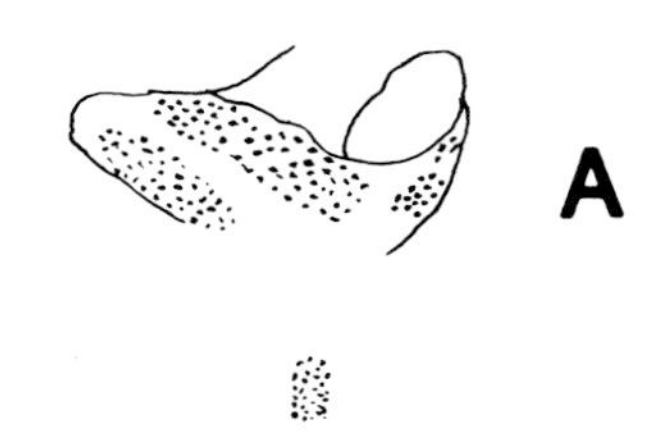

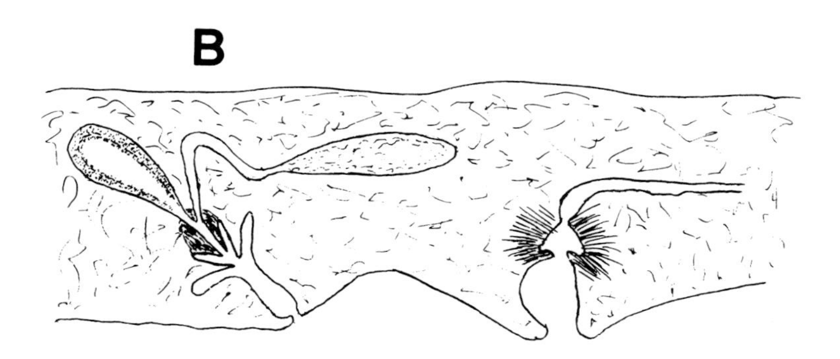

Fig. 117 *Yungia sasakii.* **A**, arrangement of eyes; **B**, sagittal section of copulatory organs (after Kaburaki).

Genus ***Licheniplana*** Heath & McGregor, 1913

PSEUDOCEROTIDAE Dorsal surface of body with small papillae. Intestinal branches numerous. Eyes between tentacles. Male copulatory complex single. Uterine canals directed posteriorly, with anastomosing branches.

TYPE-SPECIES *L. lepida* Heath & McGregor, 1913. Type by original designation.

TYPE-LOCALITY Under stones on southern shore of Monterey Bay, California, U.S.A.

NOTE Heath and McGregor did not differentiate this genus from either *Pseudoceros* or *Thysanozoon.* Hyman (1953*a*), however, points out that there is no mention in the original description of eyes in the tentacles, although they occur between them, and she suggests that the genus is possibly justified by the presence of small dorsal papillae and the absence of tentacular eyes.

Genus ***Thysanozoon*** Grube, 1840 Fig. 118

SYNONYMY *Acanthozoon* Collingwood, 1876; *Thysanoplana* Plehn, 1896.

PSEUDOCEROTIDAE Elongate or broadly-oval forms, variably coloured dorsal surface bearing papillae or tubercles, into which diverticula of intestinal branches may extend. Cerebral eyes in rounded cluster which may be longitudinally bisected. One or two symmetrically-placed male copulatory complexes. Seminal vesicle muscular and elongate; prostatic organ relatively small. One female complex.

TYPE-SPECIES *T. [Tergipes] brocchii* (Risso, 1818). Type by subsequent designation.

TYPE-LOCALITY Palermo, Sicily, Mediterranean.

SYSTEMATIC NOTE Lang (1884) regarded *Acanthozoon* as a synonym of the genus *Pseudoceros,* but Marcus (1950) recommended the resurrection of *Acanthozoon* as a valid genus on account of its short dorsal papillae, and du Bois-Reymond Marcus (1955*a*) accepted *Acanthozoon* as a subgenus of *Pseudoceros.* Hyman (1959*a*) considers that '*Acanthozoon* cannot become a subgenus of *Pseudoceros,* but would logically be made a subgenus of *Thysanozoon*'. Nevertheless, Miss Hyman defines *Acanthozoon* as a valid genus on the basis of 'dorsal surface covered with small papillae or tubercles',

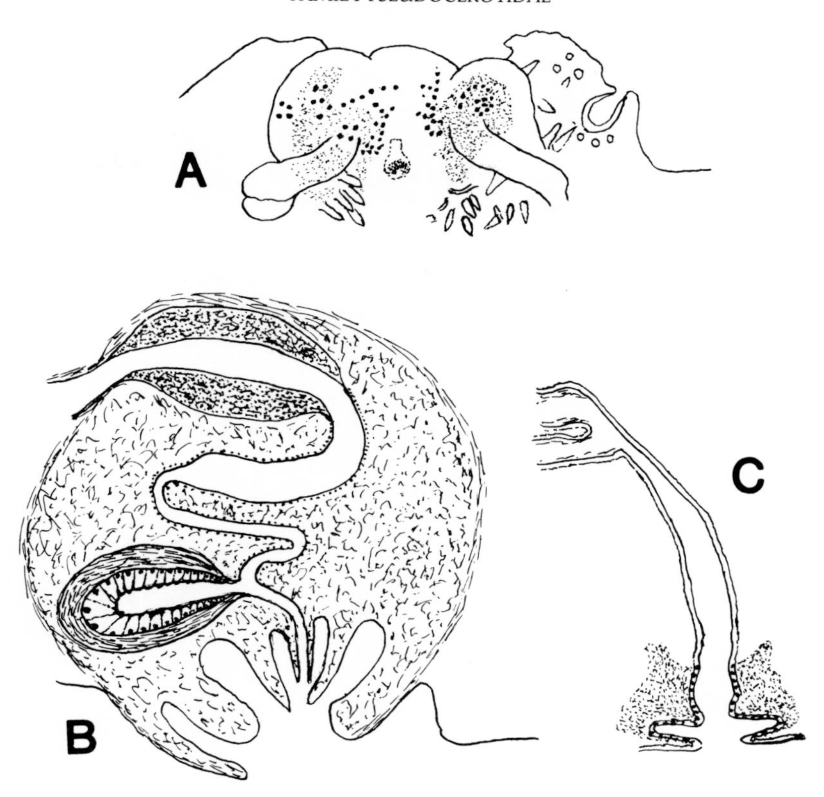

Fig. 118 *Thysanozoon*. **A**, tentacles and eyes of *T. brocchii* (after Palombi); **B**, sagittal section of ♂ copulatory organs of *T. nigrum*; **C**, sagittal section of ♀ copulatory organs (after Hyman).

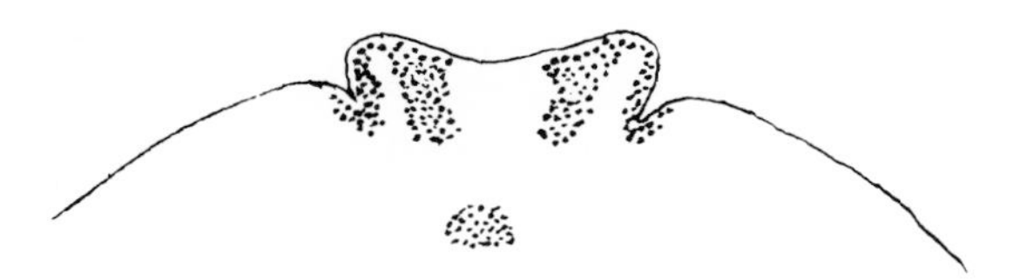

Fig. 119 *Simpliciplana marginata*. Tentacles and eyes (after Kaburaki).

and further points to the difficulty of differentiating small from large papillae. Stummer-Traunfels (1895), in a review of *Thysanozoon*, figures specimens of various species, each of which bears papillae of different shapes and sizes varying from small to large. Moreover, since the body of a pseudocerotid is likely to contract considerably at fixation, it is conceivable that at that time large papillae on the dorsal surface might contract to mere bosses. It seems, at present, that there are no sure grounds for recognizing *Acanthozoon* as a valid genus, and it is here accepted as a synonym of *Thysanozoon*.

Genus ***Simpliciplana*** Kaburaki, 1923

Fig. 119

PSEUDOCEROTIDAE Dorsal surface smooth. Ventral sucker apparently absent. Cerebral eyes in rounded group. Intestinal branches without anal pores. Seminal vesicle and prostatic organ not observed. Female pore in middle region of body.

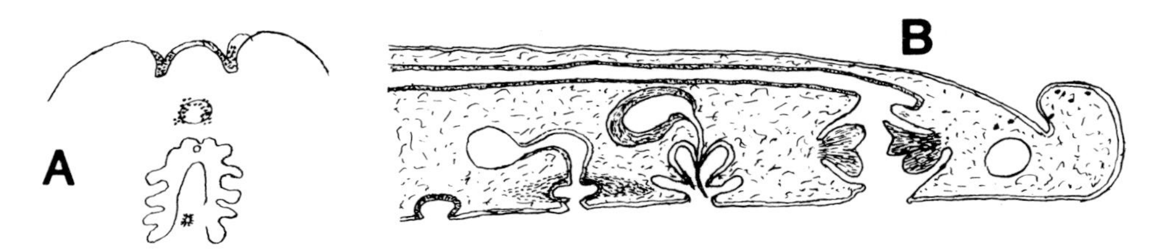

Fig. 120 *Dicteros pacificus*. **A**, tentacles and eyes; **B**, sagittal section of copulatory organs (after Plehn).

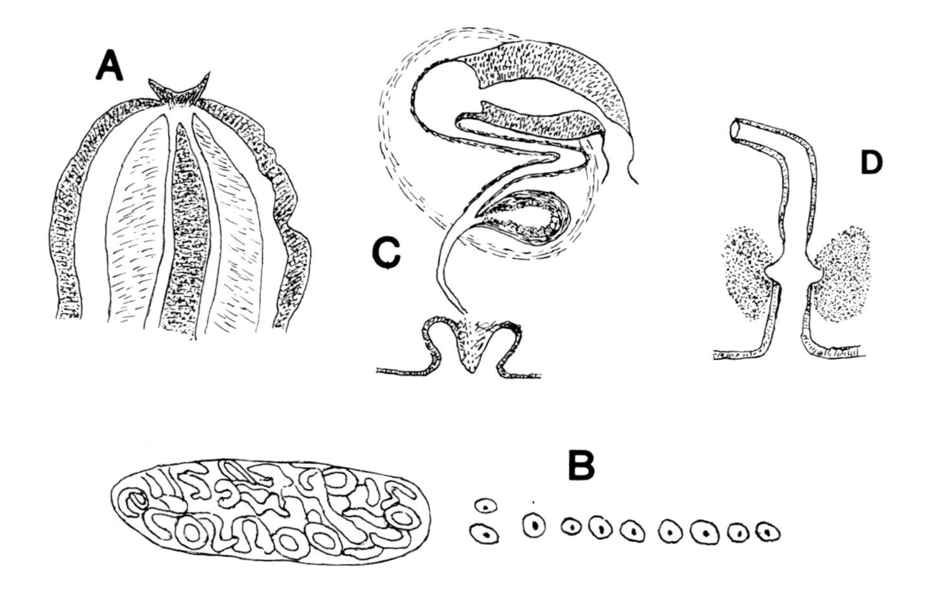

Fig. 121 *Nymphozoon bayeri*. **A**, anterior region of body (diagrammatic); **B**, pharynx and genital pores; **C**, sagittal section of ♂ copulatory organ; **D**, sagittal section of ♀ copulatory organ (after Hyman).

TYPE-SPECIES *S. marginata* Kaburaki, 1923. Type by original designation.

TYPE-LOCALITY Tawi Tawi, Tominado I., Philippines.

SYSTEMATIC NOTE Because of the lack of a ventral sucker and of a seminal vesicle and prostatic organ, Kaburaki (1923*b*) considered *Simpliciplana marginata* not to be a pseudocerotid, but related to the genus *Diplopharyngeata* and provisionally assigned *Simpliciplana* to the family Diplopharyngeatidae. Subsequent writers have, however, placed the genus in the family Pseudocerotidae, to which *Simpliciplana* bears a very strong superficial resemblance.

Genus *Dicteros* Jacubowa, 1906 Fig. 120

PSEUDOCEROTIDAE This genus appears to differ from *Pseudoceros* merely in the lack of a prostatic organ. The male antrum and the penis-pocket are, however, said to be lined with a glandular epithelium producing an eosinophilic substance very similar in appearance to that secreted by a prostatic organ.

TYPE-SPECIES *D. pacificus* Jacubowa, 1906. Type by original designation.

TYPE-LOCALITY Isle of Pines, New Caledonia, Pacific Ocean.

Genus ***Nymphozoon*** Hyman, 1959

Fig. 121

PSEUDOCEROTIDAE Large, broadly-oval forms with smooth dorsal surface. Ventral sucker apparently absent. Eyes not mentioned in original description. Elongate seminal vesicle muscular. Small, pyriform prostatic organ, which together with seminal vesicle embedded in thick mass of muscle fibres. Feebly-developed penis-papilla without stylet or penis-pocket. Two male complexes symmetrically disposed immediately posterior to highly ruffled pharynx. Close behind male pores lies median row of eight female complexes, each of which is composed of a deep tubular antrum opening into wide dorso-ventrally compressed 'shell'-chamber. Internally, each vagina runs towards a median uterus.

TYPE-SPECIES *N. bayeri* Hyman, 1959. Type by original designation.

TYPE-LOCALITY Iwayama Bay, Palau Is, Pacific Ocean.

Family **EURYLEPTIDAE** Stimpson, 1857
(Provisionally includes GRAFFIZOIDAE Heath, 1925)

COTYLEA Small to large, oval or somewhat discoid forms, sometimes highly coloured. Dorsal surface smooth, rarely papillate (*Cycloporus*). Anterior marginal tentacles of variable development, sometimes difficult to detect. Ventral sucker well developed in middle third of body and posterior to copulatory organs. Cerebral eyes in two elongate clusters; tentacular eyes (absent in *Anciliplana*) within or at bases of tentacles. Mouth at anterior end of relatively short pharyngeal chamber closely posterior to cerebral organ; pharynx cylindrical or bell-shaped, directed anteriorly. Intestinal trunk extends posteriorly, with variable number of paired lateral branches, usually anastomosing. Male genital pore posterior to mouth or united with it. Male copulatory complex simple in anterior third of body, ventral or immediately posterior to pharynx. From hind region of body vasa deferentia extend anteriorly to open into well-developed seminal vesicle. Prostatic organ relatively small, independent, anterior or dorsal to seminal vesicle. Penis-papilla enclosed in penis-pocket and often bearing a stylet. Female copulatory complex between male pore and ventral sucker. Vagina short and simple; 'shell'-chamber dilated, dorso-ventrally compressed. Uterine canals open into proximal end of vagina and extend posteriorly to hind region of intestinal trunk; uterine vesicles sometimes present.

SYSTEMATIC NOTES The family Graffizoidae was erected by Heath (1928) for a neotenic Müller's larva (*Graffizoon lobatum* Heath) which, apart from the development of reproductive organs, differs from the typical Müller's larva in the possession of lateral branches to the intestinal trunk. The erection of a distinct family for this larva was hasty, as it shows features of both young pseudocerotids and young euryleptids. Marcus (1950) agrees with Bresslau (1928–33) in regarding *Graffizoon lobatum* as a sexually mature larva of a euryleptid, and this determination is here provisionally accepted.

Hallez (1913) regarded the genera *Laidlawia* Herzig, *Enterogonimus* Hallez and *Leptoteredra* Hallez as members of the Euryleptidae, and erected the subfamily Laidlawiinae for their reception. Bock (1922), in discussing the relationships of these genera, raised the subfamily to family rank as Laidlawiidae. Subsequently, this was approved by some authorities, who appear to have based their acceptance merely on the absence of a penis-stylet. Among polyclads, the presence or absence of such a stylet can scarcely be so highly important taxonomically as to be the only feature by which two families might be separated. Nevertheless, the absence of a penis-stylet seems to be sufficient to warrant, for practical reasons, acceptance of Laidlawiinae as a subfamily of the Euryleptidae.

KEY TO EURYLEPTID SUBFAMILIES

1. Penis-papilla bearing stylet . . . . EURYLEPTINAE
   Penis-papilla without stylet . . . . LAIDLAWIINAE

Subfamily **EURYLEPTINAE** Hallez, 1913

EURYLEPTIDAE Penis-papilla with stylet.

KEY TO EURYLEPTINE GENERA

1.	Marginal tentacles small anterior prominences or not apparent	2
	Marginal tentacles large, lappet-like	4
2.	Intestinal branches with small terminal vesicles	***Cycloporus***
	No such vesicles	3
3.	Marginal eyes in distinct band anteriorly	***Euryleptides***
	Marginal eyes in two separate groups anteriorly (absent in *Stylostomum typhlum*)	***Stylostomum***
4.	Intestinal branches anastomosing	5
	Intestinal branches not anastomosing	8
5.	No tentacular eyes	***Anciliplana***
	Eyes in or about tentacles	6
6.	With uterine vesicles	7
	Without uterine vesicles	***Euryleptodes***
7.	Several pairs of uterine vesicles	***Prostheceraeus***
	One pair of such vesicles	***Oligoclado***
8.	Mouth posterior to cerebral organ	***Eurylepta***
	Mouth anterior to cerebral organ	***Oligocladus***

APPENDIX
Graffizoon = neotenic Müller's larva.

Genus ***Eurylepta*** Ehrenberg, 1831 Fig. 122

EURYLEPTIDAE: EURYLEPTINAE Oval forms of moderate size, with long marginal tentacles, containing many eyes. Cerebral eyes in two distinct elongate clusters. Ventral sucker in mid-third of body. Intestinal trunk with from three to six pairs of lateral non-anastomosing branches. At its junction with pharynx, intestinal trunk throws off three anteriorly-directed branches, one median and two passing laterally to pharynx and cerebral organ to reach near marginal tentacles. Uterine canals united posteriorly, usually with one or more pairs of uterine vesicles.

TYPE-SPECIES *E. [Planaria] cornuta* (Müller, 1776). Type by subsequent designation.

TYPE-LOCALITY On rocks, Kristiansand, southern Norway.

Genus ***Anciliplana*** Heath & McGregor, 1913 Fig. 123

EURYLEPTIDAE: EURYLEPTINAE Broadly-oval forms of moderate size. Tentacles well developed. Eyes apparently not within or about tentacles, but only in paired cerebral group. Intestinal trunk with about eight pairs of lateral anastomosing branches. Vasa deferentia in an intricate network of anastomosing canals. Uterine canals anastomosing.

TYPE-SPECIES *A. graffi* Heath & McGregor, 1913. Type by original designation.

TYPE-LOCALITY Monterey Bay, California, U.S.A.

Genus ***Cycloporus*** Lang, 1884 Fig. 124

EURYLEPTIDAE: EURYLEPTINAE Oval forms with dorsal surface smooth or papillate, variably coloured. Marginal tentacles distinct, although small, with eyes. Cerebral eyes in two closely associated elongate clusters. Intestinal trunk divides into three branches anteriorly, median branch passes between cerebral eye-clusters, side branches lateral to clusters. Six to ten pairs of lateral intestinal branches forming loose anastomosing system, except laterally, where they terminate in small vesicles opening externally by minute pores on periphery of body. Uterine canals united posteriorly, with six to eleven pairs of uterine vesicles.

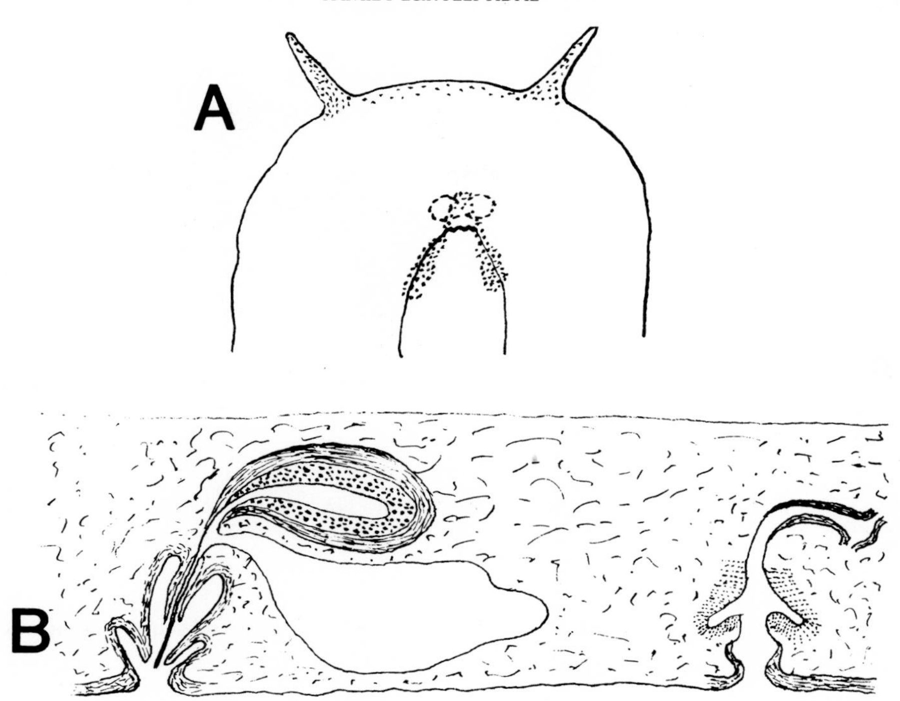

Fig. 122 *Eurylepta cornuta.* **A,** anterior region of body; **B,** sagittal section of copulatory organs (after Lang).

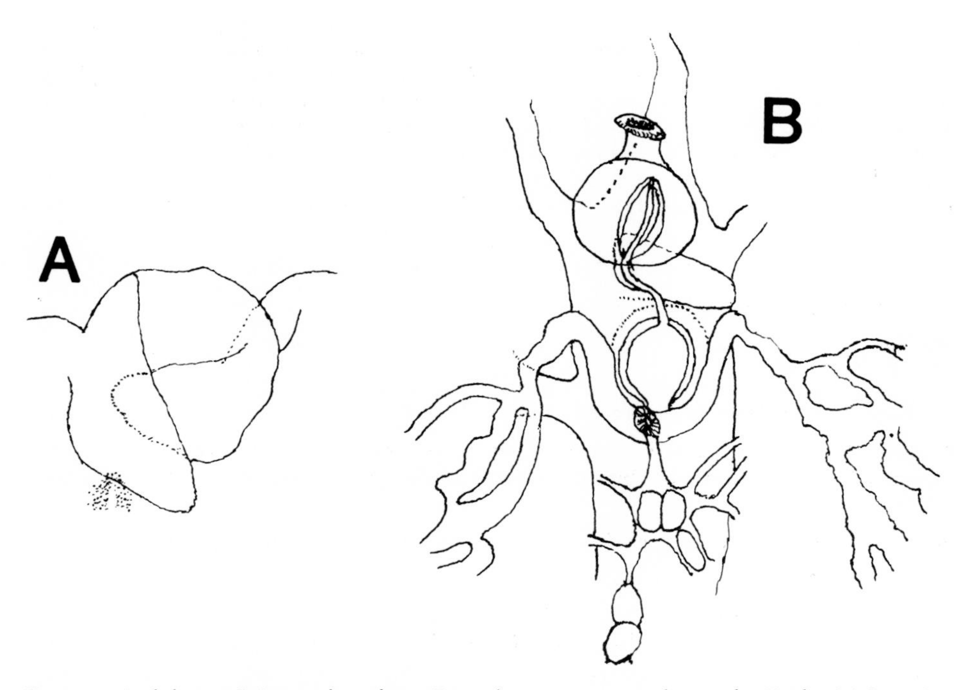

Fig. 123 *Anciliplana graffi.* **A,** tentacles and eyes; **B,** copulatory organs, ventral view (after Heath & McGregor).

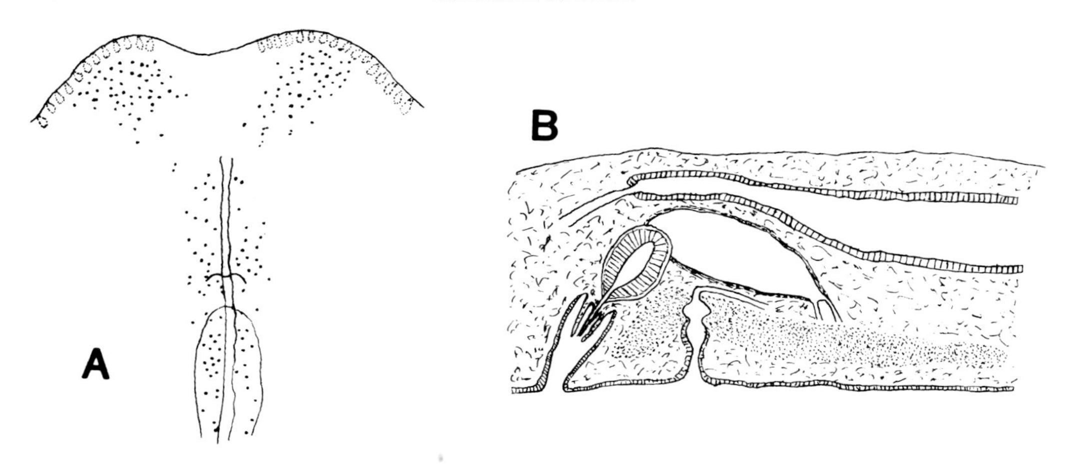

Fig. 124 *Cycloporus.* **A,** anterior region of *C. papillosus* (modified, after Lang); **B,** sagittal section of copulatory organs of *C. japonicus* (after Kato).

TYPE-SPECIES *C. [Thysanozoon] papillosus* (Sars, in Jensen, 1878). Type by subsequent designation.

TYPE-LOCALITY Florö, Norway.

Genus ***Euryleptides*** Palombi, 1924 Fig. 125

EURYLEPTIDAE: EURYLEPTINAE Small, oval forms without recognizable tentacles. Numerous eyes in complete marginal band anteriorly; cerebral eyes in two rows alongside pharynx. Intestinal trunk with about five pairs of lateral anastomosing branches. Male copulatory complex ventral to pharynx and posterior to male pore. Vasa deferentia open into muscular seminal vesicle. Independent prostatic organ and two small accessory vesicles lead into ejaculatory duct. Large penis-papilla with stylet, but no penis-pocket. Female complex posterior to female opening, with two large uterine canals forming H-shaped figure, limbs united posteriorly, and separated anteriorly; central branch of uterine figure opens into large glandular vesicle, from which two symmetrically-placed canals open into wide female antrum.

TYPE-SPECIES *E. brasiliensis* Palombi, 1924. Type by original designation.

TYPE-LOCALITY Off coast of Brazil, Atlantic Ocean.

Genus ***Euryleptodes*** Heath & McGregor, 1913 Fig. 126

EURYLEPTIDAE: EURYLEPTINAE This genus closely resembles *Eurylepta,* from which it differs in possessing lateral anastomosing intestinal branches and in anastomosis of branches of uterine canals. No uterine vesicles.

TYPE-SPECIES *E. cavicola* Heath & McGregor, 1913. Type by subsequent designation.

TYPE-LOCALITY Monterey Bay, California, U.S.A.

NOTE Heath & McGregor do not designate a type-species, but from their definition of the genus it appears very likely that they intended it to be *E. cavicola,* and this is accepted by Hyman (1953*a*). The authors also include in this genus the species *E. pannulus* and *E. phyllulus,* the latter of which is said not to possess a prostatic organ.

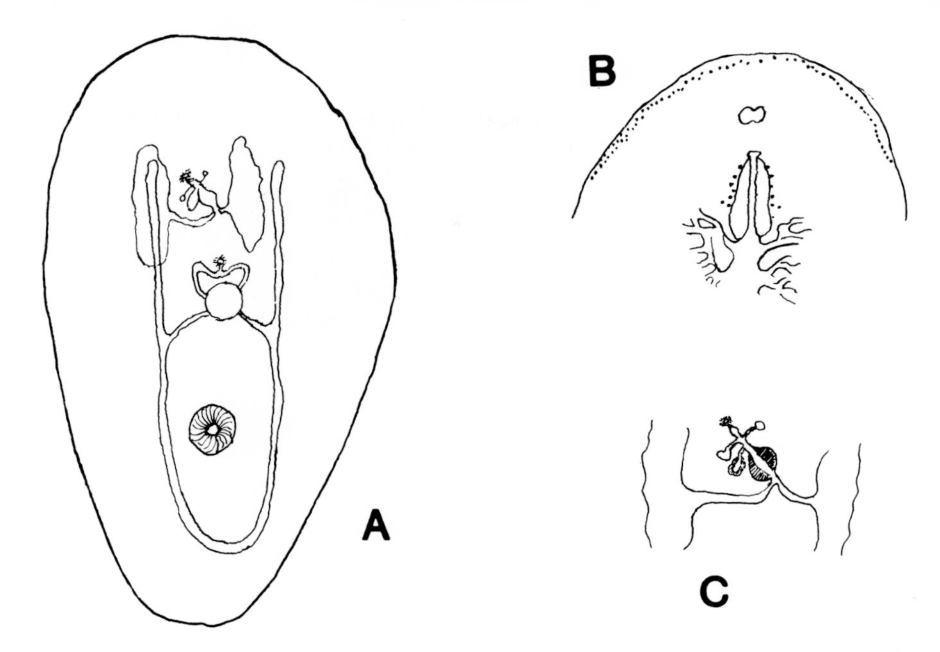

Fig. 125 *Euryleptides brasiliensis.* **A,** entire worm (cleared, ventral view); **B,** anterior region of body; **C,** male copulatory complex, ventral view (after Palombi).

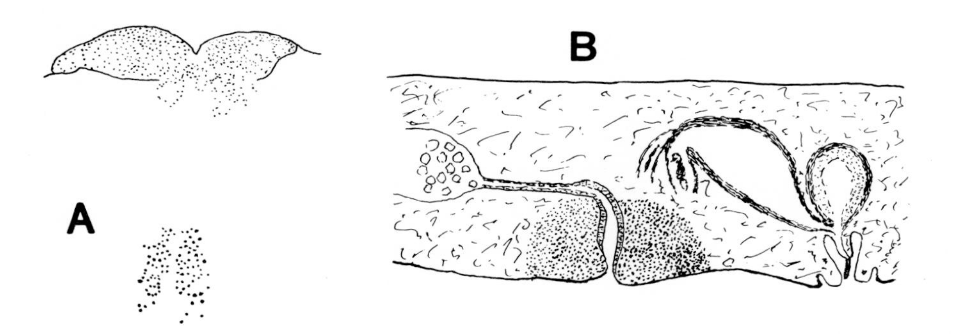

Fig. 126 *Euryleptodes.* **A,** anterior region of *E. cavicola* (after Heath & McGregor); **B,** sagittal section of copulatory organs of *E. insularis* (after Hyman).

Genus *Oligoclado* Pearse, 1938 — Fig. 127

SYNONYMY *Hymania* Pearse & Littler, 1938

EURYLEPTIDAE: EURYLEPTINAE Broadly-oval forms of moderate size, with distinct tentacles containing many eyes. Intestinal trunk with three to seven pairs of partially anastomosing lateral branches. Vasa deferentia confluent posteriorly to intestinal trunk. Penis-papilla slender; seminal vesicle and prostatic organ elongate and pyriform. Uterine canals simple, each with anterior uterine vesicle. From each vesicle a narrow canal runs posteriorly to unite with its opposite member and open externally through ventral pore in hind region of body, a feature apparently seen only at breeding time.

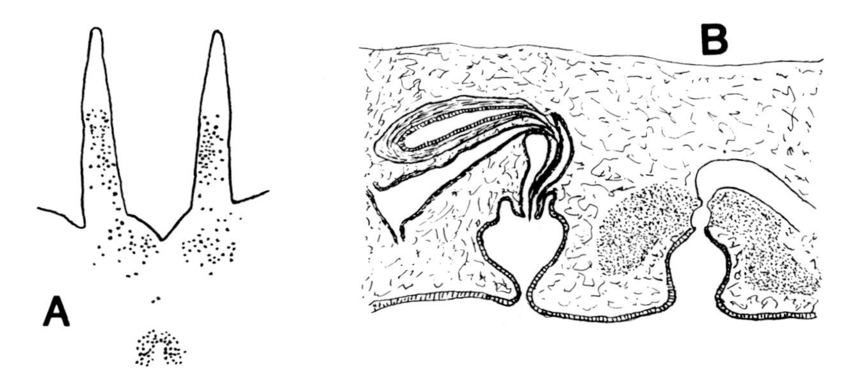

Fig. 127 *Oligoclado floridanus.* **A**, anterior region of body (after Pearse); **B**, sagittal section of copulatory organs (after Hyman).

TYPE-SPECIES *O. floridanus* Pearse, 1938. Type by original designation.

TYPE-LOCALITY Crooked Island Sound, Farndale, Florida, U.S.A.

Genus ***Oligocladus*** Lang, 1884 Fig. 128

EURYLEPTIDAE: EURYLEPTINAE Small, oval forms with anterior marginal tentacles of variable development, with eyes. Cerebral eyes in two sharply-defined clusters. Pharyngeal sheath with posterior appendage extending posteriorly beyond ventral sucker. Intestinal trunk trifurcates anteriorly, and opens to exterior posteriorly by dorsal pore; three or four pairs of non-anastomosing lateral branches. Genital pores ventral to pharynx. Uterine canals separated posteriorly; one to four pairs of uterine vesicles.

TYPE-SPECIES *O. [Proceros] sanguinolentus* (Quatrefages, 1845). Type by subsequent designation.

TYPE-LOCALITY St Malo, Atlantic coast of France

Genus ***Prostheceraeus*** Schmarda, 1859 Fig. 129

EURYLEPTIDAE: EURYLEPTINAE Large, elongate-oval forms, strikingly marked dorsally with narrow longitudinal stripes or spotted with dark brown. Tentacles well developed, containing many eyes. Cerebral eyes in two oval or rounded clusters. Bell-shaped pharynx short. Intestinal trunk extends posteriorly, several pairs of lateral branches ramifying and anastomosing freely. Uterine canals united posteriorly, with several pairs of uterine vesicles.

TYPE-SPECIES *P. [Planaria] vittatus* (Montagu, 1813). Type by subsequent designation.

TYPE-LOCALITY Kingsbridge estuary, Devon, England

Genus ***Stylostomum*** Lang, 1884 Fig. 130

SYNONYMY *Aceros* Lang, 1884, *nec* Hodgson, 1844, *nec* Boeck, 1861; *Acerotisa* Strand, 1928

EURYLEPTIDAE: EURYLEPTINAE Small, or very small, broadly-oval forms in which marginal tentacles are more or less inconspicuous. Tentacular and cerebral eyes variable in number, but each type disposed in paired cluster. Intestinal trunk with up to six pairs of lateral branches, which may or may not anastomose; no anterior branch over pharynx. Male pore and mouth may be separated or unite to form a common antrum. Male copulatory complex ventral or anterior to pharynx. Seminal vesicle ventral to prostatic organ. Penis-papilla usually with stylet in penis-pocket. Uteri represented by two wide elongate sacs, separated posteriorly. Pair of uterine vesicles often present.

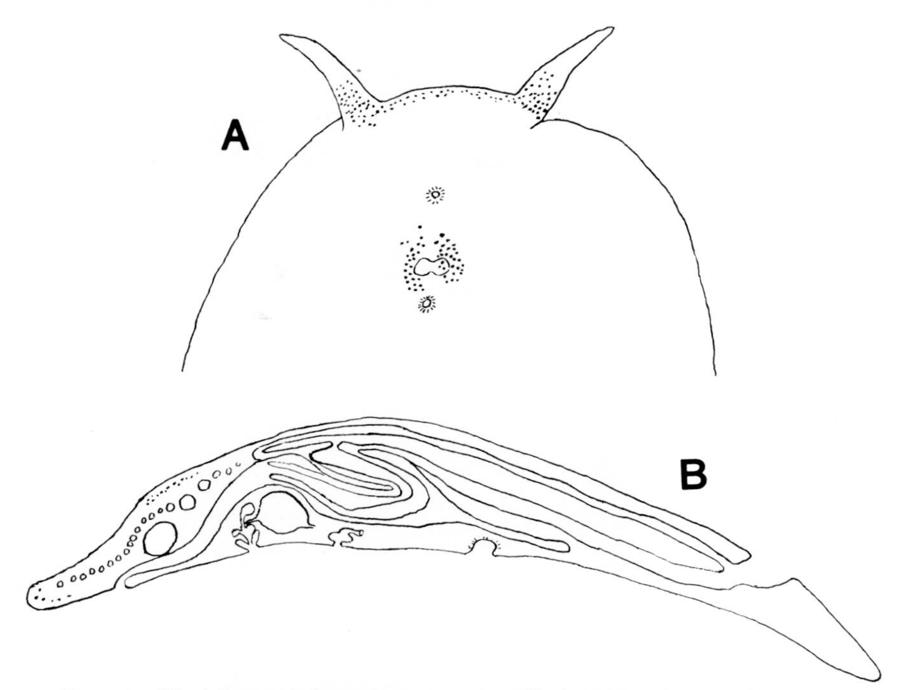

Fig. 128 *Oligocladus sanguinolentus.* **A,** anterior region of body; **B,** sagittal section of body (after Lang).

TYPE-SPECIES *S. [Planaria] ellipse* (Dalyell, 1853) [=*S. variabilis* Lang, 1884]. Type by subsequent designation.

TYPE-LOCALITY Between tide-marks, St Andrews, Fife, Scotland.

SYSTEMATIC NOTE Judging from Lang's (1884) diagnoses of *Stylostomum* and *Aceros,* these genera might be separated by two characters: (*a*) the absence of marginal tentacles in *Aceros [=Acerotisa],* and their presence in *Stylostomum*; and (*b*) the fusion of the mouth and male genital pore in *Stylostomum* and the separation of these apertures in *Aceros.* The tentacles in *Stylostomum* are often much reduced, and the presence or apparent absence of these organs can scarcely be regarded as characters of generic importance. In fact, at least two species of *Stylostomum, S. hozawai* Kato and *S. maculatum* Kato, are said to be without recognizable tentacles. Regarding the relative positions of the mouth and the male genital pore, in *Acerotisa* they may be very near each other and contraction of the body might cause them to lie in an apparent antrum. Ritter-Zahony (1907) found this to be the case in *Aceros meridianus,* and, according to Marcus (1954*1a,* fig. 109) in *Stylostomum felinum* a very thin fold of tissue separates the mouth from the male aperture. In other species that have been assigned to *Acerotisa* the space between the mouth and the male pore is most variable, thus it would appear that there are no reliable characters by which *Stylostomum* and *Acerotisa* may be differentiated.

Genus *Graffizoon* Heath, 1928

Fig. 131

EURYLEPTIDAE External form of Müller's larva. Mouth about middle of ventral surface. Pair of anterior marginal eyes and three pairs of cerebral eyes. Intestinal trunk relatively large with few lateral branches. Male genital pore anterior to mouth. Several testes; vas deferens terminates in seminal vesicle; prostatic organ independent; penis-papilla tipped with stylet in penis-pocket. Three pairs of ovaries; female pore posterior to mouth.

TYPE-SPECIES *G. lobatum* Heath, 1928. Type by original designation.

TYPE-LOCALITY In plankton, Bay of Monterey, California, U.S.A.

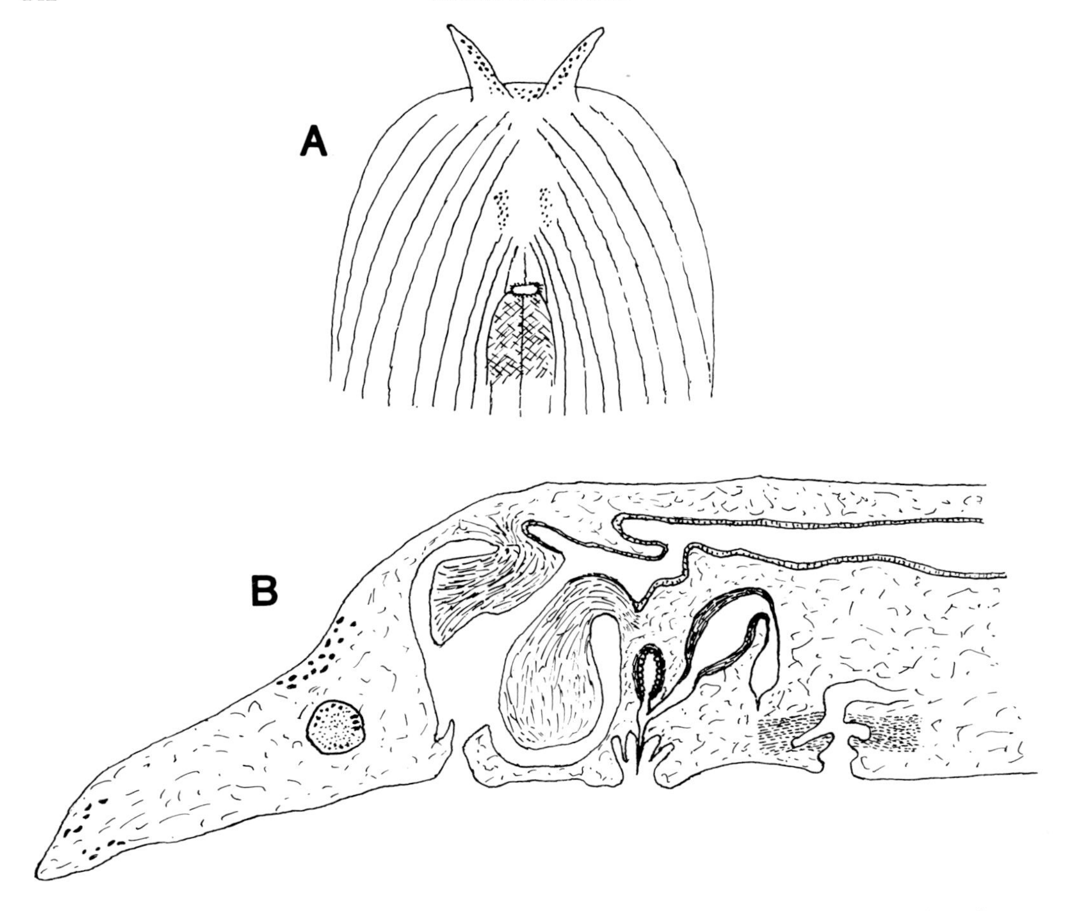

Fig. 129 *Prostheceraeus.* **A,** anterior region of *P. vittatus* (original); **B,** sagittal view of anterior region of *P. albocinctus* showing pharyngeal and copulatory complexes (after Lang).

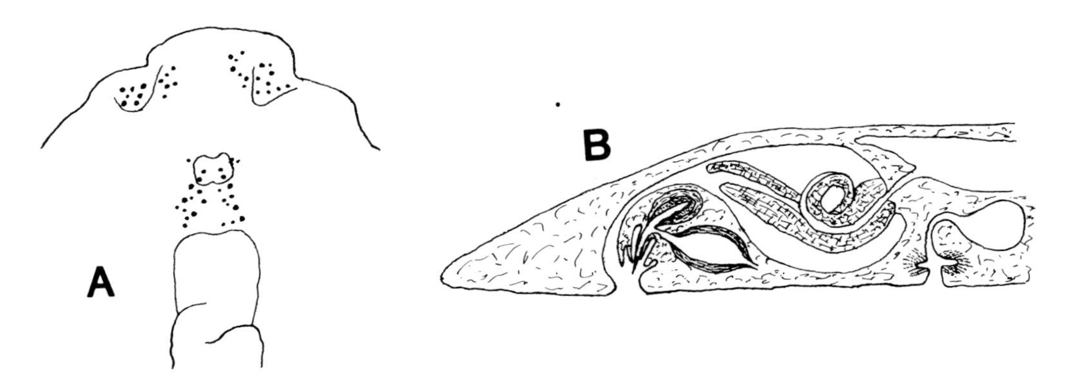

Fig. 130 *Stylostomum ellipse.* **A,** anterior region of body; **B,** sagittal view of anterior half of body showing pharyngeal and copulatory complexes (after Bock).

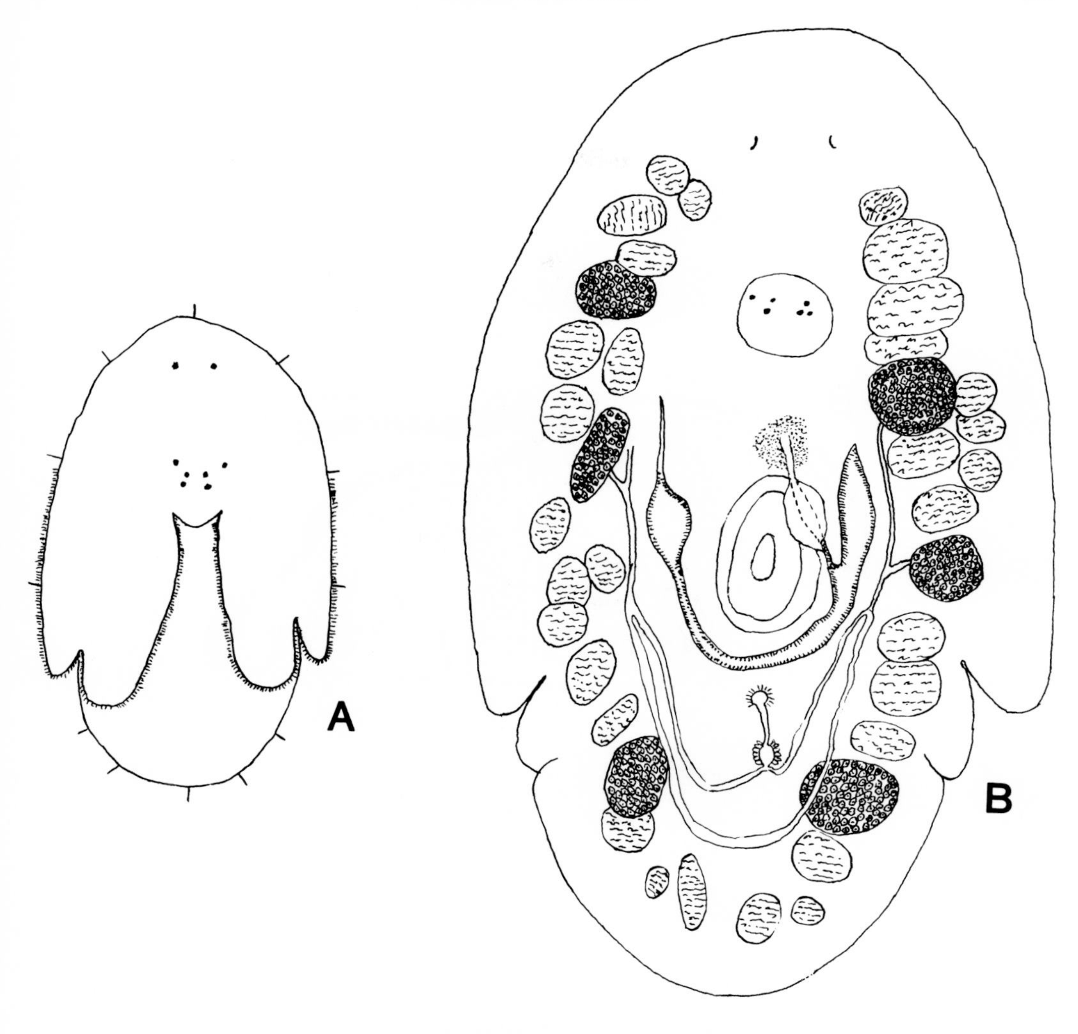

Fig. 131 *Graffizoon lobatum.* **A,** larva; **B,** reproductive system of larva (after Heath).

Subfamily **LAIDLAWIINAE** Hallez, 1913

EURYLEPTIDAE Without penis stylet.

KEY TO LAIDLAWIINE GENERA

1.	Dorsal anal pore at hinder end of intestinal trunk	***Leptoteredra***
	Intestinal trunk without external opening	2
2.	Median uterine vesicle opens dorsally	. ***Laidlawia***
	No dorsal opening to uterine vesicles	. ***Enterogonimus***

Genus ***Laidlawia*** Herzig, 1905 Fig. 132

EURYLEPTIDAE: LAIDLAWIINAE Small, broadly-oval forms with marginal tentacles inconspicuous, sometimes difficult to detect. Eyes few. Intestinal trunk with six or seven pairs of non-anastomosing

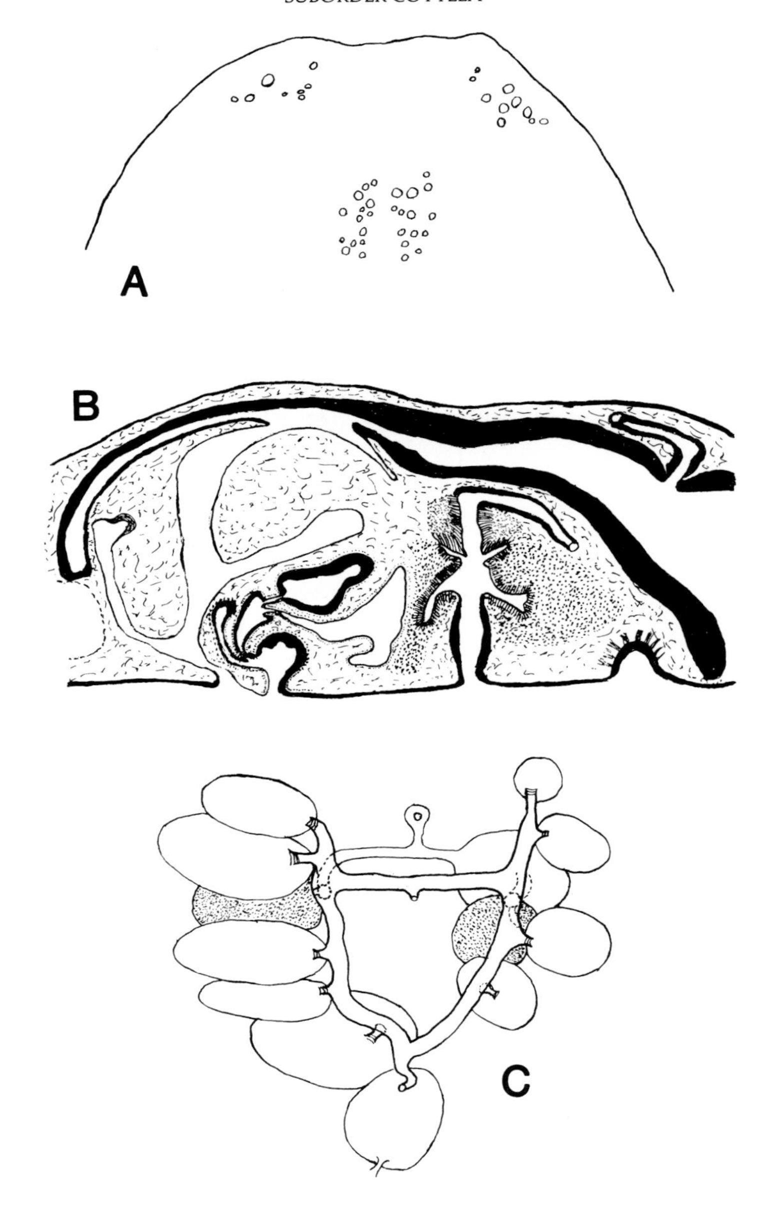

Fig. 132 *Laidlawia trigonopora.* **A**, anterior region of body (after Bock); **B**, sagittal section of copulatory organs (after Ritter-Záhony); **C**, dorsal view of ♀ copulatory complex (after Bock).

lateral branches, unpaired anterior branch. Uterine canals unite posteriorly and form large vesicle opening on dorsal surface. Behind female complex uteri connected by transverse canal leading into genito-intestinal canal. Near opening of transverse connecting canal, each uterine canal with large vesicle.

TYPE-SPECIES *L. trigonoporus* Herzig, 1905. Type by original designation.

TYPE-LOCALITY Punta Arenas, Magellan Straits, South America.

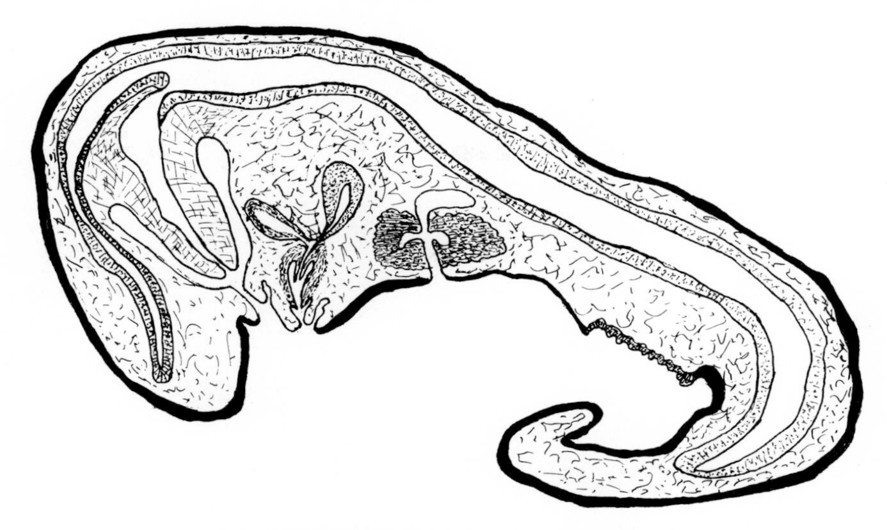

Fig. 133 *Enterogonimus aureus.* Sagittal view of complete worm (after Hallez).

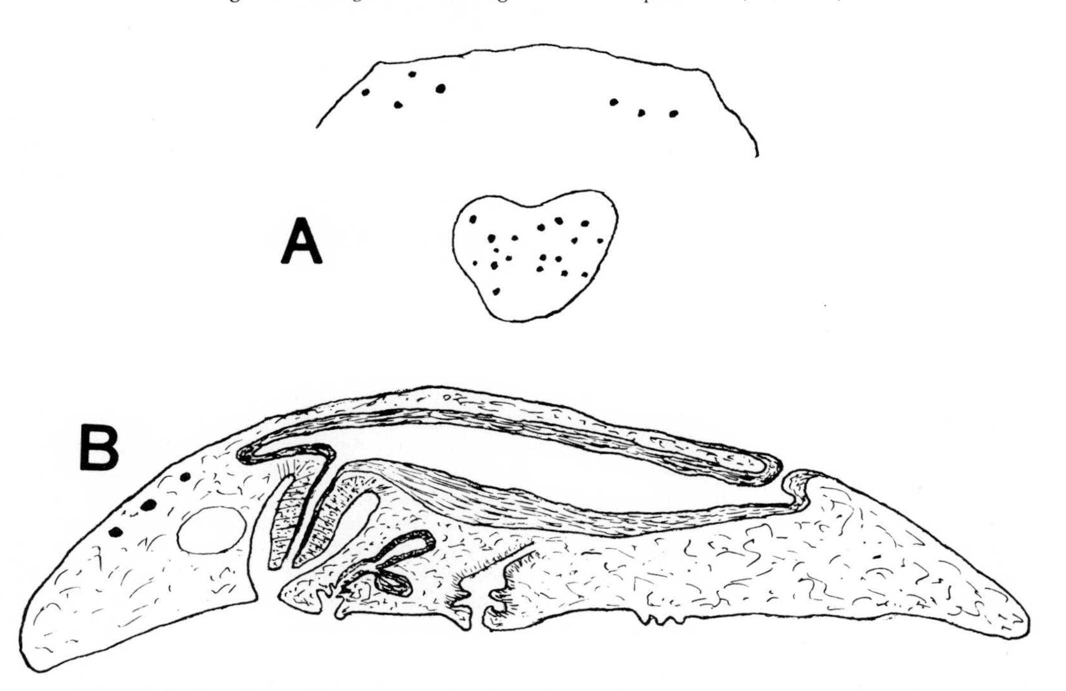

Fig. 134 *Leptotetredra maculata.* **A,** arrangement of eyes; **B,** sagittal section of complete worm (after Hallez).

Genus ***Enterogonimus*** Hallez, 1911

Fig. 133

EURYLEPTIDAE: LAIDLAWIINAE Small, oval forms, apparently without tentacles. Eyes few. Intestinal trunk with six or seven pairs of non-anastomosing lateral branches and an unpaired anterior branch. Uteri wide elongate sacs, each with uterine vesicle. Two genito-intestinal canals connect uterine canals separately with lateral walls of intestinal trunk.

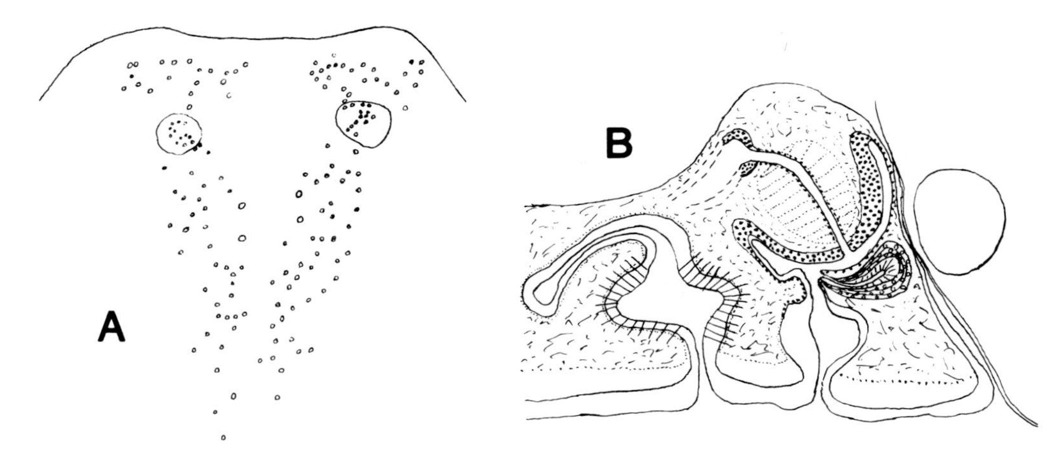

Fig. 135 *Stylochoides albus.* **A**, eyes and tentacles; **B**, sagittal section of copulatory organs (after Bock).

TYPE-SPECIES *E. aureus* Hallez, 1911. Type by original designation.

TYPE-LOCALITY Admiralty Bay, King George I., Antarctica.

Genus ***Leptoteredra*** Hallez, 1913 — Fig. 134

EURYLEPTIDAE: LAIDLAWIINAE Small, elongate oval forms with or without tentacles. Eyes few. Intestinal trunk with five pairs of non-anastomosing lateral branches and unpaired anterior branch; posterior end of trunk opens externally through a pore in median line on dorsal surface. Penis-papilla elongate and pyriform. Uterine canals united posteriorly, without vesicles.

TYPE-SPECIES *L. [Aceros] maculata* (Hallez, 1905). Type by original designation.

TYPE-LOCALITY In 18–20 metres, Bay of Carthage, Antarctica.

Family **STYLOCHOIDIDAE** Bock, 1913

COTYLEA Small, oval forms with two tentacles lying between cerebral organ and anterior margin of body. Eyes within, and anterior to, tentacles; cerebral eyes in two clusters merging with tentacular eyes. Centrally-situated ventral sucker well developed. Mouth at anterior end of pharyngeal chamber; pharynx short and tubular, closely posterior to cerebral organ; intestinal trunk with five pairs of lateral branches and unpaired anterior branch; secondary branches not anastomosing. Genital pores separated, posterior to mouth. Copulatory organs ventral to pharynx. Testes large, scattered throughout lateral regions of body; ovaries not numerous. Male copulatory complex with pair of spermiducal bulbs; no seminal vesicle. From spermiducal bulbs, vasa deferentia extend into large bulbous penis-papilla and unite at its apex. Between this union and spermiducal bulbs, vasa deferentia enclosed in mass of extracapsular gland-cells opening into lumen of each vas deferens. Pyriform glandular organ anterior to male antrum into which it opens. Vagina short; 'shell'-chamber dilated and dorso-ventrally compressed. Two wide uterine canals extend posteriorly from proximal end of vagina.

Genus ***Stylochoides*** Hallez, 1907 — Fig. 135

SYNONYMY *Nuchenoceros* Gemmill & Leiper, 1907; *Cotylocera* Ritter-Zahony, 1907.

STYLOCHOIDIDAE With characters of family.

TYPE-SPECIES *S. [Stylochus] albus* (Hallez, 1905). Type by monotypy.

TYPE-LOCALITY Dredged at 40 metres, Bay of Carthage, west coast of Graham Land, Antarctic Peninsula.

Family **PROSTHIOSTOMIDAE** Lang, 1884

COTYLEA Elongate or oval forms without tentacles. Well-developed ventral sucker generally present. Eyes marginal or submarginal, disposed anteriorly or arranged in continuous series round body; additional eyes in two elongate cerebral clusters or scattered fanwise anteriorly from cerebral organ. Mouth at anterior end of pharyngeal chamber; pharynx tubular, strongly muscular and directed anteriorly; intestinal trunk extends into posterior region of body and provided with several pairs of lateral limbs with non-anastomosing branches. Genital pores separated, between pharynx and ventral sucker. Vasa deferentia arise in posterior half of body and extend anteriorly to open independently into large, muscular seminal vesicle lying alongside pair of thick-walled accessory vesicles. Ejaculatory duct and ducts from accessory vesicles enter penis-papilla bearing strong stylet enclosed in penis-pocket, which forms a conical penis-sheath in roof of male antrum. Numerous unicellular prostatic glands invest penis-pocket and open into its lumen. Vagina short, looped anteriorly; 'shell'-chamber dilated, but dorso-ventrally compressed; no Lang's vesicle. Uterine canals form H-shaped figure, with transverse limb opening into proximal end of vagina and hinder limbs often communicating posteriorly.

KEY TO PROSTHIOSTOMID GENERA

1. Marginal eyes confined to anterior region of body 2
   Marginal eyes in continuous series round body . . . . ***Enchiridium***
2. With ventral sucker; additional eyes in one or two cerebral clusters 3
   No ventral sucker; additional eyes scattered fanwise over cephalic region . . ***Amakusaplana***
3. Ventral sucker well posterior to middle of body . . . . . . ***Euprosthiostomum***
   Ventral sucker more or less central 4
4. Muscular sheath enclosing male accessory vesicles . . . . ***Lurymare***
   Male accessory organs not enclosed in muscular sheath . . . . . ***Prosthiostomum***

Genus ***Prosthiostomum*** Quatrefages, 1845 — Fig. 136

SYNONYMY *Mesodiscus* Minot, 1877

PROSTHIOSTOMIDAE Elongate or ribbon-like forms, broadly-rounded or truncate anteriorly and tapering posteriorly. Ventral sucker centrally placed, closely posterior to female genital pore. Cerebral eyes in one or two elongate clusters; marginal eyes in anterior band; occasionally a few frontal eyes. Intestinal trunk with unpaired anterior branch lying dorsally to pharynx. Male accessory vesicles spherical, muscular, separate, one on either side of ejaculatory duct or seminal vesicle, not enclosed in muscular sheath.

TYPE-SPECIES *P. [Planaria] siphunculus* (Delle Chiaje, 1828). Type by subsequent designation.

TYPE-LOCALITY Bay of Naples, Mediterranean.

Genus ***Lurymare*** de B.-R. Marcus & Marcus, 1968 — Fig. 137

PROSTHIOSTOMIDAE Male accessory organs enclosed in muscular sheath, which may envelop seminal vesicle. Otherwise, similar to *Prosthiostomum.*

TYPE-SPECIES *L. [Prosthiostomum] drygalskii* (Bock, 1931). Type by original designation.

TYPE-LOCALITY Simonstown Bay, Cape Province, South Africa.

NOTE Poulter (1975) is probably justified in reducing *Lurymare* to subgeneric rank in the genus *Prosthiostomum.*

Genus ***Euprosthiostomum*** Bock, 1925 — Fig. 138

PROSTHIOSTOMIDAE Small, elongate forms with large ventral sucker in posterior region of body. Intestinal trunk with short anterior branch, and extends posteriorly to region of ventral sucker. Ovaries form pair of longitudinal submarginal bands, extending from pharyngeal level to copulatory complexes; distribution of testes less extensive. Arrangement of eyes and disposition of male accessory vesicles as in *Prosthiostomum.*

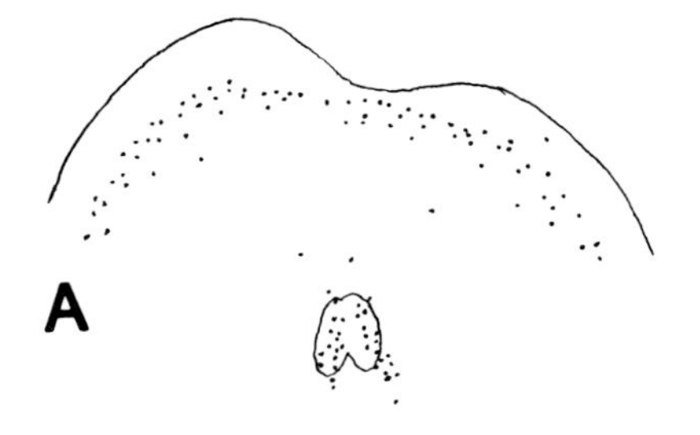

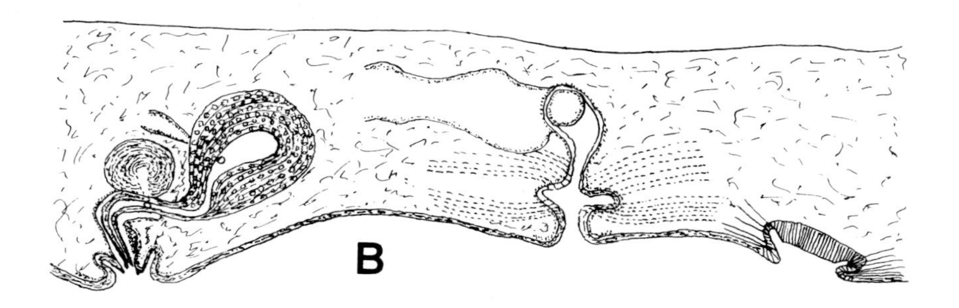

Fig. 136 *Prosthiostomum.* **A,** anterior region of *P. siphunculus* (after Palombi). **B,** sagittal section of copulatory organs of *P. lineatum* (after Meixner).

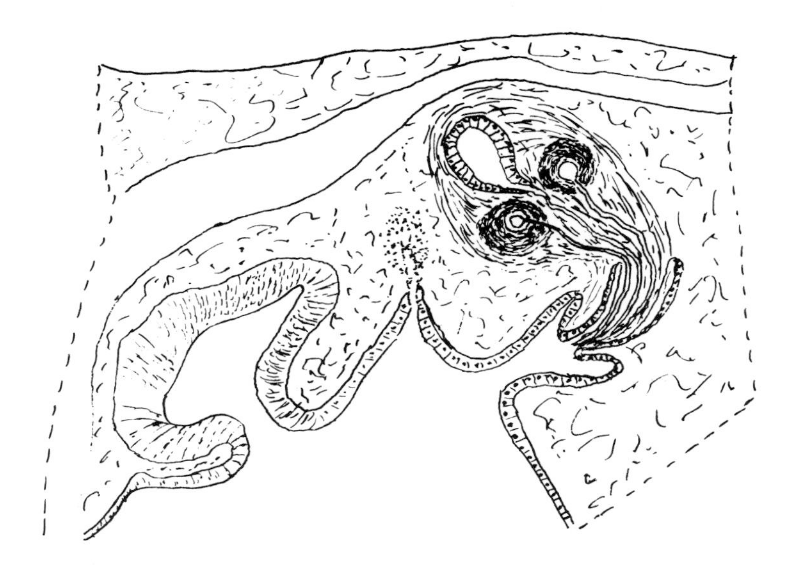

Fig. 137 *Lurymare drygalskii.* Sagittal section of copulatory organs (after Bock).

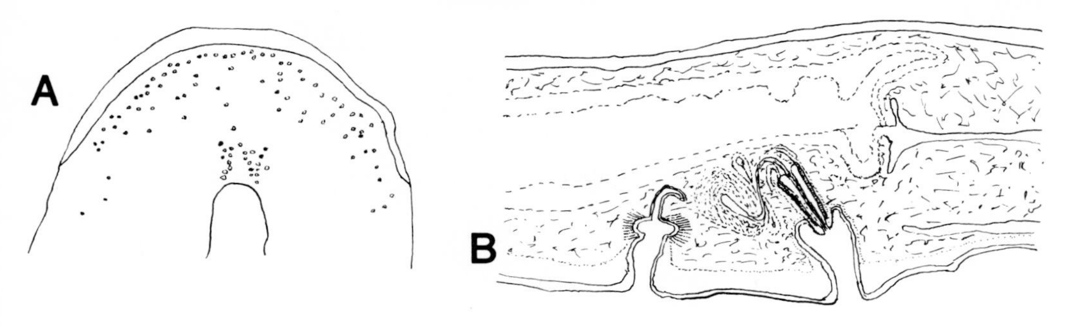

Fig. 138 *Euprosthiostomum adhaerens.* **A,** anterior region of body; **B,** sagittal section of copulatory organs (after Bock).

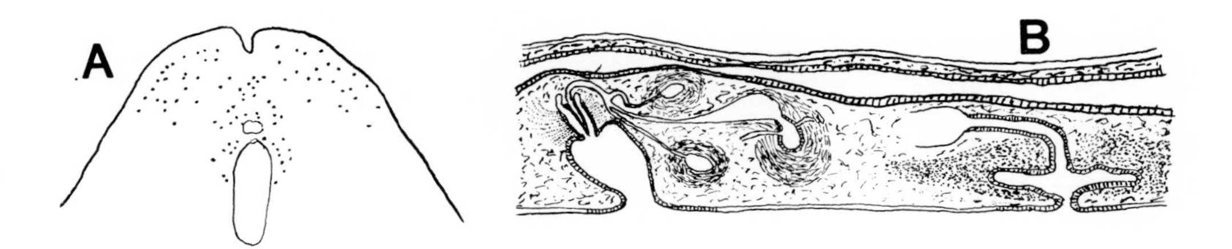

Fig. 139 *Amakusaplana ohshimai.* **A,** anterior region of body; **B,** sagittal section of copulatory organs (after Kato).

TYPE-SPECIES *E. adhaerens* Bock, 1925. Type by original designation.

TYPE-LOCALITY Gulf of Panama; at 5 to 13 metres in shells inhabited by the pagurid *Petrochirus californiensis.*

Genus *Amakusaplana* Kato, 1938 Fig. 139

PROSTHIOSTOMIDAE Elongate-oval forms of moderate size. Ventral sucker apparently absent. Eyes scattered somewhat fanwise over anterior region of body. Otherwise, similar to *Prosthiostomum.*

TYPE-SPECIES *A. ohshimai* Kato, 1938. Type by monotypy.

TYPE-LOCALITY On madreporarians at Magari-zaki, Amakusa, Japan.

Genus *Enchiridium* Bock, 1913 Fig. 140

PROSTHIOSTOMIDAE Elongate forms of moderate size, broadly rounded anteriorly and tapering posteriorly. Ventral sucker centrally situated. Cerebral eyes in two elongate clusters; marginal eyes in complete series round body. Male accessory vesicles enclosed in thick muscular envelope anteriorly to seminal vesicle. Otherwise similar to *Prosthiostomum.*

TYPE-SPECIES *E. periommatum* Bock, 1913. Type by original designation.

TYPE-LOCALITY Thatch I., U.S. Virgin Is., Caribbean.

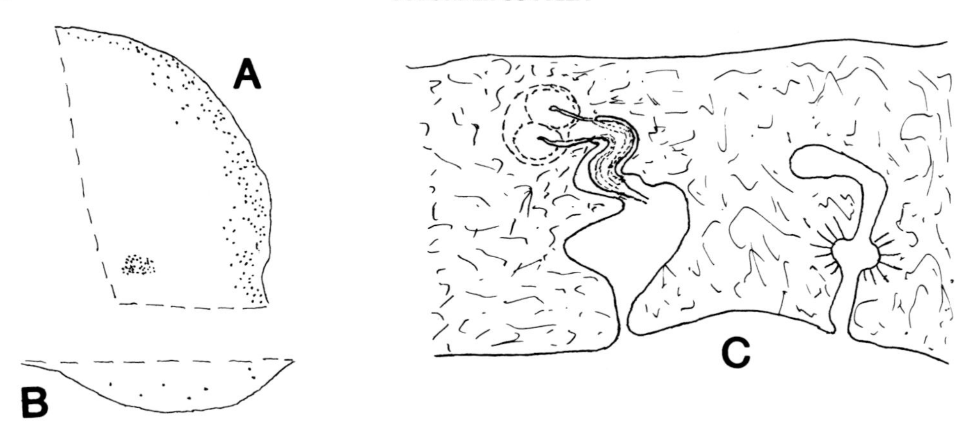

Fig. 140 *Enchiridium periommatum.* **A,** anterior arrangement of eyes; **B,** eyes at hindend of body; **C,** sagittal section of copulatory organs (after Bock).

Bibliography of suprageneric, generic, specific and infraspecific names used in Polycladida

ACANTHOZOON Collingwood, 1876, 86, 87, 95 (type *armatum* by subsequent designation); Lang, 1884, 538 (? s.o. *Pseudoceros*); du B.-R. Marcus, 1955, 42 (subgen. of *Pseudoceros*); Hyman, 1959*a*, 581 (defined). [See p. 132]

albopapillosus Hyman, 1959*a*, 583, figs (Palau Is).

albopunctatus Prudhoe, 1977, 600, figs (W. Australia).

armatum (Kelaart, 1858) Collingwood, 1876, 87, 95, fig. (redescr.); Lang, 1884, 545 (to *Pseudoceros*).

hispidus (du B.-R. Marcus, 1955*a*) Hyman, 1959*a*, 583.

micropapillosus (Kato, 1934*b*) Hyman, 1959*a*, 583.

nigropapillosus Hyman, 1959*a*, 581, fig. (Caroline Is).

papilio (Kelaart, 1858) Collingwood, 1876, 87, 95, fig. (redescr.); Lang, 1884, 546 (to *Pseudoceros*).

ACARENA Schmarda, 1859, 13 (subord. of Dendrocoela, comprising polyclads & triclads).

ACELIS Plehn, 1896*a*, 140, 146, 149 [*nec* Diesing, 1862] (type *arctica*); Benham, 1901, 31; Laidlaw, 1903*a*, 8 (systematic position); Bock, 1913, 69 (renamed *Plehnia*).

species Yeri & Kaburaki, 1918*a*, 1 (no descr.; Japan); Bock, 1923*a*, 1, 2 (= *Plehnia japonica*); Hyman, 1953*a*, 279 (to *Paraplehnia*).

arctica Plehn, 1896*a*, 140, 146, figs (79°N, 22°E – east of Spitzbergen); Bock, 1913, 70 (to *Plehnia*).

ACEPHALOLEPTIDEA Diesing, 1862, 492, 523 (family of Digonopora).

ACERIDEA Diesing, 1850, 185 (subtribe of Planariea); Lang, 1884, 446 (*ex parte* = Leptoplanidae).

ACEROIDEA Schmarda, 1859, 13 (family of Acarena); Lang, 1884, 466 (*ex parte* = Leptoplanidae).

ACEROS Lang, 1884, ix, 432, 589 (type: *inconspicuus*); Carus, 1885, 157; Gamble, 1896, 19; Benham, 1901, 31; Laidlaw, 1903*a*, 7; Heath & McGregor, 1913, 478 (defined); Bock, 1913, 273; 1922, 14; Strand, 1926, 36 (*Aceros* preoccupied, renamed *Acerotisa*).

baeckstroemi Bock, 1923*c*, 342, 362, figs (Juan Fernandez I., off Chilean coast); Stummer-Traunfels, 1933, 3584, fig.; Marcus, 1947, 141 (to *Acerotisa*).

inconspicuus Lang, 1884, 589, figs (Bay of Naples); Carus, 1885, 157 (some descr.); Marcus, 1947, 141 (to *Acerotisa*).

langi Heath & McGregor, 1913, 458, 478, figs (California); Bock, 1922, 16 ('not an *Aceros*'); Marcus, 1947, 141 (*Acerotisa*); Hyman, 1953*a*, 381, 382 (systematic position doubtful).

maculatus Hallez, 1905*b*, 125 (Bay of Carthage, Antarctica); 1907, 3, 11, figs (descr.); 1913, 38 (to *Leptoteredra*); Bock, 1922, 14 (morphological note).

meridianus Ritter-Záhony, 1907, 5, figs (Smyth Channel, Chile); Marcus, 1947, 141 (to *Acerotisa*).

nationalis Plehn, 1896*b*, 7, fig. (off Cape St Vincent, Portugal); Marcus, 1947, 141 (to *Acerotisa*).

stylostomoides Gemmill & Leiper, 1907, 819, figs (S. Orkney Is, Antarctica); Bock, 1922, 15, 16 (to *Leptoteredra*; ? = *L. maculata*); Marcus, 1947, 141 (to *Acerotisa*).

typhlus Bock, 1913, 273, figs (Norway); Marcus, 1947, 141 (to *Acerotisa*); Hendelberg, 1974*b*, 17 (Sweden).

ACEROTISA Strand, 1926, 36 (nom. nov. pro *Aceros* Lang preocc.); Hyman, 1940*a*, 489 (defined); 1953*a*, 378 (defined); Marcus, 1947, 141 (systematic note; list of spp.). [See p. 140]

species Hyman, 1952, 199, fig. (Florida).

alba (Freeman, 1933) Hyman, 1953*a*, 378, fig. (Puget Sound).

arctica Hyman, 1953*a*, 379, figs (Alaska).

baeckstroemi (Bock, 1923*c*) Marcus, 1947, 141.

baiae Hyman, 1939*e*, 153; 1940*a*, 489, fig. (Florida); 1952, 199 (Massachusetts).

bituna Marcus, 1947, 100, 138, 166, (Brazil); du B.-R. Marcus, 1955*b*, 286; du B.-R. Marcus & Marcus, 1968, 80, fig. (redescr.; Curaçao).

californica Hyman, 1953*a*, 379, figs (California).

inconspicua (Lang, 1884) Marcus, 1947, 141; Hyman, 1959*a*, 587.

langi (Heath & McGregor, 1913) Marcus, 1947, 141.

leuca Marcus, 1947, 100, 139, 166, figs (Brazil); du B.-R. Marcus, 1955*b*, 286.
maculata (Hallez, 1905*b*) Marcus, 1947, 141.
meridiana (Ritter-Záhony, 1907) Marcus, 1947, 141; Hyman, 1953*a*, 379.
multicelis Hyman, 1955*b*, 139, figs (Florida & Puerto Rico).
nationalis (Plehn, 1896*b*) Marcus, 1947, 141.
notulata (Bosc, 1802) Hyman, 1939*d*, 21, figs (redescr.; Sargasso Sea); du B.-R. Marcus & Marcus, 1968, 78, figs (redescr.; Curaçao).
pellucida Pearse, 1938, 90, figs (Florida); Hyman, 1940*a*, 482 (to *Enantia*).
piscatoria Marcus, 1947, 99, 136, 166, figs (Brazil); du B.-R. Marcus, 1955*b*, 286; du B.-R. Marcus & Marcus, 1968, 82, figs (redescr.; Bonaire I., Caribbean).
rugosa Hyman, 1959*a*, 585, 593, 594, figs (Palau Is).
stylostomoides (Gemmill & Leiper, 1907) Marcus, 1947, 141 (*Leptoteredra*).
typhla (Bock, 1913) Marcus, 1947, 141; Hyman, 1953*a*, 579.

ACOTYLEA Lang, 1884, xi, 3, 433 ('Tribus' of Polycladidea); Gamble, 1893*a*, 500, 521 (key to families & genera found in Britain); Hallez, 1893, 147 (defined); 1894, 200 (key to families) 201 (defined); Verrill, 1893, 461 (=tribe of subord. Digonopora); Laidlaw, 1902, 307 (differentiated from Cotylea); 1903*e*, 9 (key to genera); Bock, 1913, 59 (defined); Heath & McGregor, 1913, 455 (defined); Poche, 1926, 98 (Tribe of Planoceridea); Bresslau, 1928–33, 284 (sub-ord. of Polycladida); Pearse, 1936, 1ff (Acotylina); Hyman, 1939*b*, 128 (defined); 1940*a*, 450 (defined); 1953*a*, 275 (defined); de Beauchamp, 1961, 56 (classification of 'Acotyles'); Prudhoe, 1982*c*, 27 (defined).

Acotyles indeterminables de Beauchamp, 1951*b*, 248 (Mauritania).

ACOTYLINA Pearse, 1936, 1 (pro Acotylea).

ADENOPLANA Stummer-Traunfels, 1933, 3486, 3510 (type: *obovata*); Hyman, 1955*b*, 116 (defined).
antillarum Hyman, 1955*b*, 116, figs (Virgin Is, Caribbean); du B.-R. Marcus & Marcus, 1968, 62 (to *Boninia*).
evelinae Marcus, 1950, 5, 76, 114, figs (Brazil); du B.-R. Marcus, 1955*b*, 282.
obovata (Schmarda, 1859) Stummer-Traunfels, 1933, 3486, 3505, figs (type redescr.; Jamaica).
platae Hyman, 1955*d*, 9, figs (Uruguay).

AEOLIDICEROS [pro *Eolidiceros* Quatrefages] Moseley, 1877, 29.

ALLEENA Marcus, 1947, 99, 102, 162 (type: *callizona*); Hyman, 1953*a*, 275 (defined). [See p. 48]
callizona Marcus, 1947, 99, 103, figs (Brazil); du B.-R. Marcus, 1955*b*, 282.
mexicana Hyman, 1953*a*, 275, figs (=*Latocestus* sp. Steinbeck & Ricketts, 1941; Gulf of California); Brusca, 1973, 66 (Gulf of California).

ALLOIOPLANA Plehn, 1896*a*, 140, 142, 143 (type: *delicata*); Laidlaw, 1903*e*, 10 (morphological note); Bock, 1913, 185 (defined); Marcus, 1949, 76 (Leptoplaninae); Hyman, 1953*a*, 346 (planocerid). [See p. 90]
californica (Heath & McGregor, 1913) Hyman, 1953*a*, 346, figs (lectotype; California & Lower California); Morris *et al.*, 1980, 79, fig. (coloured) (California).
delicata Plehn, 1896*a*, 140, 142, figs (nr. Payta, Peru); Bock, 1913, 185.
sandiegensis (Boone, 1929) Hyman, 1953*a*, 349, figs (California & Gulf of California); Brusca, 1973, 68 (Gulf of California).

AMAKUSAPLANA Kato, 1938*a*, 560, 573, 575 (type: *ohshimai*).
ohshimai Kato, 1938*a*, 560, 573, figs (on madreporarians, Amakusa, Japan); Kato, 1944, 313; Hyman, 1959*a*, 593 (doubts validity).

AMBLYCERAEUS Plehn, 1897, 94 (type: *luteus*); Laidlaw, 1902, 307.
luteus Plehn, 1897, 94, figs (Monterey Bay, California); Hyman, 1953*a*, 366 (to *Pseudoceros*), 370.

AMEMIYAIA Kato, 1944, 271 (type: *pacifica*); Hyman, 1953*a*, 300 (systematic position).
pacifica Kato, 1944, 271, figs (Japan).

AMYELLA Bock, 1922, 20, 30 (type: *lineata*).
lineata Bock, 1922, 20, figs (Japan); Bresslau, 1928–33, 292, fig.; Kato, 1944, 306.

AMYRIS du Bois-Reymond Marcus & Marcus, 1968, 52 (type: *hummelincki*).
favis Sopott-Ehlers & Schmidt, 1975, 195, fig. (Galapagos Is).
hummelincki du B.-R. Marcus & Marcus, 1968, 53, figs (Curaçao & Bonaire, Caribbean).
ujara du B.-R. Marcus & Marcus, 1968, 55, figs (Klein Bonaire I., Caribbean).

ANANDROPLANA Hyman, 1955*b*, 125 (type: *muscularis*).
muscularis Hyman, 1955*b*, 125 figs (Puerto Rico).

portoricensis Hyman, 1955*b*, 127, figs (Puerto Rico). [see p. 78]

ANCORATHECA Prudhoe, 1982*a*, 223; 1982*b*, 363 (defined; type: *australiensis*).

australiensis Prudhoe, 1982*a*, 223 (S. Australia); 1982*b*, 363, 364, fig. (S. Australia).

pacifica (Bock, 1923*c*), Prudhoe, 1982*b*, 363 (Juan Fernandez I.)

ANCILIPLANA Heath & McGregor, 1913, 458, 479 (type: *graffi*); Hyman, 1953*a*, 381.

graffi Heath & McGregor, 1913, 458, 480, figs (California); Hyman, 1953*a*, 381.

ANONYMIDAE Lang, 1884, xi, 430, 521; Hallez, 1893, 168; 1894, 218 (defined); Gamble, 1896, 19; Benham, 1901, 31; Bock, 1922, 10 (diagnostic note); Bresslau, 1928–33, 290; Stummer-Traunfels, 1933, 3576.

ANONYMUS Lang, 1884, viii, 430, 522 (type: *virilis*).

virilis Lang, 1884, 62, 522, figs (Secca di Benda Palummo, Italy); Gamble, 1896, 18, fig.; Laidlaw, 1906, 712, fig. (Cape Verde Is), Martin, 1914, 251, 262 (nematocysts); Bresslau, 1928–33, 141, figs.; Stummer-Traunfels, 1933, 3573, fig. (some descr.).

APIDIOPLANA Bock, 1926*a*, 133 (type: *mira*).

mira Bock, 1926*a*, 133, figs (Fiji); 1927*a*, 6, figs (redescr.); Bresslau, 1928–33, 89, fig.; Hyman, 1951, 116, figs.

okadai Kato, 1944, 287, figs (Japan).

similis Bock, 1927*a*, 87, 97, figs (Fiji).

APIDIOPLANIDAE Bock, 1926*a*, 133; 1927*a*, 100 (defined); Bresslau, 1928–33, 289; Stummer-Traunfels, 1933, 3567, 3586.

APODA Örsted, 1843, 545 (Order of Class Vermes); Diesing, 1850, 179 (=Order Turbellariea).

APOROCEPHALES Blanchard, 1847, 143 (Order=Dendrocoela Ehrenberg); Lang, 1884, 433 (*ex parte*=Polycladidea).

APROSTATUM Bock, 1913, 149, 152, 157 (type: *stiliferum*); Marcus, 1952, 81, 114 (a stylochid).

stiliferum Bock, 1913, 152, figs (Chile); Marcus, 1954*a*, 52, figs (redescr.; Chile).

AQUAPLANA Hyman, 1953*b*, 191 (type: *oceanica*).

oceanica Hyman, 1953*b*, 191, figs (Galapagos Is).

pacifica Hyman, 1959*a*, 555, 594, figs (Palau Is).

ASOLENIA Hyman, 1959*a*, 562 (type: *deilogyna*).

deilogyna Hyman, 1959*a*, 562, 593, figs (Palau Is).

ASTHENOCEROS Laidlaw, 1903*c*, 302, 515 (type: *woodworthi*); Bock, 1923*b*, 26 (features).

woodworthi Laidlaw, 1903*c*, 303, 315, fig. (Malaysia); Bock, 1923*b*, 26 (features).

BERGENDALIA Laidlaw, 1903*c*, 302, 310, 312 (type: *anomala*); Bock, 1913, 86, 148 (diagnosis); Yeri & Kaburaki, 1918*a*, 433 (features); 1918*b*, 10 (defined); Bock, 1925*a*, 20; 1925*b*, 172 (systematic position); Kato, 1944, 261 (s.o. *Trigonoporus*).

species Dawydoff, 1952, 82 (Vietnam).

anomala Laidlaw, 1903*c*, 302, 310, figs (Straits of Malacca, Malaysia); Bock, 1913, 148; 1925*a*, 27; 1925*b*, 177; Kato, 1944, 20 (a *Trigonoporus*).

diversa Yeri & Kaburaki, 1918*a*, 431, 434 (diagnosis; Japan); 1918*b*, 2, 8, 46, figs (diagn.; Japan); Bock, 1925*a*, 27; 1925*b*, 177; Kato, 1944, 261 (to *Trigonoporus*).

mirabilis Kato 1938*b*, 577, 578, figs (Middle Japan); 1944, 261 (to *Trigonoporus*).

BONINIIDAE Bock, 1923*b*, 29; Bresslau, 1928–33, 290; Stummer-Traunfels, 1933, 3576; Prudhoe, 1944, 333 (redefined; key to genera); du B.-R. Marcus & Marcus, 1968, 63 (key to spp.); du Beauchamp, 1961, 64 (genera).

BONINIA Bock, 1923*b*, 1 (type: *mirabilis*); Prudhoe, 1944, 334 (generic features).

antillarum (Hyman, 1955*b*) du B.-R. Marcus & Marcus, 1968, 62, figs (Curaçao, Caribbean).

divae du Bois-Reymond Marcus & Marcus, 1968, 64, figs (Curaçao).

mirabilis Bock, 1923*b*, 1 (Bonin Is & possibly Philippines); Bresslau, 1928–33, 145 (fig.); Stummer-Traunfels, 1933, 3583, fig.; Kato, 1944, 294.

CALLIOPLANA Stimpson, 1857, 22 (type: *marginata*); Diesing, 1862, 569 (s.o. *Stylochus*); Lang, 1884, 434 (s.o. *Planocera*); Meixner, 1907*b*, 397 (distinct from *Stylochus*); Yeri & Kaburaki, 1918*a*, 438; 1918*b*, 25 (systematic note), 31; Stummer-Traunfels, 1933, 3561 (syn. *Diplosolenia*); Hyman, 1959*c*, 10 (defined).

evelinae Marcus, 1954*b*, 476, figs (Brazil); du B.-R. Marcus, 1955*b*, 284.
marginata Stimpson, 1857, 22, 29 (Japan); Diesing, 1862, 569 (to *Stylochus*); Lang, 1884, 445 (to *Planocera*); Meixner, 1907*b*, 397, 403; Yeri & Kaburaki, 1918*a*, 438 (diagn.; Japan); 1918*b*, 32, 48, figs (Japan); Stummer-Traunfels, 1933, 3487, 3561, 3585 (syns. *Stylochus oxyceraeus* & *Diplosolenia johnstoni*); Kato, 1934*b*, 127 (Japan); 1944, 288 (descr.; Japan); Dakin, 1953, 144 (Australia); Utinomi. 1956, 29 (coloured fig.; Japan); Hyman, 1959*c*, 12 (N.S.W., Australia).
CALLIOPLANIDAE Hyman, 1953*a*, 357 (=Diplosoleniidae); 1959*c*, 10 (defined); Marcus, 1954*b*, 475 (systematic note); de Beauchamp, 1961, 62.
CANDIMBA Marcus, 1949, 7, 74, 103 (type: *divae*); du B.-R. Marcus & Marcus, 1968, 35 (defined).
divae Marcus, 1949, 7, 74, 103, figs (Brazil); du B.-R. Marcus, 1955*b*, 284.
rabita du Bois-Reymond Marcus & Marcus, 1968, 2, 25, figs (Caribbean); Prudhoe, 1982*b*, 371 (to *Candimboides*).
CANDIMBOIDES Prudhoe, 1982*b*, 362, 371 (type: *rabita*).
cuneiformis Prudhoe, 1982*b*, 362, 371, fig. (Victoria, Australia).
rabita (du B.-R. Marcus & Marcus, 1968) Prudhoe, 1982*b*, 371
CARENOCERAEUS Schmarda, 1859, 14 (type: *oceanicus*); Diesing, 1862, 546 (s.o. *Nautiloplana*); Bock, 1913, 277 (s.o. *Planocera*).
oceanicus (Darwin, 1844) Schmarda, in Diesing, 1862, 546 (to *Nautiloplana*); Bock, 1913, 277 (s.o. *Planocera pellucida*).
CARENOTA Schmarda, 1859, 14, 36 (subord. of Dendrocoela).
CENTROSTOMUM Diesing, 1850, 199 (type: *lichenoides*); Schmarda, 1859, 13, 24; Stimpson, 1857, 21; Diesing, 1862, 492, 543.
bilobatum (Leuckart, 1828) Schmarda, 1859, 24; Lang, 1884, 607 (*Planaria*).
dubium Schmarda, 1859, 25, figs (Ceylon); Diesing, 1862, 544; Lang, 1884, 499 (to *Leptoplana*); Stummer-Traunfels, 1933, 3487, 3536 (type re-examd.; to *Notoplana*).
gigas (Schmarda, 1859) Diesing, 1862, 544; Lang, 1884, 508 (*Leptoplana*).
incisum (Darwin, 1844) Diesing, 1850, 200; 1862, 543; Lang, 1884, 609 (*Planaria*).
jaltensis Czerniavsky, 1881, 220, figs (Black Sea); Lang, 1884, 500 (to *Leptoplana*); Jacubowa, 1909, 1 (s.o. *Leptoplana tremellaris*).
lichenoides (Mertens, 1832) Diesing, 1850, 199; 1862, 543; Stimpson, 1857, 21; Lang, 1884, 469 (to *Discocelis*).
mertensii Claparède, 1861, 147, figs (Arran I., Scotland); Diesing, 1863, 174; Lang, 1884, 499 (to *Leptoplana*).
ocellatum (Kelaart, 1858) Collingwood, 1876, 97, fig. (some descr.); Lang, 1884, 511 (to *Penula*).
polycyclium Schmarda, 1859, 24, fig. (Ceylon); Diesing, 1862, 543; Lang, 1884, 498 (to *Leptoplana*).
polysorum Schmarda, 1859, 25, figs (New Zealand); Diesing, 1862, 544; Hutton, 1879, 315; Lang, 1884, 499 (to *Leptoplana*); Stummer-Traunfels, 1933, 3486, 3528 (type re-examd.; to *Leptostylochus*).
punctatum (Kelaart, 1858) Collingwood, 1876, 88, 97, fig. (some descr.); Lang, 1884, 616 (to *Penula*).
taenia Schmarda, 1859, 24, figs (Peru); Diesing, 1862, 543; Lang, 1884, 498 (to *Leptoplana*).
CEPHALOCERIDEAE Diesing, 1862, 202 (family); Lang, 1884, 523 (*ex parte*=Pseudoceridae), 553 (*ex parte*=Euryleptidae).
CEPHALOCEROIDEA Schmarda, 1859, 14, 30 (family of subord. Acarena); Lang, 1884, 553 (*ex parte*=Euryleptidae).
CERATOPLANA Bock, 1925*a*, 30 (type: *colobocentroti*); Marcus, 1949, 76 (Leptoplaninae).
colobocentroti Bock, 1925*a*, 4, figs (Krakatau I., Indonesia).
colobocentroti var. *hawaiiensis* Bock, 1925*a*, 6, figs (Hawaii).
CERIDEA Diesing, 1850, 202 (subtribe of Planariea).
CESTOPLANA Lang, 1884, viii, 430, 516 (type: *rubrocincta*); Bock, 1913, 250 (distribution of spp.); 1925*a*, 72; Kato, 1937*a*, 227 (might be regarded as a cotylean); Marcus, 1949, 81 (key to spp.).
australis Haswell, 1907*b*, 479, fig. (N.S.W., Australia); Kato, 1937*a*, 226, 227 (var. of *rubrocincta*); 1944, 293 (s.o. *rubrocincta*)
ceylanica Laidlaw, 1902, 302 (Ceylon).
cuneata Sopott-Ehlers & Schmidt, 1975, 210, figs (Galapagos Is).
faraglionensis Lang, 1884, 219, 520, figs (Italy); Carus, 1885, 153 (some descr.); Bresslau, 1928–33, 105 (fig.).

filiformis Laidlaw, 1903*d*, 110 (East Africa); Kato, 1937*a*, 226, 227 (var. of *rubrocincta*); 1944, 293 (s.o. *rubrocincta*).
lactea Kato, 1937*a*, 211, 223, figs (Japan); 1944, 293.
maldivensis Laidlaw, 1902, 290, fig. (Laccadive, Is, Arabian Sea); 1903*c*, 313 ('a *Latocestus*'); Bock, 1913, 64 (*Latocestus*).
marina Kato, 1938*a*, 559, 567, figs (Japan); 1944, 293.
meridionalis Prudhoe, 1982*a*, 223; 1982*b*, 362, 378, figs (South Australia).
microps (Verrill, 1903) Hyman, 1939*d*, 14, figs (redescr., Bermuda & Sargasso Sea).
nexa Sopott-Ehlers & Schmidt, 1975, 213, fig. (Galapagos Is).
polypora Meyer, 1922, 138, 149, figs (Red Sea); Bock, 1925*a*, 73.
raffaelei Ranzi, 1928, 3, figs (Naples, Italy).
rubrocincta (Grube, 1840) Lang, 1884, 49, 516, figs (redescr.; Italy); Carus, 1885, 153 (some descr.); Hallez, 1893, 167 (Pas-de-Calais); 1894, 218 (NW. France); Gamble, 1896, 18, fig.; Laidlaw, 1906, 712 (Cape Verde Is); Bock, 1913, 250; Stummer-Traunfels, 1933, 3573, fig. (some descr.); Kato, 1937*a*, 225, figs (descr.; synonymy; Japan); 1938*b*, 586, fig. (Japan); 1944, 293; Utinomi, 1956, 29 (coloured fig., Japan); Prudhoe, 1982*c*, 53, fig. (English Channel).
salar Marcus, 1949, 7, 79, 104, figs (Brazil); du Bois-Reymond Marcus, 1955*b*, 283.
techa du Bois-Reymond Marcus, 1957, 174, figs (Brazil).
species Child, 1905*b*, 48 (form regulation); 1905*b*, 261 (regeneration); 1905*d*, 157 (form regulation); 1907, 357 (functional regulation in intestine).
species Jacubowa, 1906, 149 (Lifu, I., New Caledonia).
species Yeri & Kaburaki, 1918*b*, 1 (Japan).
Cestoplanid juvenile Hyman, 1959*a*, 565 (Caroline Is, Pacific Ocean).
CESTOPLANIDAE Lang, 1884, 430, 516; Hallez, 1894, 216 (defined).
CESTOPLANIDES Poche, 1926, 100 (superfam. of Acotylea = Emprosthommata Bock, 1913); Prudhoe, 1982*c*, 52 (emend. to Cestoplanoidea).
Cestoplanoids Laidlaw, 1902, 305 (Malaysia).
CHROMOPLANA Bock, 1922, 1, 29 (type: *bella*).
bella Bock, 1922, 1 figs (Japan); Kato, 1938*b*, 588, fig. (Japan); 1944, 306; Utinomi, 1956, 30 (coloured fig.; Japan).
CHROMOPLANIDAE Bock, 1922, 20, 29; Bresslau, 1933, 293; Stummer-Traunfels, 1933, 3576; de Beauchamp, 1961, 67 (genera).
CHROMYELLA Corrêa, 1958, 81 (type: *saga*).
saga Corrêa, 1958, 81, figs (Brazil).
CIRROPOSTHIA Steinbock, 1937, 5 (type: *steueri*); Marcus, 1947, 133, 166 (s.o. *Neoplanocera*).
steueri Steinböck, 1937, 5, figs (Alexandria, Egypt); Marcus, 1947, 133 (*Neoplanocera*).
COMPROSTATUM Hyman, 1944*a*, 7 (type: *insularis*); 1953*a*, 297, 300 (s.o. *Phaenocelis*).
insularis Hyman, 1944*a*, 7, figs (Florida); 1952; 195 (s.o. *P. purpurea*); 1953*a*, 299 (s.o. *Phaenocelis purpurea*).
veneris Hyman, 1944*a*, fig. 16 [err. pro *insularis*].
CONJUGUTERUS Pearse, 1938, 80 (type: *parvus*); Hyman, 1939*c*, 5 (s.o. *Euplana*); 1940*a*, 470 (s.o. *Euplana*).
parvus Pearse, 1938, 81, fig. (Florida, N. Carolina & Prince Edward I.); Hyman, 1939*c*, 5 (s.o. *Euplana gracilis*); 1940*a*, 470 (s.o. *Euplana gracilis*).
CONOCEROS Lang, 1884, ix, 428, 446 (type: *conoceraeus*); Gamble, 1896, 19; Benham, 1901, 31; Meixner, 1907*b*, 398 (= *Stylochoplana*).
conoceraeus (Schmarda, 1859) Lang, 1884, 446 (Ceylon); Meixner, 1907*b*, 398, 403, fig. (to *Stylochoplana*); Stummer-Traunfels, 1933, 3487, 3563 (type redescr.; to *Stylochoplana*).
COPIDOPLANA Bock, 1913, 55, 187, 214, 218 (type: *paradoxa*); Hyman, 1953*a*, 339 (defined).
paradoxa Bock, 1912, 214, figs (Gulf of Siam).
tripyla Hyman, 1953*a*, 339, figs (California); Stasek, 1966, 7 (location of type-specimen).
virgae Sopott-Ehlers & Schmidt, 1975, 202, figs (Galapagos Is).
CORONADENA Hyman, 1939*e*, 153; 1939*f*, 238; 1940*a*, 450 (type: *mutabilis*).
mutabilis (Verrill, 1873) Hyman, 1939*e*, 155; 1939*f*, 238, figs; 1940*a*, 450, figs (synonymy; Florida to Massachusetts); 1940*b*, 15 (descr.); 1951, 116, figs. (New England coast); Pearse & Williams, 1951, 137 (N. Carolina); Lawler, 1969, 65 (associating with oysters and barnacles in Chesapeake Bay, Virginia).

COTYLEA Lang, 1884, xi, 430, 521 ('Tribus'); Carus, 1885, 153 ('Tribus'); Verrill, 1893, 494 (Tribe of suborder Digonopora); Gamble, 1893*a*, 522 (key to genera found in Britain); Hallez, 1893, 168; 1894, 200 (key to families), 218 (definition); Benham, 1901, 31; Laidlaw, 1902, 306, 307 (relationships within, and differentiation from, Acotylea); 1903*e*, 12 (key to families); Heath & McGregor, 1913, 474; Bock, 1913, 251 (defined); Poche, 1926, 100 (Tribus of Planoceridea); Bresslau, 1928–33, 289; Freeman, 1933, 136; Pearse, 1936, 1ff. (Cotylina); Hyman, 1939*b*, 148; 1940*a*, 484; 1940*b*, 19; 1951, 170; 1953*a*, 362 (defined); 1959*c*, 14 (defined); Marcus, 1952, 101, 117 (juveniles, Brazil); de Beauchamp, 1961, 64 ('Cotyles', classification); Marcus & Marcus, 1966, 333 (synopsis); Prudhoe, 1982*c*, 54 (defined).

COTYLINA Pearse, 1936, 1ff. (pro Cotylea).

COTYLOCERCA Ritter-Záhony, 1907, 3 (type: *michaelseni*); Hallez, 1913, 40, 41; Bock, 1913, 40 (s.o. *Stylochoides*).

michaelseni Ritter-Záhony, 1907, 3, figs (Falkland Is & Tierra del Fuego); Hallez, 1913, 40 (s.o. *Stylochoides albus*); Bock, 1913; 276 (s.o. *Stylochoides albus*).

CRASPEDOMMATA Bock, 1913, 46, 56, 59 (section of Acotylea); Poche, 1926, 98 (= superfam. Stylochides); Bresslau, 1928–33, 284; Freeman, 1933, 112; Hyman, 1939*b*, 129; 1940*a*, 450; 1940*b*, 15; 1951, 166; 1953*a*, 275 (defined); 1959*c*, 1 (defined); Marcus & Marcus, 1966, 321 (superfam. Craspedommatidea; key to families).

CRASPEDOMMATIDEA Marcus & Marcus, 1966, 321 (superfam. of Acotylea).

CRASSANDROS Hyman, 1955*b*, 132 (type: *dominicanus*).

dominicanus Hyman, 1955*b*, 132, figs (Dominica, Caribbean).

CRASSIPLANA Hyman, 1955*d*, 14 (type: *albatrossi*).

albatrossi Hyman, 1955*d*, 14, figs (Uruguay).

CRYPTOCELIDAE Laidlaw, 1903*c*, 12; Bock, 1913, 149 (defined); Poche, 1926, 99 (emended to Cryptocelididae); Bresslau, 1928–33, 288 (defined); Stummer-Traunfels, 1933, 3567, 3586; Hyman, 1953*a*, 294 (defined); 301 (constitution); Marcus, 1947, 106 (key to genera); 1952, 81 (key to genera); de Beauchamp, 1961, 58 (generic constitution); Marcus & Marcus, 1966, 326 (key to genera).

CRYPTOCELIDES Bergendal, 1890, 327 (type: *loveni*); 1893*a*, 237; Bock, 1913, 99, 108 (diagnosis); 1923*a*, 14; Palombi, 1924*b*, 7 (redefined).

loveni Bergendal, 1890, 327 (Sweden; not descr.); 1893*a*, 237 (descr.); 1893*d*, 1 (redescr.); Laidlaw, 1904*a*, 5 (Victoria, Australia & Firth of Clyde, Scotland; doubts validity of former locality); 1906, 706; Bock, 1913, 100, figs (redescr.; Iceland, Denmark, Sweden & Norway); Bresslau, 1928–33, 61, 83, figs; Steinböck, 1938, 17 (Iceland); Westblad, 1955, 5 (W. Norway); de Beauchamp, 1951*b*, 244 (7°33'–70°45'W, 33°33'–34°1'N, off Moroccan coast); Hendelberg, 1974*b*, 9, figs (Sweden); Prudhoe, 1982*b*, 361; 1982*c*, fig. (Scotland).

samoensis Palombi, 1924*a*, 33 (14°32'15"S, 167°43'W, off Samoa, Pacific Ocean); 1924*b*, 4, figs.

'n.sp.' Bock, 1927*a*, 75 (Samoa).

CRYPTOCELIDIDAE Bergendal, 1893*a*, 239; 1893*b*, 367 (changed to Polypostiadae); 1893*c*, 5 (type: *Cryptocelides* Berg.); Bock, 1913, 85 (s.o. Polyposthiidae).

CRYPTOCELIDIDAE Poche, 1926, 99 (= Cryptocelidae of Laidlaw, 1903*e*).

CRYPTOCELIS Lang, 1884, viii, 101, 429, 471 (type: *alba*); Carus, 1885, 151; Hallez, 1893, 149; Gamble, 1896, 19; Bock, 1913, 150 (defined); 1923*a*, 27 (comparison of spp.); Hyman, 1953*a*, 294 (defined).

alba (Lang, in Schmidtlein, 1880) Lang, 1884, 101, 471, figs (descr.; Mediterranean), 473 (*'lactea'*); Carus, 1885, 151 (some descr.); Lo Bianco, 1888, 398 (Naples); 1899*a*, 477 (Naples); Bock, 1913, 150; Clark & Milne, 1955, 175 (Firth of Clyde, Scotland); Prudhoe, 1982*c*, 34, figs.

amakusaensis Kato, 1936*a*, 17, figs (Japan); 1938*a*, 564 (Japan); 1944, 268.

arenicola Hallez, 1893, 150, 197, figs (Pas-de-Calais; nom. nov. pro *equiheni*); 1894, 204, figs (NW. France): Bock, 1913, 151 (footnote).

compacta Lang, 1884, 249, 474, figs (Italy); Carus, 1885, 151 (some descr.): Jacubowa, 1909, 18 (some descr.; Black Sea); Bock, 1913, 150; Stummer-Traunfels, 1933, 3571, fig.; Levetzow, 1939, 780 (Naples; regeneration).

equiheni Hallez, 1888, 104 (NW. France); 1893, 150 (renamed *arenicola*); Bock, 1913, 150, 151 (correct name); Prudhoe, 1982*c*, 36, figs.

glandulata Jacubowa, 1909, 2, 14, figs (Black Sea); Bock, 1913, 150; du Bois-Reymond Marcus & Marcus, 1968, 15 (note).

ijimai Bock, 1923*a*, 15, figs (Japan); Kato, 1944, 268.
insularis Hyman, 1953*b*, 186, figs (Galapagos Is).
lactea Lang, 1884, 473 [err. pro *alba*].
lilianae du Bois-Reymond Marcus & Marcus, 1968, 2, 3, 13, figs (Brazil).
littoralis Kato, 1937*f*, 347, 355, figs (Japan); 1944, 268.
occidentalis Hyman, 1953*a*, 294, figs (California).
orientalis Kato, 1939*b*, 141, 142, figs (Japan); 1944, 268.
(?) species Jacubowa, 1906, 139, figs (Isle of Pines, New Caledonia); Bock, 1913, 151; 1923*a*, 35, 36 (possibly a juvenile *Latocestus*).
species Jacubowa, 1906, 149 (New Britain); Bock, 1913, 151.
species Haswell, 1907*a*, 644 (N.S.W., Australia).
species Bock, 1913, 151 (east coast of N. America).

CRYPTOCOELA Örsted, 1844*a*, 44 (family; constitution); Lang, 1884, 433 (= Polycladidea).

CRYPTOCOELIDES [pro *Cryptocelides*] Benham, 1901, 31.

CRYPTOCOELIS [pro *Cryptocelis*] Benham, 1901, 31.

CRYPTOCOELUM Stimpson, 1857, 21, 26 (type: *opacum*); Diesing, 1862, 521 (s.o. *Typhlolepta*).
opacum Stimpson, 1857, 21, 26 (in echinoderm; Hong Kong); Diesing, 1862, 522 (renamed *Typhlolepta stimpsoni*); Lang, 1884, 612.

CRYPTOPHALLINAE Bresslau, 1928–33, 287 (subfam. of Stylochidae).

CRYPTOPHALLUS Bock, 1913, 49, 120, 124 (type, *wahlbergi*); 1925*b*, 172; Freeman, 1933, 113.
aegypticus Melouk, 1940, 125, figs (Red Sea).
bartschi Kaburaki, 1923*b*, 635, 636, figs (Philippines).
eximius Kato, 1937*f*, 347, 353, figs (Japan); 1944, 267.
japonicus Kato, 1944, 267, figs (nom. nov. pro *C. sondaicus* of Kato, 1938*b*, *nec* Bock, 1925*b*).
magnus Freeman, 1933, 111, 113, figs (Washington State, U.S.A.); Kato, 1937*f*, 355 (s.o. *Kaburakia excelsa*); Ricketts & Calvin, 1948, 175, 300 (California); 1952, 229 (Alaska); MacGinitie & MacGinitie, 1949, 152 (California); Hyman, 1953*a*, 285, 287 (*C. magnus* of MacGinitie & MacGinitie s.o. *Stylochus californicus*), 290 (*C. magnus* of Freeman s.o. *Kaburakia excelsa*).
sondaicus Bock, 1925*b*, 120, 177, figs (Amboina, Indonesia); Kato, 1938*b*, 580, figs (redescr.; Japan); 1944, 267 (Japanese specimens renamed *C. japonicus*).
wahlbergi Bock, 1913, 120, figs (Durban, S. Africa); 1925*b*, 177; Stummer-Traunfels, 1933, 3571, fig. (some descr.).

CRYSTOPHALLUS [err. pro *Cryptophallus*] Bock, 1925*b*, 122.

CYCLOPORUS Lang, 1884, vii, 431, 568 (type: *papillosus* (Sars, in Jensen).); Carus, 1885, 156; Hallez, 1893, 171; 1894, 231 (defined); Gamble, 1893*a*, 505; 1896, 19; Benham, 1901, 31; Yeri & Kaburaki, 1918*a*, 440 (features); 1918*b*, 49 (features); du B.-R. Marcus & Marcus, 1968, 77 (key to spp.); Prudhoe, 1982*c*, 62 (defined).
australis Prudhoe, 1982*a*, 222, 223; 1982*b*, 362, 381, fig. (South Australia).
gabriellae Marcus, 1950, 89, 115, figs (Brazil); 1952, 96, 116, figs. (Brazil); du B.-R. Marcus, 1955*b*, 286; du B.-R. Marcus & Marcus, 1968, 77 (Caribbean).
japonicus Kato, 1944, 305 (= *C. papillosus* of Yeri & Kaburaki, 1918*b*; of Kato, 1937*a*, 1938*a* & *b*, 1939*b*); du B.-R. Marcus & Marcus, 1968, 77 (s.o. *misakiensis*).
maculatus Hallez, 1893, 171, 197, figs (NW. France); 1894, 222, figs (NW. France); Hallez, 1905*a*, xlix (Boulogne; some descr.); Prudhoe, 1982*c*, 64, fig. (English Channel).
misakiensis (Kato, 1938*a*) du Bois-Reymond Marcus & Marcus, 1968, 77 (*papillosus* (Sars) var. *misakiensis* raised to specific status).
papillosus (Sars, *in* Jensen, 1878) Bock, 1913, 262, fig. (synonymy; Sweden & Norway); Yeri & Kaburaki, 1918*b*, 40, 49, fig. (Japan); Bresslau, 1928–33, 90, fig.; Steinböck, 1933, 19 (Gulf of Trieste); Stummer-Traunfels, 1933, 3574, 3579, figs; Kato, 1937*a*, 229, figs (Japan); 1939*b*, 149, fig. (Japan); Hyman, 1951, 169, figs; Dawydoff, 1952, 81 (Vietnam); Laverack & Blackler, 1974, 32 (St Andrews Bay, Scotland); Hendelberg, 1974*b*, 15, figs (Sweden); Hill, 1974, 263 (Channel Is, U.K.); Prudhoe, 1982*c*, 62, fig. (U.K.).
papillosus Lang, 1884, 568, figs (nom. nov. pro *Proceros tuberculatus* Lang, *in* Schmidtlein); Carus, 1885, 156 (some descr.); Lo Bianco, 1888, 398 (Naples); Gamble, 1893*a*, 506 (descr.; Britain); 1893*b*, 46 (south Devon); 1893*c*, 168, figs (Liverpool Bay); 1896, 19 (Britain); Pruvot, 1897, 21 (NW. France); Francotte, 1897, 30, figs (embryology); 1898, 250, figs (= *Planaria schlosseri*; Roscoff; development of egg); Lo Bianco, 1899*a*, 477 (Naples); Gamble, 1900, 812 (Ireland);

Benham, 1901, 32, fig.; Bock, 1913, 263 (s.o., *papillosus* (Sars)); Southern, in Farran, 1915, 36 (Ireland); Yeri & Kaburaki, 1918*a*, 440 (diagn., Japan); Southern, 1936, 71 (Ireland); Eales, 1939, 55, fig.; Kato, 1944, 305 (*papillosus* of Yeri & Kaburaki, 1918*b*, s.o. *japonicus*).

papillosus Lang var. *laevigatus* Lang, 1884, 570 (Mediterranean); Gamble, 1893*b*, 46, fig. (south Devon); Jameson, 1897, 163 (NW. England); Hallez, 1905, xlix (Boulogne); Laidlaw, 1906, 713 (Cape Verde Is); Southern, in Farran, 1915, 36 (Ireland); Southern, 1936, 71 (Ireland).

papillosus (Sars) var. *misakiensis* Kato, 1938*a*, 571 (Japan); 1938*b*, 588 (Japan); 1939*b*, 149, fig. (Japan); 1944, 305 (to *japonicus*); du B.-R. Marcus & Marcus, 1968, 77 (*misakiensis* raised to specific status).

tuberculatus (Lang, *in* Schmidtlein, 1880) Vaillant, 1889, 656 (NW France); Bock, 1913, 263 (s.o. *papillosus* (Sars)).

variegatus Kato, 1934*b*, 123, 133, figs (Japan); 1937*a*, 229 (Japan); 1944, 304; Dawydoff, 1952, 81 (Vietnam).

DENDROCOELA Ehrenberg, 1831, 54 (order of Turbellaria); Örsted, 1844*a*, 26; Diesing, 1850, 180 (suborder); Leuckart, 1856, 109 (family); Stimpson, 1857, 19 (subtribe); Schmarda, 1859, 13 (order); Lang, 1884, 433 (*ex parte* = Polycladidea); Vaillant, 1889, 645, 646, 647 (subord. of Planariaea).

DENDROCOELA DIGONOPORA Diesing, 1862, 492, 521; Lang, 1884, 433 (= Polycladidea).

DICELIS Schmarda, 1859, 13, 15 (type: *megalops*) [*nec Dicelis* Dujardin, 1845 (Nematode); *nec* Stimpson, 1857 (Nemertine)]; Diesing, 1862, 523 (renamed *Diopis*).

megalops Schmarda, 1859, 15, figs (Jamaica); Diesing, 1862, 523 (to *Diopis*); Lang, 1884, 510; Stummer-Traunfels, 1933, 3486, 3488, figs (to *Stylochus*; type-specimen re-examined).

DICTEROS Jacubowa, 1906, 146 (type: *pacificus*); Bock, 1913, 39 ('*Dicterus*').

gamblei (Laidlaw, 1902) Marcus, 1950, 85, 115 (*Dicteros*).

pacificus Jacubowa, 1906, 146, figs (New Caledonia).

DIGONOPORA Stimpson, 1857, 19 (subtribe of Turbellaria Dendrocoela); Diesing, 1862, 492, 521 (section of subord. Turbellaria Dendrocoela); Vaillant, 1889, 645, 651 (tribe of Dendocoela), 653 (key to families); Verrill, 1893, 460, 461 (= Polycladidea).

DIGYNOPORA Hyman, 1939*e*, 153; 1940*a*, 473 (type: *americana*); Marcus, 1949, 76 (Leptoplaninae).

americana Hyman, 1939*e*, 153; 1940*a*, 473, figs (nom. nov. pro *Leptoplana angusta* of Pearse & Littler, 1938, pl. 10, fig. 4; Florida).

DIONCUS Stimpson, 1855*b*, 389 (type: *badius*); Diesing, 1862, 528 (s.o. *Leptoplana*); Haswell, 1907*a*, 471; Bock, 1913, 205 (s.o. *Notoplana*).

badius Stimpson, 1855*b*, 389 (N.S.W., Australia), 1857, 22, 27 (descr.); Diesing, 1862, 528 (to *Leptoplana*); Lang, 1884, 511; Whitelegge, 1890, 206 (N.S.W.); Haswell, 1907*a*, 471 (? = *Leptoplana australis* Laidlaw); Bock, 1913, 205 (? = s.o. *Notoplana australis* (Laidlaw)).

oblongus Stimpson, 1855*b*, 389 (N.S.W., Australia); 1857, 22, 28 (descr.); Diesing, 1862, 528 (renamed *Leptoplana stimpsoni*); Lang, 1884, 511; Whitelegge, 1890, 206 (N.S.W.); Haswell, 1907*a*, 471(? = *Leptoplana australis* Laidlaw); Bock, 1913, 205 (? = s.o. *Notoplana australis* (Laidlaw).)

DIOPSIS Diesing, 1862, 492, 523 (type: *megalops*) (nom. nov. pro *Dicelis* Schmarda).

borealis Diesing, 1862, 524 (Arctic Ocean).

megalops (Schmarda, 1859) Diesing, 1862, 523; Lang, 1884, 510 (*Dicelis*).

DIPLANARIA Darwin, 1844, 249 (type; *notabilis*); Diesing, 1850, 202; Schmarda, 1859, 13; Diesing, 1862, 542 (s.o. *Leptoplana*); Lang, 1884, 475 ('Diplanariae' s.o. *Leptoplana*).

notabilis Darwin, 1844, 249, fig. (Chonos Archipelago, Chile); Diesing, 1850, 202; 1862 542 (to *Leptoplana*).

DIPLANDROS Hyman, 1953*a*, 341 (type: *singularis*).

singularis Hyman, 1953*a*, 341 (California); Stasek, 1966, 7 (location of type-specimen).

DIPLONCHUS Stimpson, 1857, 22 (type: *marmoratus*); Diesing, 1862, 492, 545; Lang, 1884, 429, 462 (diagn.); Gamble, 1896, 19.

marmoratus Stimpson, 1857, 22, 29 (Oho-sima, Japan); Diesing, 1862, 545; Lang, 1884, 463.

DIPLOPHARYNGEATA Plehn, 1986*a*, 140, 167 (type: *filiformis*).

filiformis Plehn, 1896*a*, 140, 167, figs (off north coast of Sumatra).

DIPLOPHARYNGEATIDAE Plehn, 1896*a*, 167, 169; Benham, 1901, 31; Bresslau, 1928–33, 293; Stummer-Traunfels, 1933, 3576, 3586; de Beauchamp, 1961, 67 (genera).

DIPLOSOLENIA Haswell, 1907*b*, 466, 469, 470 (type: *johnstoni*); Bock, 1913, 248; Stummer-Traunfels, 1933, 3561 (s.o. *Callioplana*).

johnstoni Haswell, 1907*b*, 469, figs (Australia); Bock, 1913, 248; Pope, 1943, 246 (N.S.W.); Kato, 1944, 289 (s.o. *Callioplana marginata*); Dakin *et al.*, 1948, 208 (N.S.W.); Dakin, 1953, 144, figs (east coast of Australia).

DIPLOSOLENIDAE Bock, 1913, 248; Yeri & Kaburaki, 1918*b*, 25 (redefined); Poche, 1926, 100 (emend. to Diplosoleniidae); Bresslau, 1928–33, 289; Stummer-Traunfels, 1933, 3576, 3585; Freeman, 1933, 133. Hyman, 1935*a*, 357 (renamed Callioplanidae).

DIPOSTHIDAE Woodworth, 1898, 65; Laidlaw, 1903*c*, 307 (relationship); 1903*e*, 13 ('Diposthiidae'); Bock, 1913, 19; Bresslau, 1928–33, 290; Stummer-Traunfels, 1933, 3576; Hyman, 1959*c*, 15 (defined).

DIPOSTHUS (Woodworth, 1898, 64 (type: *corallicola*); Bock, 1922, 19 (relationship); Hyman, 1959*c*, 15 (defined).

corallicola Woodworth, 1898, 64, figs (Hope I., Great Barrier Reef); Bock, 1923*b*, 25 (Note).

popeae Hyman, 1959*c*, 15, figs. (N.S.W., Australia).

DISCOCELIDAE Laidlaw, 1903*e*, 12; Bock, 1913; 59; Poche, 1926, 98 (emend. to Discocelididae); Bresslau, 1928–33, 284; Stummer-Traunfels, 1933, 3567, 3586; Hyman, 1940*a*, 450; 1940*b*, 15; Marcus, 1950, 75 (systematic note; key to genera); Hyman, 1955*b*, 115 (defined); 1955*c*, 66 (defined); 1959*c*, 1 (defined); de Beauchamp, 1961, 56 (constitution).

DISCOCELIDES Bergendal, 1893*a*, 240, 241 (type: *langi*); 1893*b*, 1; 1893*a*, 240; Benham, 1901, 31 ('Discocoelides'); Bock, 1913, 73, 85 (defined); Hyman, 1940*a*, 454 (defined); 1953*a*, 279 (defined).

ellipsoides (Girard, 1853) Hyman, 1939*e*, 153; 1940*a*, 454, fig. (descr.; synonymy; New England & Canada); 1940*b*, 16 (descr.); 1952, 195 (to *Plehnia*); 1953*a*, 279 (*Plehnia*); Linkletter *et al.*, 1977, 11 (New Brunswick).

langi Bergendal, 1893*a*, 241; 1893*b*, 367; 1893*c*, 1, 28; Bock, 1913, 73, figs (descr.; Scandinavia & Shetland Is); Remane, 1929, 78 (Kiel Bay), Steinböck, 1932, 333, 337 (distribution); Westblad, 1955, 5 (W. Norway); Pedersen, 1966, 94 (Denmark); Hendelberg, 1974*b*, 10, figs (Sweden); Prudhoe, 1982*c*, 32, fig.

DISCOCELIDIDAE (Discocelidae emended) Poche, 1926, 98.

DISCOCELIS Ehrenberg, 1836, 67 (type: *lichenoides*); Diesing, 1862, 543 (? s.o. *Centrostomum*); Lang, 1884, 429, 466 (diagn.); Carus, 1885, 151; Verrill, 1893, 492; Gamble, 1896, 19; Bock, 1913; 60 (syn. *Thalamoplana*) (defined); Yeri & Kaburaki, 1918*a*, 433 (features); 1918*b*, 46 (features); Hyman, 1955*c*, 66 (defined); 1959*c*, 1 (defined).

australis Hyman, 1959*c*, 1, figs (N.S.W., Australia); Prudhoe, 1982*a*, 223, 224 (South Australia); 1982*b*, 362 (some descr.).

binoculata Verrill, 1903, 43, figs (Bermuda); Bock, 1913, 61 (? *Discocelis*); Hyman, 1939*d*, 8 (to *Notoplana*).

cyclops Verrill, 1903, 44, fig. (Bermuda); Bock, 1913, 61 (? *Discocelis*); Hyman, 1939*d*, 19 (to *Prosthiostomum*).

fulva (Kato, 1944, 260 (figs (Japan).

grisea Pearse, 1938, 67, fig. (Florida); Pearse & Littler, 1938, 235, fig. (descr.; North Carolina); Hyman, 1940*a*, 450. (s.o. *Coronadena mutabilis*).

herdmani (Laidlaw, 1904*b*) Bock, 1913, 61.

insularis Hyman, 1955*c*, 66, figs (Tuamoto Is).

japonica Yeri & Kaburaki, 1918*a*, 431, 433 (diagn.); 1918*b*, 2, 3, 46, figs (descr.; Japan); Kato, 1937*a*, 212, figs (descr.; Japan); 1938*a*, 561 (Japan); 1938*b*, 578 (Japan); 1944, 259; Dawydoff, 1952, 81 (Vietnam).

lactea (Stimpson, 1857) Lang, 1884, 470; Bock, 1913, 61 (? *Discocelis*).

lichenoides (Mertens, 1832) Ehrenberg, 1836, 67; Örsted, 1844*a*, 49 (to *Leptoplana*); Diesing, 1850, 199 (to *Centrostomum*); Lang, 1884, 469; Gamble, 1896, 23 (fig.); Bock, 1913, 61.

mutabilis (Verrill, 1873); Verrill, 1893, 493, figs (descr.; Connecticut); Sumner *et al.*, 1913, 580 (Massachusetts); Bock, 1913, 61 (? *Discocelis*) Pearse & Walker, 1939, 16, fig. (North Carolina; '*mutabalis*'); Hyman, 1940*a*, 450 (to *Coronadena*).

pusilla Kato, 1938*a*, 559, 560, figs (Japan); 1944, 260; Kikuchi, 1968, 167 (Japan).

tigrina (Blanchard, 1847) Lang, 1884, 71, 467, figs (Italy); Carus, 1885, 151 (descr.); Lo Bianco, 1888, 599 (Naples); 1899*a*, 477 (Naples); Pruvot, 1897, 20 (Gulf of Lion, Mediterranean); Micoletzky, 1910, 179 (Trieste); Bock, 1913, 61; Palombi, 1939*b*, 96, fig. (Rio de Oro, NW. Africa).

species Plehn, 1896*b*, 9 (Ascension I.; no description).

DISCOPLANA Bock, 1913, 219 (type: *gigas*); Hyman, 1939*b*, 136 (s.o. *Euplana*); 1939*c*, 5 (s.o. *Euplana*); 1940*a*, 470 (s.o. *Euplana*).

species Stummer-Traunfels, 1933, 3566, fig. (Indonesia).

concolor (Meixner, 1907*a*) Bock, 1913, 221 (diagn.); Hyman, 1954*b*, 33 (to *Euplana*).

gigas Schmarda, 1859) Stummer-Traunfels, 1933, 3486, 3492, figs (type-specimen, redescribed; syn. *D. subviridis*), 3565, fig. (Amboina [Ambon], Indonesia); Kato, 1944, 277.

inquieta (Heath & McGregor, 1913) Stummer-Traunfels, 1933, 3504; Hyman, 1953*a*, 333 (to *Notoplana*).

longipenis Kato 1943*c*, 79, 83, figs (Palau Is); 1944, 278.

malayana (Laidlaw, 1903*c*) Bock, 1913, 221 (diagn.).

pacificola (Plehn, 1896*a*) Bock, 1913, 220, figs (diagn.); Hyman, 1953*a*, 332 (to *Euplana*).

subviridis (Plehn, 1896*c*) Bock, 1913, 220, figs (diagn.; Indonesia, Maldive Is & Somalia); 1925*a*, 2 (Bonin & Gilbert Is); 1925*b*, 120 (Indonesia); Stummer-Traunfels, 1933, 3487, 3494 (s.o. *gigas*).

takewakii Kato, 1935*b*, 149, figs (in genital bursa of ophiurid, *Ophioplocus japonicus*, Japan); 1938*b*, 582 (Japan); 1944, 278; Hyman, 1953*a*, 333 (to *Euplana*).

DISCOSOLENIA Freeman, 1933, 111, 112, 133, (type; *washingtonensis*); Hyman, 1953*a*, 358 (s.o. *Pseudostylochus*). [See p. 114]

washingtonensis Freeman, 1933, 111, 112, 133, figs (Washington State); Hyman, 1953*a*, 358 (s.o. *Pseudostylochus burchami*).

DISCOSTYLOCHUS Bock, 1925*a*, 31, 47 (type; *parcus*); 1925*b*, 172.

parcus Bock, 1925*a*, 31, figs (Hawaii Is); 1925*b*, 177; Bresslau, 1930, 147, fig.

yatsui Kato, 1937*f*, 347, 351, figs (Japan); 1944, 267.

DISPAROPLANA Laidlaw, 1903*d*, 103, 105 (type: *dubia*); 1903*e*, 14; Bock, 1913, 239 (diagn.).

dubia Laidlaw, 1903*d*, 103, figs (East Africa); Bock, 1913, 239; Steinböck, 1937, 9, 10, fig.

DITREMAGENIA Palombi, 1928, 614 (type: *macropharynx* – Tricladida Maricola); Bresslau, 1928–33, 290 (Polycladida, Cotylea). [See p. 122]

macropharynx Palombi, 1928, 614, figs (Suez Canal).

DITREMAGENIIDAE Bresslau, 1928–33, 290.

DORIS Linnaeus, 1758 (now a mollsucan genus).

electrina Pennant, 1777, 36, fig. (Anglesea, Wales); Lang, 1884, 572 (? s.o. *Eurylepta cornuta*).

ECHINOPLANA Haswell, 1907*b*, 466, 475, 477 (type: *celerrima*); Bock, 1913, 247 (diagn.); Prudhoe, 1982*b*, 377 (diagn.).

celerrima Haswell, 1907*b*, 475, figs (N.S.W., Australia); Bock, 1913, 247; Galleni, 1978*a*, 214 (Tuscan coast, Italy); 1978*b*, 139 (Tuscan coast, Italy; redescribed); Prudhoe, 1982*a*, 223, 224 (South Australia); 1982*b*, 377 (redescr.; S. Australia).

ELASMODES Le Conte, 1851, 319 (type: *discus*); Stimpson, 1857, 21; Schmarda, 1859, 13 (s.o. *Leptoplana*); Schmidt, 1861, 5 (s.o. *Leptoplana*); Diesing, 1862, 527 (s.o. *Leptoplana*).

acutus (Stimpson, 1855*a*) Stimpson, 1857, 21, 26 (descr.).

discus Le Conte, 1851, 319 (Panama Isthmus); Stimpson, 1857, 21; Diesing, 1862, 527 (to *Leptoplana*); Lang, 1884, 611.

flexilis (Dalyell, 1814) Stimpson, 1857, 21; Lang, 1884, 477 (s.o. *Leptoplana tremellaris*).

gracilis (Girard, 1850*b*) Stimpson, 1857, 21; Diesing, 1862, 541 (to *Leptoplana*); Lang, 1884, 610 (*Prosthiostomum*).

modestus (Quatrefages, 1845) Stimpson, 1857, 21; Lang, 1884, 490 (s.o. *Leptoplana pallida*); Palombi, 1928, 589 (s.o. *Stylochoplana pallida*).

obtusus Collingwood, 1876, 88, 93, fig. (Singapore); Lang, 1884, 512; Laidlaw, 1903*a*, 2 (probably a leptoplanid); 1903*c*, 302 (to *Leptoplana*).

pallidus (Quatrefages, 1845) Stimpson, 1857, 21; Lang, 1884, 490 (*Leptoplana*); Palombi, 1928, 589 (s.o. *Stylochoplana pallida*).

tenellus Stimpson, 1857, 21, 26 (Oshima, Japan); Diesing, 1862, 528 (to *Leptoplana*).

tigrinus (Blanchard, 1847) Stimpson, 1857, 21; Lang, 1884, 467 (to *Discocelis*).

EMPROSTHOMMATA Bock, 1913, 57, 249 (section of Acotylea); Poche, 1926, 99 (= superfamily Planocerides); Bresslau, 1928–33, 289; Hyman, 1951, 170; Marcus & Marcus, 1966, 321 (superfamily Emprosthommatidea; key to families); de Beauchamp, 1961, 63 (superfamily).

EMPROSTHOPHARYNGIDAE Bock, 1913, 161; Bresslau, 1928–33, 288; Stummer-Traunfels, 1933, 3567, 3586; de Beauchamp, 1961, 59.

EMPROSTHOPHARYNX Bock, 1913, 51, 161, 166 (type: *opisthoporus*).

opisthoporus Bock, 1913, 161, figs (Galapagos Is, Pacific Ocean); 1925*a*, 61, figs (descr.; in houses of pagurid, *Petrochinus californiensis*, Panama).

rasae Prudhoe, 1968, 408 (associating with hermit-crab, Hawaii).

vanhoeffeni Bock, 1931, 261, 262, 268, figs (Cape Verde Is & Morocco); de Beauchamp, 1951*b*, 245 (Mauritania).

'n.sp.' Bock, 1925*a*, 61, 67, 77 (Gilbert Is, Pacific Ocean); 1926*a*, 134; 1927*a*, 7, 104.

ENANTIA Graff, 1890, 1 (type: *spinifera*); Benham, 1901, 31; Laidlaw, 1902, 307; 1903*e*, 15 (affinity with *Haploplana*).

pellucida (Pearse, 1938) Hyman, 1939*e*, 153; 1940*a*, 482, figs. (descr.).

spinifera Graff, 1890, 1, figs (Trieste); Benham, 1901, 32, fig.; Micoletzky, 1910, 180 (Trieste); Hyman, 1951, 73 (figs of spines).

ENANTIADAE Graff, 1890, 14; Hallez, 1894, 217 (defined); Gamble, 1896, 19 (emended to Enantiidae).

ENANTIIDAE Gamble, 1896, 19; Benham, 1901, 31; Bresslau, 1928–33, 289; Stummer-Traunfels, 1933, 3567, 3586; Hyman, 1940*a*, 481.

ENCHIRIDIUM Bock, 1913, 48, 287 (type: *periommatum*); Hyman, 1953*a*, 386 (defined).

species Correa, 1958, 81 (Brazil).

evelinae Marcus, 1949, 7, 91, 105, figs (Brazil); 1952, 100, 117; du Bois-Reymond Marcus, 1955*b*, 287; du B.-R. Marcus & Marcus, 1968, 92, figs (Curaçao).

japonicum Kato, 1943*b*, 69, 75, figs (Taiwan); 1944, 313; Poulter, 1975, 328, 337, figs (Oahu, Hawaii Is).

periommatum Bock, 1913, 287, figs (Thatch I., Caribbean); Hyman, 1955*b*, 146, figs (Gulf of Mexico); 1955*e*, 266 (Jamaica).

punctatum Hyman, 1953*a*, 386, figs (Californian coast & Gulf of California); Morris *et al.*, 1981, 81, fig. (coloured) (Calif.).

ENDOCELIS Schmankewitsch, 1873, 275 (type: *ovata*); Jacubowa, 1909, 1 (s.o. *Stylochus*).

ovata Schmankewitsch, 1873, 275 (Black Sea); Jacubowa, 1909, 1 (a *Stylochus*).

ENTEROGONIA Haswell, 1907*b*, 478 (type *orbicularis*); 1907*a*, 644; Bock, 1913, 152 (defined); 1925*b*, 172; Hyman, 1959*c*, 5 (defined); Prudhoe, 1982*b*, 362 (redefined).

orbicularis (Schmarda, 1859) Stummer-Traunfels, 1933, 3486, 3511, figs (type-specimen descr.; New Zealand); Prudhoe, 1982*a*, 223, 224 (Australia); 1982*b*, 362 (descr.).

orbicularis pigrans Hyman, 1959*c*, 5; Prudhoe, 1982*b*, 363 (s.o. *orbicularis*).

pigrans Haswell, 1907*a*, 644; 1907*b*, 478, fig. (N.S.W., Australia); Bock, 1925*b*, 177; Bresslau, 1928–33, 146, fig.; Hyman, 1959*c*, 5, figs (subspecies of *orbicularis*).

pigrans novaezeylandiae Bock, 1925*b*, 142, 177, figs (New Zealand); 1927*b*, 7 (genito-intestinal duct); Stummer-Traunfels, 1933, 3486, 3487, 3514 (s.o. *orbicularis*).

ENTEROGONIINAE Bresslau, 1928–33, 287 (subfam. of Stylochidae).

ENTEROGONIMUS Hallez, 1911*b*, 141 (type: *aureus*); 1913, 33, 36, 41.

aureus Hallez, 1911*b*, 141 (Antarctica); 1913, 1, 24, 34, figs (Admiralty Bay & George Ross I., Antarctica); Bresslau, 1928–33, 292, fig.

EOLIDICEROS Quatrefages, 1845, 139 (type: *brocchii*), 174 (s.o. *Thysanozoon*); Diesing, 1850, 211 (s.o. *Thysanozoon*); Schmarda, 1859, 29 (subgenus of *Thysanozoon*); Moseley, 1877, 29 ('*Aeolidiceros*'); Lang, 1884, 524 (s.o. *Thysanozoon*).

brocchii (Risso, 1818) Quatrefages, 1845, 140, figs (Naples & Toulon); Diesing, 1850, 213 (*Thysanozoon*); Lang, 1884, 526 (*Thysanozoon*).

cruciatum Schmarda, 1859, 30, figs (*Thysanozoon (Eolidiceros)*; N.S.W., Australia, & Auckland, New Zealand); Diesing, 1862, 557; Lang, 1884, 526 (? s.o. *Thysanozoon brocchii*); Laidlaw, 1906, 713 (var. of *brocchii*).

ovale Schmarda, 1859, 29, fig. (*Thysanozoon (Eolidiceros)*; Ceylon); Diesing, 1862, 557; Lang, 1884, 526 (s.o. *Thysanozoon brocchii*).

panormus Quatrefages, 1845, 142, figs (Palermo, Sicily); Diesing, 1850, 213 (to *Thysanozoon*); Stimpson, 1857, 20 (to *Planeolis*); Lang, 1884, 525 (s.o. *Thysanozoon brocchii*); Palombi, 1928, 604.

EUPLANA Girard, 1893, 198 (type: *gracilis*); Hyman, 1939*b*, 136 (syn. *Discoplana*); 1939*c*, 5 (syns. *Discoplana* and *Conjuguterus*); 1940*a*, 470 (synonymy); 1953*a*, 332 (defined); 333 (constitution).

carolinensis Hyman, 1939*e*, 153; 1940*a*, 472, figs (nom. nov. pro *Leptoplana angusta* of Pearse & Littler, 1938, 237, *nec* fig. (North Carolina).)

clippertoni Hyman, 1939*c*, 4, figs (Clipperton I., Galapagos).

concolor (Meixner, 1907*b*) Hyman, 1954*b*, 333.

gigas (Schmarda, 1859) Hyman, 1955*c*, 76 (synonymy; Gilbert Is, Pacific Ocean); 1959*a*, 553 (Micronesia).

gracilis (Girard, 1850*b*) Girard, 1893, 198, fig.; Pearse & Littler, 1938, 238, fig. (descr.; North Carolina); Pearse & Walker, 1939, 18, figs; Hyman, 1939*b*, 136, fig. (Massachusetts); 1939*c*, 5 (syn. *Conjuguterus parvus*); 1940*a*, 470 (descr.; Florida to Prince Edward I.); 1940*b*, 18 (descr. synonymy); 1951, 164, fig. (New England); Christensen, 1971, 457 (early development; Maryland); Prudhoe, 1982*a*, 223 (Victoria, Australia); 1982*b*, 365, fig. (Victoria).

hymanae Marcus, 1947, 99, 129, 165, figs (Brazil); Hyman, 1953*a*, 333; du Bois-Reymond Marcus, 1955*b*, 285.

inquieta (Heath & McGregor, 1913) Marcus, 1947, 131, 165; Hyman, 1953*a*, 333 (to *Notoplana*).

longipenis (Kato, 1943*c*) Hyman, 1953*a*, 333.

pacificola (Plehn, 1896*a*) Hyman, 1953*a*, 332, figs (Mexico).

takewakii (Kato, 1935*b*) Hyman, 1953*a*, 333.

tropicalis Hyman, 1954*b*, 331, fig. (Hawaiian Is); Marcus & Marcus, 1966, 330 (to *Phylloplana*).

EUPLANINA Sopott-Ehlers & Schmidt, 1975, 204 (type: *horrida*).

horrida Sopott-Ehlers & Schmidt, 1975, 204, figs (Galapagos Is).

EUPLANINAE Marcus, 1947, 129, 165 (=3rd Series of Leptoplanidae of Bock, 1913, 169); Hyman, 1953*a*, 343 (criticism).

EUPROSTHIOSTOMUM Bock, 1925*a*, 49, 61 (type: *adhaerens*); Palombi, 1936, 39; Kato, 1937*f*, 369; Hyman, 1953*a*, 386 (defined); du B.-R. Marcus & Marcus, 1968, 91 (key to spp.).

adhaerens Bock, 1925*a*, 3, 48, 49, figs (associating with hermit-crabs, Gulf of Panama).

molle (Freeman, 1930) Hyman, 1953*a*, 386 (California).

mortenseni Marcus, 1948, 111, 184, 196, figs (Brazil); du B.-R. Marcus, 1955*b*, 287.

pakium du Bois-Reymond Marcus & Marcus, 1968, 91, figs (Florida).

viscosum Palombi, 1936, 1, 32 (associating with hermit-crab in Bay of Naples).

EURYLEPTA Ehrenberg, 1831, 56 (type: *cornuta*); Örsted, 1844*a*, 46, 50 (list of spp.); Diesing, 1850, 208 (diagn.); Stimpson, 1857, 20; Schmarda, 1859, 13, 26; Johnston, 1865, 7; Lang, 1884, 431, 572; Carus, 1885, 156; Verrill, 1893, 495; Hallez, 1893, 174; 1894, 225 (defined); Gamble, 1893*a*, 507; 1896, 19; Benham, 1901, 31; Heath & McGregor, 1913, 481; Freeman, 1933, 138; Hyman, 1939*b*, 148; 1940*a*, 486; Marcus, 1952, 95–96 (list of spp.); Hyman, 1953*a*, 370 (defined); du Bois-Reymond Marcus, 1955*b*, 45 (list of spp.); Prudhoe, 1982*c*, *56 (defined)*.

affinis Kelaart, in Collingwood, 1876, 87, 96, fig. (Ceylon); Lang, 1884, 593; Stummer-Traunfels, 1933, 3566, fig. (to *Pseudoceros*).

argus (Quatrefages, 1845) Diesing, 1850, 209; Keferstein, 1869, 8, figs (NW. France); Diesing, 1862, 553 (*Proceros*); Lang, 1884, 564 (*Prostheceraeus*).

atraviridis Kelaart, in Collingwood, 1876, 87, 95, fig. (Ceylon); Lang, 1884, 594; Hyman, 1959*a*, 565 (to *Pseudoceros*).

aurantiaca Heath & McGregor, 1913, 458, 481, figs (California); Hyman, 1953*a*, 370, figs (San Diego to Vancouver I.); du B.-R. Marcus, 1955*a*, 45; Marcus, 1952, 96 (note); Hyman, 1955*a*, 10 (Puget Sound & California); Stasek, 1966, 7 (location of type-specimen); Morris *et al.*, 1981, 80, coloured fig. (California).

auriculata (Müller, 1788*b*) Diesing, 1850, 211; 1862, 550; Hallez, 1878*a*, 193 (embryology); 1879*b*, 118, figs (NW. France); Lang, 1884, 583 (*auriculata* of Hallez, s.o. *Oligocladus auritus*), 584 (*Planaria auriculata* of Müller a rhabdocoele); Hallez, 1893, 176 (*E. auriculata* of Hallez, 1878*c*, s.o. *Oligocladus auritus*).

aurita Claparède, 1861, 144, figs (Wales & west coast of Scotland); Diesing, 1863, 175 (to *Proceros*); Lang, 1884, 583 (to *Oligocladus*).

californica Hyman, 1959*b*, 11, figs (California); Morris *et al.*, 1981, 80, coloured fig. (California).

cardiosora Schmarda, 1859, 28, figs (Ceylon); Diesing, 1862, 552 (to *Proceros*); Lang, 1884, 546 (to *Pseudoceros*); Stummer-Traunfels, 1933, 3487, 3540, 3544 (type-specimen re-examined; *Pseudoceros*).

cerebralis (Kelaart, 1858) Collingwood, 1876, 87, 96, fig. (descr.); Lang, 1884, 546 (to *Pseudoceros*).

coccinea Stimpson, 1857, 20, 25 (Luchu I., Japan); Diesing, 1862, 549; Lang, 1884, 592; Kato 1944, 298 (to *Pseudoceros*).

cornuta (Müller, 1776) Ehrenberg, 1831, 56; Örsted, 1844*a*, 50; Diesing, 1850, 208; Leuckart, 1859, 183; Diesing, 1862, 548; Johnston, 1865, 7; Lankester, 1866, 388 (Herm, Channel Is); Keferstein, 1869, 9, figs (descr., synonymy, NW. France); Jensen, 1878, 78 (descr.; Norway); Lang, 1884, 572 (synonymy); Carus, 1885, 156 (descr.); Koehler, 1885, 37, 49, 56 (Channel Is); Lo Bianco, 1888, 399 (Naples); Vaillant, 1889, 656 (NW. France); Hallez, 1893, 174 (NW. France); 1894, 224 (NW. France); Bergendal, 1893*a*, 237 (Sweden); Gamble, 1893*a*, 507 (Britain); 1893*b*, 47 (Devon); 1896, 19 (Britain); Pruvot, 1897, 21 (Brittany & Gulf of Lion); Lo Bianco, 1899, 478 (Naples); Gamble, 1900, 812 (Ireland); Bock, 1913, 264, figs (descr.); Southern, in Farran, 1915, 36 (Ireland); Southern, 1936, 71 (Ireland); Marcus, 1952, 95, 96 (Note); Williams, 1954, 55 (NE. Ireland); du Bois-Reymond Marcus, 1955*a*, 45; Hendelberg, 1974*b*, 17, figs (Sweden); Prudhoe, 1982*c*, 56, fig. (Britain).

cornuta var. *melobesiarum* (Schmidtlein, 1880) Lang, 1884, 73, 576, figs (Naples); Carus, 1885, 156 (descr.); Lo Bianco, 1909, 568 (Naples); Stummer-Traunfels, 1933, 3582, fig.; Marcus, 1952, 95 (Note); du B.-R. Marcus, 1955*a*, 45.

cornuta var. *wandeli* Hallez, 1907, 2, 7, figs (Wanda I., Antarctica); Bock, 1913, 264 (s.o. *cornuta* s.s.); Marcus, 1952, 96 (Note). du B.-R. Marcus, 1955*a*, 45 (distinct from *cornuta*).

cristata (Quatrefages, 1845) Diesing, 1850, 210; Selenka, 1881*b*, 229 (NW. France); 1881*c*, 1 (NW. France); Lang, 1884, 554 (s.o. *Prostheceraeus vittatus*).

dalyellii Johnston, 1865, 7 (nom. nov. pro *Planaria cornuta* of Dalyell); Lang, 1884, 572 (s.o. *cornuta*).

dulcis (Kelaart, 1858) Collingwood, 1876, 87, 96, fig. (descr.); Hyman, 1959*b*, 565 (*Pseudoceros*).

flavomarginata Ehrenberg, 1831, 56 (Red Sea); Örsted, 1844*a*, 27, 50; Diesing, 1850, 208; Stimpson, 1857, 20; Diesing, 1862, 548; Lang, 1884, 544 (? s.o. *Pseudoceros limbatus*), 563 (to *Prostheceraeus* (?)).

fulminata Stimpson, 1855*a*, 380 (Loo Choo I. (Ryu Kyu I.), Japan); 1857, 20, 25 (descr.); Diesing, 1862, 548; Lang, 1884, 592 (? = *Prostheceraeus* ? *flavomarginata*); Kato, 1944, 298 (to *Pseudoceros*).

fulvolimbata Grube, 1868, 46 (Samoa, Pacific Ocean); Lang, 1884, 592.

fusca (Kelaart, 1858) Collingwood, 1876, 87, 95, fig. (descr.); Hyman, 1959*a*, 565 (*Pseudoceros*).

guttatomarginata Stimpson, 1855*a*, 380 (Loo Choo I. (Ryu Kyu I.), Japan); 1857, 20, 26 (descr.); Diesing, 1862, 549; Lang, 1884, 592; Kato, 1944, 298 (to *Pseudoceros*).

herberti Kirk, 1882, 267 (Wellington, New Zealand); Lang, 1884, 617 (? *Prostheceraeus*).

interrupta Stimpson, 1855*a*, 380 (Loo Choo I. (Ryu Kyu I.), Japan); 1857, 20, 26 (descr.); Diesing, 1862, 550; Lang, 1884, 591; Kato, 1944, 298 (to *Pseudoceros*).

japonica Stimpson, 1857, 20, 26 (Jesso I., Japan); Diesing, 1862, 549; Lang, 1884, 565 (? *Prostheceraeus*); Stummer-Traunfels, 1933, 3566; Kato, 1944, 298 (to *Pseudoceros*).

kelaartii Collingwood, 1876, 87, 92, figs (Singapore); Lang, 1884, 568 (to *Prostheceraeus*); Laidlaw, 1903*a*, 302 (to *Pseudoceros*).

leoparda Freeman, 1933, 111, 112, 138, figs (Washington State, U.S.A.); Marcus, 1952, 96 (note); Hyman, 1953*a*, 372, fig.; du B.-R. Marcus 1955*a*, 46; Hyman, 1959*b*, 12, fig. (California); Ching, 1977, 338 (British Columbia & Washington State).

limbata (Leuckart, 1828) Örsted, 1844*a*, 50; Diesing, 1850, 210; Stimpson, 1857, 20; Schmarda, 1859, 26; Diesing, 1862, 554 (to *Proceros*); Lang, 1884, 544 (to *Pseudoceros*).

lobianchii (Lang, 1879) Lang, 1884, 78, 578 (*'lobiancoi'*), figs; Carus, 1885, 156 (descr.); Hallez, 1893, 175 (Pas-de-Calais); 1894, 227 (NW. France); Micoletzky, 1910, 179 (var. of *cornuta*; Adriatic); Bock, 1913, 266, 267 (var. of *cornuta*); Stummer-Traunfels, 1933, 3575, 3582 (descr.); du B.-R. Marcus, 1955*a*, 45 (*'lobianchoi'*); Marcus, 1952, 95 (note).

maculata (Gray, *in* Pease, 1860) Diesing, 1862, 548; Lang, 1884, 547 (to *Pseudoceros*).

maculosa Verrill, 1893, 495, figs (Massachusetts); Sumner *et al.*, 1913, 580 (Massachusetts); Pearse, 1938, 87; Hyman, 1938, 87, 97; Pearse & Walker, 1939, 19, fig. (Massachusetts); Hyman, 1939*b*, 150, figs (co-type descr.); 1940*a*, 486 (descr.); 1940*b*, 20 (descr.); 1952, 197 (to *Prostheceraeus*); Marcus, 1952, 96 (note); du B.-R. Marcus, 1955*a*, 45.

miniata Schmarda, 1859, 27, figs (Ceylon); Diesing, 1862, 554 (to *Proceros*); Lang, 1884, 551 (to *Yungia*); Stummer-Traunfels, 1933, 3539, 3544 (type-specimen re-examined; to *Pseudoceros*).

neptis du Bois-Reymond Marcus, 1955*a*, 42, figs (Brazil); 1955*b*, 286.
nigra Stimpson, 1857, 20 (*'niger'*), 26 (Loo Choo I. (Ryu Kyu I.), Japan); Diesing, 1862, 549; Lang, 1884, 565 (to *Prostheceraeus* ?); Stummer-Traunfels, 1933, 3566 (to *Pseudoceros*).
nigrocincta Schmarda, 1859, 26, figs (Ceylon); Diesing, 1862, 551 (to *Proceros*); Lang, 1884, 547 (to *Pseudoceros*); Stummer-Traunfels, 1933, 3487, 3537, 3538 (type-specimen re-examined; *Pseudoceros*).
oceanica (Darwin, 1844) Diesing, 1850, 211; Stimpson, 1857, 20 (to *Nautiloplana*); Lang, 1884, 608 (*Planaria*).
orbicularis Schmarda, 1859, 28, figs (Jamaica); Diesing, 1862, 553 (to *Proceros*); Lang, 1884, 551: Stummer-Traunfels, 1933, 3487, 3545, figs (type-specimen re-exmd.; to *Pericelis*).
pantherina Grube, 1868, 46 (Samoa, Pacific Ocean); Lang, 1884, 592 (poss. a *Cycloporus*).
praetexta Ehrenberg, 1831, 56 (Red Sea); Örsted, 1844*a*, 27, 50; Diesing, 1850, 208; 1862, 547; Lang, 1884, 544 (? s.o. *Pseudoceros limbatus*); 591 (*Eurylepta*).
pulchra Örsted, 1845, 415 (Norway); Diesing, 1862, 550; Lang, 1884, 572 (? s.o. *cornuta*); Bock, 1913, 267 (s.o. *Oligocladus sanguinolentus*).
punctata Kaburaki, 1923*a*, 199, figs (Japan); Kato, 1937*e*, 131; 1944, 302, du B.-R. Marcus, 1955*a*, 45; Marcus, 1952, 96 (note).
purpurea (Kelaart, 1858) Collingwood, 1876, 87, 96, fig. (descr.); Hyman, 1959*a*, 565 (*Pseudoceros*).
rubrocincta Schmarda, 1859, xiii, 26, fig. (Ceylon); Diesing, 1862, 546 (*Schmardea*); Lang, 1884, 550 (to *Yungia*); Stummer-Traunfels, 1933, 3487, 3537, 3538 (type-specimen re-examined; to *Pseudoceros*).
sanguinolenta (Quatrefages, 1845) Diesing, 1850, 209; Schmarda, 1859, 26; Diesing, 1862, 552 (to *Proceros*); Johnston, 1865, 7 (Britain); Lang, 1884, 580 (to *Oligocladus*); Bock, 1913, 267 (*Oligocladus*).
striata (Kelaart, 1858), Collingwood, 1876, 87, 97, fig. (descr.); Lang, 1884, 546 (to *Pseudoceros*).
striata Schmarda, 1859, 27, figs (Ceylon); Diesing, 1862, 551 (to *Proceros*); Lang, 1884, 552; Stummer-Traunfels, 1933, 3487, 3540, 3544, figs (type-specimen re-examined; to *Pseudoceros*); Hyman, 1959*a*, 566 (s.o. *Pseudoceros gratus*).
superba Schmarda, 1859, 28, figs (Ceylon); Diesing, 1862, 552 (to *Proceros*); Lang, 1884, 552 (s.o. *Planaria undulata*); Stummer-Traunfels, 1933, 3487, 3541, 3544, figs (type-specimen re-examined; to *Pseudoceros*).
susakiensis (Kato, 1934*b*) Kato, 1944, 302, figs (Japan), du B.-R. Marcus, 1955*a*, 46, Marcus, 1952, 96 (note).
turma Marcus, 1952, 5, 94, 116, figs (Brazil); du B.-R. Marcus, 1955*a*, 46; 1955*b*, 286.
undulata (Kelaart, 1858) Collingwood, 1876, 87, 95, fig. (descr.); Lang, 1884, 552 (*Planaria*); Marcus, 1950, 88 (to *Pseudoceros*).
velutina (Blanchard, 1847) Diesing, 1850, 210; Stimpson, 1857, 20; Schmarda, 1859, 26; Diesing, 1862, 548; Lang, 1884, 538, figs. (to *Pseudoceros*); Lo Bianco, 1909, 569.
violacea (Kelaart, 1858) Collingwood, 1876, 87; Hyman, 1959*a*, 566 (to *Pseudoceros*).
violacea Schmarda, 1859, 27, figs (Ceylon); Diesing, 1862, 553 (to *Proceros*); Lang, 1884, 540 (to *Pseudoceros velutinus* var. *violaceus*); Stummer-Traunfels, 1933, 3487, 3539, 3544, figs type-specimen re-examined; to *Pseudoceros*); Hyman, 1959*a*, 566 (renamed *Pseudoceros perviolaceus*).
viridis (Kelaart, 1858) Collingwood, 1876, 87, 96, fig. (descr.); Lang, 1884, 567 (s.o. *Prostheceraeus viridis* Schmarda); Stummer-Traunfels, 1933, 3487, 3544 (to *Pseudoceros*); Hyman, 1959*a*, 569 (distinct from Schmarda's form).
vittata (Montagu, 1813) Diesing, 1850, 209; Schmarda, 1859, 26; Diesing, 1862, 548; Johnston, 1865, 8; Jensen, 1878, 78 (descr.; Norway); Lang, 1884, 554 (to *Prostheceraeus*).
zebra (Leuckart, 1828), Diesing, 1850, 211; Stimpson, 1857, 20; Schmarda, 1859, 26; Diesing, 1862, 554 (to *Proceros*); Lang, 1884, 544 (to *Pseudoceros*).
zeylanica (Kelaart, 1858) Collingwood, 1876, 87, 97, fig. (descr.); Lang, 1884, 546 (to *Pseudoceros*).
species Bergendal, 1893*a*, 327 (Sweden).
species Morton & Miller, 1968, 172, fig. (New Zealand).
species Morris *et al.*, 1980, 81, fig. (coloured) (Pacific Grove, Monterey, California).

EURYLEPTIDAE Stimpson, 1857, 19; Lang, 1884, 431, 553 (diagn.); Carus, 1885, 155; Vaillant, 1889, 155 (key to genera); Hallez, 1893, 170 (key to genera); 1894, 219 (defined); 220 (key to genera); Verrill, 1893, 495; Gamble, 1893*a*, 503; 1896, 19; Benham, 1901, 31; Hallez, 1907, 4 (key to genera); 1913, 43 (diagn.); 44 (new subfams.: Prostheceraeinae & Laidlawiinae); Heath & McGregor, 1913, 476; Bock, 1922, 10 (note); Freeman, 1933, 136; Bresslau, 1928–33, 291; Stummer-Traunfels, 1933, 3576; Hyman, 1939*b*, 148; 1940*a*, 486; 1940*b*, 19; 1953*a*, 370 (defined); de Beauchamp, 1961, 66 (genera); Prudhoe, 1982*c*, 56 (defined).

EURYLEPTIDEA Diesing, 1862, 493, 547; Lang, 1884, 524 (*ex parte* = Pseudoceridae), 553.

EURYLEPTIDES Palombi, 1924*a*, 37 (type: *brasiliensis*).

brasiliensis Palombi, 1924*a*, 37 (Brazil); 1924*b*, 21, figs (off Brazilian coast – 14°14′5″S, 38°39′W).

EURYLEPTINAE Hallez, 1913, 44.

EURYLEPTODES Heath & McGregor, 1913, 458, 483 (type: *cavicola*); Hyman, 1953*a*, 375, 382 (defined).

cavicola Heath & McGregor, 1913, 458, 483, figs (California); Hyman, 1953*a*, 376, 381.

insularis Hyman, 1953*a*, 375, figs (California).

pannulus Heath & McGregor, 1913, 458, 484, figs (California); Hyman, 1953*a*, 376, 381.

phyllulus Heath & McGregor, 1913, 458, 484, figs (California); Hyman, 1953*a*, 376, 381.

EUSTYLOCHUS Verrill, 1893, 467 (type: *ellipticus*); Laidlaw, 1903*e*, 13, 14 (note); Meixner, 1907*b*, 399 (s.o. *Stylochus*); Pearse, 1938, 72 (distinct from *Stylochus*); Hyman, 1939*b*, 131 (s.o. *Stylochus*).

ellipticus (Girard, 1850*b*), Verrill, 1893, 467, figs (Connecticut to Maine); Van Name, 1899, 263 (in tubes of nemertine, Connecticut; embryology); Verrill, 1895, 532 ('spermatophores' in dorsal integument); Meixner, 1907*b*, 399; Sumner *et al.*, 1913, 579 (Massachusetts); Pearse, 1938, 73 (Massachusetts & Rhode I.); Pearse, & Walker, 1939, 17, figs (Massachusetts & Prince Edward I.); Hyman, 1939*b*, 130 (to *Stylochus*); 1940*a*, 459 (to *Stylochus*); Pearse, 1949, 26 (in burrows of shrimp, *Upogebia affinis*, North Carolina).

meridionalis Pearse, 1938, 73, fig. (Atlantic coast of U.S.A.); Pearse, & Littler, 1938, 236, fig. (descr.); Hyman, 1940*a*, 459 (s.o. *ellipticus*).

FASCIOLA Linnaeus, 1758 (now a genus of trematodes).

tremellaris Müller, 1774, 72 (Norway); 1776, 223 (to *Planaria*); Örsted, 1843, 569 (to *Leptoplana*); Diesing, 1850, 198 (*Leptoplana*); Lang, 1884, 476 (*Leptoplana*).

FREEMANIA Hyman, 1953*a*, 336 (type: *litoricola*); 1955*a*, 10.

litoricola (Heath & McGregor, 1913), Hyman, 1953*a*, 336, figs (California, Puget Sound & British Columbia); 1955*a*, 9 (California).

GENNEOCEROS [err. pro *Gnesioceros*] Diesing, 1862, 494.

GLOSSOSTOMA Le Conte, 1851, 319 (type: *nematoideum*).

nematoideum Le Conte, 1851, 319 (Isthmus of Panama); Diesing, 1862, 573; Lang, 1884, 611.

GNESIOCERINAE Marcus & Marcus, 1966, 331; du B.-R. Marcus & Marcus, 1968, 53; Prudhoe, 1982*b*, 375 (raised to family status).

GNESIOCEROS Diesing, 1862, 571, 575 (type: *sargassicola*); Lang, 1884, 434, 446; Meixner, 1907*b*, 398; Hyman, 1939*b*, 144 (syn. *Pelagoplana*); Hyman, 1939*d*, 11; 1940*a*, 478; 1955*b*, 134 (defined).

floridana (Pearse, 1938) Hyman, 1940*a*, 478 (Florida & North Carolina; syns. *Imogine oculifera* of Verrill, 1893; *Stylochoplana oculifera* of Pearse & Walker, 1939; *Gnesioceros verrilli* Hyman, 1939*b*); Hyman, 1940*b*, 19 (descr.; synonymy); 1954*c*, 301 (Gulf of Mexico), 302 (littoral variant of *Gnesioceros sargassicola*); 1955*b*, 135 (Virgin Is); du Bois-Reymond Marcus & Marcus, 1968, 48 (s.o. *sargassicola*).

mertensi (Diesing, 1850) Diesing, 1862, 572; Lang, 1884, 454 (s.o. *Stylochus sargassicola*); Meixner, 1907*b*, 398, 402; Hyman, 1939*b*, 146 (s.o. *sargassicola*).

pellucidus (Mertens, 1832) Diesing, 1862, 571; Lang, 1884, 437 (s.o. *Planocera pellucida*); Meixner, 1907*b*, 398, 401; Bock, 1931, 277 (s.o. *Planocera pellucida*).

sargassicola (Mertens, 1832) Hyman, 1939*b*, 146 (synonymy; distribution); 1939*d*, 11, figs (amongst *Sargassum*, Caribbean Sea, Bermuda, Gulf of Mexico, western & central North Atlantic); Prudhoe, 1944, 325 (distribution); Hyman, 1951, 88, 114, figs (Bermuda); 1954*c*, 302; Lenhoff, 1964, 841 (Florida); du B.-R. Marcus & Marcus, 1968, 48, figs (Caribbean, Florida & central Atlantic; systematic note); Cheng & Lewin, 1975, 518 (as *Stylochoplana sargassicola*, Baja California).

sargassicola (Mertens) var. *lata* Hyman, 1939*d*, 13, fig. (Bermuda).

verrilli Hyman, 1939*b*, 146, figs (Massachusetts; *Imogine oculifera* of Verrill, 1892, renamed); 1940*a*, 478 (s.o. *floridana*); du B.-R. Marcus & Marcus, 1968, 48, 51 (s.o. *sargassicola*).

GNESIOCEROTIDAE Prudhoe, 1982*b*, 375 (Gnesiocerinae raised to family-rank).

GRAFFIZOIDAE Heath, 1928, 203.

GRAFFIZOON Heath, 1928, 187, 197, 202 (type; *lobatum*); Marcus, 1950, 92.

lobatum Heath, 1928, 187, 203, figs (California); Bock, 1931, 282 (neotenic larva); Stummer-Traunfels, 1933, 3588, figs.

HAPLOPLANA Laidlaw, 1903*d*, 109, 110 (type: *elioti*); 1903*e*, 15 (affinity with *Enantia*); Bock, 1913, 222 (diagn.).

elioti Laidlaw, 1903*d*, 109, figs (East Africa); Bock, 1913, 222.

HETEROPLANA *newtoni* Willey, 1898, 203 [a mutilated planarian ?].

HETEROSTYLOCHUS Verrill, 1893, 462, 467 (type: *maculatus*).

maculatus (Quatrefages, 1845) Verrill, 1893, 467, 468 (footnote).

HIRUDO Linnaeus, 1758 (now genus of Hirudinea).

plana Strøm, 1768, 365 (Norway); Müller, 1779, 72 (s.o. *Planaria tremellaris*); Diesing, 1850, 198 (s.o. *Leptoplana tremellaris*); Graff, 1904, 1736 (unrecognizable polyclad).

HOPLOPLANA Laidlaw, 1902, 303 (nom. nov. pro *Planocera* Group B of Lang, 1884; (type: insignis); Bock, 1913, 224 (diagn.); Yeri & Kaburaki, 1918*a*, 435 (features); 1918*b*, 47 (features); Hyman, 1939*b*, 142; Smith, 1960, 387 (key to spp.); du B.-R. Marcus & Marcus, 1968, 47 (systematic note); Marcus, 1952, 87 (key to spp.).

californica Hyman, 1953*a*, 344, figs (California); Morris *et al.*, 1980, 74, figs (coloured), (California).

cupida Kato, 1938*b*, 577, 582, figs (Japan); 1944, 280; Sandô, 1964, 27 (Japan).

deanna Kato, 1939*b*, 141, 144, figs (Japan); 1944, 280.

divae Marcus, 1950, 79, 114, figs (Brazil); du B.-R. Marcus, 1955*b*, 283 (Brazil); du B.-R. Marcus & Marcus, 1968, 47 (Caribbean).

grubei (Graff, 1892) Bock, 1913, 225 (descr.; off St Thomas I., Caribbean); Hyman, 1939*b*, 142 (Massachusetts); 1939*d*, 10 (descr.; Gulf of Mexico, Bermuda & central N. Atlantic); Prudhoe, 1944, 325 (Cayman Is, Caribbean); Hyman, 1954*c*, 302.

inquilina (Wheeler, 1894) Bock, 1913, 228; Pearse, 1938, 79 (in shells of *Busycon*, Massachusetts); Pearse & Walker, 1939, 18, fig. (Massachusetts); Hyman, 1939*b*, 143; figs (Massachusetts); 1940*a*, 476 (descr.; commensal in mantle-cavity of *Busycon*); 1940*b*, 18 (descr.); Stauber, 1941, 541 (descr.; in mantle-cavity of *Urosalpina cinerea* & *Eupleura caudata*; New Jersey); Schechter, 1943, 362 (in mantle-cavity of *Thais floridana*, Florida & Louisiana); Hyman, 1944*a*, 7 (in mantle-cavity of *Urosalpinx cinerea*, New Jersey & in *Thais floridana haysae*, Louisiana); 1954*c*, 301.

inquilina thaisana (Pearse) Hyman, 1940*a*, 477; 1944*a*, 7 (subsp. not valid); 1951, 169, fig.; Smith, 1960, 389 (subsp. valid).

insignis (Lang, 1884) Laidlaw, 1902, 303; Bock, 1913, 225; Steinböck, 1933, 19 (Gulf of Trieste).

luracola Smith, 1961, 69, figs (in pallial groove of *Nerita (Ritena) scabricosta ornata*, Panama City, 9°0′N, 79°30′W).

ornata Yeri & Kaburaki, 1918*a*, 432, 435 (diagn.; Japan); 1918*b*, 2, 15, 47, figs (Japan); Bock, 1924, 22 (Japan); Kato, 1938*a*, 565 (Japan); 1944, 280; Utinomi, 1956, 29 (coloured fig.); Kukuchi, 1966, 68, table 21 (Japan); 1968, 167 (Japan).

papillosa (Lang, 1884) Bock, 1913, 225.

rosea Prudhoe, 1977, 588, fig. (South Australia); 1982*b*, 224, 373 (southern Australia).

rubra Kato, 1944, 281, figs (Japan).

schizoporellae Kato, 1944, 280, figs (Japan).

thaisana Pearse, 1938, 79, fig. (in *Thais floridana*, Florida); Hyman, 1940*b*, 477 (subsp. of *inquilina*); 1944*a*, 7 (s.o. *inquilina*).

usaguia Smith, 1960, 385, figs (in mantle-cavity of prosobranch molluscs, Brazil).

villosa (Lang, 1884), Bock, 1913, 225; Stummer-Traunfels, 1933, 3579, fig.; Kato, 1937*a*, 216, figs (Japan); 1940, 557, figs (Japan; development); 1944, 280; Dawydoff, 1952, 82 (Vietnam); Kikuchi, 1968, 167 (Japan).

HOPLOPLANIDAE Stummer-Traunfels, 1933, 3576, 3585; Hyman, 1940*a*, 476; 1940*b*, 18; Marcus, 1952, 86 (defined); Hyman, 1953*a*, 343 (defined); de Beauchamp, 1961, 60 (genera).

HYMANIA Pearse & Littler, 1938, 239 (type: *prytherci*); Hyman, 1940*a*, 487 (s.o. *Oligocladus*).
prytherci Pearse & Littler, 1938, 239, figs (North Carolina); Hyman, 1940*a*, 487, (s.o. *Oligocladus floridanus*).

IDIOPLANA Woodworth, 1898, 63 (type: *australiensis*); Laidlaw, 1903*e*, 10, 12; Meixner, 1907*b*, 440; Haswell, 1907*b*, 471; Bock, 1913, 141 (diagn.); 1924*b*, 347, 348; 1925*a*, 40 (remarks on Woodworth's description); 1925*b*, 171.
australiensis Woodworth, 1898, 63, figs (Great Barrier Reef); Meixner, 1907*b*, 440, figs; Bock, 1913, 142; 1925*b*, 177; Palombi, 1928, 586, figs (Suez); Prudhoe, 1952, 178 (Palombi's specimens appear to belong to *Idioplanoides*); Kato, 1943*c*, 81 (Palombi's specimens not *australiensis*).
pacifica Kato, 1943*c*, 79, 80, figs (Palau Is); 1944, 266 (Palau).
IDIOPLANINAE Bresslau, 1933, 286 (subfamily of Stylochidae).
IDIOPLANOIDES Barbour, 1912, 187 (nom. nov. pro *Woodworthia* Laidlaw, 1903*e*); Bock, 1924*b*, 347, 348; 1925*b*, 172
atlanticum (Bock, 1913) Bock, 1925*a*, 23; 1925*b*, 177.
insignis (Laidlaw, 1904*b*) Barbour, 1912, 187; Bock, 1925*b*, 177.
IGLUTA Marcus & Marcus, 1966, 326; du B.-R. Marcus & Marcus, 1968, 2, 18 (type: *tipuca*).
tipuca Marcus & Marcus, 1966, 327 (Caribbean); du B.-R. Marcus & Marcus, 1968, 2, 18, figs (Caribbean).
ILYPLANA Bock, 1925*b*, 97, 102, 111, 171 (type: *aberrans*).
aberrans Bock, 1925*b*, 97, 102, 177, figs (New Zealand).
ILYPLANOIDES Kato, 1944, 268 (type: *mitsuii*).
mitsuii Kato, 1944, 269, figs. (Japan).
IMOGINE Girard, 1853, 367 (type: *oculifera*); Stimpson, 1857, 22; Schmarda, 1859, 14, 35; Diesing, 1862, 570 (s.o. *Stylochus*); Lang, 1884, 428, 445 (diagn.); Girard, 1893, 192; Gamble, 1896, 19; Meixner, 1907*b*, 396 (? s.o. *Stylochus*); du B.-R. Marcus & Marcus, 1968, 6 (subgen. of *Stylochus*; list of spp.).
conoceraea Schmarda, 1859, 35, figs (Ceylon); Diesing, 1862, 568 (to *Stylochus*); Lang, 1884, 446 (to *Conoceros*); Meixner, 1907*b*, 397, 398 (a *Stylochoplana*); Stummer-Traunfels, 1933, 3487, 3563, figs type re-examined; to *Stylochoplana*).
oculifera Girard, 1853, 367 (South Carolina); Diesing, 1862, 570 (to *Stylochus*); Lang, 1884, 446; Girard, 1893, 193 (descr.); Verrill, 1893, 475, fig. (descr.; New England); Meixner, 1907*b*, 396; Sumner *et al.*, 1913, 580 (Massachusetts); Pearse & Walker, 1939, 18, fig. (*Stylochoplana oculifera* of Verrill); Hyman, 1939*b*, 146 (*I. o.* of Verrill, 1893, renamed *Gnesioceros verrilli*); 1940*a*, 464, fig. (*Stylochus*).
truncata Schmarda, 1859, 35, fig. (Ceylon); Diesing, 1862, 567 (to *Stylochus*); Lang, 1884, 465; Meixner, 1907*b*, 397 (not a *Stylochus*); Stummer-Traunfels, 1933, 3487, 3562 (type-specimen re-examined; s.o. *Notoplana dubia*).
INDIPLANA Stummer-Traunfels, 1933, 3486, 3527 (type: *oosora*); Marcus, 1949, 76 (Leptoplaninae).
oosora (Schmarda, 1859) Stummer-Traunfels, 1933, 3486, 3522, figs (type-specimen re-examd.; Ceylon).
INDISTYLOCHUS Hyman, 1955*b*, 123 (type: *hewatti*).
hewatti Hyman, 1955*b*, 123, figs (Puerto Rico).
ITANNIA Marcus, 1947, 99, 134, 166 (type: *ornata*).
ornata Marcus, 1947, 99, 135, 166, figs (Brazil); Marcus, 1952, 88, 115, figs (Brazil); du B.-R. Marcus, 1955*b*, 283.
ornata forma *murna* du Bois-Reymond Marcus, 1957, 174, fig. (Brazil).

KABURAKIA Bock, 1925*b*, 97, 132, 142, 172 (type: *excelsa*); Hyman, 1953*a*, 290 (defined).
excelsa Bock, 1925*b*, 98, 132, 142, figs (British Columbia & Puget Sound); Bresslau, 1930, 146, 177, fig.; Kato, 1937*f*, 355 (syn. *Cryptophallus magnus*); Hyman, 1953*a*, 290 (descr., synonymy; Puget Sound to Alaska); 1955*a*, 7 (California); Morris *et al.*, 1980, 78, fig. (coloured) (California).
gloriosa Kato, 1938*a*, 559, 562, figs (Japan); 1944, 267.

LAIDLAWIA Herzig, 1905, 329 (type: *trigonopora*); Hallez, 1913, 36, 41 (compared with *Stylochoides*); Bock, 1923*b*, 27 (note); 1927*a*, 10 (relationship); 1931, 287 (systematic note).
polygenia Palombi, 1938, 329, 360, figs (South Africa).

trigonopora Herzig, 1905, 329, fig. (Punta Arenas, Chile); Ritter-Záhony, 1907, 9, figs (redescr.; Punta Arenas); Bock, 1927*b*, 10 (uterine vesicle); 1931, 287, figs (descr.; Antarctica); Westblad, 1952, 11 (South Georgia).

LAIDLAWIIDAE Bock, 1922, 20; Bresslau, 1928–33, 292; de Beauchamp, 1961, 67 (genera).

LAIDLAWIINAE Hallez, 1913, 44 (subfam. of Euryleptidae; key to genera); Bock, 1922, 14, 18, 20 (raised to family rank – [by accident or design?]); Poche, 1926, 100 (= Stylochoididae).

LA PELLICULE ANIMÉE Dicquemare, 1781, 142, figs; Bosc, 1803, 63 ('La Planaire pellicule'); Lang, 1884, 476 (? s.o. *Leptoplana tremellaris*).

LA PLANAIRE PELLICULE Bosc, 1803, 63 (nom. nov. pro 'La pellicule animée' of Dicquemare, 1781); [Lang (1884, 6, 476) and Graff (1904–05, 1740) indicate that Bosc renamed Dicquemare's 'La pellicule animée' as *Planaria pellucida*, but this appears to be incorrect.]; Prudhoe, 1950*b*, 712.

LATOCESTIDAE Laidlaw, 1903*a*, 7; Bock, 1913, 62 (diagn.); Stummer-Traunfels, 1933, 3567, 3586; Bresslau, 1928–33, 284 (classification); Hyman, 1940*a*, 457; 1944*b*, 73; 1953*a*, 275 (defined), 277 (juvenile specimen, Lower California); Marcus, 1947, 100 (key to genera); 1949, 68 (key to genera); 1950, 78 (key to genera); 1952; 77 (key to genera); Hyman, 1955*c*, 70, figs. (juvenile specimen from the Marianas).

LATOCESTUS Plehn, 1896*a*, 140, 161 (type: *atlanticus*); Benham, 1901, 31; Laidlaw, 1903*a*, 7; Meixner, 1907*b*, 461; Bock, 1913, 63 (defined); Hyman, 1940*a*, 457 (syn. *Oculoplana*); 1953*a*, 275 (defined); 1953*b*, 183 (defined); du B.-R. Marcus & Marcus, 1968, 4 (key to West Atlantic species).

argus Laidlaw, 1903*c*, 302, 312, figs (Malaysia); Bock, 1913, 64; Prudhoe, 1950*a*, 41, fig. (Burma).

atlanticus Plehn, 1896*a*, 140, 159, figs (Cape Verde Is & Brazil); Bock, 1913, 64; du B.-R. Marcus & Marcus, 1968, 5.

brasiliensis Hyman, 1955*d*, 12, figs (Brazil); du B.-R. Marcus & Marcus, 1968, 5.

caribbeanus Prudhoe, 1944, 323, figs (Cayman Is, Caribbean); du B.-R. Marcus & Marcus, 1968, 4.

galapagensis Hyman, 1953*b*, 183, figs (Galapagos Is).

maldivensis (Laidlaw, 1902) Laidlaw, 1903*c*, 313; Bock, 1913, 64.

marginatus Meixner, 1907*a*, 168 (Somalia); 1907*b*, 461, figs.; Bock, 1913, 64.

ocellatus Marcus, 1947, 99, 100, 162, figs (Brazil); 1949, 67 (Brazil); Marcus & Marcus, 1951, 13; du B.-R. Marcus, 1955*b*, 282; du B.-R. Marcus & Marcus, 1968, 5.

pacificus Laidlaw, 1903*a*, 3, 8, fig. (Rotuna I., Pacific Ocean); Bock, 1913, 64; Hyman, 1959*d*, 543, 593, figs (Palau Is).

plehni Laidlaw, 1906, 711, fig. (Cape Verde Is); Bock, 1913, 64; Palombi, 1940, 110, fig. (Angola).

viridis Bock, 1913, 64, figs (Panama); Stummer-Traunfels, 1933, 3570, fig. (descr.).

whartoni (Pearse, 1938) Hyman, 1939*e*, 153; 1940*a*, 458, fig. (Florida & North Carolina); 1951, 164, fig. (Florida); du B.-R. Marcus & Marcus, 1968, 4 (Florida).

species Laidlaw, 1903*a*, 9 (Malaysia); Bock, 1913, 64.

species Bock, 1913, 151 (= *Cryptocelis* ? Jacubowa, 1906, 139); 1923*e*, 35, 36.

species Steinbeck & Ricketts, 1941, 336 (Lower California); Hyman, 1953*a*, 277 (to *Alleena mexicana*).

species Prudhoe, 1982*a*, 225 (Victoria, Australia); 1982*b*, 362 (Victoria).

LEPTOCERA Jacubowa, 1906, 135 (type: *delicata*); Bock, 1913, 185 (defined); Marcus, 1949, 76 (systematic position).

delicata Jacubowa, 1906, 135, figs (New Britain, Pacific Ocean); Bock, 1913, 186.

LEPTOPLANA Ehrenberg, 1831, 56 (type: *hyalina*); Örsted, 1843, 568 (diagn.); 1844*a*, 46, 48 (list of spp.); Diesing, 1850, 194 (diagn.); Maitland, 1851, 187; Stimpson, 1857, 21; Schmarda, 1859, 13, 16 (syn. *Elasmodes*); Schmidt, 1861, 5; Diesing, 1862, 532; Johnston, 1865, 5; Lang, 1884, 429, 475; Carus, 1885, 151; Hallez, 1893, 154 (key to spp. on NW. coast of France); 1894, 209 (defined; key to spp.); Verrill, 1893, 477; Gamble, 1893*a*, 498; 1896, 19; Laidlaw, 1903*c*, 307 (key to spp.); 1903*e*, 15 (note); 1904*a*, 5 (key to spp. with penis-stylet); Meixner, 1907*b*, 451; Bock, 1913, 181 (defined); Hyman, 1953*a*, 310 (defined); 310 (constitution); Marcus, 1949, 76; Prudhoe, 1982*c*, 38.

acticola Boone, 1929, 38, figs. (California); Hyman, 1939*a*, 437 (not a *Leptoplana*); Ricketts & Calvin, 1948, 26, 300 (California); 1952, 106, fig., 146; Hyman, 1953*a*, 321 (to *Notoplana*); Stasek, 1966, 7 (location of type-specimen).

acuta Stimpson, 1855*a*, 381 (China); Stimpson, 1857, 21, 26 (descr.; to *Elasmodes*); Diesing, 1862, 527; Lang, 1884, 498.

affinis (Stimpson, 1857) Diesing, 1862, 539; Lang, 1884, 603 (? s.o. *Prosthiostomum grande*); Yeri & Kaburaki, 1918*b*, 42 (s.o. *P. grande*).

alba (Kelaart, 1858) Diesing, 1862, 542 (possibly a *Leptoplana*); Lang, 1884, 613.

alba Lang, in Schmidtlein, 1880, 172 (Mediterranean); Lang, 1884, 101, 471, figs (to *Cryptocelis*).

alcinoi Schmidt, 1861, 5, figs (Corfu, Greece); Diesing, 1862, 541; Selenka, 1881*b*, 229 (Naples); 1881*c*, 2 (France); Lang, 1884, 486, figs (descr.; synonymy; Naples); Carus, 1885, 151 (descr.); Graff, 1886, 342 (Lesina, Adriatic); Lo Bianco, 1888, 399 Naples); 1899*a*, 478 (Naples); Monti, 1900, 1; Laidlaw, 1903*c*, 307; 1904*a*, 5; 1906, 707 (Cape Verde Is); Jacubowa, 1909, 19, fig. (descr.; Black Sea); Micoletzky, 1910, 179 (Trieste); Bock, 1913, 187, 210 (to *Notoplana*); Bresslau, 1928–33, 105, fig.; Kensler, 1964, 961 (Banyuls-sur-Mer); 1965, 856, 875 (coasts of Spain & NW. Africa).

angusta Verrill, 1893, 485, figs (Massachusetts or North Carolina); Laidlaw, 1906, 711 (? congeneric with *Zygantroplana verrilli*); Bock, 1913, 223; Palombi, 1928, 590, figs (Suez Canal; to *Stylochoplana*); Pearse, 1938, 76, fig (Florida & North Carolina & Baffin Bay); Pearse & Littler, 1938, 237, fig. (N. Carolina); Pearse & Walker, 1939, 17, fig. (Florida & N. Carolina); Hyman, 1939*b*, 139 (type-specimens redescribed; to *Stylochoplana*; *S. a.* of Palombi, 1928, *Leptoplana angusta* of Pearse, 1938, and of Pearse & Littler, 1938, are not representatives of Verrill's species); Hyman, 1940*a*, 467 (to *Stylochoplana*); 1940*a*, 468 (*L. a.* of Pearse, 1938, s.o. *Notoplana atomata*); 1940*a*, 472 (*L. a.* of Pearse & Littler, 1938, 237 *nec.* fig. s.o. *Euplana carolinensis*); 1940*a*, 473 (*L. a.* of Pearse & Littler, pl. 20, fig. 4 = *Digynopora americana*); Corrêa, 1949, 175 (to *Zygantroplana*).

arcta (Quatrefages, 1845) Diesing, 1850, 196; 1862, 538; Lang, 1884, 595 (s.o. *Prosthiostomum siphunculus*); Yeri & Kaburaki, 1918*b*, 41 (s.o. *P. siphunculus*).

atomata (Müller, 1776) Örsted, 1843, 569; 1844*a*, 49, fig.; 1844*b*, 79 (Öresund, Denmark/Sweden); Diesing, 1850, 197; Maitland, 1851, 188 (Netherlands); Stimpson, 1857, 21; Leuckart, 1859, 183 (= *Planaria maculata* Dalyell); Diesing, 1862, 532; Johnston, 1865, 6 (Scotland); McIntosh, 1874, 150 (Scotland); 1875, 107; Möbius, 1875, 154; Lang, 1884, 514 (*Planaria*); Gamble, 1896, 19 (Britain); Schultz, 1901, 527; Bock, 1913, 195 (Schultz's material = *Leptoplana tremellaris*), 195, figs (to *Notoplana*).

aurantiaca Collingwood, 1876, 88, 94, fig. (Singapore); Lang, 1884, 615; Laidlaw, 1903*c*, 301, 302, 317 (to *Prosthiostomum*).

australis (Schmarda, 1859) Diesing, 1862, 529; Lang, 1884, 505 (to *Polycelis*); Stummer-Traunfels, 1933, 3486 (to *Notoplana*; Australia and New Zealand).

australis Laidlaw, 1904*a*, 3, 5, fig. (Port Phillip Bay, Australia); Stead, 1907, 1 (N.S.W., Australia); Haswell, 1907*b*, 471, figs (N.S.W., Tasmania & New Zealand); Bock, 1913, 205, fig. (to *Notoplana*; synonymy); Dakin *et al.*, 1948, 208, 218 (N.S.W.); Pope, 1943, 246 (N.S.W.); Dakin, 1953, 144 (eastern coast of Australia).

bacteoalba [err.pro lactoalba Verrill] Palombi, 1939*b*, 108 (s.o. *Stylochoplana pallida*).

badia (Stimpson, 1855*b*) Diesing, 1862, 528; Lang, 1884, 511 (*Dioncus*).

brunnea Cheeseman, 1883, 214 (Auckland Harbour, New Zealand); Lang, 1884, 618; Powell, 1947, 15, fig. (Auckland).

californica Plehn, 1897, 93, fig. (Monterey Bay, California); 1899, 451, fig. (Chatham Is, New Zealand); Laidlaw, 1903c, 308; Bock, 1913, 173, 180 (renamed *Stylochoplana plehni*); Heath & McGregor, 1913, 458, 470 (descr.); Hyman, 1953*a*, 305, 319 (poss. synonymous with *Notoplana inquieta*).

capensis (Schmarda, 1859) Diesing, 1862, 530; Lang, 1884, 506 (*Polycelis*).

chierchiae Plehn, 1896*a*, 155, figs (Peru); Laidlaw, 1903*c*, 308; Ritter-Zahony, 1907, 15, figs (Chile); Bock, 1913, 211 (diagn.; to *Notoplana*; *L. chierchiae 'ex parte* Zahony, 1907: 16' = *Notoplana gardineri*).

chilensis Schmarda, 1859, 17, figs (Chile); Diesing, 1862, 538; Lang, 1884, 509; Stummer-Traunfels, 1933, 3486, 3494 (type-specimen re-examined, to *Stylochoplana*).

chloranota (Boone, 1929) Hyman, 1953*a*, 310, figs (synonymy; British Columbia & California); du B.-R. Marcus & Marcus, 1968, 23 (to *Stylochoplana*).

collaris Stimpson, 1855*a*, 381 (Lu Chu I. (Ryu Kyu I.), Japan); 1857, 22 (to *Prosthiostomum*); Diesing, 1862, 536; Lang, 1884, 612.

concolor Meixner, 1907*a*, 167 (Somalia); 1907*b*, 452, figs; Bock, 1913, 221 (to *Discoplana*).

constipata (Stimpson, 1857) Diesing, 1862, 537; Lang, 1884, 604 (to *Prosthiostomum*).

crassiuscula (Stimpson, 1857) Diesing, 1862, 537; Lang, 1884, 604 (to *Prosthiostomum*).

cribraria (Stimpson, 1857) Diesing, 1862, 537; Lang, 1884, 604 (to *Prosthiostomum*).

delicatula Stimpson, 1857, 22, 27 (Hong Kong); Diesing, 1862, 534; Lang, 1884, 496.

diaphana Stummer-Traunfels, 1933, 3531, figs (Naples); du B.-R. Marcus & Marcus, 1968, 28 (to *Stylochopana*).

discus (Le Conte, 1851) Diesing, 1862, 527; Lang, 1884, 611 (*Elasmodes*).

droebachensis Örsted, 1845, 415 (Norway); Diesing, 1862, 526; Jensen, 1878, 76, figs (Norway); Lang, 1884, 494; Bergendal, 1890, 326 (Sweden); Gamble, 1893*a*, 502 (features; possibly =*Planaria atomata*); 1893*b*, 46 (south Devon); 1896, 19 (Britain); Laidlaw, 1903*c*, 308; 1904*b*, 3, 5 (from 'Godthal' [Godthab, Greenland]; descr.); Bock, 1913, 196 (s.o. *Notoplana atomata*), 202 (*L. d.* of Verrill, 1893, ?=*Notoplana kuekenthalii*); Southern, in Farran, 1915, 35 (Ireland); Southern, 1936, 70 (Ireland).

drowbachensis [err. pro *droebachensis*] Hyman, 1940*b*, 468 (s.o. *Notoplana atomata*).

dubia (Schmarda) Lang, 1884, 499; Stummer-Traunfels, 1933, 3487 (to *Notoplana*).

ellipsis (Dalyell, 1853) Diesing, 1862, 542; Johnston, 1865, 7 (Scotland); McIntosh, 1874, 150 (Scotland); 1875, 107 (Scotland); Gamble, 1893*a*, 511 (possibly =*Stylostomum variabile* Lang); Bock, 1913, 270 (*Stylostomum*).

ellipsoides Girard, *in* Stimpson, 1853, 27, fig. (New Brunswick); Stimpson, 1857, 21; Diesing, 1862, 533; Verrill, 1879, 13 (Cape Cod); Girard, 1893, 200, figs; Verrill, 1893, 483, figs (New England); Kingsley, 1901, 165 (Maine); Laidlaw, 1903*c*, 308; Bock, 1913, 196 (s.o. *Notoplana atomata*); Palombi, 1928, 595 (s.o. *Notoplana atomata*); Hyman, 1940*a*, 454, fig. (to *Discocelides*; synonymy), 468 (*L. e.* of Verrill, 1893, s.o. *Notoplana atomata*); 1952, 195 (to *Plehnia*); 1953*a*, 279 (to *Plehnia*).

elongata (Quatrefages, 1845) Diesing, 1850, 196; 1862, 538; Lang, 1884, 595 (s.o. *Prosthiostomum siphunculus*); Yeri & Kaburaki, 1918*b*, 41 (s.o. *P. siphunculus*).

erythrotaenia (Schmarda, 1859) Diesing, 1862, 529; Lang, 1884, 505 (*Polycelis*); Stummer-Traunfels, 1933, 3486 (to *Notoplana*).

fallax (Quatrefages, 1845) Diesing, 1850, 198; Stimpson, 1857, 31; Diesing, 1862, 533; Lang, 1884, 492; Carus, 1885, 152 (descr.); Vaillant, 1889, 654; Hallez, 1893, 165, 197, fig. (Pas-de-Calais); 1894, 214, fig. (NW. France); Gamble, 1896, 19 (Channel Is); Laidlaw, 1903*a*, 308; Bock, 1913, 204 (to *Notoplana*); Southern, *in* Farran, 1915, 35 (Ireland); Palombi, 1928, 595 (? s.o. *Notoplana atomata*); Southern, 1936, 70 (Ireland).

ferruginea (Schmarda, 1859) Diesing, 1862, 530; Lang, 1884, 506 (*Polycelis*); Stummer-Traunfels, 1933, 3521 (*Notoplana*).

flexilis (Dalyell, 1814) Diesing, 1850, 194; Stimpson, 1857, 3; Leuckart, 1859, 183; Johnston, 1865, 6 (Britain); Lankester, 1866, 388 (Channel Isles); Keferstein, 1869, 6 (s.o. *tremellaris*); McIntosh, 1874, 150 (Scotland); 1875, 107 (Scotland); Lang, 1884, 476 (s.o. *L. tremellaris*).

folium Verrill, 1873, 487, 632 (Long I. Sound to Eastport, Maine); Lang, 1884, 512; Verrill, 1893, 487, figs (to *Trigonoporus*); Girard, 1893, 201; Hyman, 1940*a*, 454 (s.o. *Discocelides ellipsoides*).

formosa (Darwin, 1944) Diesing, 1850, 199; 1862, 541; Lang, 1884, 609 (*Planaria*).

fulva (Kelaart, 1858) Lang, 1884, 613.

fusca Stimpson, 1857, 22, 26 (Hong Kong); Diesing, 1862, 531; Lang, 1884, 497.

gardineri Laidlaw, 1904*b*, 134, fig. (Ceylon); Ritter-Záhony, 1907, 15 (s.o. *chierchiae*); Bock, 1913, 211 (to *Notoplana*).

genicotyla (Palombi, 1939*b*); Marcus, 1949, 76 (combination not made).

gigas Schmarda, 1859, xii, 17, fig. (Ceylon); Diesing, 1862, 544 (to *Centrostomum*); Lang, 1884, 508; Stummer-Traunfels, 1933, 3486, 3492 (type-specimens re-examined; to *Discoplana*).

gracilis (Girard, 1850*b*) Diesing, 1862, 541; Lang, 1884, 610 (to *Prosthiostomum*).

graffii Laidlaw, 1906, 708, fig. (Cape Verde Is); Bock, 1913, 179 (descr.; to *Stylochoplana graffi*).

grandis (Stimpson, 1857) Diesing, 1862, 539; Lang, 1884, 603 (to *Prosthiostomum*).

haloglena (Schmarda, 1859) Diesing, 1862, 528; Lang, 1884, 505 (*Polycelis*); Stummer-Traunfels, 1933, 3486 (to *Notoplana*).

hamata (Schmidt, 1861) Diesing, 1862, 538; Lang, 1884, 595 (s.o. *Prosthiostomum siphunculus*); Yeri & Kaburaki, 1918*b*, 41 (s.o. *P. siphunculus*).

humilis Stimpson, 1857, 22, 27 ('Jesso I.' [? Hokkaido], Japan); Diesing, 1862, 533; Lang, 1884, 496; Yeri & Kaburaki, 1918*b*, 11 (to *Notoplana*).

hyalina Ehrenberg, 1831, 56, figs (Red Sea); Örsted, 1844*a*, 28, 48; Diesing, 1850, 197; Stimpson, 1857, 21; Diesing, 1862, 532; Lang, 1884, 476 (s.o. *tremellaris*); Palombi, 1928, 594 (s.o. *tremellaris*).

inconspicua (Gray, *in* Pease, 1860) Diesing, 1862, 536; Lang, 1884, 615 (*Peasia*).

inquieta Heath & McGregor, 1913, 456, 458, 470, figs (California); Stummer-Traunfels, 1933, 3504 (*Discoplana*); Hyman, 1939*a*, 437 (not a *Leptoplana*); 1953*a*, 319 (to *Notoplana*); Marcus, 1947, 131, 165 (*Euplana*).

irrorata (Gray, *in* Pease, 1860) Diesing, 1862, 536; Lang, 1884, 605 (*Peasia*).

jaltensis (Czerniavsky, 1881) Lang, 1884, 500; Jacubowa, 1909, 1 (s.o. *tremellaris*).

kuekenthalii Plehn 1896*a*, 149, figs (Spitzbergen); Laidlaw, 1903*c*, 307; Bock, 1913, 202, figs (to *Notoplana*; synonymy).

lactea (Stimpson, 1857) Diesing, 1862, 531; Lang, 1884, 470 (to *Discocelis*).

lacteoalba [err. pro *lactoalba* Verrill] Laidlaw, 1903*a*, 308.

lactoalba Verrill, 1900, 595, fig. (Bermuda); 1904, 595; Bock, 1913, 173, 179 (to *Stylochoplana 'lacteoalba'*); Palombi, 1939*b*, 108 (*'bacteoalba'* s.o. *Stylochoplana pallida*); Hyman, 1939*d*, 6, figs (to *Notoplana*).

lactoalba var. *tincta* Verrill, 1903, 46, fig. (Bermuda); Bock, 1913, 179 (s.o. *Stylochoplana lacteoalba [sic]*); Crozier, 1918, 379 (*tincta* apparently regarded as a distinct species); Palombi, 1939*b*, 108 (*L. bacteoalba [sic]* var. *tincta* s.o. *Stylochoplana pallida*).

lanceolata Schmarda, 1859, 19, figs (Chile); Diesing, 1862, 540; Lang, 1884, 510; Stummer-Traunfels, 1933, 3496 (s.o. *Stylochoplana chilensis*; type-specimen re-examined).

levigata (Quatrefages, 1845) Diesing, 1850, 198 (*'laevigata'*); Stimpson, 1857, 21; Schmidt, 1861, 8, figs (*'laevigata'*; Cephalonia, Greece); Diesing, 1862, 532 (*'laevigata'*); Keferstein, 1869, 6 (s.o. *tremellaris*); Czerniavsky, 1881, 218 (Black Sea); Stossich, 1882, 225 (Trieste; *'laevigata'*); Lang, 1884, 476 (s.o. *tremellaris*); Palombi, 1928, 594 (s.o. *tremellaris*).

lichenoides (Mertens 1832) Örsted, 1844*a*, 49; Diesing, 1850, 200 (to *Centrostomum*); Lang, 1884, 469 (to *Discocelis*).

limnoriae Hyman, 1953*a*, 313, figs (in burrows of *Limnoria*; California); du B.-R. Marcus & Marcus, 1968, 24 (to *Stylochoplana*).

littoralis [Author ?] Morgan, 1905, 187 (California); Bock, 1913, 223.

lutea [err. pro *luteola* (Delle Chiaje)] Örsted, 1844*a*, 49; Diesing, 1850, 199; 1862, 542.

luteola (Delle Chiaje, 1822) Örsted, 1844*a*, 49 (to *Leptoplana* (?) *'lutea'*).

lyrosora (Schmarda, 1859) Diesing, 1862, 535; Lang, 1884, 507 (to *Polycelis*).

macrorhyncha (Schmarda, 1859) Diesing, 1862, 531; Lang, 1884, 614 (*Polycelis*; perhaps a *Prosthiostomum*); Stummer-Traunfels, 1933, 3486 (to *Prosthiostomum*).

macrosora Schmarda, 1859, 18, figs (Jamaica); Diesing, 1862, 538; Lang, 1884, 615; Hyman, 1955*b*, 148 (prob. unrecognizable).

maculosa Stimpson, 1857, 22, 27 (California); Diesing, 1862, 534; Lang, 1884, 496; Heath & McGregor, 1913, 455, 472 (descr.); Hyman, 1953*a*, 361 (? = *Parviplana californica* or *Diplandros singularis*).

malayana Laidlaw, 1903*c*, 302, 306, 308, figs (Malaysia); Bock, 1913, 221 (descr.; to *Discoplana*).

mertensii (Claparède, 1861) Lang, 1884, 499; Gamble, 1893*a*, 501.

microsora (Schmarda, 1859) Diesing, 1862, 529; Lang, 1884, 506 (to *Polycelis*).

modesta (Quatrefages, 1845) Diesing, 1850, 195; 1862, 527; Schmidt, 1861, 10; Lang, 1884, 490 (s.o. *pallida*).

monosora Schmarda, 1859, 16, fig. (Ceylon); Diesing, 1862, 535; Lang, 1884, 508; Stummer-Traunfels, 1933, 3486, 3490 (type-specimen examined; to *Prosthiostomum*).

moseleyi Lang, 1884, 500 (2°55′N, 124°53′E, off Siao I., Indonesia; nom. nov. pro *Leptoplana* (?) sp. Moseley, 1877).

nadiae Melouk, 1941, 41, figs (Red Sea); du B.-R. Marcus & Marcus, 1968, 23 (to *Stylochoplana*).

nationalis Plehn, 1896*b*, 6, figs (8°S, 14°W, off Ascension I., Atlantic); Laidlaw, 1903*c*, 308; 1904*a*, 5; 1904*b*, 135; Bock, 1913, 207 (renamed *Notoplana atlantica*).

nigripunctata Örsted, 1843, 569 (Scandinavia); 1844*a*, 49; 1844*b*, 79 (Öresund, Denmark/Sweden); Diesing, 1850, 198 (*'nigropunctata'*); Lang, 1884, 513.

notabilis (Darwin, 1844) Diesing, 1862, 542; Lang, 1884, 501.

oblonga Stimpson, 1857, 22, 27 (Port Simoda, Honshu, Japan); Diesing, 1862, 533; Lang, 1884, 496; Kato, 1944, 276 ('may be referable to *Notoplana humilis*').

obovata (Schmarda, 1859) Diesing, 1862, 528; Lang, 1884, 504 (*Polycelis*); Stummer-Traunfels, 1933, 3505 (to *Adenoplana*).

obscura Stimpson, 1855*a*, 381 (China); 1857, 22 (to *Prosthiostomum*); Diesing, 1862, 539; Lang, 1884, 604 (*Prosthiostomum*).

obtusum (Collingwood, 1876) Laidlaw, 1903*c*, 302; Dawydoff, 1952, 81 (Vietnam).

oosora (Schmarda, 1859) Diesing, 1862, 530; Lang, 1884, 506 (*Polycelis*); Stummer-Traunfels, 1933, 3486 (to *Indiplana*).

ophryoglena (Schmarda, 1859) Diesing, 1862, 526; Lang, 1884, 613 (*Polycelis*; perhaps a *Prosthiostomum*).

orbicularis (Schmarda, 1859) Diesing, 1862, 527; Lang, 1884, 504 (*Polycelis*); Stummer-Traunfels, 1933, 3486 (to *Enterogonia*).

otophora Schmarda, 1859, xiii, 18, figs (Ceylon); Diesing, 1862, 527; Lang, 1884, 509, fig.; Stummer-Traunfels, 1933, 3486, 3497 (type-specimen re-examined; to *Notoplana*).

pacificola Plehn, 1896*a*, 140, 153, figs (var. *chilensis* & var. *peruensis* Chile & Peru, respectively); Laidlaw, 1903*c*, 308; Bock, 1913, 220 (to *Discoplana*); Hyman, 1953*a*, 332 (to *Euplana*).

pallida (Quatrefages, 1845) Diesing, 1850, 195; 1862, 527; Lang, 1884, 489, figs (synonymy; Bay of Naples); Carus, 1885, 152 (descr.); Graff, 1886, 342 (Adriatic); Lo Bianco, 1888, 399 (Naples); Francotte, 1898, 239, figs (Roscoff); Lo Bianco, 1899*a*, 478 (Naples); Benham, 1901, 32, fig.; Laidlaw, 1903*c*, 308; 1906, 708 (Cape Verde Is); Bock, 1913, 172, 179 (to *Stylochoplana*); Riedl, 1953, 132 (Sicily); 1959, 204 (Tyrrhenian Sea); Euzet & Poujol, 1963, 826 (S. France).

panamensis Plehn, 1896*a*, 151, figs (Gulf of Panama); Laidlaw, 1903*c*, 308; Bock, 1913, 173, 179 (to *Stylochoplana*).

pardalis Laidlaw, 1902, 287, figs (Maldives Is, Indian Ocean, also Funafuti I., Pacific Ocean); 1903*a*, 7 (Funafuti); 1903*c*, 307 (=*subviridis*); 1903*b*, 580 (=*subviridis*); Bock, 1913, 220 (s.o. *Discoplana subviridis*).

patellarum Stimpson, 1855*b*, 389 (under limpets on rocks, Simon's Bay, South Africa); 1857, 22, 27 (descr.); Diesing, 1862 ,534; Lang, 1884, 496; Palombi, 1936, 18 (?=*Notoplana ovalis* Bock); 1939*a*, 128, fig. (East London, S. Africa; to *Notoplana*; synonymy).

patellensis Collingwood, 1876, 88, 93, figs (under mantle of limpet, Simon's Bay, South Africa); Lang, 1884, 511; Palombi, 1936, 18 (?=*Notoplana ovalis* Bock); 1939*a*, 128 (s.o. *Notoplana patellarum*).

pellucida Grube, 1840, 53 (Sicily); Delle Chiaje, 1841, pt. 5, iii (s.o. *Planaria syphunculus [sic]*); Örsted, 1844*a*, 48; Diesing, 1850, 196; 1862, 532; Lang, 1884, 605 (to *Prosthiostomum*); Carus, 1885, 152; Bock, 1913, 283 (? *Prosthiostomum*).

polycyclia (Schmarda, 1859) Lang, 1884, 498.

polysora (Schmarda, 1859) Lang, 1884, 499, fig.; Stummer-Traunfels, 1933, 5486 (*Leptostylochus*).

punctata Stimpson, 1857, 22, 27 (Oho-sima I., Japan); Diesing, 1862, 534; Lang, 1884, 497.

punctata (Kelaart, 1858) Diesing, 1862, 542 (? *Leptoplana*); Lang, 1884, 616 (=*Penula punctata*).

purpurea Schmarda, 1859, 18, figs (Jamaica); Diesing, 1862, 540; Lang, 1884, 509; Stummer-Traunfels, 1933, 3486, 3499 (to *Phaenocelis*; type-specimens re-examined, also includes a species of *Phylloplana*).

rupicola Heath & McGregor, 1913, 455, 457, 464, figs (California); Hyman, 1939*a*, 437 (prob. a *Notoplana*); 1953*a*, 315 (to *Notoplana*).

saxicola Heath & McGregor, 1913, 456, 457, 467, figs (California); Hyman, 1939*a*, 437 ('evidently a *Notoplana*'); 1953*a*, 318 (to *Notoplana*); Stasek, 1966, 8 (location of type-specimen).

schizoporellae Hallez, 1893, 155, 156, 197, figs (NW France); 1894, 212, figs (NW. France); Bock, 1913, 223; Prudhoe, 1982*c*, 42, fig.

schoenbornii Stimpson, 1857, 22, 26 (Cape of Good Hope, South Africa); Diesing, 1862, 530; Lang, 1884, 497.

sciophila Boone, 1929, 40, figs (California); Hyman, 1939*a*, 437 (*Stylochoplana* or *Notoplana*); 1953*a*, 323, figs (to *Notoplana*); Stasek, 1966, 8 (location of type-specimen).

sparsa Stimpson, 1855*a*, 381 (Kikaisima, Japan); 1857, 22, 29 (to *Prosthiostomum*); Diesing, 1862, 540; Lang, 1884, 603 (*Prosthiostomum*).

stimpsoni Diesing, 1862, 528 (nom. nov. pro *Dioncus oblongus* Stimpson); Lang, 1884, 511 (s.o. *Dioncus oblongus*).

striata Schmarda, 1859, 17, figs (Peru); Diesing, 1862, 536; Lang, 1884, 508.

subauriculata (Johnston, 1836) Diesing, 1850, 195; Maitland, 1851, 188 (Netherlands); Diesing, 1862, 527; Johnston, 1865, 6; Lankester, 1866, 388 (Channel Is); McIntosh, 1874, 150 (Scotland); 1875, 106 (Scotland); Lang, 1884, 459 (s.o. *Stylochoplana maculata*).

subviridis Plehn, 1896*c*, 330, figs (Indonesia); Laidlaw, 1903*b*, 580 (syn. *pardalis*); Meixner, 1907*a*, 167 (African coast of Arabian Sea); 1907*b*, 457, figs; Bock, 1913, 220, figs (to *Discoplana*; Somali

coast, Maldive Is, the Moluccas, Timor & Funafuti).
suteri Jacubowa, 1906, 150, figs (New Zealand); Bock, 1913, 205 (s.o. *Notoplana australis*).
taenia (Schmarda, 1859) Lang, 1884, 498.
tenebrosa (Stimpson, 1857); Diesing, 1862, 538; Lang, 1884, 604 (*Prosthiostomum*).
tenella (Stimpson, 1857) Diesing, 1862, 528; Lang, 1884, 498.
tigrina (Blanchard, 1847) Diesing, 1850, 195; 1862, 527; Lang, 1884, 467 (to *Discocelis*).
timida Heath & McGregor, 1913, 455, 457, 466, figs (California); Bresslau, 1928–33, 286, fig.; Hyman, 1939*a*, 437 (prob. not a *Leptoplana*); 1953*a*, 361, 362 (? = *Notoplana sanguinea*); 1954*d*, 56, fig. (NW. U.S.A.).
trapezoglena (Schmarda, 1859) Diesing, 1862, 531; Lang, 1884, 507 (*Polycelis*); Stummer-Traunfels, 1933, 3486, 3528, figs (type-specimen re-examined).
tremellaris (Müller, 1774) Örsted, 1843, 569; 1844*a*, 49 (descr.); Diesing, 1850, 197; Maitland, 1851, 187 (Netherlands); Stimpson, 1857, 21; Diesing, 1862, 532; Johnston, 1865, 6 (Scotland); Keferstein, 1869, 6, figs (NW. France; synonymy); Uljanin, 1870, 106 (Black Sea); Mobius, 1875, 104 (Keil Bay); Giard, 1877, 812 (parasites); Jensen, 1878, 77; Hallez, 1878*a*, 193 (embryology); 1878*c*, 250 (embryology); 1879*b*, 95, figs (embryology); Mereschkowsky, 1879, 53 (White Sea); Levinsen, 1879, 199 (Greenland); Selenka, 1881*b*, 229 (NW. France); 1881*c*, 1 (NW. France). Francotte, 1883, 465 (Belgium); Lang, 1884, 476, figs (Italy; synonymy); Koehler, 1885, 14, 37, 49, 56 (Channel Is); Carus, 1885, 151 (descr.); Wagner, 1885, 50 (White Sea); Marenzeller, 1886 (Jan Mayen I.); Graff, 1886, 342; Heape, 1888, 168 (S. Devon); Lo Bianco, 1888, 399 (Naples); Hoyle, 1889, 458 (West Scotland); Vaillant, 1889, 654, (NW. France); Graff, 1892*b*, 217; Gamble, 1893*a*, 498 (descr.; synonymy); 1893*b*, 45 (S. Devon); 1893*c*, 166, fig. (Liverpool Bay); Hallez, 1893, 155, 197, figs (Pas-de-Calais); 1894, 210 (NW France); Francotte, 1894, 382, figs; Gamble, 1896, 7, 19, figs (Britain); Francotte, 1897, 5, figs; Pruvot, 1897, 20 (NW. France & Gulf of Lion); Francotte, 1898, 189 (embryology); Lo Bianco, 1899*a*, 399 (Naples); Sabussow, 1897, 1 (U.S.S.R); Gamble, 1900, 812 (Ireland); Gérard, 1901, 143, fig.; Colgan, 1907, 323 (Ireland); Jacubowa, 1909, 20, 21, fig. (Black Sea); Micoletzky, 1910, 179 (Trieste); Hallez, 1911*a*, xx; Bock, 1913, 181, fig. (Scandinavia); 202 (*L. t.* of Mereschkowsky, 1879, of Levinsen, 1879, of Wagner, 1885, and of Marenzeller, 1886, all probably = *Notoplana kuekenthalii*); Southern, *in* Farran, 1915, 35 (Ireland); Chumley, 1918, 50, 69, 116, 160 (Clyde Sea area, Scotland); Palombi, 1928, 594 (Port Said); Wesenberg-Lund, 1928, 56 (distribution in Greenland); Remane, 1929, 78 (Kiel Bay); Steinbock, 1933, 18 (Gulf of Trieste); Stummer-Traunfels, 1933, 3572, fig.; Palombi, 1936, 16 (associating with hermit-crab, Bay of Naples); Fleming, 1936, 265 (Outer Hebrides); Southern, 1936, 70 (Ireland); Eales, 1939, 54, fig.; Jones, 1939, 21 (Port Erin, Isle of Man); Arndt, 1943, 2 (Naples); Bassindale & Barrett, 1957, 251 (S. Wales); Purchon, 1948, 296, 300, 308 (S. Wales); 1957, 219 (Bristol Channel); Williams, 1954, 55 (NE. Ireland); Crothers, 1966, 22 (S. Wales); Marinescu, 1971, 41 (Black Sea); Laverack & Blackler, 1974, 32 (St Andrews Bay, Scotland); Hendelberg, 1974*b*, 13, figs (Sweden); Galleni & Puccinelli, 1981, 37 (Gt Britain); Prudhoe, 1982*c*, 39, figs (Britain).
tremellaris forma *mediterranea* Bock, 1913, 184.
tremellaris var. *taurica* Jacubowa, 1909, 21, figs (Black Sea).
trullaeformis Stimpson, 1855*a*, 381 (China); 1857, 22, 27 (descr.; Hong Kong); Diesing, 1862, 535; Lang, 1884, 497.
tuba Grube, 1871, 28 (Fiji); Lang, 1884, 603 (? s.o. *Prosthiostomum grande*).
variabilis (Girard, 1850*b*) Diesing, 1862, 542; Verrill, 1879, 13 (Massachusetts); Lang, 1884, 610 (*Polycelis*); Girard, 1893, 199 (descr.); Kingsley, 1901, 165 (Maine); Laidlaw, 1903*c*, 308; Sumner *et al.*, 1913, 580 (Massachusetts); Bock, 1913, 196 (s.o. *Notoplana atomata*); Palombi, 1928, 595 (s.o. *Notoplana atomata*); Hyman, 1939*b*, 135 (? s.o. *Notoplana atomata*); 1940*a*, 468 (s.o. *N. atomata*).
vesiculata Hyman, 1939*a*, 434, fig. (Washington, State, U.S.A.); 1953*a*, 313 (note).
virilis Verrill, 1893, 478, figs (Massachusetts); Laidlaw, 1903*c*, 308; Bock, 1913, 187, 208 (to *Notoplana*); Pearse & Walker, 1939, 17, fig. (Baffin Bay); Hyman, 1940*a*, 468 (s.o. *Notoplana atomata*).
vitrea Lang, 1884, 71, 493, figs (Trieste & Naples); Carus, 1885, 152 (descr.); Laidlaw, 1903*c*, 307; 1904*a*, 5; Micoletzky, 1910, 179 (Trieste); Bock, 1913, 207 (to *Notoplana*).
species (Schultze, 1854) of von Kennel, 1879, 120; Lang, 1884, 477 (s.o. *Leptoplana tremellaris*).
species Uljanin, 1870, 107 (near *alcinoi*; Black Sea).

species (*Dicelis*) Studer, 1876, 7 (Kerguelen Is); 1879, 123; Lang, 1884, 515.
species Moseley, 1877, 27, fig. (2°55′N, 124°53′E Indonesia); Lang, 1884, 500 (*Leptoplana moseleyi*).
species Andrews, 1892, 75 (Jamaica); Bock, 1913, 223; Hyman, 1955*b*, 148 (probably unrecognizable).
species Bock, 1913, 224 (Heligoland).
species Wilson, 1898, 63 (Washington State); Bock, 1913, 224.
species Shelford, 1901, 21 (in freshwater, Borneo). [cf. *Limnostylochus*]
species Laidlaw, 1902, 289, 289 (Maldive Is, Indian Ocean); Bock, 1913, 223.
species Child, 1904*a*, 95; 1904*b*, 463; 1904*c*, 513; 1905*a*, 253
species Jacubowa, 1906, 148 (New Caledonia, Pacific Ocean); Bock, 1913, 224.
species Graw & Darsie, 1918, 68 (Laguna Beach, California).
species Dakin *et al*., 1948, 224 (N.S.W., Australia).

LEPTOPLANEA Ehrenberg, 1831, 56 (family of order Rhabdocoela); Orsted, 1843, 535 (family & generic name); 1844*a*, 27; Lang, 1844, 466 (*ex parte* = Leptoplanidae).

LEPTOPLANID Bock, 1923*c*, 356, figs (Juan Fernandez Is, Pacific Ocean); Palombi, 1928, 603, 604 (to *Notoplanides* sp.?); Marcus, 1947, 129, 165 (? a *Pucelis*).

LEPTOPLANIDAE Stimpson, 1957, 21; Schmidt, 1861, 5; Lang, 1884, 429, 466 (diagn.); Carus, 1885, 151; Hallez, 1893, 149 (key to genera); Verrill, 1893, 475 (morphological notes); Gamble, 1893*a*, 498; Hallez, 1894, 203 (key to genera); Benham, 1901, 31; Meixner, 1907*b*, 451; Bock, 1913, 167 (taxonomy; defined); 1925*a*, 23; Stummer-Traunfels, 1933, 3567, 3585, 3586; Bresslau, 1928–33, 288; Freeman, 1933, 116; Hyman, 1939*b*, 134; 1940*a*, 467; 1953*a*, 301 (defined); 343 (systematic note); Marcus, 1947, 109, 163 (systematic note); Hyman, 1959*a*, 554, 593, 594, figs (juveniles from Micronesia); 1959*c*, 7 (defined).

LEPTOPLANIDEA Diesing, 1862, 492, 524 (family of Digonopora).

LEPTOPLANINAE Marcus, 1947, 110, 163 (= 1st series of Bock, 1913, 168); 1949, 75 (constitution); Hyman, 1953*a*, 343 (criticism).

LEPTOSTYLOCHUS Bock, 1925*b*, 98, 111, 120, 171 (type: *elongatus*); Palombi, 1938, 341 (defined); Hyman, 1959*c*, 3 (defined).
capensis Palombi, 1938, 329, 334, figs (= *Leptostylochus* sp. ? Palombi, 1936; South Africa); 1939*a*, 125, fig. (S. Africa); Day *et al*., 1970 (False Bay, S. Africa).
elongatus Bock, 1925*b*, 98, 111, 177, figs (New Zealand).
gracilis Kato, 1934*a*, 374, figs (Japan); 1944, 266.
novacambrensis Hyman, 1959*c*, 3, figs (N.S.W., Australia).
ovatus Kato, 1937*f*, 347, 349, figs (Japan); 1944, 266.
polysorus (Schmarda, 1859) Stummer-Traunfels, 1933, 3486, 3533, figs (type-specimen re-examd.; New Zealand).
species Palombi, 1936, 1, 11, figs (S. Africa); 1938, 334 (= *capensis*).
species Kato, 1937*b*, 233, fig. (Korea).
species nov. Bock, 1925*b*, 120, 177 (N.S.W., Australia).

LEPTOTEREDRA Hallez, 1913, 38, 40, 41 (type: *maculata*); Bock, 1922, 14 (relationships).
maculata (Hallez, 1905*b*) Hallez, 1913, 38, 40 (affinities & specific definition); Bock, 1922, 14; Westblad, 1952, 10 (South Georgia).
stylostomoides (Gemmill & Leiper, 1907) Bock, 1922, 15 (probably identical with *maculata*); Marcus, 1947, 141.
tentaculata Kato, 1943*a*, 48 figs (Japan); Kato, 1944, 306.

LICHENIPLANA Heath & McGregor, 1913, 474 (type: *lepida*), 458 ('*Lichenoplana*').
lepida Heath & McGregor, 1913, 458, 474, figs (California); Hyman, 1953*a*, 370; du B.-R. Marcus, 1955*a*, 42 (to *Pseudoceros (Acanthozoon)*.)

LIMNOSTYLOCHUS Bock, 1913, 344 corrigendum (pro *Shelfordia* Stummer-Traunfels) (type: *borneensis*); Bock, 1923*c*, 348; 1925*b*, 172.
amara (Kaburaki, 1918) Bock, 1925*b*, 177.
annandalei (Kaburaki, 1918) Bock, 1925*b*, 177.
borneensis (Stummer-Traunfels, 1902) Bock, 1923*c*, 346; 1925*b*, 177.

LOBOPHORA Dawydoff, 1940, 448, 450 (nom. nov. pro Müller's larva).
actinotrocha Dawydoff, 1940, 460, 461 (Vietnam).
albonigra Dawydoff, 1940, 457 (Vietnam).

flavofusca Dawydoff, 1940, 465, 466 (Vietnam).
gargantua Dawydoff, 1940, 461 (Vietnam).
gigantea Dawydoff, 1940, 448, 457, 460 (Vietnam).
lineata Dawydoff, 1940, 464, 465 (Vietnam).
rubra Dawydoff, 1940, 462 (Vietnam).
LONGIPROSTATUM Hyman, 1953*a*, 300 (type; *rickettsi*).
rickettsi Hyman, 1953*a*, 300, figs (Gulf of California).
LURYMARE du Bois-Reymond Marcus & Marcus, 1968, 88 (type: *drygalskii*); Poulter, 1975, 317 (subgenus of *Prosthiostomum*).
delicatum (Palmobi, 1939*a*) du B.-R. Marcus & Marcus, 1968, 88.
drygalskii (Bock, 1931) du B.-B. Marcus & Marcus, 1968, 88.
gabriellae (Marcus, 1949) du B.-R. Marcus & Marcus, 1968, 89.
matarazzoi (Marcus, 1950) du B.-R. Marcus & Marcus, 1968, 89 (Brazil & Curaçao).
purum (Kato, 1937*f*) du B.-R. Marcus & Marcus, 1968, 88.
russoi (Palombi, 1939*a*) du B.-R. Marcus & Marcus, 1968, 89.
utarum (Marcus, 1952) du B.-R. Marcus & Marcus, 1968, 90 (Brazil & Florida).

MACGINITIELLA Hyman, 1953*a*, 337 (type: *delmaris*).
delmaris Hyman, 1953*a*, 337, figs (*California*).
MARCUSIA Hyman, 1953*a*, 296 (type: *ernesti*); 1955*e*, 262 (s. o. *Pericelis*).
ernesti Hyman, 1953*a*, *296*, figs (Mexico); 1955*e*, 263 (to *Pericelis*); Brusca, 1973, 66 (Gulf of California).
MARINA Leuckart, 1856, 109 (section of family Dendrocoela to include polyclads).
MEIXNERIA Bock, 1913, 36, 112, 119 (type: *furva*); 1925*b*, 171.
furva Bock, 1913, 112, figs (Gulf of Siam); 1925*b*, 177.
MESOCELA Jacubowa, 1906, 141 (type: *caledonica*); Bock, 1913, 159 (diagnosis and systematic position).
caledonica Jacubowa, 1906, 141, figs (New Caledonia, Pacific Ocean).
MESODISCUS Minot, 1877*a*, 451 (type: *inversiporus*); Lang, 1884, 594 (s.o. *Prosthiostomum*).
inversiporus Minot, 1877*a*, 451 (Trieste); Lang, 1884, 595 (s.o. *Prosthiostomum siphunculus*); Yeri & Kaburaki, 1918*b*, 41 (s.o. *P. siphunculus*).
METAPOSTHIA Palombi, 1924*a*, 35 (type: *norfolkensis*); 1924*b*, 12, 15.
norfolkensis Palombi, 1924*a*, 35 (off Norfolk I., Pacific Ocean); 1924*b*, 12, figs (28°20′S, 170°5′E, Pacific Ocean).
MEXISTYLOCHUS Hyman, 1953*a*, 291 (type: *tuberculatus*); Hyman, 1955*a*, 9 (s.o. *Ommatoplana*); 1955*c*, 74 (s.o. *Ommatoplana*).
levis Hyman, 1953*a*, 293, fig. (Mexico); 1955*a*, 9 (to *Ommatoplana*); Brusca, 1973, 66 (Gulf of California).
tuberculatus Hyman, 1953*a*, 291, figs (California & Mexico); 1955*a*, 9 (referred to *Ommatoplana* & renamed *O. mexicana*); Brusca, 1973, 64, 66 (Gulf of California).
MICROCELIS Plehn, 1899, 449 (type: *schauinslandi*); Bock, 1913, 151.
schauinslandi Plehn, 1899, 449, fig. (Tasmania); Haswell, 1907*b*, 474 (descr.; Tasmania): Bock, 1913; 151; Prudhoe, 1982*a*, 225 (descr.).
MICROCOELA Örsted, 1843, 548, 568 (family of tribe Planariea containing *Leptoplana* & *Typhlolepta*); Lang, 1884, 433 (= Polycladidea).
MIROSTYLOCHUS Kato, 1937*e*, 124, 127 (type: *akkashiensis*).
akkashiensis Kato, 1937*e*, 124, figs (Japan); 1944, 267.
MONOSOLENIA Hyman, 1853*a*, 359 (type: *asymmetrica*).
asymmetrica Hyman, 1953*a*, *359*, figs (Lower California).
MUCROPLANA Sopott-Ehlers & Schmidt, 1975, 207 (type: *caelata*).
caelata Sopott-Ehlers & Schmidt, 1975, 208, figs (Galapagos Is).
MUCROPLANINAE Sopott-Ehlers & Schmidt, 1975, 207.
MÜLLER'S LARVA Müller, 1850, 485, figs (Mediterranean); 1854, 75, fig. (Mediterranean); Claparède, 1863, 22, fig. (Normandy); Steinböck, 1937, 14 (Alexandria); Dawydoff, 1940, 450 (= *Lobophora*).

NAUTILOPLANA Stimpson, 1857, 20 (type: *oceanica*); Diesing, 1862, 493, 546.

oceanica (Darwin, 1844) Stimpson, 1857, 20; Diesing, 1862, 546; Lang, 1884, 608 (*Planaria*).

NAUTILOPLANIDAE Stimpson, 1857, 20.

NAUTILOPLANIDEA Diesing, 1862, 493, 545.

NEOPLANOCERA Yeri & Kaburaki, 1918*a*, 432, 436 (type: *elongata*); 1918*b*, 2, 17, 47 (features); Kato, 1937*a*, 223 (defined); Marcus, 1947, 133 (syn. *Cirroposthia*).

elongata Yeri & Kaburaki, 1918*a*, 432, 436 (diagn.; Japan); 1918*b*, 2, 17, 47, figs (Japan); Kato, 1937*a*, 220, figs (descr.; Japan); 1944, 286.

steueri (Steinböck, 1937) Marcus, 1947, 133 (combination not made).

NEOSTYLOCHUS Yeri & Kaburaki, 1920, 591, 595 (type: *fulvopunctatus*); Bock, 1923*c*, 346, 348; 1925*b*, 171.

fulvopunctatus Yeri & Kaburaki, 1920, 591, figs (Japan); Bock, 1925*b*, 177; Kato, 1944, 266.

pacificus Bock,1923*c*, 342, figs (Juan Fernandez I., Pacific Ocean); 1925*b*, 177; Prudhoe, 1982*b*, 363 (to *Ancoratheca*).

viridis Bock, 1925*a*, 3, 4, [MS. name]; 1927*a*, 20.

species Bock, 1923*c*, 346 (Japan).

'sp. nov.' Day *et al.*, 1970, 19 (False Bay, South Africa).

NONATONA Marcus, 1952, 5, 77, 114 (type: *euscopa*).

euscopa Marcus, 1952, 5, 77, 114, figs (Brazil); du B.-R. Marcus, 1955*b*, 282.

NOTOCERIDEAE Diesing, 1850, 215 (family); Schmarda, 1859, 14, 33 (emended to Notoceroidea); Lang, 1884, 433 (=Planoceridae).

NOTOCEROIDEA Schmarda, 1859, 14, 33 (=Notocerideae Diesing, family of suborder Acarena); Lang, 1884, 433 (=Planoceridae).

NOTOPLANA Laidlaw, 1903*c*, 301, 302, 305 (type: *evansii*); Meixner, 1907*b*, 447; Bock, 1913, 185 (genus divided into groups A, B & C); Yeri & Kaburaki, 1918*a*, 434 (features); 1918*b*, 46 (features); Freeman, 1933, 120; Hyman, 1939*b*, 134; 1939*c*, 4; 1940*a*, 468; 1953*a*, 315 (defined); 1959*c*, 8 (defined); Marcus, 1948, 180, 182 (systematic notes); 1954*a*, 61 (systematic notes); du B.-R. Marcus & Marcus, 1968, 37 (notes on constitution).

alcinoi (Schmidt, 1861) Bock, 1913, 187, 210 (diagn.); Palombi, 1930, 2 (Mediterranean); Marinescu, 1971, 45 (external features; Rumania); Domenici *et al.*, 1975, 239 (Italy); Galleni *et al.*, 1976, 62 (Tuscan coast); Galleni & Puccinelli, 1981, 33 (Tuscan coast); Lanfranchi *et al.*, 1981, 267 (Ligurian Sea).

acticola (Boone, 1929) Hyman, 1953*a*, 321, figs. (California); 1955*a*, 9; Thum, 1970, 553 (California; reproductive ecology); 1974, 431 (California); Chien & Koopowitz, 1972, 277 (southern California); Morris *et al.*, 1981, 78, fig. (coloured).

annula du Bois-Reymond Marcus & Marcus, 1968, 43, figs. (Curaçao & Florida).

atlantica Bock, 1913, 207 (nom. nov. pro *Leptoplana nationalis* Plehn, off Ascension I., Atlantic).

atomata (Müller, 1776) Bock, 1913, 195, figs (synonymy; Scandinavia, Faroes, Shetlands, Newfoundland & Massachusetts); Palombi, 1928, 594 (Port Said; note on distribution & synonymy); Hadenfeldt, 1929, 596, figs; Remane, 1929, 73, fig. (Kiel Bay; pelagic stage); Bresslau, 1930, 146, fig.; Steinböck, 1932, 333, 337 (distribution); 1938, 17 (Iceland); Pearse, 1938, 75 (Maine & Newfoundland); Pearse & Walker, 1939, 17, fig. (SE Canada); Hyman, 1939*b*, 135, figs (Maine); 1939*d*, 18 (descr.; synonymy); 1940*a*, 468 (descr.; distribution & synonymy); 1953*a*, 315 (Maine & Alaska); 387 (distribution); Westblad, 1955, 6 (W. Norway); Hyman, 1951, 95, 164, figs (New England); Hartog, 1968, 215 (English Channel); 1977, 29 (Netherlands); Holleman, 1972, 410 (Puget Sound); Mettrick & Boddington, 1972, 1 (New Brunswick); Hendelberg, 1974*b*, 17–18, figs (Sweden); Prudhoe, 1982*c*, 48, fig. (Britain).

australis (Laidlaw, 1904*a*) Bock, 1913, 205, fig. (synonymy); 1925*b*, 143 (*'australiensis'*); Stummer-Traunfels, 1933, 3487, 3518 (s.o. *N. australis* (Schmarda).)

australis (Schmarda, 1859) Stummer-Traunfels, 1933, 3486, 3487, 3517, figs (type-specimen re-examined); Dakin, 1953, 1ff. (Australia); Hyman, 1959*c*, 8, fig. (N.S.W., Australia); Prudhoe, 1982*a*, 225 (S. Australia); 1982*b*, 366 (descr.; S. Australia); Anderson, 1977, 303 (N.S.W., development).

australis (Schmarda, 1859) forma *huina* Marcus, 1954*a*, 56, figs (Chile); Prudhoe, 1982*b*, 366 (=*australis* s.s.).

bahamensis Bock, 1913, 187, 208, figs (Bahamas); Stummer-Traunfels, 1933, 3487 (s.o. *ferruginea*);

Hyman, 1939*c*, 3 (s.o. *binoculata*); 1939*d*, 8 (s.o. *binoculata*, but not of *ferruginea*); 1955*b*, 131 (s.o. *ferruginea*).

binoculata (Verrill, 1903) Hyman, 1939*c*, 3 (syn. *bahamensis*); 1939*d*, 8, figs (syn. *bahamensis*); 1955*b*, 131 (s.o. *ferruginea*).

caribbeana Hyman, 1939*c*, 2, figs (Old Providence I., W. Indies); 1939*d*, 9 (s.o. *ferruginea*).

celeris Freeman, 1933, 111, 112, 125, figs (Puget Sound); Hyman, 1953*a*, 329 (lectotype descr.).

chierchiae (Plehn, 1896*a*) Bock, 1913, 211 (diagn.); Marcus, 1954*a*, 63, figs (Chile); du B.-R. Marcus, 1957, 173 (Peru).

cotylifera Meixner, 1907*a*, 167; 1907*b*, 448, figs (Somalia); Bock 1913, 190 (diagn.); Prudhoe, 1952, 177 (Red Sea).

delicata (Yeri & Kaburaki, 1918*a*, 432, 435 (diagn.; Japan); 1918*b*, 2, 13, 47, figs (Japan); Kato, 1934*b*, 124 (Japan); 1938*a*, 564 (Japan); 1938*b*, 582 (Japan); 1940, 538 (development); 1944, 276.

distincta Prudhoe, 1982*b*, 367, fig. (South Australia).

divae Marcus, 1948, 111, 178, 195, figs (Brazil); du B.-R. Marcus, 1955*b*, 284.

dubia (Schmarda, 1859) Stummer-Traunfels, 1933, 3487, 3536, figs (type-specimen re-examined; syns: *Imogene truncatus* & *Notoplana evansii*); Prudhoe, 1950*a*, 44 (Burma).

erythrotaenia (Schmarda, 1859) Stummer-Traunfels, 1933, 3486, 3487, 3518, figs (type-specimen re-examined; syns: *Polycelis lyrosora* & *Notoplana ovalis*); Palombi, 1939*a*, 128 (?s.o. patellarum).

evansii Laidlaw, 1903*c*, 301, 302, figs (Malaysia); Bock, 1913, 187, figs (Gulf of Siam); Bresslau, 1928–33, 146, fig.; Stummer-Traunfels, 1933, 3487, 3537, (s.o. *dubia*).

fallax (Quatrefages, 1845) Bock, 1913, 204 (synonymy); Palombi, 1928, 595 (?s.o. *atomata*).

ferruginea (Schmarda, 1859) Stummer-Traunfels, 1933, 3486, 3487, 3521, figs (type-specimen re-examined; syn. *bahamensis*); Hyman, 1939*d*, 9 (syn. *caribbeana*, but not *bahamensis*); 1955*b*, 131 (synonymy; Puerto Rico); 1955*e*, 260, fig. (Jamaica); du B.-R. Marcus & Marcus, 1968, 40, fig. (Caribbean).

gardineri (Laidlaw, 1904) Bock, 1913, 211 (note); Prudhoe, 1952, 176; fig. (Red. Sea).

haloglena (Schmarda, 1859) Stummer-Traunfels, 1933, 3486, 3514, figs (type-specimen re-examined; ? var of *australis*).

humilis (Stimpson, 1857) Yeri & Kaburaki, 1918*a*, 432, 435 (diagn.; Japan); 1918*b*, 11, 47, figs (Japan); Kaburaki, 1923*a*, 191, fig. (Japan); Kato, 1934*b*, 124, (Japan); 1937*c*, 35 (Japan); 1937*e*, 127 (Japan); 1938*a*, 564 (Japan); 1938*b*, 582 (Japan); 1939*a*, 72 (Japan); 1939*b*, 144 (descr.); 1940, 554, figs, (development); 1944, 276 (? syn. *Leptoplana oblonga*); Utinomi, 1956, 29, fig. (coloured); Okada *et al.*, 1971, 62 (Japan).

igiliensis Galleni, 1974, 395, figs (Ligurian Sea); Galleni & Puccinelli, 1975, 375 (Tuscan coast).

inquieta (Heath & McGregor, 1913) Hyman, 1953*a*, 319, figs (California & British Columbia).

inquieta Freeman, 1933, 111, 120, figs (Puget Sound); Hyman, 1953*a*, 310 (s.o. *Leptoplana chloranota*).

inquilina Hyman, 1955*a*, 1, figs (Puget Sound).

insularis Hyman, 1939*c*, 1, figs (Old Providence I., W. Indies); 1955*b*, 130 (Florida & Caribbean).

japonica Kato, 1937*a*, 211, 215, figs (Japan); 1938*b*, 582 (Japan); 1944, 276, Sandô, 1964, 27 (Japan).

koreana Kato, 1937*b*, 233, 234, figs (Korea); 1939*a*, 72, figs; 1944, 277.

kuekenthalii Plehn, 1896*a*) Bock, 1913, 202, figs (synonymy); Steinböck, 1932, 333, 337 (distribution); Steven, 1938, 61 (W. Greenland).

lactoalba (Verrill, 1900) Hyman, 1939*d*, 6, figs (Bermuda); du B.-R. Marcus & Marcus, 1968, 41, fig. (Curaçao & Florida).

lapunda du Bois-Reymond Marcus & Marcus, 1968, 46, figs (Curaçao).

libera Kato, 1939*a*, 65, 68, figs (Japan); 1944, 277.

longastyletta (Freeman, 1933) Hyman, 1953*a*, 325, figs (Puget Sound & Aleutian Is).

longicrumena Prudhoe, 1982*a*, 222, 225 (S. Australia); 1982*b*, 362, 369, fig. (S. Australia).

longiducta Hyman, 1959*c*, 10, figs (N.S.W., Australia).

longisaccata Hyman, 1959*c*, 8, figs (N.S.W., Australia).

martae Marcus, 1948, 111, 180, 195, figs (Brazil); du B.-R. Marcus, 1955*b*, 285.

megala Marcus, 1952, 5, 85, 115, figs (Brazil); du B.-R. Marcus, 1955*b*, 285.

micheli Marcus, 1949, 7, 78, 104, figs (Brazil); du B.-R. Marcus, 1955*b*, 285.

micronesiana Hyman, 1959*a*, 551, figs (Caroline Is, Pacific).

microsora (Schmarda, 1859) Stummer-Traunfels, 1933, 3486, 3520, figs (type-specimen re-examd.).

mortenseni Bock, 1913, 40 ,187, 191, figs (Thailand).

natans Freeman, 1933, 111, 123, figs (Puget Sound); Hyman, 1953*a*, 328, fig. (Puget Sound).
nationalis (Plehn, 1896*b*) Bock, 1913, 187, 207 (diagn.).
otophora (Schmarda, 1859) Stummer-Traunfels, 1933, 3486, 3497, figs (type-specimen re-examd.).
ovalis Bock, 1913, 212, figs (Mauritius, Indian Ocean); Stummer-Traunfels, 1933, 3487, 3519 (s.o. *erythrotaenia*); Palombi, 1936, 18, figs (synonymy; in mantle of *Patella* spp., South Africa); 1938, 341 (in *Patella oculus*, S. Africa); 1939*a*, 128 (s.o. *Notoplana patellarum*); Marcus, 1947, 127, 165 (distinguished from *patellensis* & *patellarum*).
palaoensis Kato, 1943*c*, 79, 81, figs (Palau Is); 1944, 277.
palta Marcus, 1954*a*, 65, figs (Chile).
parvula Palombi, 1924*a*, 37 (Sunda Sea); 1924*b*, 19, figs (90 miles south of Borneo – 4°21′S, 110°26′E).
patellarum (Stimpson, 1855*a*) Palombi, 1939*a*, 123, 128, fig. (synonymy; South Africa); Marcus, 1947, 127, 165 (note); Day *et al.*, 1970, 19 (S. Africa).
pegnis [err. pro *segnis*] Nicoll, 1934, 76.
plecta Marcus, 1947, 99, 124, 165, figs (Brazil); du B.-R. Marcus, 1955*b*, 285.
puma Marcus, 1954*a*, 59, figs (Chile).
queruca du Bois-Reymond Marcus & Marcus, 1968, 44, figs (Caribbean & Florida).
robusta Palombi, 1928, 579, 596, figs (Suez Canal); 1939*b*, 107 (fig. 176 & fig. 180 in 1928 paper must be transposed); du B.-R. Marcus & Marcus, 1968, 31 (to *Stylochoplana*).
rupicola (Heath & McGregor, 1913) Hyman, 1939*a*, 437 (prob. a *Notoplana*); 1953*a*, 315, figs (Oregon & California).
sanguinea Freeman, 1933, 111, 122, figs (Puget Sound); Hyman, 1953*a*, 327, figs (Puget Sound).
sanjuania Freeman, 1933, 111, 112, figs (Puget Sound); Hyman, 1953*a*, 330, fig. (Puget Sound & Alaska, mostly in association with crab, *Paralithodes camtschatica*).
sanpedrensis Freeman, 1930, 337, figs (California); Hyman, 1953*a*, 325 (possibly identical with *acticola*).
sawayai Marcus, 1947, 99, 121, 164, figs (Brazil); du B.-R. Marcus, 1955*b*, 285.
saxicola (Heath & McGregor, 1913) Hyman, 1939*a*, 437 ('evidently a *Notoplana*'); 1953*a*, 318, figs (California).
sciophila (Boone, 1929) Hyman, 1953*a*, 321, 323, figs (California).
segnis Freeman, 1933, 111, 112, 130, figs (Puget Sound); Ricketts & Calvin, 1948, 175, 300 (California); Hyman, 1953*a*, 326 (s.o. *Freemania litoricola*).
septentrionalis Kato, 1937*e*, 127, figs (Japan); 1944, 277.
serica Kato, 1939*a*, 559, 564, figs (Japan); 1944, 277.
'*similis* Stimps.' Dawydoff, 1952, 81 (Vietnam).
sophia Kato, 1939*a*, 65, 70, figs (Japan); 1944, 277.
stilifera Bock, 1923*c*, 341, 348, figs (Juan Fernandez Is, Pacific Ocean).
syntoma Marcus, 1947, 99, 123, 164, figs (Brazil); 1948, 183, 195, figs (Brazil); du B.-R. Marcus, 1955*b*, 285.
tavoyensis Prudhoe, 1950*a*, 44, figs (Burma).
virilis (Verrill, 1893) Bock, 1913, 187, 208 (diagn.); Hyman, 1940*b*, 468 (s.o. *atomata*).
vitrea (Lang, 1884) Bock, 1913, 207 (diagn.).
willeyi Jacubowa, 1906, 131, figs (New Britain, Pacific Ocean); Bock, 1913, 190 (diagn.).
species Miller & Batt, 1973, 64 (New Zealand).

NOTOPLANELLA Bock, 1931, 261, 271 (type: *inarmata*).
inarmata Bock, 1931, 261, 272, figs (Simon's Bay, South Africa); Day *et al.*, 1970, 19 (S. Africa; '*Notoplana*').

NOTOPLANIDES Palombi, 1928, 579, 599, 603 (type: *opisthopharynx*); Marcus, 1947, 110 (systematic position); 1949, 76 (systematic position); Hyman, 1953*a*, 343 (?validity). [s.o. *Stylochoplana* – see p. 90]
alcha du Bois-Reymond Marcus & Marcus, 1968, 2, 33 (Caribbean).
opisthopharynx Palombi, 1928, 579, 599, figs (Suez Canal).
species Palombi, 1928, 603, 604 (= Leptoplanid of Bock, 1923*c*, 356); Marcus, 1947, 129, 165 (poss. = *Pucelis*).

NOTOPLANINAE Marcus, 1947, 120, 164 (= 2nd leptoplanid series of Bock, 1913, 168); 1954*a*, 55 (note); Hyman, 1953*a*, 343 (criticism); Marcus & Marcus, 1966, 329 (key to genera).

NOTOPLANOIDES [err. pro *Notoplanides*] Nicoll, 1929, 46.

NUCHENCEROS Gemmill & Leiper, 1907, 823 (type: *orcadensis*); Bock, 1913, 276 (s.o. *Stylochoides*); Hallez, 1913, 40 (s.o. *Stylochoides*).
orcadensis Gemmill & Leiper, 1907, 823, figs (S. Orkney Is, Antarctica); Bock, 1913, 276 (s.o. *Stylochoides albus*); Hallez, 1913, 23 (s.o. *Stylochoides albus*).
NYMPHOZOON Hyman, 1959*a*, 575 (type: *bayeri*).
bayeri Hyman, 1959*a*, 578, 594, figs (Palau Is, Pacific Ocean).

OCULOPLANA Pearse, 1938, 83 (type: *whartoni*); Hyman, 1940*b*, 458 (s.o. *Latocestus*).
whartoni Pearse, 1938, 83, fig. (Florida & North Carolina); Pearse & Littler, 1938, 238, fig.; Hyman, 1940*a*, 458, fig. (to *Latocestus*).
OLIGOCLADO Pearse, 1938, 87 (type: *floridanus*); Hyman, 1940*a*, 487 (syn. *Hymania*).
floridanus Pearse, 1938, 88, fig. (Florida & North Carolina); Pearse & Littler, 1938, 241, fig.; Hyman, 1940*a*, 487, fig. (syn. *Hymania prytherchi*).
OLIGOCLADUS Lang, 1884, vii, 431, 580 (type: *sanguinolentus*); Carus, 1885, 156, Hallez, 1893, 176; 1894, 228 (defined); Gamble, 1893*a*, 509; 1896, 19; Benham, 1901, 31; Freeman, 1933, 140; Prudhoe, 1982*c*, 66 (defined).
albus Freeman, 1933, 111, 112, 140, figs (Puget Sound); Hyman, 1953*a*, 378 (to *Acerotisa*); 1959*a*, 587 (prob. an. *Acerotisa*).
auritus (Claparède, 1861) Lang, 1884, 583; Vaillant, 1889, 656 (NW. France); Hallez, 1893, 176 (descr.; Pas-de-Calais); 1894, 228 (NW. France); Gamble, 1893a, 510 (west coast of Scotland); 1896, 19 (Britain); Francotte, 1897, 28, figs; Bock, 1913, 269 (variety of *sanguinolentus*).
sanguinolentus (Quatrefages, 1845) Lang, 1884, 127, 580, figs (Naples); Carus, 1885, 156 (descr.); Koehler, 1885, 14, 56 (Channel Is, U.K.); Graff, 1886, 342 (Adriatic); Vaillant, 1889, 656, fig. (NW. France); Gamble, 1893*a*, 509, fig. (Britain); 1893*b*, 47 (Devon); 1893*c*, 170, figs (Isle of Man, U.K.); 1896, 19, 22 (Britain); Pruvot, 1897, 21 (Brittany); Gamble, 1900, 812 (Ireland); Laidlaw, 1906, 707, 714 (Cape Verde Is, Atlantic Ocean); Bock, 1913, 267, fig. (synonymy; Sweden & Norway); Bresslau, 1928–33, 105, fig.; 1930, 114, fig.; Steinböck, 1932, 334, 337 (distribution); Southern, 1936, 71 (Ireland); Eales, 1939, 54, fig.; Bruce, 1948, 46 (Isle of Man); Crothers, 1966, 22 (S. Wales); Hendelberg, 1974*b*, 17, fig. (Sweden); Prudhoe, 1982*c*, 66, fig. (U.K.).
species Bergendal, 1890, 327 (Sweden).
OMMATOPLANA Laidlaw, 1903*d*, 111, 113 (type: *tuberculata*); Bock, 1913, 157 (defined); Hyman, 1955*c*, 72 (defined); 74 (systematic note).
levis (Hyman, 1953) Hyman, 1955*a*, 9.
mexicana Hyman, 1955*a*, 9 (nom. nov. pro *Mexistylochus tuberculatus*); 1955*c*, 74 (n. n. pro *M. tuberculatus*).
oceanica Hyman, 1955*c*, 72, figs (Marianas Is, Pacific Ocean).
tuberculata Laidlaw, 1903*d*, 111, figs (East Africa); Bock, 1913, 157.
OPISTHOGENIA Palombi, 1928, 579, 608, 613 (type: *tentaculata*).
tentaculata Palombi, 1928, 579, 608, figs (Suez Canal); Bresslau, 1928–33, 291, figs.
OPISTHOGENIIDAE Palombi, 1928, 579, 608, 613; Stummer-Traunfels, 1933, 3576, 3587.
OPISTHOPORUS Minot, 1877*a*, 451 (type: *tergestinus*); Lang, 1884, 475 ('Opisthopori' s.o. *Leptoplana*).
tergestinus Minot, 1877*a*, 451, figs (Trieste); Lang, 1884, 486 (s.o. *Leptoplana alcinoi*).
ORTHOSTOMUM Grube, 1840, 56 (type: *rubrocinctum*).
rubrocinctum Grube, 1840, 56 (Mediterranean); Örsted, 1843, 556; 1844*a*, 75; Diesing, 1850, 238; Grube, 1855, pl. vi, fig. 6; Stimpson, 1857, 3 (to *Typhlolepta*); Lang, 1884, 516, fig. (to *Cestoplana*).

PACHYPLANA Stimpson, 1857, 22, 28 (type: *lactea*); Lang, 1884, 466 (?s.o. *Discocelis*).
lactea Stimpson, 1857, 22, 28 (Ryukyu Is, Japan); Diesing, 1862, 531 (to *Leptoplana*); Lang, 1884, 470 (to *Discocelis* (?).).
PARABONINIA Prudhoe, 1944, 326 (type: *caymanensis*).
caymanensis Prudhoe, 1944, 326, figs (Cayman Is, Caribbean).
PARAPLANOCERA Laidlaw, 1903*a*, 3, 4 (type: *langii*); 1903*e*, 13 (note); Haswell, 1907*b*, 471, 477; Bock, 1913, 246 (diagn.); Yeri & Kaburaki, 1918*a*, 437 (features); 1918*b*, 48 (features); Kato, 1936*b*, 21 (validity of spp.); Prudhoe, 1945*a*, 195 (review of spp.); Hyman, 1953*a*, 353 (defined); 355 (review of spp.).

aurora Laidlaw, 1903*d*, 102, figs (Zanzibar); 1904b, 128 (Ceylon); Bock, 1923, 246; Kato, 1936*b*, 28 (? = *langii*); Prudhoe, 1945*a*, 200.

discus (Willey, 1897) Bock, 1913, 246 (New Britain, Pacific Ocean; syn. *laidlawi*); Stummer-Traunfels, 1933, 3488, 3556 (s.o. *oligoglena*), 3565, fig. (Indonesia); Kato, 1936*b*, 28 (prob. syns.); 1943*b*, 73 (s.o. *oligoglena*); 1944, 286 (s.o. *oligoglena*); Prudhoe, 1945*a*, 199, 201 (s.o. *oligoglena*).

fritillata Hyman, 1959*a*, 557, figs (Micronesia, Pacific).

laidlawi Jacubowa, 1906, 115, figs (New Britain, Pacific Ocean; apparently a new name for *Planocera discus*); Bock, 1913, 246 (s.o. *discus*); Kato, 1943*b*, 73 (s.o. *oligoglena*); 1944, 286 (s.o. *oligoglena*); Prudhoe, 1945*a*, 199, 201 (s.o. *oligoglena*).

langii (Laidlaw, 1902) Laidlaw, 1903*a*, 3, 4; 1903*e*, 14 (note); Bock, 1913, 246 (*langi*); Kato, 1936*b*, 28 (prob. synonyms); 1943*b*, 73 (s.o. *oligoglena*); 1944, 286 (s.o. *oligoglena*); Prudhoe, 1945*a*, 199, 201 (s.o. *oligoglena*); Hyman, 1953*a*, 355 (a distinct species).

marginata Meyer, 1922, 138, 139, figs (Red Sea); Kato, 1936*b*, 28 (? = *discus*); 1943*b*, 73 (s.o. *oligoglena*); 1944, 286 (s.o. *oligoglena*); Prudhoe, 1945*a*, 195, fig. (Kenya); Gallagher, 1976, 36 (Persian Gulf).

misakiensis Yeri & Kaburaki, 1918*a*, 432, 437 (features; Japan); 1918*b*, 2, 23, 48, figs (Japan); Kato, 1936*b*, 22, figs (Japan; ? = *discus*); 1943*b*, 73 (s.o. *oligoglena*); 1944, 286 (s.o. *oligoglena*) Prudhoe, 1945*a*, 199, 201 (s.o. *oliglogl ena*); Hyman, 1953*a*, 357 (? = *langii*).

oligoglena (Schmarda, 1859) Stummer-Traunfels, 1933, 3487, 3488, 3552, 3585, figs (type-specimen re-examined; syn. *discus*); Kato, 1943*b*, 73 (synonymy); 1944, 286 (synonymy); Prudhoe, 1945*a*, 195, fig. (synonymy; Christmas I. & Tahiti); Hyman, 1953*a*, 353, 357 (Gulf of California); 1954*b*, 333 (Hawaii); 1955*c*, 77, fig. (synonymy; Marshall Is, Pacific Ocean); 1960, 308 (Hawaii); Brusca, 1973, 68 (Gulf of California); Prudhoe, 1977, 590, fig. (W. Australia).

rotumanensis Laidlaw, 1903*a*, 3, 4 (Rotuma I., Pacific Ocean); Bock, 1913, 246; Kato, 1936*b*, 28 (? = *langii*); 1943*b*, 73 (s.o. *oligoglena*); 1944, 286 (s.o. *oligoglena*); Prudhoe, 1945*a*, 199, 201 (s.o. *oligoglena*); Hyman, 1953*a*, 357 (s.o. *langii*).

rubrifasciata Kato, 1937*f*, 347, 360, figs (Japan); 1944, 286; Prudhoe, 1945*a*, 198, 201 (s.o. *marginata*).

PARAPLEHNIA Hyman, 1953*a*, 279 (type: *japonica*).

japonica (Bock, 1923*a*) Hyman, 1953*a*, 279.

pacifica (Kato, 1939*a*) Hyman, 1953*a*, 279.

PARASTYLOCHUS Bock, 1913, 125, 127 (type: *astis*); 1925*b*, 171.

astis Bock, 1913, 125, figs (Java Sea); 1925*b*, 177.

PARVIPLANA Hyman, 1953*a*, 318 (type: *californica*).

californica (Woodworth, 1894) Hyman, 1953*a*, 314, figs (California).

PEASIA Gray, *in* Pease, 1860, 37 (type not designated); Lang, 1884, 434, 524; Girard, 1893, 190; Stimpson, 1861, 134.

inconspicua Gray, *in* Pease, 1860, 37, figs (Hawaii); Stimpson, 1861, 134 (a *Prosthiostomum*); Diesing, 1862, 536 (to *Leptoplana*); Lang, 1884, 615.

irrorata Gray, *in* Pease, 1860, 38, figs (Hawaii); Stimpson, 1861, 134 (a *Prosthiostomum*); Diesing, 1862, 536 (to *Leptoplana*); Lang, 1884, 605; Poulter, 1975, 327.

maculata Gray, *in* Pease, 1860, 38, figs (Hawaii); Stimpson, 1861, 134 (*Eurylepta*); Diesing, 1862, 548 (*Eurylepta*); Lang, 1884, 547 (to *Pseudoceros*).

reticulata Gray, in Pease, 1860, 37, figs (Hawaii); Stimpson, 1861, 134 (*Stylochus*); Diesing, 1862, 561 (to *Planocera*); Lang, 1884, 440 (*Planocera*); Hyman, 1960, 308 (s.o. *Paraplanocera oligoglena*).

tentaculata Gray, *in* Pease, 1860, 37, figs (Hawaii); Stimpson, 1861, 134 (a *'Physanozoum'*); Diesing, 1862, 556 (to *Thysanozoon*); Lang, 1884, 536 (var. of *Thysanozoon brocchii*); Plehn, 1899, 448 (= *Thysanozoon brocchii*); Hyman, 1960, 309 (*Thysanozoon*).

PELAGOPLANA Bock, 1913, 173, 232 (type: *sargassicola*); Hyman, 1939*b*, 144 (s.o. *Gnesioceros*); 1939*d*, 11 (s.o. *Gnesioceros*).

sargassicola (Mertens, 1832) Bock, 1913, 233; Hyman, 1939*b*, 146 (to *Gnesioceros*).

PENTAPLANA Marcus, 1949, 7, 67, 102 (type: *divae*).

divae Marcus, 1949, 7, 68, 102, figs (Brazil); du B.-R. Marcus, 1955*b*, 283.

PENULA Kelaart, 1858, 134, 138 (type not designated).

alba Kelaart, 1858, 139 (Ceylon); Diesing, 1862, 542 (? *Leptoplana*); Lang, 1884, 613.

fulva Kelaart, 1858, 139 (Ceylon); Diesing, 1862, 542 (? *Leptoplana*); Lang, 1884, 613.

ocellata Kelaart, 1858, 138 (Ceylon); Diesing, 1862, 542 (? *Leptoplana*); Lang, 1884, 511; Collingwood, 1876, 88, 87, fig.

punctata Kelaart, 1858, 138 (Ceylon); Diesing, 1862, 542 (? *Leptoplana*); Lang, 1884, 616; Collingwood, 1876, 88, 97, fig. (to *Centrostomum*).

PERICELIDAE Laidlaw, 1902, 291; Bock, 1923*b*, 28 (note); Poche, 1926, 100 (emended to Pericelididae); Bresslau, 1933, 290; Stummer-Traunfels, 1933, 3576; Hyman, 1955*e*, 262 (defined).

PERICELIS Laidlaw, 1902, 291, 306 (type: *byerleyana*); Bock, 1922, 19 (morphological note); Hyman, 1955*e*, 263 (comparison of spp.).

byerleyana (Collingwood, 1876) Laidlaw, 1902, 291, figs (Laccadive Is & Rotuma); 1903*a*, 3, 9 (Rotuma I., Pacific Ocean); Meixner, 1907*a*, 171 (Somali coast); 1907*b*, 473, figs (Somalia & Java); Bock, 1923*b*, 5 (Fiji; Gilbert & Ellis Is); Bresslau, 1930, 143, fig.; 1933, 290, fig.; Palombi, 1938, 351, figs (Mauritius); Kato, 1943*c*, 84, figs (Palau Is); 1944, 294 (distribution).

cata du Bois-Reymond Marcus & Marcus, 1968, 59, figs (Curaçao).

ernesti (Hyman, 1953) Hyman, 1955*e*, 263 (Gulf of California).

hymanae Poulter, 1974, 93 (Hawaii).

orbicularis (Schmarda, 1859) Stummer-Traunfels, 1933, 3487, 3545, 3583, 3595, figs (type-specimen re-examined); du B.-R. Marcus & Marcus, 1968, 56, fig. (Florida and Texas).

species Poulter, 1970, 553 (Hawaii).

PHAENOCELIDAE Stummer-Traunfels, 1933, 3501, 3586; Marcus, 1952, 81, 114 (=Cryptocelidae); Hyman, 1953*a*, 300 (=Cryptocelidae).

PHAENOCELIS Stummer-Traunfels, 1933, 3486, 3501 (type: *purpurea*) Hyman, 1952, 115 (syn. *Comprostatum*); 1953*a*, 297 (defined); du B.-R. Marcus & Marcus, 1968, 17 (key to spp.).

medvedica Marcus, 1952, 5, 81, 114, figs (Brazil); du B.-R. Marcus, 1955*b*, 282.

mexicana Hyman, 1953*a*, 299, figs (Mexico & California); Brusca, 1973, 66 (Gulf of California).

peleca du Bois-Reymond Marcus & Marcus, 1968, 2, 16, figs (Curaçao).

purpurea (Schmarda, 1859) Stummer-Traunfels, 1933, 3485, 3499, figs (type-specimen re-examined); Hyman, 1952, 195 (Florida; syn. *Comprostatum insularis*); 1954*c*, 302; 1955*e*, 259 (Jamaica); du B.-R. Marcus & Marcus, 1968, 15, fig. (Curaçao).

PHARYNGEATIDAE [err. pro Diplopharyngeatidae] Bock, 1913, 50.

PHARYNGOCOELA Leuckart, 1856, 108 (subord. of Turbellaria, includes polyclads).

PHYLLOPLANA Laidlaw, 1903*d*, 100, 107, 109 (type: *lactea*); Bock, 1913, 222 (defined); Heath & McGregor, 1913, 472 (defined); Hyman, 1953*a*, 333 (defined).

chloranota Boone, 1929, 43, figs (California); Hyman, 1953*a*, 310, 337 (to *Leptoplana*); Stasek, 1966, 8 (location of type-specimen); du B.-R. Marcus & Marcus, 1968, 23 (to *Stylochoplana*).

lactea Laidlaw, 1903*d*, 107, fig. (East Africa); Bock, 1913, 222.

litoricola Heath & McGregor, 1913, 458, 472, figs (California); Stummer-Traunfels, 1933, 3504 (not a *Phylloplana*); Hyman, 1953*a*, 336 (to *Freemania*); Stasek, 1966, 8 (location of type-specimen).

tropicalis (Hyman, 1954) Marcus & Marcus, 1966, 330.

viridis (Freeman, 1933) Hyman, 1953*a*, 334, figs (Puget Sound); Morris *et al.*, 1980, 78, fig. (coloured) (California).

species Stummer-Traunfels, 1933, 3502, 3504 (Jamaica; *ex parte*=*Leptoplana purpurea* Schmarda, 1859).

PLAGIOPLANA [err. pro *Plagiotata*] Benham, 1901, 31.

PLAGIOTATA Plehn, 1896*a*, 140, 144, 146 (type: *promiscua*); Laidlaw, 1903*e*, 10; Bock, 1913, 218 (diagn.).

promiscua Plehn, 1896*a*, 140, 144, figs (Hong Kong); Bock, 1913, 219.

PLANARIA Müller, 1774 [Now a valid genus of Tricladida]

armata Kelaart, 1858, 135 (Ceylon); Diesing, 1862, 560; Collingwood, 1876, 87, 95 (to *Acanthozoon*); Lang, 1884, 545 (to *Pseudoceros*).

atomata Müller, 1776, 282 (nom. nov. pro *punctata* Müller, 1776, 223; Norway); 1779, 73 (*drobachiensi*); 1788*a*, 37, figs; Bosc, 1802, 260; Fleming, 1823, 297 (east coast of Scotland); Blainville, 1826, 217; Delle Chiaje, 1831, 179, 196 (descr.; Naples); Forbes & Goodsir, 1840, 80 (Orkney & Shetland Is); Delle Chiaje, 1841, pt. III, 133, 135; pt. V, 112, pt. CIX, fig. 16 (Sicily); Örsted, 1843, 569 (to *Leptoplana*); 1844*a*, 25; Johnston, 1846, 436; Frey & Leuckart, 1847, 146, 149 (descr.; Heligoland); Diesing, 1850, 197 (*Leptoplana*); Lang, 1884, 514 (synonymy); Carus,

1885, 152; Gamble, 1893*a*, 501 (descr.; ? = *Leptoplana droebachensis*); 1893*b*, 17; Bock, 1913, 195, figs (to *Notoplana*).

aurantiaca Delle Chiaje, 1830, figs Pl. 78; 1841, pt. III, 131; 135, pt. V, 111, pt. XXXIX (Sicily), figs 1–13; Guérin-Meneville, 1844, 14, fig. Örsted, 1844*a*, 47 (to *Thysanozoon*); Verany, 1846, 9 (Italy); Diesing, 1850, 214 (*Thysanozoon*); Milne-Edwards, 1859, 455, 456; Lang, 1879, 459 (to *Proceros*); 1884, 548, figs (to *Yungia*).

aurea Kelaart, 1858, 137 (Ceylon); Diesing, 1862, 562 (to *Planocera*); Lang, 1884, 466.

auriculata Müller, 1788*b*, 37, figs (Norway); Bosc, 1802, 261; Diesing, 1850, 211 (to *Eurylepta*); Lang, 1884, 584 (a rhabdocoel).

bilobata Leuckart, 1828, 11–15 (Red Sea); Diesing, 1850, 281, 522 (to *Typhloleptà*?) Lang, 1884, 607.

bituberculata Leuckart, 1828, 13, figs (Red Sea); Ehrenberg, 1831, 55 (? = *Stylochus suesensis*); Diesing, 1850, 215 (*Planaria 'bitentaculata'* Leuck. ?s.o. *Stylochus suesensis*); Schmarda, 1859, 33 (a *Planocera*); Lang, 1884, 451 (s.o. *Stylochus suesensis*); Palombi, 1928, 582 (s.o. *St. suesensis*).

brocchii (Risso, 1818) Risso, 1826, 264 (descr.; S. France); Blainville, 1826, 218; Grube, 1840, 55 (related to *Thysanozoon diesingii*); Delle Chiaje, 1841, pt. V, 112; Örsted, 1844*a*, 47 (*Thysanozoon*); Quatrefages, 1845, 140 (to *Eolidiceros*); Diesing, 1850, 213 (to *Thysanozoon*); Lang, 1884, 525 (*Thysanozoon brocchii*).

cerebralis Kelaart, 1858, 135 (Ceylon); Diesing, 1862, 558; Collingwood, 1876, 87, 96, fig. (to *Eurylepta*); Lang, 1884, 546 (to *Pseudoceros*).

corniculata Dalyell, 1853, 101, figs (Scotland); Leuckart, 1859, 183 (*Polycelis*); Diesing, 1862, 571 (to *Stylochus*?); Johnston, 1865, 6 (s.o. *Leptoplana subauriculata*); Lang, 1884, 459 (?s.o. *Stylochoplana maculata*).

cornuta Müller, 1776, 221 (Norway); 1788*a*, 37, figs (Norway); Bosc, 1802, 260; Blainville, 1826, 210, figs; Johnston, 1832, 344, figs (Scotland); 429 (= *Planaria vittata*), 678 (*Planaria cornuta*); Guerin-Méneville, 1844, 14, figs; Thompson, 1845, 320 (Ireland); 1846, 392 (British records are those of *Proceros sanguinolentus*); Johnston, 1846, 436; Diesing, 1850, 208 (*Eurylepta*); Dalyell, 1853, 97, figs; Schmarda, 1859, 30 (to *Prostheceraeus*); Leuckart, 1859, 183 (*Proceros* or *Eurylepta*); Diesing, 1862, 553 (*P. cornuta* of Thompson, 1845, s.o. *Proceros sanguinolentus*); Lang, 1884, 572 (*Eurylepta*).

dicquemari (Risso, 1818) Risso, 1826, 263 (descr.; S. France); Delle Chiaje, 1822, Pl. cviii, figs 1, 4 & 5; 1841, pt. III, 132, 135; pt. V, 112, pl. CIX, fig. 20 (Sicily); Blainville, 1826, 217; Örsted, 1844*b*, 47 (to *Thysanozoon*); Verany, 1846, 9 (Italy); Diesing, 1850, 213 (*Thysanozoon*); Lang, 1884, 525 (*Planaria dicquemaris* of Delle Chiaje, 1841, s.o. *T. brocchii*; of Örsted, 1844*a*, s.o. *brocchii*); 550, (*Pl. dicq*. Risso to *Yungia*); Palombi, 1928, 604 (s.o. *T. brocchii*).

dicquemaris var. *verrucosa* Delle Chiaje, 1841, pt. III, 132, pt. V, 112 (Sicily); Lang, 1884, 525 (s.o. *Thysanozoon brocchii*); Palombi, 1928, 604 (s.o. *T. brocchii*).

dubia Blainville, 1826, 218; Lang, 1884, 436 (s.o. *Planocera gaimardi*).

dulcis Kelaart, 1858, 137 (Ceylon); Diesing, 1862, 559; Collingwood, 1876, 87, 96, fig. (to *Eurylepta*); Lang, 1884, 593.

elegans Kelaart, 1858, 136 (Ceylon); Diesing, 1862, 562 (to *Planocera*); Lang, 1884, 465.

ellipsis Dalyell, 1853, 101, figs (Scotland); Leuckart, 1859, 183 (*Polycelis*); Diesing, 1862, 542 (to *Leptoplana*); Lang, 1884, 588 (to *Stylostomum*); Gamble, 1893*a*, 511 (? = *Stylostomum variable*); Bock, 1913, 270 (*Stylostomum*).

flava Delle Chiaje, 1822, Pl. CVIII, fig. 11 (no text) (Naples); 1841, V, 112, pl. XXXVI, fig. 11; Örsted, 1844*a*, 47 (to *Thysanozoon*); Lang, 1884, 548 (s.o. *Yungia aurantiaca*).

flexilis Dalyell, 1814, 5, figs (Scotland); Johnston, 1836, 17; Johnston, 1846, 436; Thompson, 1849, 354 (Ireland; = *Planaria subauriculata*); Diesing, 1850, 194 (to *Leptoplana*); Dalyell, 1853, 102, figs; Leuckart, 1859, 183 (*Leptoplana*); Keferstein, 1869, 6 (s.o. *Leptoplana tremellaris*); McIntosh, 1874, 140, 150 (Scotland); Lang, 1884, 476 (s.o. *Leptoplana tremellaris*).

formosa Darwin, 1844, 247 (Tierra del Fuego I., S. America); Diesing, 1850, 199 (to *Leptoplana*); Lang, 1884, 609.

fusca Kelaart, 1858, 136 (Ceylon); Diesing, 1862, 559; Collingwood, 1876, 87, 95, fig. (to *Eurylepta*); Lang, 1884, 593.

gigas Leuckart, 1828, 11–15 (Red Sea); Ehrenberg, 1831, 55 (? = *Stylochus suesensis*); Örsted, 1844*a*, 48 (? = *Planocera suesensis*); Diesing, 1850, 215 (? = *Stylochus suesensis*); Lang, 1884, 606. Meixner, 1907*b*, 422 (a pseudocerid).

incisa Darwin, 1844, 248, fig. (Cape Verde Is); Diesing, 1850, 200 (to *Centrostomum*); Lang, 1884, 609.
lichenoides Mertens, 1832, 4, figs (Isle of 'Sitcha' [Sitka], NW. America); Ehrenberg, 1836, 67 (to *Discocelis*); Örsted, 1844*a*, 49 (to *Leptoplana*); Diesing, 1850, 199 (to *Centrostomum*); Lang, 1884, 469 (*Discocelis*).
limbata Leuckart, 1828, 11–15, figs (Red Sea); Ehrenberg, 1831, 56 (?*Eurylepta*); Örsted, 1844*a*, 50 (to *Eurylepta*); Diesing, 1862, 554 (to *Proceros*); Lang, 1884, 544 (to *Pseudoceros*).
luteola Delle Chiaje, 1822, Pl. XXXV, fig. 28, (Italy); 1828, 81, 118, 120; 1841, Pt. 3, 131, 135; Pt. 5, 111, Pl. XXXIV, fig. 20 (Sicily); Örsted, 1844*a*, 49 (to *Leptoplana* (?) *lutea*); Verany, 1846, 9 (Italy; '*lutea*'); Diesing, 1850, 199 (*Leptoplana lutea*); Lang, 1884, 513; Carus, 1885, 152.
maculata Dalyell, 1853, 104, figs (? = *atomata*; Scotland); Leuckart, 1859, 183 (= *Leptoplana atomata*); Diesing, 1862, 532 (s.o. *Leptoplana atomata*); Lang, 1884, 514, (?s.o. *Planaria atomata*); Palombi, 1928, 595 (s.o. *Notoplana atomata*).
meleagrina Kelaart, 1858, 137, fig. (Ceylon); Diesing, 1862, 572; Collingwood, 1876, 88, 98, fig. (to *Stylochoplana*); Lang, 1884, 613; Kaburaki, 1923*a*, 646, figs (to *Prosthoceraeus*; Philippines).
muelleri Audouin, in Savigny, 1827, 247, fig. (Red Sea); Ehrenberg, 1831, 55 (? = *Stylochus suesensis*); Örsted, 1844*a*, 48 (? = *Planocera suesensis*); Diesing, 1850, 215 (?s.o. *Stylochus suesensis*; Schmarda, 1859, 33 (to *Planocera*); Lang, 1884, 451 (s.o. *St. suesensis*); Palombi, 1928, 582 (s.o. *St. suesensis*).
muelleri Delle Chiaje, 1831, 179, 196 (Naples); 1841, Pt. III, 132, 135, Pt. V, 112, pl. cxxxix, figs (Sicily); Örsted., 1844*a*, 47 (to *Thysanozoon*); Diesing, 1850, 215 (*Thysanozoon*); Lang, 1884, 545 (to *Pseudoceros*).
neapolitana Delle Chiaje, 1841, Pt. III, 133, 135; Pt. V, 112, pl. cix, figs; Goette, 1878, 75; 1882*a*, 1 (*Pl. neapolitana* of Goette, 1878, renamed *Stylochopsis pilidium*); Lang, 1884, 447, figs (to *Stylochus*; Mediterranean).
nesidensis Delle Chiaje, 1822, pl. xci, figs (Naples); 1841, Pt. III, 133, 134; Pt. V, 112; Pt. XXI, figs (Sicily); Lang, 1884, 513; Carus, 1885, 152.
notulata Bosc, 1802, 254, figs (Atlantic Ocean); Blainville, 1826, 217; Lang, 1884, 513; Graff, 1892*b*, 217 (Sargasso Sea); Hyman, 1939*d*, 21, figs (Atlantic Ocean; to *Acerotisa*).
oceanica Darwin, 1844, 246, fig. (Atlantic Ocean, 5°S, 35°W); Diesing, 1850, 211 (to *Eurylepta*); Stimpson, 1857, 20 (to *Nautiloplana*); Schmarda, 1859, 14 (type of *Carenoceraeus*); Lang, 1884, 608; Graff, 1892*b*, 199 (? = *Planocera pellucida*); Bock, 1913, 240 (prob. = *Planocera pellucida*); 1931, 277 (? = *Planocera pellucida*); Kato, 1944, 283 (s.o. *Planocera pellucida*).
ornata Focke, *in litt.*, Diesing, 1850, 213 (renamed *Thysanozoon fockei*; Adriatic Sea).
papilionis Kelaart, 1858, 136 (Ceylon); Diesing, 1862, 560; Collingwood, 1876, 87, 95, fig. (to *Acanthozoon*); Lang, 1884, 546 (to *Pseudoceros*); Kaburaki, 1923*a*, 646, fig. (to *Prostheceraeus*).
pellucida Bosc. 1803, of Lang, 1884, 6, 476; & of Graff, 1904–05 – La Planaire pellicule Bosc, *supra* – s.o. *Leptoplana tremellaris*; Prudhoe, 1950*b*, 712 (footnote).
pellucida Mertens, 1832, 8, figs (Atlantic Ocean, 7°48″N, 23°–56″W); Ehrenberg, 1836, 67 (to *Stylochus*); Örsted, 1844*a*, 48 (to *Planocera*); Diesing, 1850, 216 (*Stylochus*); 1862, 571 (to *Gnesioceros*); Prudhoe, 1950*b*, 710 (taxonomic note).
punctata Müller, 1776, 223 (Greenland), 282 (renamed *atomata*); Diesing, 1850, 197 (s.o. *Leptoplana atomata*); Palombi, 1928, 594 (s.o. *Notoplana atomata*).
purpurea Kelaart, 1858, 136 (Ceylon); Diesing, 1862, 559; Collingwood, 1876, 87, 96, fig. (to *Eurylepta*); Lang, 1884, 593.
retusa Viviani, 1805, 13, figs (near St Nazzaro, Italy); Diesing, 1850, 200 (to *Typhlolepta*); Lang, 1884, 606.
sargassicola Mertens, 1832, 11, 13, figs (Atlantic Ocean, 21°–35°N, 36°–38°W); Ehrenberg, 1836, 67 (to *Stylochus*); Örsted, 1844*a*, 48 (to *Planocera*); Diesing, 1850, 216 (renamed *Stylochus mertensi*); Lang, 1884, 454 (*Stylochus*); Graff, 1892*a*, 146 (to *Stylochoplana*); Bock, 1913, 173, 233 (to *Pelagoplana*); Hyman, 1939*b*, 146 (to *Gnesioceros*); Prudhoe, 1950*b*, 710 (*sargassicola* of Graff probably not *sargassicola* of Mertens).
schlosseri Girard, 1873, 488, fig. (NW. France); Lang, 1884, 590; Francotte, 1898, 250 (= *Cycloporus papillosus* Lang); Bock, 1913, 262 (?s.o. *Cycloporus papillosus* (Sars).)
siphunculus Delle Chiaje, 1822, Pl. XXXV, figs (Naples); 1828, 81, 118, 120; 1841, Pt. III, 131, Pt. V, 111, pl. cxii, figs ('*syphunculus*'; Sicily); Verany, 1846, 9 (Italy; '*syphunculus*'); Lang, 1884, 595, fig. (to *Prosthiostomum*; synonymy).

striata Kelaart, 1858, 137 (Ceylon); Diesing, 1862, 559; Lang, 1884, 546 (to *Pseudoceros*).

subauriculata Johnston, 1836, 16, figs (Britain); 1846, 436; Thompson, 1849, 355 (= *Pl. flexilis*); Diesing, 1850, 195 (to *Leptoplana*); Lang, 1884, 459 (?s.o. *Stylochoplana maculata*).

thesea Kelaart, 1858, 136 (Ceylon); Diesing, 1862, 562; Collingwood, 1876, 88, 98, fig. (to *Planocera*); Lang, 1884, 465.

tremellaris (Müller, 1774) Müller, 1776, 223; 1779, 72 (Norway); 1788*a*, 36, figs; Bosc, 1802, 262; 1803, 63 (Baltic); Blainville, 1826, 217, fig.; Ehrenberg, 1836, 67; Grube, 1840, 52 (descr.; Sicily); Thompson, 1840, 247 (Ireland); Örsted, 1843, 569; Johnston, 1846, 436; Diesing, 1850, 198 (*Leptoplana*); 1862, 532 (*Leptoplana*); Lang, 1884, 476 (*Leptoplana*), 607 (*tremellaris* of Grube, 1840, *nec* Müller).

tuberculata Delle Chiaje, 1822, Pl. XXV, figs. (Naples); 1828, 81, 119, 120; Grube, 1840, 55 (related to *Thysanzoon diesingii*); Delle Chiaje, 1841, Pt. III, 132, 135, Pt. V, 112, pl. cxii, figs (Sicily; syns. *Tisanozoon [sic] diesingii* Grube, & *Planaria brocchii* ?Risso); Örsted, 1844*a*, 48 (? = *Planocera suesensis*); Diesing, 1850, 272 (*Thysanozoon*); Schmarda, 1859, 33 (a *Planocera*); Lang, 1884, 525 (s.o. *Thysanozoon brocchii*); Palombi, 1928, 604 (s.o. *Thysanozoon brocchii*).

undulata Kelaart, 1858, 137 (Ceylon); Diesing, 1862, 559; Collingwood, 1876, 87, 95, fig. (to *Eurylepta*); Lang, 1884, 552 (syn. *Eurylepta superba*; to *Prostheceraeus*); Marcus, 1950, 88 (to *Pseudoceros*).

velellae Lesson, 1830, 453 (Atlantic Ocean); Lang, 1884, 607; Graff, 1892*b*, 199 (prob. = *Planocera pellucida*); Bock, 1913, 240 (prob. = *Planocera pellucida*); 1931, 277 (? = *Planocera pellucida*); Kato, 1944, 283 (s.o. *Planocera pellucida*).

verrucosa Delle Chiaje, 1831, 180, 197 (Naples); Diesing, 1850, 212 (s.o. *Thysanozoon diesingii*); Lang, 1884, 525 (s.o. *Thysanozoon brocchii*); Palombi, 1928, 604, (s.o. *Thysanozoon brocchii*).

violacea Delle Chiaje, 1830, pl. cviii (Naples); 1841, Pt. III, 132, 135; V, 112, pl. xxxvi, fig. (Sicily); Örsted, 1844*a*, 47 (to *Thysanozoon*); Diesing, 1850, 214 (*Thysanozoon*); Lang, 1884, 563 (to *Prostheceraeus*).

violacea Kelaart, 1858, 135 (Ceylon); Diesing, 1862, 558; Collingwood, 1876, 87, 96, fig. (to *Eurylepta*); Lang, 1884, 544 (s.o. *Pseudoceros zebra*); Hyman, 1959*a*, 566 (a *Pseudoceros*).

viridis Kelaart, 1858, 135 (Ceylon); Diesing, 1862, 559; Collingwood, 1876, 87, 96, fig. (to *Eurylepta*); Lang, 1884, 567 (s.o. *Prostheceraeus viridis* Schmarda); Hyman, 1959*a*, 569 (to *Pseudoceros*).

vittata Montagu, 1813, 25, figs (England); Blainville, 1826, 217; Thompson, 1840, 247 (Ireland); 1846, 392 (syn. *Proceros ?cristatus* Quatref.); Johnston, 1846, 436; Diesing, 1850, 209 (to *Eurylepta*); Harvey, 1857, 157 (Ireland); Lang, 1884, 554, figs (to *Prostheceraeus*).

zebra Leuckart, 1828, 11, figs (Red Sea); Diesing, 1850, 211 (to *Eurylepta*); 1862, 554 (to *Proceros*); Lang, 1884, 544 (synonymy; to *Pseudoceros*).

zeylanica Kelaart, 1858, 138 (Ceylon); Diesing, 1862, 559; Collingwood, 1876, 87, 97, fig. (to *Eurylepta*); Lang, 1884, 546 (to *Pseudoceros*).

PLANARIEA Ehrenberg, 1931, 54 (family of order Dendrocoela); Örsted, 1844*a*, 26, 41 (tribe of order Apoda); 1843, 545; Diesing, 1850, 180 (tribe of Dendrocoela).

PLANCTOPLANA Graff, 1892*b*, 190, 213, 216 (type: *challengeri*); Gamble, 1896, 19; Benham, 1901, 32 (*'Planknoplana'*); Laidlaw, 1903*c*, 11; Bock, 1913, 231 (defined); Prudhoe, 1950*b*, 715 (s.o. *Stylochoplana*).

challengeri Graff, 1892*b*, 190, 213, figs (New Guinea); Bock, 1913, 231; Prudhoe, 1950*b*, 713, fig. (original material re-examined; to *Stylochoplana*).

PLANCTOPLANELLA Hyman, 1940*a*, 479 (type: *atlantica*).

atlantica Hyman, 1940*a*, 479, figs (off North Carolina).

PLANCTOPLANINAE Bresslau, 1933, 289 (subfam. of Planoceridae).

PLANEOLIS Stimpson, 1857, 20 (type: *panormus*); Diesing, 1862, 493, 554; Lang, 1884, 524 (s.o. *Thysanozoon*).

panormus (Quatrefages, 1845) Stimpson, 1857, 20; Diesing, 1862, 554 (synonymy); Lang, 1884, 526 (s.o. *Thysanozoon brocchii*).

PLANOCERA Blainville, 1828, 578 (type: *gaimardi*); Örsted, 1844*a*, 47 (list of spp.); Diesing, 1850, 217 (diagn.); Schmarda, 1859, 14; Diesing, 1862, 493, 560; Johnston, 1865, 8; Lang, 1884, 428, 434 (diagn.; synonymy); Carus, 1885, 149; Verrill, 1893, 471; Gamble, 1893*a*, 496; Girard, 1893, 188; Benham, 1901, 31; Laidlaw, 1902, 303; 1903*d*, 101 (table of spp.); Heath & McGregor, 1913, 459; Yeri & Kaburaki, 1918*a*, 436 (features); 1918*b*, 47 (features); Hyman, 1953*a*, 188 (defined).

amphibolus (Schmarda, 1859) Lang, 1884, 444; Stummer-Traunfels, 1933, 3487, 3552 (s.o. *Paraplanocera oligoglena*); Kato, 1944, 286 (s.o. *oligoglena*); Prudhoe, 1945*a*, 201 (s.o. *oligoglena*).

armata Laidlaw, 1902, 282, 305 (Laccadive Is., Indian Ocean); 1903*d*, 101; Bock, 1913, 245.

aurea (Kelaart, 1858) Diesing, 1862, 562.

bituberculata (Leuckart, 1828) Schmarda, 1859, 33; Lang, 1884, 451 (s.o. *Stylochus suesensis*).

burchami Heath & McGregor, 1913, 457, 461, figs (California); Hyman, 1953*a*, 358 (to *Pseudostylochus*).

californica Heath & McGregor, 1913, 455, 457, figs (California); Ricketts & Calvin, 1948, 16, 300 (California); 1952, 180, fig.; Hyman, 1953*a*, 346 (to *Alloioplana*); Stasek, 1966, 9 (location of type-specimens). [See p. 104]

corniculata (Dalyell, 1853) Leuckart, 1859, 183.

crosslandi Laidlaw, 1903*d*, 100 (East Africa); Bock, 1913, 245; Prudhoe, 1952, 175, fig. (Red Sea).

dictyotus (Schmarda, 1859) Lang, 1884, 443.

discoidea Willey, 1897, 156, figs (New Britain); Jacubowa, 1906, 121, figs; Bock, 1913, 245.

discus Willey, 1897, 155, fig. (New Caledonia); Jacubowa, 1906, 115 (renamed *Paraplanocera laidlawi*); Bock, 1913, 246 (to *Paraplanocera*); Kato, 1944, 286 (s.o. *Paraplanocera oligoglena*).

edmondsi Prudhoe, 1982*a*, 222, 225 (southern Australia); 1982*b*, 373, figs (Tasmania & South Australia).

elegans (Kelaart, 1858) Diesing, 1862, 562.

elliptica Girard, 1850*a*, 398; 1850*b*, 251 (Massachusetts); 1851*a*, 348; 1852*a*, 300; 1854*b*, 27; Stimpson, 1857, 23; Diesing, 1862, 561; Lang, 1884, 463; Verrill, 1893, 467, figs (to *Eustylochus*); Girard, 1893, 190, figs; Laidlaw, 1903*d*, 107 (= *Stylochus littoralis*); Hyman, 1939*b*, 130, figs (to *Stylochus*); 1940*b*, 459 (to *Stylochus*).

folium (Grube, 1840) Örsted, 1844*a*, 48; Diesing, 1850, 216 (*Stylochus*); Johnston, 1865, 8 (Britain); Lang, 1884, 440; Carus, 1885, 149 (descr.); Gamble, 1893*a*, 496 (descr.); 1896, 6, 19 (Britain); Laidlaw, 1903*d*, 102; Bock, 1913, 244; Prudhoe, 1982*c*, 50.

gaimardi Blainville, 1828, 578, 579, fig.; Ehrenberg, 1836, 67 (to *Planoceros*); Diesing, 1850, 217; Stimpson, 1857, 23; Schmarda, 1859, 33 (*Stylochus*); Diesing, 1862, 561; Lang, 1884, 436 (syns.); Graff, 1892*b*, 199 (prob. = *P. pellucida*); Girard, 1893, 189; Bock, 1913, 240 (prob. = *P. pellucida*); Kato, 1944, 283 (s.o. *P. pellucida*); Prudhoe, 1950*b*, 712 (status; syn. *P. pellucida* (Mertens).)

gilchristi Jacubowa, 1906 ,153; 1908, 145, figs (Cape Town, South Africa); Bock, 1913, 245; Palombi, 1936, 23 (descr.; Still Bay, S. Africa); 1939*a*, 129, fig. (East London, South Africa); Day *et al.*, 1970, 19 (False Bay, S. Africa).

graffii Lang, 1879, 470, figs (Gulf of Salerno); 1884, 434, figs (Naples); Carus, 1885, 149 (descr.); Gamble, 1896, 18, fig.; Benham, 1901, 32, fig.; Laidlaw, 1903*d*, 102; 1906, 706 (Cape Verde Is); Bock, 1913, 244; Bresslau, 1928, 76, 77, 105, figs; Riedl, 1959, 204 (Tyrrhenian Sea).

grubei Graff, 1892*b*, 190, 205, figs (Atlantic & Indian Oceans); Plehn, 1896*b*, 10, fig. (off south coast of Newfoundland); Bock, 1913, 225 (19°N, 65°W, off St Thomas I., Caribbean; to *Hoploplana*).

hawaiiensis Heath, 1907, 145, figs (Hawaii); Bock, 1913, 245.

heda Kato, 1944, 285, figs (Japan).

heteroglena (Schmarda, 1859) Lang, 1884, 444; Stummer-Traunfels, 1933, 3486, 3490, 3558 (s.o. *Stylochus megalops*).

inquilina Wheeler, 1894, 195 (in *Sycotypus canaliculatus*; Massachusetts); Surface, 1907, 732; 1908, 514, figs (in *S. canaliculatus*; Mass.); Sumner *et al.*, 1913, 580 (Massachusetts); Bock, 1913, 288 (to *Hoploplana*).

insignis Lang, 1884, 236, 442, figs (Naples); Carus, 1885, 149 (descr.); Laidlaw, 1902, 303 (to *Hoploplana*); Bock, 1913, 225.

langii Laidlaw, 1902, 286, figs (Minikoi Is, Indian Ocean); 1903*a*, 4 (to *Paraplanocera*); Kato, 1944, 286 (s.o. *Paraplanocera oligoglena*); Prudhoe, 1945*a*, 201 (s.o. *oligoglena*).

marginata (Stimpson, 1857) Lang, 1884, 445.

muelleri (Audouin, in Savigny, 1827) Schmarda, 1859, 33; Lang, 1884, 451 (s.o. *Stylochus suesensis*); Meixner, 1907*b*, 421 (s.o. *S. suesensis*); Palombi, 1928, 582 (s.o. *Stylochus suesensis*).

multitentaculata Kato, 1940, 559, figs (Japan); 1944, 284 (Japan); Kikuchi, 1968, 167 (Japan). [See *Planocera reticulata* multitentaculate form – Yeri & Kaburaki, 1918*b*, 47.]

nebulosa Girard, 1853, 367 (South Carolina); Diesing, 1862, 561; Verrill, 1873, 325, 332, 632, fig.; Lang, 1884, 463; Verrill, 1893, 472, figs (New England; type of subgenus *Planoceropsis*); Girard,

1893, 191 (descr.); Van Name, 1899, 263; Meixner, 1907*b*, 399 (to *Stylochus*); Sumner *et al.*, 1913, 580 (Massachusetts); Laidlaw, 1903*d*, 102 (not a *Planocera*); Pearse & Walker, 1939, 18, fig.; Hyman, 1940*a*, 459, (s.o. *Stylochus ellipticus*).

oligoglena (Schmarda, 1859) Lang, 1884, 444 (*'olygoglena'*); Prudhoe, 1945*a*, 201 (s.o. *Paraplanocera oligoglena*).

oxyceraea (Schmarda, 1859) Lang, 1884, 445; Kato, 1944, 288 (s.o. *Callioplana marginata*).

pacifica Hyman, 1954*b*, 333, figs (Hawaiian Is).

papillosa Lang, 1884, 236, 442, figs (Isle of Capri, Italy); Carus, 1885, 149 (descr.); Bock, 1913, 225 (to *Hoploplana*).

pelagica (Moseley, 1877) Lang, 1884, 236, 439; Laidlaw, 1903*d*, 102; 1904*a*, 1 (Atlantic Ocean; distinct from *pellucida*); Bock, 1913, 233 (s.o. *Pelagoplana sargassicola*).

pellucida (Mertens, 1832) Örsted, 1844*a*, 48; Diesing, 1850, 216 (*Stylochus*); Lang, 1884, 437; Graff, 1892*b*, 195, figs (Atlantic & Pacific Oceans); Woodworth, 1894, 49 (pelagic, North Pacific, 13°33′30″N, 97°57′30″W); Plehn, 1896*a*, 170 (Atlantic, 5°N, 25°W & Pacific, south of Galapagos Is); 1896*b*, 3, 4, 11, fig. (between Ascension & Cape Verde Is & 42°N, 55°W (off Newfoundland)); Laidlaw, 1903*d*, 102; Bock, 1913, 173, 240, fig. (synonymy; Durban, S. Africa, Pacific Ocean (29°50′S, 175°0′E) & various localities in Atlantic Ocean); 1923*c*, 341, 357 (Juan Fernandez I., off Chilean coast); 1931, 277, figs (synonymy; various localities in Atlantic, Indian & Pacific Oceans); Bresslau, 1933, 234, fig.; Stummer-Traunfels, 1933, 3568, fig. (6°S, 8°E); Kato, 1938*c*, 231, fig. (Japan); 1944, 282 (synonymy); Prudhoe, 1950*b*, 710, 712 (s.o. *gaimardi*); Westblad, 1952, 9 (distribution); Dawydoff, 1952, 82 (Vietnam); Prudhoe, 1982*c*, 50, fig. (North Sea).

profunda Kato, 1937*f*, 347, 359, figs (Japan); 1944, 284.

purpurea Yeri & Kaburaki, 1918*a*, 432, 437 (diagn.; Japan); 1918*b*, 2, 22, 47, figs (Japan); Kato, 1937*a*, 220 (Japan); 1944, 283 (Japan); Dawydoff, 1952, 82 (Vietnam).

rainieri Belloc, 1961, 297 [*sine descr.*]; (Villefranche, Mediterranean).

reticulata (Stimpson, 1855*a*) Lang, 1884, 445; Yeri & Kaburaki, 1918*a*, 436 (descr.; Japan); 1918*b*, 19, 47, figs (multitentaculate form; Japan); Bresslau, 1928–33, 83, fig.; Kato, 1934*b*, 127 (Japan); 1937*b*, 235 (Korea); 1938*a*, 565 (Japan); 1938*b*, 584 (Japan); 1939*b*, 146 (multi-tentacular forms; Japan); 1940, 565, figs (Japan); 1944, 283, figs; Dawydoff, 1952, 82 (Vietnam); Utinomi, 1956, 30, fig. (coloured) (Japan); Okada *et al.*, 1971, 62 (Japan); Gallagher, 1978, 36 (Persian Gulf).

reticulata (Gray, *in* Pease, 1860) Diesing, 1862, 561; Lang, 1884, 440; Laidlaw, 1903*d*, 102.

sandiegensis Boone, 1929, 33, figs (California); Freeman, 1930, 339 (morphological note; California); Hyman, 1953*a*, 349 (to *Alloioplana*); Stasek, 1966, 9 (location of type-specimens).

sargassicola (Mertens, 1832) Örsted, 1844*a*, 48; Diesing, 1850, 216 (renamed *Stylochus mertensi*); Lang, 1884, 454 (*Stylochus*); Bock, 1913, 233 (to *Pelagoplana*).

simrothi Graff, 1892*b*, 190, 200, figs (pelagic, in shell of *Janthina communis*, north of Ascension I.); Plehn, 1896*b*, 3, 11 (descr.; between Equator & Ascension I.); Laidlaw, 1903*d*, 102; Bock, 1913, 244.

species Laidlaw, 1903*c*, 301, 302, fig. (Straits of Malacca); Bock, 1913, 245.

species Laidlaw, 1903*a*, 4 (between Dauer & Murray Is, Torres Strait, Australia); Bock, 1913, 245 (?*Planocera*).

species Haswell, 1907*a*, 643 (N.S.W., Australia).

species Steinbeck & Ricketts, 1941, 337 (Lower California).

species Prudhoe, 1945*b*, 381 (Victoria, Australia; host of trematode metacercaria).

species Morton & Miller, 1968, 172 (New Zealand).

suesensis (Ehrenberg, 1831) Orsted, 1844*a*, 48 (? = *Planaria muelleri*, *Pl. gigas* & *Pl. tuberculata*); Diesing, 1850, 215 (*Stylochus*); Lang, 1884, 451 (*Stylochus*); Palombi, 1928, 582 (*Stylochus*).

thesea (Kelaart, 1858) Collingwood, 1876, 88, 98, fig. (Ceylon); Lang, 1884, 465.

tridentata Hyman, 1953*a*, 188, figs (Galapagos Is).

uncinata Palombi, 1939*a*, 123, 129, figs (South Africa).

villosa Lang, 1884, 62, 441, figs (Bay of Naples); Carus, 1885, 1885, 149 (descr.); Bock, 1913, 225 (to *Hoploplana*).

PLANOCERATIDAE Stechow, 1922 – see Poche, 1926, 99 (= Planoceridae).

PLANOCERIDAE Stimpson, 1857, 23; Johnston, 1865, 5; Lang, 1884, 428, 433; Carus, 1885, 148; Verrill, 1893, 462; Hallez, 1893, 147 (key to genera); 1894, 201 (defined; key to genera); Gamble, 1893*a*, 496; 1896, 19; Benham, 1901, 31; Meixner, 1907*b*, 390; Bock, 1913, 228, 231 (diagn., key to genera); Heath & McGregor, 1913, 459; Bresslau, 1933, 288; Stummer-Traunfels, 1933, 3576,

3586; Steinböck, 1937, 9; Hyman, 1940*b*, 19; 1953*a*, 346 (defined); Marcus, 1947, 131, 135, 165 (systematic note; key to genera); de Beauchamp, 1961, 60 (genera).

PLANOCERIDEA Diesing, 1862, 493, 560; Lang, 1884, 433 (=Planoceridae); Poche, 1926, 98 (Order of Polycladida).

PLANOCERIDES Poche, 1926, 99 (=Schematommata; superfam. of Acotylea).

PLANOCERIDIUM Poche, 1926, 99 (sedis incertae Planocerides – *Hoploplana*).

PLANOCERINAE Meixner, 1907*b*, 385, 394 (subfam. of Planoceridae); Bresslau, 1933, 289.

PLANOCERODES Palombi, 1936, 1, 25, 30 (type: *ceratommata*).

ceratommata Palombi, 1936, 1, 25, figs (Still Bay, S. Africa).

PLANOCEROIDEA Nicoll, 1936, 74 (emendation of Planocerides Poche).

(PLANOCEROPSIS) Verrill, 1893, 471 (subgen. of *Planocera*; type: nebulosa). Laidlaw, 1903*e*, 14 (affinities); Meixner, 1907*b*, 399 (s.o. *Stylochus*).

nebulosa (Girard, 1853) Verrill, 1893, 471; Meixner, 1907*b*, 399 (to *Stylochus*).

PLANOCEROS [pro *Planocera*] Ehrenberg, 1831, 54 (type: *gaymardi*).

gaymardi [err. pro *gaimardi*] (Blainville, 1828) Ehrenberg, 1831, 54; Örsted, 1844*a*, 26.

PLATYENDRON Simonetta & Delle Cave, 1978 (type: *ovale* Simonetta & Delle Cave). [Fossil].

ovale Simonetta & Delle Cave, 1978, 45, figs (Middle Cambrian, British Columbia).

PLEHNIA Bock, 1913, 36, 69 (nom. nov. pro *Acelis* Plehn; type: *arctica*); Hyman, 1953*a*, 277 (defined), 279 (systematic note).

arctica (Plehn, 1896*a*) Bock, 1913, 70, fig. (Jan Mayen I. & east Greenland); Bock, 1923*a*, 13, 15 (=*Polyporus coeca*); 1926*b*, 293 (prob. =*Polyporus coeca*); Steinböck, 1932, 332, 337 (Arctic distribution).

caeca Hyman, 1953*a*, 277, figs (California).

caeca var. *oculifera* Hyman, 1953*a*, 279, fig. (California).

ellipsoides (Girard, 1853) Hyman, 1952, 195.

japonica Bock, 1923*a*, 1, figs (Japan; =*Acelis* sp. Yeri & Kaburaki, 1918); Bock, 1925*b*, 110 ('new genus'); Kato, 1944, 261; Hyman, 1953*a*, 279 (to *Paraplehnia*).

pacifica Kato, 1939*a*, 65, 66, figs (Japan); 1944, 261; Hyman, 1953*a*, 279 (to *Paraplehnia*).

tropica Hyman, 1959*a*, 546, figs (Palau Is). [See p. 47]

PLEHNIIDAE Bock, 1913, 69; 1923*a*, 14 (systematic relationship); Stummer-Traunfels, 1933, 3567, 3586; Bresslau, 1933, 285; Hyman, 1940*a*, 453; 1940*b*, 16; 1953*a*, 277 (defined).

POLYCELIS Ehrenberg, 1831 [Now a valid genus of Tricladida] Quatrefages, 1845, 131 (subgenera *Polycelis* and *Prosthiostomum*); Schmarda, 1859, 13, 19; Lang, 1884, 475 (subgen. (*Polycelis*) of Quatrefages s.o. *Leptoplana*).

australis Schmarda, 1859, 21 fig. (N.S.W., Australia & Auckland, New Zealand); Diesing, 1862, 529 (to *Leptoplana*); Hutton, 1879, 315 (New Zealand); Lang, 1884, 505; Whitelegge, 1890, 206 (N.S.W.); Haswell, 1907*a*, 471 (? =*Leptoplana australis* Laidlaw); Bock, 1913, 205 (? =*Notoplana australis* (Laidlaw)); Stummer-Traunfels, 1933, 3486, 3517 (to *Notoplana*; type-specimen re-examined).

capensis Schmarda, 1859, xiii, 22, figs (off Cape of Good Hope, South Africa); Diesing, 1862, 530 (to *Leptoplana*); Lang, 1884, 506.

ellipsis (Dalyell, 1853) Leuckart, 1859, 183; Diesing, 1862, 542 (to *Leptoplana*); Lang, 1884, 588 (to *Stylostomum*); Bock, 1913, 270 (*Stylostomum*).

erythrotaenia Schmarda, 1859, xiii, 21, fig. (Table Bay, South Africa); Diesing, 1862, 529 (to *Leptoplana*); Lang, 1884, 505; Stummer-Traunfels, 1933, 3486, 3518 (to *Notoplana*; type-specimen re-examd.; syns: *Polycelis lyrosora & Notoplana ovalis*); Palombi, 1939*a*, 128 (? s.o. *Notoplana patellarum*).

fallax Quatrefages, 1845, 135, fig. (NW. France); Bock, 1913, 204 (to *Notoplana*).

ferruginea Schmarda, 1859, 22, fig. (Jamaica, W.I.); Diesing, 1862, 530 (to *Leptoplana*); Lang, 1884, 506; Stummer-Traunfels, 1933, 3486, 3521 (to *Notoplana*; type-specimen re-examd.).

haloglena Schmarda, 1859, 21, figs (Viña del Mar, Chile); Diesing, 1862, 528 (to *Leptoplana*); Lang, 1884, 505; Stummer-Traunfels, 1933, 3486, 3514 (to *Notoplana*; type-specimen re-examd.).

levigatus Quatrefages, 1845, 134, figs (NW. France); Diesing, 1850, 198 (to *Leptoplana laevigata*); Beneden, 1861, 42, fig. (Belgium); Keferstein, 1869, 6 (s.o. *Leptoplana tremellaris*); Stossich, 1882, 226 (Trieste); Lang, 1884, 476 (s.o. *tremellaris*); Koehler, 1885, 14, 37, 46, 56 (Channel Is).

lineoliger, Blanchard, 1849, 610 (San Cárlos de Chiloe, Chile); Lang, 1884, 610.

lyrosora Schmarda, 1859, xiii, 24, figs (Cape of Good Hope, South Africa); Diesing, 1862, 535 (to

Leptoplana); Lang, 1884, 507; Stummer-Traunfels, 1933, 3486, 3532, fig. (s.o. *Notoplana erythrotaenia*; type-specimen re-exmd.); Palombi, 1936, 18 (? = *Notoplana ovalis*); 1939*a*, 128 (s.o. *Notoplana patellarum*).

macrorhyncha Schmarda, 1859, 23, figs (Ceylon); Diesing, 1862, 531 (to *Leptoplana*); Lang, 1884, 614 (perhaps a *Prosthiostomum*); Stummer-Traunfels, 1933, 3527 (to *Prosthiostomum*; type-specimen re-examd.).

microsora Schmarda, 1859, 22, fig. (Ceylon); Diesing, 1862, 529 (to *Leptoplana*); Lang, 1884, 506; Stummer-Traunfels, 1933, 3486, 3520 (to *Notoplana*; type-specimen re-examd.).

modestus Quatrefages, 1845, 133, fig. (Naples); Diesing, 1850, 195 (to *Leptoplana*); Stimpson, 1857, 3 (to *Elasmodes*); Lang, 1884, 490 (s.o. *Leptoplana pallida*); Palombi, 1928, 589 (s.o. *Stylochoplana pallida*).

mutabilis Verrill, 1873, 746 (Connecticut); Lang, 1884, 616; Verrill, 1893, 493 (to *Discocelis*); Hyman, 1939*e*, 153; 1939*f*, 238 (to *Coronadena*); 1940*a*, 450, figs (*Coronadena*).

obovata Schmarda, 1859, 20 fig. (Jamaica); Diesing, 1862, 528 (to *Leptoplana*); Lang, 1884, 504; Stummer-Traunfels, 1933, 3486, 3505 (to *Adenoplana*; type-specimen re-examd.).

oosora Schmarda, 1859, 22, figs (Ceylon); Diesing, 1862, 530 (to *Leptoplana*); Lang, 1884, 506; Stummer-Traunfels, 1933, 3486, 3522 (to *Indiplana*; type-specimen re-examd.).

ophryoglena Schmarda, 1859, 20, figs (Peru); Diesing, 1862, 613 (to *Leptoplana*); Lang, 1884, 613 (perhaps a *Prosthiostomum*).

orbicularis Schmarda, 1859, 20, fig. (Chile); Diesing, 1862, 527 (to *Leptoplana*); Lang, 1884, 504; Stummer-Traunfels, 1933, 3486, 3511, 3517 (to *Enterogonia*; type-specimen re-examd.; evidence that type-locality is New Zealand).

pallidus Quatrefages, 1845, 133, figs (Sicily); Diesing, 1850, 195 (to *Leptoplana*); Stimpson, 1857, 21 (to *Elasmodes*); Knappert, 1865, 2; Lang, 1884, 489 (*Leptoplana*); Bock, 1913, 172, 179 (to *Stylochoplana*).

roseimaculata Blanchard, 1849, 72 (San Cárlos de Chiloe, Chile); Lang, 1884, 610.

tigrinus Blanchard, 1847, 271, figs (Genoa); Diesing, 1850, 195 (to *Leptoplana*); Stimpson, 1857, 21 (to *Elasmodes*); Lang, 1884, 71, 467, figs (to *Discocelis*).

trapezoglena Schmarda, 1859, 23, figs (Ceylon); Diesing, 1862, 531 (to *Leptoplana*); Lang, 1884, 507; Stummer-Traunfels, 1933, 3486, 3528 (to *Leptoplana*; type-specimen re-examd.).

variabilis Girard, 1850*b*, 251 (Massachusetts); 1851*c*, 3; Diesing, 1862, 542 (to *Leptoplana*); Lang, 1884, 610; Bock, 1913, 196 (?s.o. *Notoplana atomata*); Hyman, 1939*b*, 135 (?s.o. *Notoplana atomata*); 1940*a*, 468 (?s.o. *N. atomata*); Palombi, 1928, 595 (?s.o. *N. atomata*).

species Schultze, 1854, 222; Kennel, 1879, 120 (to *Leptoplana*); Lang, 1884, 477 (?s.o. *Leptoplana tremellaris*).

POLYCLADA Hallez, 1893, 146, 165 (suborder of Dendrocoela); (Class of Turbellaria; key to families); 1894, 200 (Class of Turbellaria; division of families); Poche, 1926, 95 (classification).

POLYCLADEN Lang, 1881*a*, 549; 1884, 433 (= Polycladidea); Poche, 1926, 9 (Subsubclass of Turbellaria).

POLYCLADIDA Gamble, 1893*a*, 520 (suborder of Turbellaria).

POLYCLADIDEA Lang, 1884, 3, 433 (sub-order of Dendrocoela = Cryptocoela of Örsted); Verrill, 1893, 460 (= Digonopora of Stimpson); Gamble, 1893*a*, 520 (emend. to Polycladida); Gamble, 1896, 19 (classification); Benham, 1901, 31 (Order); Bresslau, 1933, 284 (Order).

POLYPHALLOPLANA Bock, *in* de Beauchamp, 1951*b*, 239 (type: *bocki*).

bocki de Beauchamp, 1951*b*, 239, figs (off NW. coast of Africa, 9°51′10″–9°50′40″W and 30°28′10″–30°25′30″N).

POLYPORUS Plehn, 1897, 90 [*nec* Grube, 1840 ?Vermes, ?Mollusca] (type: *caecus*); Benham, 1901, 31; Bock, 1913, 250 (diagn.); 1923*a*, 15 (prob. s.o. *Plehnia*); Marcus & Marcus, 1966, 324.

caecus Plehn, 1897, 90, figs (Spitzbergen); Bock, 1913, 251; 1923*a*, 15 (= *Plehnia arctica*); 1925*a*, 22; 1926*b*, 302 (prob. = *Plehnia arctica*); Steinböck, 1932, 334, 337 (Arctic distribution).

POLYPOSTHIA [emend. pro *Polypostia*] Bock, 1913, 87 (diagn.). 1923*a*, 14; Bresslau, 1933, 285; Prudhoe, 1982*c*, 27 (diagn.).

similis Bergendal, 1892, 551 (Sweden); 1893*b*, 366 (Sweden); 1893*c*, 1 (descr.); Bock, 1913, 87, figs (Sweden & Skagerrak); Palombi, 1929, 196 (genital complex); Hendelberg, 1974*b*, 9, figs (Sweden); Prudhoe, 1982*c*, 28, fig. (Scotland).

POLYPOSTHIDES Palombi, 1924*a*, 34 (type: *karimatensis*); 1924*b*, 7, 12 (defined).

affinis Palombi, 1924*a*, 35 (Karimata Straits, Indonesia); 1924*b*, 9, figs (1°20′S, 107°57′E); Bresslau, 1933, 240, fig.

caraibica Palombi, 1924*a*, 35 (Caribbean); 1924*b*, 10, figs (17°5′N, 85°14′W).

karimatensis Palombi, 1924*a*, 34 (Karimata Straits, Indonesia); 1924*b*, 7, figs (Karimata Straits – 1°20′S, 107°57′E).

POLYPOSTHIIDAE [emend. pro Polypostiidae] Bock, 1913, 85; 1923*a*, 14 (relationship with Plehniidae); Palombi, 1929, 196 (genital organs & affinities of family); Bresslau, 1933, 285; Stummer-Traunfels, 1933, 3567, 3586; de Beauchamp, 1961, 57 (generic constitution).

POLYPOSTIA Bergendal, 1892, 551; 1893*b*, 366; 1893*c*, 1; (type: *similis*); Gamble, 1896, 19; Graff, 1896, 93 (emend. to *Polyposthia*); Benham, 1901, 31; Bock, 1913, 87 (*Polyposthia*).

similis Bergendal, 1892, 551 (Sweden); 1893*b*, 366 (Sweden); 1893*c*, 1 (descr.; Sweden); 1902, 750; Laidlaw, 1903*e*, 5; Bock, 1913, 87, figs (Skagerrak & Sweden); Palombi, 1929, 196 (genital organs); Hendelberg, 1974*b*, 9, figs (Sweden).

POLYPOSTIADAE Bergendal, 1893*a*, 367; 1893*b*, 367 (nom. nov. pro Cryptocelididae Bergendal, 1893); 1893*c*, 13; Bock, 1913, 85 (emended to Polyposthiidae).

POLYPOSTIIDAE [emend. pro Polypostiadae] Benham, 1901, 31.

PROCEROS Quatrefages, 1845, 129, 137 (type: *argus*); Diesing, 1850, 208 (s.o. *Eurylepta*); Stimpson, 1857, 20; Schmarda, 1859, 14, 30 (renamed *Prosthoceraeus*); Diesing, 1862, 493, 550; Grube, 1864, 97; Laidlaw, 1903*a*, 2 (prob. = *Pseudoceros*).

albicornis Stimpson, 1857, 20, 25 (Japan); Diesing, 1862, 551; Lang, 1884, 564 (to *Prostheceraeus*); Kato, 1944, 298 (to *Pseudoceros*).

argus Quatrefages, 1845, 137, figs (NW. France); Diesing, 1850, 209 (to *Eurylepta*); Stimpson, 1857, 20; Schmarda, 1859, 30 (to *Prostheceraeus*); Diesing, 1862, 553; Lang, 1884, 563 (*Prostheceraeus*); Koehler, 1885, 37, 56 (Channel Is); Pruvot, 1897, 21 (Brittany & Gulf of Lion).

aurantiacus (Delle Chiaje, 1830) Lang, 1879, 459; Schmidtlein, 1880, 172; Lang, 1881*c*, 308; 1884, 548 (to *Yungia*).

auritus (Claparède, 1861) Diesing, 1863, 175.

buskii Collingwood, 1876, 87, 91, figs (Singapore); Lang, 1884, 547 (to *Pseudoceros*).

cardiosorus (Schmarda, 1859) Diesing, 1862, 552; Lang, 1884, 546 (to *Pseudoceros*).

concinnus Collingwood, 1876, 87, 90, figs (North Borneo); Lang, 1884, 593; Laidlaw, 1903*c*, 315 (a Euryleptid); Kaburaki, 1923*a*, 642, figs (to *Pseudoceros*).

cristatus Quatrefages, 1845, 139, figs (NW. France); Thompson, 1846, 392 (s.o. *Planaria vittata*); Diesing, 1850, 210 (to *Eurylepta*); Schmarda, 1859, 14, 30, (to *Prostheceraeus*); Diesing, 1862, 546; Lang, 1884, 554 (s.o. *Prostheceraeus vittatus*).

hancockanus Collingwood, 1876, 87, 91, figs (Singapore); Lang, 1884, 567 (to *Prostheceraeus*); Laidlaw, 1903*c*, 302 (to *Pseudoceros 'hancockianus'*; Straits of Malacca); Bock, 1913, 258 (? s.o. *Pseudoceros malayensis*); Stummer-Traunfels, 1933, 3566 (s.o. *Pseudoceros niger*).

limbatus (Leuckart, 1828) Diesing, 1862, 554; Lang, 1884, 544 (to *Pseudoceros*).

lobianchii Lang, 1879, 486 (Mediterranean); 1884, 578, figs (to *Eurylepta*).

melobesiarum Schmidtlein, 1880, 172 (Mediterranean); Lang, 1884, 576, figs (var. of *Eurylepta cornuta*).

miniatus (Schmarda, 1859) Diesing, 1862, 554; Lang, 1884, 551 (to *Yungia*).

nigrocinctus (Schmarda, 1859) Diesing, 1862, 551; Lang, 1884, 547 (to *Pseudoceros*).

orbicularis (Schmarda, 1859) Diesing, 1862, 553; Lang, 1884, 551 (to *Eurylepta*).

sanguinolentus Quatrefages, 1845, 138, figs (NW. France); Thompson, 1846, 392 (syn. *Planaria cornuta* of British authors); Diesing, 1850, 209 (to *Eurylepta*); Stimpson, 1857, 20; Diesing, 1862, 552; Grube, 1864, 97 (Adriatic Sea); Stossich, 1882, 226 (Trieste); Lang, 1884, 580, figs (to *Oligocladus*); Bock, 1913, 267 (*Oligocladus*).

striatus (Schmarda, 1859) Diesing, 1862, 551; Lang, 1884, 552 (to *Eurylepta*); Stummer-Traunfels, 1933, 3487 (to *Pseudoceros*).

superbus (Schmarda, 1859) Diesing, 1862, 552; Lang, 1884, 552 (s.o. *Planaria undulata*); Stummer-Traunfels, 1933, 3487 (to *Pseudoceros*).

tuberculatus Lang, *in* Schmidtlein, 1880, 172 (Mediterranean); Lang, 1881*b*, 225, fig.; 1884, 568 (renamed *Cycloporus papillosus* sp. nov.); Bock, 1913, 263 (s.o. *Cycloporus papillosus* (Sars.).)

velutinus Blanchard, 1847, 273, figs (Genoa); Diesing, 1850, 210 (to *Eurylepta*); Schmarda, 1859, 30; Lang, 1879, 459 (anatomy); 1884, 49, 538, figs (to *Pseudoceros*).

violaceus (Schmarda, 1859) Diesing, 1862, 553; Lang, 1884, 540 (var. of *Ps. velutinus*).

zebra (Leuckart, 1828) Diesing, 1862, 554; Lang, 1884, 544 (to *Pseudoceros*).

PROSTHECERAEINAE Hallez, 1913, 44 (subfam. of Euryleptidae = Euryleptidea of Lang, 1884).

PROSTHECERAEUS Schmarda, 1959, 14, 30 (nom. nov. pro *Proceros* Quatrefages *nec* Rafinesque) (type: *cristatus* [= *vittatus*]; Lang, 1884, 431, 553 (diagn.); Carus, 1885, 155; Gamble, 1893*a*, 503; Hallez, 1894, 220 (defined); Gamble, 1896, 19; Benham, 1901, 31; Steinböck, 1937, 11, 12, fig.; Hyman, 1953*a*, 373 (defined); Prudhoe, 1982*c*, 58 (defined).

albicornis (Stimpson, 1857) Lang, 1884, 104, 564.

albocinctus Lang, 1884, 557, figs (Mediterranean); Carus, 1885, 155 (descr.).

anomalus Haswell, 1907*b*, 481, fig. (N.S.W., Australia).

argus (Quatrefages, 1845) Schmarda, 1859, 30 (footnote); Diesing, 1862, 553 (*Proceros*); Lang, 1884, 563; Gamble, 1893*a*, 505 (descr.); 1896, 19 (Channel Is); Prudhoe, 1982*c*, 58, fig.

bellostriatus Hyman, 1953*a*, 373, figs (California); 1959*b*, 14, figs (California); Morris *et al.*, 1980, 81, fig. (coloured), (California).

clavicornis Schmarda, 1859, 32, figs (Ceylon); Diesing, 1862, 564; Lang, 1884, 566; Stummer-Traunfels, 1933, 3487, 3543, 3544, figs (type-specimen re-examd.; to *Pseudoceros*).

cornutus (Müller, 1776) Schmarda, 1859, 30 (footnote); Diesing, 1862, 548 (*Eurylepta*); Lang, 1884, 572 (*Eurylepta*).

cristatus (Quatrefages, 1845) Schmarda, 1859, 14, 30 (footnote); Lang, 1884, 554 (s.o. *vittatus*).

flavomaculatus Graff, *in* Saville-Kent, 1893, 362, fig. (Great Barrier Reef); 1904–05, 1888 (*Pseudoceros* – error?)

flavomarginatus (Ehrenberg, 1831) Lang, 1884, 563.

floridanus Hyman 1955*b*, 138, fig. (Gulf of Mexico).

giesbrechtii Lang, 1884, 558, fig. (Naples); Carus, 1885, 155 (descr.); Dawydoff, 1952, 81 (Vietnam).

hancockanus (Collingwood, 1876) Lang, 1884, 567 (syn. *Stylochopsis malayensis*); Bock, 1913, 258 (*ex parte* s.o. *Pseudoceros malayensis*).

japonicus (Stimpson, 1857) Lang, 1884, 565; Kato, 1944, 298 (to *Pseudoceros*).

kelaartii (Collingwood, 1876) Lang, 1884, 568; Laidlaw, 1903*c*, 302 (to *Pseudoceros*).

latissimus Schmarda, 1859, 31, figs (Ceylon); Diesing, 1862, 563; Lang, 1884, 566; Stummer-Traunfels, 1933; 3487, 3544, figs (to *Pseudoceros*; type-specimen re-examd.).

maculosus (Verrill, 1893) Hyman, 1952, 197, figs (Massachusetts).

meleagrinus (Kelaart, 1858) Kaburaki, 1923*b*, 635, 646, figs. (Philippines); Dawydoff, 1952, 81 (Vietnam).

microceraeus Schmarda, 1859, 31, figs (Ceylon); Diesing, 1862, 563; Lang, 1884, 565; Stummer-Traunfels, 1933, 3487, 3541, 3544, figs (type-specimen re-examd.; to *Pseudoceros*).

moseleyi Lang, 1884, 116, 560, figs (Tyrrhenian Sea); Carus, 1885, 155 (descr.).

niger (Stimpson, 1857) Lang, 1884, 565; Stummer-Traunfels, 1933, 3566 (to *Pseudoceros*).

nigricornis Schmarda, 1859, 31, figs (Peru); Diesing, 1862, 563; Lang, 1884, 566.

panamensis Woodworth, 1894, 51, figs (off Panama).

papilio (Kelaart, 1858) Kaburaki, 1923*b*, 635, 646, fig. (Philippines); Dawydoff, 1952, 81 (Vietnam).

pseudolimax Lang, 1884, 559, figs (Mediterranean); Carus, 1885, 155 (descr.).

roseus Lang, 1884, 562, figs (Isle of Capri & Naples); Carus, 1885, 156 (descr.).

rubropunctatus Lang, 1884, 562, fig. (Mediterranean); Carus, 1885, 156 (descr.); Laidlaw, 1906, 714 (Cape Verde Is).

terricola Schmarda, 1859, 30 (a terricolous triclad).

undulatus (Kelaart, 1858) Lang, 1884, 552.

violaceus (Delle Chiaje, 1822) Lang, 1884, 563 (Naples); Carus, 1885, 156 (descr.).

viridis Schmarda, 1859, 32, figs (Ceylon); Diesing, 1862, 564; Lang, 1884, 567 (syn. *Planaria viridis* Kelaart); Stummer-Traunfels, 1933, 3487, 3543, 3544, figs (type-specimen re-examd.; to *Pseudoceros*).

vittatus (Montagu, 1913) Lang, 1884, 49, 554, fig. (Naples; syn. *Proceros vittatus*); Carus, 1885, 155 (descr.); Koehler, 1885, 14, 37, 49, 56 (Channel Is); Vaillant, 1889, 656, fig. (Atlantic coast of France); Bergendal, 1890, 327 (Sweden); Gamble, 1893*a*, 504 (descr.); 1893*b*, 46 (S. Devon); Girard, 1894*a*, 245 (NW. France); Hallez, 1894, 221 (NW. France); Klinckowström, 1896, 587; Gamble, 1896, 19 (Britain); Pruvot, 1897, 20 (Brittany & Gulf of Lion); Francotte, 1897, 36, figs; 1898, 243, figs (Concarneau); Gamble, 1900, 812 (Ireland); Gérard, 1901, 141, figs; Retzius, 1906, 41; Theel, 1907, 61 (Sweden); Lo Bianco, 1909, 569 (Naples); Bock, 1913, 261 (Norway & Sweden); Southern, *in* Farran, 1915, 35 (Ireland); Bresslau, 1928, 105, fig.; Stummer-Traunfels,

1933, 3574, fig.; Southern, 1936, 71 (Ireland); Bassindale & Barrett, 1957, 221 (S. Wales); Williams, 1954, 55 (NE. Ireland); Pasterino & Canu, 1965, 1 (Ligurian Sea); Crothers, 1966, 22 (S. Wales); Franzén, 1956, 355, 364 (Sweden); Hendelberg, 1974*b*, 15, figs (Sweden); Prudhoe, 1982*c*, 60, fig. (U.K.).

zebra Hyman, 1955*e*, 266, figs (Gulf of Mexico & Jamaica).

species Plehn, 1896*a*, 172 (Brazil).

PROSTHIOSTOMIDAE Lang, 1884, ix, 432, 594; Carus, 1885, 157; Hallez, 1893, 182; 1894, 235 (defined); Gamble, 1896, 19 (Prosthiostomatidae); Benham, 1901, 31; Laidlaw, 1902, 307 (probably derived from Euryleptidae); Bock, 1913, 281; 1922, 11 (diagn. note); Bresslau, 1933, 293; Stummer-Traunfels, 1933, 3576; Hyman, 1940*a*, 490; 1953*a*, 382 (defined); Marcus, 1948, 184 (key to genera); de Beauchamp, 1961, 67 (genera); Prudhoe, 1982*c*, 70 (defined).

PROSTHIOSTOMUM Quatrefages, 1845, 129, 132, 133, (subgenus of *Polycelis*; type: *arctum* = *siphunculus*); Diesing, 1850, 194 (s.o. *Leptoplana*); Stimpson, 1857, 22 (to generic rank); Schmidt, 1861, 5; Lang, 1884, 432, 594 (diagnosis); Vaillant, 1889, 656 (*Prosthiostoma*); Verrill, 1895, 496; Laidlaw, 1902, 302 (distribution of spp.); Bock, 1913, 281; Yeri & Kaburaki, 1918*a*, 441 (features); 1918*b*, 49 (features); Bock, 1925*a*, 58; Kato, 1937*f*, 369; Hyman, 1939*c*, 7; 1953*a*, 382 (defined); Marcus, 1949, 89 (list of spp.); 1950, 96 (list of spp.); 1952, 99 (supplement to list of spp.); Poulter, 1975, 317 (systematics; *Lurymare* a subgenus).

affine Stimpson, 1857, 22, 28 (Hong Kong); Diesing, 1862, 539 (to *Leptoplana*); Lang, 1884, 603 (?s.o. *grande*); Yeri & Kaburaki, 1918*b*, 42 (?s.o. *grande*).

angustum Bock, 1913, 282, figs (Bahamas); Marcus, 1950, 97 101 (note).

arctum Quatrefages, 1845, 135, figs (Naples); Diesing, 1850, 196 (to *Leptoplana*); Stimpson, 1857, 22; Lang, 1884, 595 (s.o. *siphunculus*).

asiaticum Kato, 1937*b*, 233, figs (Korea); Kato, 1944, 307; Marcus, 1950, 97, 99, 101 (note).

aurantiacum (Collingwood, 1876) Laidlaw, 1903*c*, 302, 317.

auratum Kato, 1937*f*, 347, 363, figs (Japan); 1938*a*, 572 (Japan); 1938*b*, 589, fig. (Japan); 1939*b*, 152 (descr.; Japan); 1944, 307; Marcus, 1950, 97, 101 (note).

awaense Yeri & Kaburaki, 1918*a*, 432, 441 (diagn.; Japan); 1918*b*, 3, 44, 50, figs (Japan); Kato, 1944, 307; Marcus, 1950, 97 (note).

bellum Kato, 1939*a*, 66, 77, figs (Japan); 1944, 308; Marcus, 1950, 97, 99, 101 (note).

capense Bock, 1931, 261, 296, figs (Simon's Bay, South Africa); Marcus, 1950, 97, 99 (note); Day *et al.*, 1970, 19 (False Bay, South Africa).

collare (Stimpson, 1855) Stimpson, 1857, 22, 28 (descr.); Diesing, 1862, 537 (*Leptoplana*); Lang, 1884, 612 (*Leptoplana*).

constipatum Stimpson, 1857, 22, 28 (Hokkaido, Japan); Diesing, 1862, 537 (to *Leptoplana*); Lang, 1884, 604; Marcus, 1950, 96 (note).

cooperi Laidlaw, 1902, 301, fig. (Maldive Is, Indian Ocean); Marcus, 1950, 97 (note); Poulter, 1975, 317 (to *P.* (*Lurymare*).)

crassiusculum Stimpson, 1857, 22, 29 (Japan); Diesing, 1862, 537 (to *Leptoplana*); Lang, 1884, 612; Marcus, 1950, 96, 101 (notes).

cribrarium Stimpson, 1857, 22, 28 (Hokkaido, Japan); Diesing, 1862, 537 (to *Leptoplana*); Lang, 1884, 604; Marcus, 1950, 96, 101 (notes).

cyclops (Verrill, 1901) Hyman, 1939*d*, 19 (type-specimen(?) re-examd.); Marcus, 1950, 97, 99, 101 (notes); du B.-R. Marcus & Marcus, 1968, 83, figs (Caribbean).

cynarium Marcus, 1950, 5, 99, 115, figs (Brazil); du Bois-Reymond Marcus, 1955*b*, 287.

delicatum Palombi, 1939*a*, 123, 135, figs (East London, South Africa); Marcus, 1950, 97, 101 (notes); du B.-R. Marcus & Marcus, 1968, 88 (to *Lurymare*); Day *et al.*, 1970, 19 (False Bay, South Africa); Poulter, 1975, 317 (to *P.* (*Lurymare*).)

dohrnii Lang, 1884, 274, 601, figs (Italy); Benham, 1901, 32, fig.; Laidlaw, 1906, 714 (descr.; Cape Verde Is); Marcus, 1950, 97, 99 (note).

drygalskii Bock, 1931, 261, 298, figs (Simon's Bay, S. Africa); Marcus, 1950, 97 (note); du B.-R. Marcus & Marcus, 1968, 88 (to *Lurymare*); Day *et al.*, 1970, 19 (False Bay, S. Africa).

elegans Laidlaw, 1902, 298, 305, figs (Maldives Is, Indian Ocean); Marcus, 1950, 97, 101 (notes); Poulter, 1975, 317 (to *P.* (*Lurymare*)).

elongatum Quatrefages, 1845, 136, figs (I. de Bréhat, NW. France); Diesing, 1850, 196 (to *Leptoplana*); Stimpson, 1857, 22; Schmidtlein, 1880, 72; Lang, 1884, 595 (s.o. *P. siphunculus*); Yeri & Kaburaki, 1918*b*, 41 (s.o. *siphunculus*).

emarginatum Leuckart, 1863, 169; Lang, 1884, 595 (?s.o. *P. siphunculus*); Yeri & Kaburaki, 1918*b*, 41 (?s.o. *siphunculus*).
exiguum Hyman, 1959*a*, 587, figs (Eniwetok Atoll, Micronesia).
formosum Kato, 1943*b*, 69, 74, figs (Taiwan); 1944, 308.
gabriellae Marcus, 1949, 7, 88, 105, figs (Brazil); Marcus, 1950, 97 (note); du B.-R. Marcus, 1955*b*, 287; du B.-R. Marcus & Marcus, 1968, 89 (to *Lurymare*); Poulter, 1975, 317 (to *P.* (*Lurymare*).)
gilvum Marcus, 1950, 5, 98, 115, figs (Brazil); du B.-R. Marcus, 1955*b*, 287; Marcus & Marcus, 1951, 25 (2 suckers).
gracile Girard, 1850*b*, 251 (Massachusetts); Diesing, 1862, 541 (to *Leptoplana*); Verrill, 1879, 13 (Massachusetts); Lang, 1884, 610; Girard, 1893, 198 (to *Euplana*); Verrill, 1893, 496, figs (Massachusetts & Connecticut); Sumner *et al.*, 1913, 580 (Massachusetts); Pearse & Walker, 1939, 19, fig.
grande Stimpson, 1857, 22, 28 (Japan); Diesing, 1862, 539 (to *Leptoplana*); Lang, 1884, 603 (synonymy); Yeri & Kaburaki, 1918*b*, 42, 49, figs (synonymy; Japan); Kato, 1934*b*, 135 (Japan); 1938*a*, 572 (Japan); 1938*b*, 589 (Japan); 1944, 307; Marcus, 1950, 96, 101 (notes); Dawydoff, 1952, 81 (Vietnam); Utinomi, 1956, 30, fig. (coloured) (Japan).
griseum Hyman, 1959*a*, 592, figs (Eniwetok Atoll, Micronesia).
hamatum Schmidt, 1861, 9, figs (Cephalonia I., Greece); Diesing, 1862, 538 (to *Leptoplana*); Leuckart, 1863, 169 (specific features); Lang, 1884, 595 (s.o. *siphunculus*); Yeri & Kaburaki, 1918*b*, 41 (s.o. *siphunculus*).
katoi (*Lurymare*) Poulter, 1975, 323, 336 (Hawaii).
komaii Kato, 1944, 311, figs (Japan).
laetum Kato, 1938*b*, 578, 590, figs (Japan); 1944, 308; Marcus, 1950, 97 (note).
latocelis Hyman, 1953*a*, 382, figs (California).
lineatum Meixner, 1907*a*, 172 (Somalia); 1907*b*, 482, figs; Marcus, 1950, 97.
lobatum Pearse, 1938, 91, fig. (Florida & North Carolina); Pearse & Littler, 1938, 242, fig.; Hyman, 1940*a*, 490 (descr.,); Marcus, 1950, 97, 99 (notes).
macrorhynchum (Schmarda, 1859) Stummer-Traunfels, 1933, 3486, 3527, figs (type-specimen re-examd.); Marcus, 1950, 97 (note).
maculatum Haswell, 1907*b*, 482, fig. (Australia); Marcus, 1950, 97, 99, 101 (notes).
marmoratum Yeri & Kaburaki, 1918a, 432, 441 (diagn.; Japan); 1918*b*, 2, 43, 49, figs. (Japan); Kato, 1937*a*, 230 (Japan); 1938*b*, 589 (Japan); 1944, 307; Marcus, 1950, 97, 101 (note).
matarazzoi Marcus, 1950, 5, 94, 99, 101, 117, figs (Brazil); du B.-R. Marcus 1955*b*, 287; du Bois-Reymond Marcus & Marcus, 1968, 89, fig. (to *Lurymare*; Caribbean); Poulter, 1975, 317 (to *P.* (*Lurymare*).).
milcum du Bois Reymond Marcus & Marcus, 1968, 87, figs (Caribbean & Florida).
molle Freeman, 1930, 334, fig. (California); Hyman, 1953*a*, 386 (to *Euprosthiostomum*); Marcus, 1950, 97, 99 (note).
monosorum (Schmarda, 1859) Stummer-Traunfels, 1933, 3486, 3490, figs. (type-specimen re-examd.); Marcus, 1950, 96 (note).
montiporae (Lurymare) Jokiel & Townsley, 1974, 361 (Hawaii); Poulter, 1975, 317, 334, figs (Hawaii).
multicelis Hyman, 1953*a*, 384 (California and Baja California).
nationalis Plehn, 1896*b*, 8, fig. (off Labrador).
notoensis Kato, 1944, 310, figs (Japan).
nozakensis Kato, 1944, 311, figs (Japan).
obscurum (Stimpson, 1855) Stimpson, 1857, 22, 28 (descr.); Diesing, 1862, 539 (to *Leptoplana*); Lang, 1884, 604; Marcus, 1950, 96 (note).
ostreae Kato, 1937*f*, 347, 365, figs (Japan); 1944, 308; Marcus, 1950, 97 (note).
pallidum Laidlaw, 1903*c*, 302, 317, fig. (Malaysia); Marcus, 1950, 97 (note).
parvicelis Hyman, 1939*c*, 6, figs (Galapagos); Marcus, 1950, 97, 101 (note); Hyman, 1953*a*, 193, figs (Galapagos).
pellucidum (Grube, 1840) Lang, 1884, 605; Bock, 1913, 283 (?*Prosthiostomum*).
pulchrum Bock, 1913, 285, figs (Bahamas); Marcus, 1950, 97, 99 (notes); Hyman, 1955*b*, 142, figs (Virgin Is, Caribbean); du B.-R. Marcus & Marcus, 1968, 85, fig. (Caribbean).
purum Kato, 1937*f*, 347, 366, figs (Japan); 1944, 308; Marcus, 1950, 97, 101 (notes); du B.-R. Marcus & Marcus, 1968, 88 (to *Lurymare*); Poulter, 1975, 317 (to P. (*Lurymare*).).

rubropunctatum Yeri & Kaburaki, 1918*a*, 432, 442 (diagn.; Japan); 1918*b*, 3, 45, 50, figs (Japan); Kato, 1944, 307; Dawydoff, 1952, 81 (Vietnam); Marcus, 1950, 97 (note).

russoi Palombi, 1939*a*, 123, 141, figs (East London, S. Africa); Marcus, 1950, 97, 99, 101 (notes); du B.-R. Marcus & Marcus, 1968, 89 (to *Lurymare*); Day *et al.*, 1970, 19 (False Bay, South Africa); Poulter, 1975, 317 (to *P.* (*Lurymare*).)

sadoensis Kato, 1944, 309, figs (Japan).

sancum du Bois-Reymond Marcus, 1965, 134, figs (Peru).

singulare Laidlaw, 1904*b*, 135 (Ceylon); Marcus, 1950, 97 (note).

siphunculus (Delle Chiaje, 1829) Lang, 1884, 595, fig. (synonymy; Naples). Carus, 1885, 157 (descr.); Graff, 1886, 342 (Adriatic Sea); Lo Bianco, 1888, 399 (Naples); Vaillant, 1889, 656 (Atlantic coast of France); Hallez, 1894, 235 (Atlantic coast of France); Gamble, 1896, 18, fig.; Pruvot, 1897, 21 (south coast of France); Francotte, 1898, 190, figs (NW. France); Lo Bianco, 1899*a*, 478 (Naples); Laidlaw, 1906, 706 (Channel Is); Meixner, 1907*a*, 172 (Somalia); 1907*b*, 481 fig.; Jacubowa, 1909, 23, figs (Black Sea); Micoletzky, 1910, 180 (Trieste); Yeri & Kaburaki, 1918*a*, 441 (diagn.; Japan); 1918*b*, 41, 49, fig. (Japan); Bresslau, 1928–33, 105, fig.; Stummer-Traunfels, 1933, 3575, fig.; Southern, 1936, 72 (Ireland); Palombi, 1936, 31, figs (Still Bay, S. Africa); Steinböck, 1937, 11, figs (Egypt); Kato, 1937*a*, 230 (Japan); 1944, 308 (*siphunculus* of Yeri & Kaburaki, 1918 = *vulgaris*); Marcus, 1950, 97, 99, 101 (note); de Beauchamp, 1951*b*, 249 (Mauritania); Dawydoff, 1952, 81 (Vietnam); Riedl, 1953, 108 (Sicily); 1959, 204 (Tyrrhenian Sea); Marinescu, 1971, 45 (Rumania); Prudhoe, 1982*c*, 70, fig. (Britain).

sonorum Kato, 1938*a*, 560, 572, figs (Japan); 1944, 308; Marcus, 1950, 97, 101 (notes).

sparsum (Stimpson, 1855) Stimpson, 1857, 22, 29 (descr.); Diesing, 1862, 540 (to *Leptoplana*); Lang, 1884, 603; Marcus, 1950, 96 (note).

susakiensis Kato, 1944, 312, figs (Japan).

tenebrosum Stimpson, 1857, 22, 29 (Hong Kong); Diesing, 1862, 538 (to *Leptoplana*); Lang, 1884, 604; Marcus, 1950, 96 (note).

trilineatum Yeri & Kaburaki, 1920, 596, figs (Japan); Kato, 1944, 307; 1943, 87, fig. (Palau Is, Pacific Ocean); Utinomi, 1956, 30, fig. (coloured) (Japan).

utarum Marcus, 1952, 5, 98, 117, figs (Brazil); du Bois-Reymond Marcus, 1955*b*, 287; du B.-R. Marcus & Marcus, 1968, 89 (to *Lurymare*), 90 (descr. from Florida); Poulter, 1975, 317 (to *P.* (*Lurymare*).)

vulgaris Kato, 1938*a*, 560, 572 (Japan; nom. nov. pro *Prosthiostomum siphunculus* of Yeri & Kaburaki, 1918*b*); 1938*b*, 578, 589, figs (Japan); 1944, 308; Marcus, 1950, 97, 101 (note); Kikuchi, 1968, 167 (Japan).

wagurenisis Kato, 1944, 309, figs (Japan).

yerii Kato, 1937*f*, 347, 367, figs (Japan); 1944, 308; Marcus, 1950, 97 (note).

species Plehn, 1896*a*, 172 (Brazil).

species Plehn, 1896*b*, 9 (Ascension I., Atlantic Ocean).

species Laidlaw, 1906, 714 (Cape Verde Is).

species Stummer-Traunfels, 1933, 3566, fig. (Indonesia; prob. = *P. awaense*).

species Prudhoe, 1950*a*, 48, fig. (Burma).

species (juvenile) Hyman, 1955*b*, 144, fig. Puerto Rico).

PSEUDOCERATIDAE Stechow, 1922 – see Poche, 1926, 100 (emended to Pseudocerotidae).

PSEUDOCERIDAE Lang, 1884, xi, 430, 523; Carus, 1885, 153; Hallez, 1893, 169; 1894, 219 (defined); Gamble, 1896, 19; Benham, 1901, 31; Heath & McGregor, 1913, 474; Bock, 1922, 10 (systematic note); Poche, 1926, 100 (emended to Pseudocerotidae); Bresslau, 1928–33, 291; Stummer-Traunfels, 1933, 3576; Hyman, 1940*a*, 484; 1953*a*, 362 (defined); 1959*c*, 14 (defined); du B.-R. Marcus & Marcus, 1968, 71 (west Atlantic forms); de Beauchamp, 1961, 64 (genera).

PSEUDOCEROIDEA Schmarda, 1859, 13, 25 (family of sub-order Acarena); Lang, 1884, 524 (= Pseudoceridae).

PSEUDOCEROS Lang, 1884, xi, 430, 538 (type: *velutinus*); Carus, 1885, 154; Gamble, 1896, 19; Benham, 1901, 31; Bock, 1913, 253; Yeri & Kaburaki, 1918*a*, 439 (features); 1918*b*, 49 (features); Stummer-Traunfels, 1933, 3544 (arrangement of eye-spots); Hyman, 1939*d*, 18; 1940*a*, 485; 1953*a*, 363 (defined); 1954*a*, 219 (distribution & taxonomy); 1959*c*, 14 (defined); Marcus, 1949, 87 (western Atlantic forms); Hyman, 1959*a*, 565 (systematic note); Marcus, 1950, 84 (list of spp.); du B.-R. Marcus, 1955*a*, 42 (*Acanthozoon* a subgen. of *Pseudoceros*); du Bois-Reymond Marcus & Marcus, 1968, 71 (note).

affinis (Kelaart, *in* Collingwood, 1876) Stummer-Traunfels, 1933, 3566, fig. (Indonesia); Hyman, 1960, 309 (Hawaii).
albicornis (Stimpson, 1857) Kato, 1944, 298.
albomarginatus Hyman, 1959*c*, 14, figs (N.S.W., Australia).
armatus (Kelaart, 1858) Lang, 1884, 545; du B.-R. Marcus, 1955*a*, 42 (to subgen. *Acanthozoon*); Marcus, 1950, 84 (note).
asamusiensis Kato, 1939*b*, 141, 148, figs (Japan); 1944, 300; Marcus, 1950, 84 (note).
ater Hyman, 1959*a*, 571, 594, fig. (Palau Is).
atraviridis (Kelaart, *in* Collingwood, 1876) Hyman, 1959*a*, 565.
atropurpureus Kato, 1934*b*, 123, 129, figs (Japan); 1938*a*, 568 (descr.; Japan); 1938*b*, 587 (Japan); 1944, 299; Marcus, 1950, 84 (note).
aureolineata Verrill, 1903, 42, fig. (Bermuda); Hyman, 1939*d*, 18, fig. (Bermuda); Marcus, 1950, 84 (note; emended to *aureolineatus*).
bajae Hyman, 1953*a*, 365, fig. (Lower California); Brusca, 1973, 64, 69 (Gulf of California).
bedfordi Laidlaw, 1903*c*, 302, 314, fig. (Singapore); Bock, 1913, 254, fig. (Mendanao, West of Billiton I., Indonesia); Bresslau, 1928–33, 59, fig.; Kato, 1943, 87 (descr.; Palau Is); 1944, 299 (distribution); Marcus, 1950, 84 (note); Hyman, 1954*a*, 220 (Great Barrier Reef); Dawydoff, 1952, 82 (Vietnam); Bates, 1956, 557, fig. (Caroline Is); Hyman, 1959*a*, 566 (Caroline Is, (?)Philippines – presumably misread Mendanao as Mindanao); Bennett, 1971, 166, photo p. 121 (Great Barrier Reef); Prudhoe, 1977, 586 (Great Barrier Reef).
bicolor Verrill, 1903, 42, fig. (Bermuda); Hyman, 1939*d*, 19 (Bermuda); du B.-R. Marcus & Marcus, 1968, 84 (note; Caribbean); Marcus, 1950, 84 (note).
bimarginatum Meixner, 1907*a*, 169 (Somalia); 1907*b*, 465, figs; Marcus, 1950, 84 (note).
buskii (Collingwood, 1876) Lang, 1884, 547; Laidlaw, 1902, 298 (descr.; Maldive & Laccadive Is, Indian Ocean); 1903*c*, 302; Kaburaki, 1923*b*, 641, fig. (Philippines; 'perhaps identical with Collingwood's *Proceros buskii*'); Marcus, 1950, 84.
caeruleocinctus Hyman, 1959*a*, 569, 593, fig. (Palau Is).
caeruleopunctatus Palombi, 1928, 579, 605, figs (Suez Canal); Marcus, 1950, 85 (note).
canadensis Hyman, 1953*a*, 368, figs (British Columbia); 1954*a*, 219 (Puget Sound); 1955*a*, 9 (Puget Sound); Ching, 1978, 1372 (British Columbia).
cardinalis Hasell, 1907*b*, 480, fig. (N.S.W., Australia); Marcus, 1950, 85 (note).
cardiosorus (Schmarda, 1859) Lang, 1884, 546; Stummer-Traunfels, 1933, 3487, 3540, 3544, figs (type-specimen re-examd.); Marcus, 1950, 85 (note).
cerebralis (Kelaart, 1858) Lang, 1884, 546; Marcus, 1950, 85 (note).
chloreus Marcus, 1949, 7, 86, 104, figs (Brazil); 1950, 85 (note); du B.-R. Marcus, 1955*b*, 286.
cinereus Palombi, 1931*a*, 3, 5, figs (New Guinea); Marcus, 1950, 85 (note).
clavicornis (Schmarda, 1859) Stummer-Traunfels, 1933, 3487, 3543, 3544, figs (type-specimen re-examd.); Marcus, 1950, 85 (note).
coccineus (Stimpson, 1857) Kato, 1944, 298 (Japan).
colemani Prudhoe, 1977, 598, figs (N.S.W., Australia).
collingwoodi Laidlaw, 1903*c*, 302, 314 (Malaysia); Marcus, 1950, 85 (note).
concinnus (Collingwood, 1876) Kaburaki, 1923*b*, 635, 642, figs (Philippines); Stummer-Traunfels, 1933, 3565, fig. (Indonesia); Marcus, 1950, 85 (note); Dawydoff, 1952, 81 (Vietnam); Hyman, 1954*a*, 200, fig. (Indonesia); 223 (doubts authenticity of determinations of Kaburaki and of Stummer-Traunfels).
corallophilus Hyman, 1954*a*, 223, fig. (Great Barrier Reef); Prudhoe, 1977, 586 (Great Barrier Reef).
crozieri Hyman, 1939*d*, 17, figs (Bermuda; nom. nov. pro *P.* sp. of Crozier); 1951, 67, fig. (Bermuda); 1952, 197 (Florida); Marcus, 1950, 85 (note); du B.-R. Marcus & Marcus, 1968, 73, fig. (Caribbean & Florida).
devisii Woodworth, 1898, 63, fig. (Queensland); Marcus, 1950, 85 (note).
dimidiatus Graff, *in* Saville-Kent, 1893, 362, fig. (Great Barrier Reef); Marcus, 1950, 85 (note).
dulcis (Kelaart, 1858) Hyman, 1959*a*, 565.
evelinae Marcus, 1950, 5, 81, 115, figs (Brazil); 1952, 93 (note); du B.-R. Marcus, 155*b*, 286.
exoptatus Kato, 1938*b*, 577, 587, figs (Japan); 1944, 300; Marcus, 1950, 85 (note); Utinomi, 1956, 30 fig. (coloured).
ferrugineus Hyman, 1959*a*, 571, 594, figs (Palau Is); Prudhoe, 1977, 586 (Great Barrier Reef).
flavomaculatus (Graff, *in* Saville-Kent, 1893) Graff, 1904–05, 1888 [? misprint for *Prostheceraeus*

flavomaculatus]; Marcus, 1950, 85 (note).
flavomarginatus Laidlaw, 1902, 298 (Laccadive Is, Indian Ocean); Kato, 1944, 298 (descr.; syn. *luteomarginatus*); Marcus, 1950, 85 (note); Prudhoe, 1977, 592 (Queensland).
fulminatus (Stimpson, 1855); Kato, 1944, 298.
fulvogriseus Hyman, 1959*a*, 573, 593, 594, figs (Palau Is).
fuscogriseus Hyman, 1959*a*, 575, figs (Palau Is).
fuscopunctatus Prudhoe, 1977, 586, 594, figs (Great Barrier Reef).
fuscus (Kelaart, 1858) Hyman, 1959*a*, 565.
gamblei Laidlaw, 1902, 297, figs (Laccadive Is, Indian Ocean); Marcus, 1950, 85, 115 (to *Dicteros*).
gardineri Laidlaw, 1902, 296 (Maldive Is, Indian Ocean); Marcus, 1950, 85 (note).
gratus Kato, 1937*a*, 211, 227, figs (Japan); 1944, 300; Marcus, 1950, 85 (note); Utinomi, 1956, 30, fig. (coloured) (Japan); Hyman, 1959*a*, 566 (synonymy); Kato, 1943*c*, 86, fig (Palau Is); Prudhoe, 1977, 593 (W. Australia).
gravieri Meixner, 1907*a*, 170 (Somalia); 1907*b*, 468, figs; Marcus, 1950, 85 (note).
griseus Hyman, 1959*b*, 10, figs (California).
guttatomarginatus (Stimpson, 1855) Kato, 1944, 298.
habroptilus Hyman, 1959*a*, 569, fig. (Solomon Is, Pacific Ocean).
haddoni Laidlaw, 1903*a*, 3, 10 (Torres Straits, Australia).
hancockanus (Collingwood, 1876) Laidlaw, 1903*c*, 301, 302, 315 (Malaysia; '*hancockianus*'); Kaburaki, 1923*b*, 635, 639 (descr.; Philippines; syns. ?*Stylochopsis malayensis* & *Ps. malayensis* Bock); Marcus, 1950, 86 (note).
hispidus du Bois-Reymond Marcus, 1955*a*, 39, figs (Brazil); 1955*b*, 286; Hyman, 1959*a*, 583 (*Acanthozoon*).
interruptus (Stimpson, 1855) Kato, 1944, 298.
izuensis Kato, 1944, 301, figs (Japan); Hyman, 1959*a*, 567, 593, figs (Palau Is); Marcus, 1952, 93 (note).
japonicus (Stimpson, 1857) Kato, 1944, 298.
kelaartii (Collingwood, 1876) Laidlaw, 1903*c*, 302, 314 (Malaysia); Marcus, 1950, 86 (note); Dawydoff, 1952, 82 (Vietnam).
kentii Graff, *in* Saville-Kent, 1893, 362, fig. (Great Barrier Reef); Marcus, 1950, 86 (note).
lacteus (Collingwood, 1876) Lang, 1884, 548; Yeri & Kaburaki, 1918*a*, 440 (diagn.; Japan); 1918*b*, 37, 49, fig.; Marcus, 1950, 86 (note); Dawydoff, 1952, 81 (Vietnam).
latissimus (Schmarda, 1859) Stummer-Traunfels, 1933, 3487, 3542, 3544, figs (type-specimen re-examd.); Marcus, 1950, 86 (note).
lepida (Heath & McGregor, 1913) du B.-R. Marcus, 1955*a*, 42 (subgen. *Acanthozoon*).
leptostictus Bock, 1913, 256, figs (W. Australia); Marcus, 1950, 86 (note), Prudhoe, 1977, 597 (descr.; W. Australia).
limbatus (Leuckart, 1828) Lang, 1884, 544 (synonymy); Marcus, 1950, 86 (note).
limbatus Haswell, 1907*b*, 480 (Queensland); Marcus, 1950, 86 (renamed *liparus*).
liparus Marcus, 1950, 86, 115 (nom. nov. pro *limbatus* Haswell *nec* Leuckart).
litoralis Bock, 1913, 259, figs (Gulf of Siam); Kaburaki, 1923*b*, 640, fig. (Philippines); Marcus, 1950, 86 (note).
lividus Prudhoe, 1982*a*, 222, fig.; 1982*b*, 381 (descr.; South Australia).
luteomarginatus Yeri & Kaburaki, 1918*a*, 432, 440 (diagn.; Japan); 1918*b*, 2, 37, 49, figs (Japan); Kato, 1937*a*, 229 (size & colour; Japan); 1937*b*, 239 (Korea); 1944, 298 (s.o. *flavomarginatus*); Marcus, 1950, 86 (note); Dawydoff, 1952, 82 (Vietnam).
luteus (Plehn, 1898) Hyman, 1953*a*, 366, figs (California), 367 (colour variant); Morris *et al.*, 1980, 80, fig. (coloured) (California).
maculatus (Gray, *in* Pease, 1860) Lang, 1884, 547; Marcus, 1950, 86 (note).
maculosus Pearse, 1938, 85, fig. (Florida); Hyman, 1940*a*, 485 (type-specimen re-examd.); 1954*a*, 219 (to *Acanthozoon*); Marcus, 1950, 86 (note); du B.-R. Marcus, 1955*a*, 42 (subgen. *Acanthozoon*).
malayensis (Collingwood, 1876) Bock, 1913, 258 (synonymy; Borneo & Gulf of Siam); Kaburaki, 1923*a*, 639 (s.o. *hancockanus*).
marmoratus Plehn, 1898, 145, fig. (Indonesia); Marcus, 1950, 86 (note).
maximus Lang, 1884, 49, 541, figs (Naples); Carus, 1885, 154 (descr.); Lo Bianco, 1888, 400 (Naples); 1899*a*, 479 (Naples); Sabussow, 1905, 489 (Mediterranean); Steinböck, 1933, 19 (Gulf of

Trieste); Levetzow, 1939, 780 (Naples); Marcus, 1950, 86 (note).
memoralis Kato, 1938*a*, 560, 569, figs (Japan); 1944, 300; Marcus, 1950, 87 (note).
mexicanus Hyman, 1953*a*, 363, figs (Gulf of California); Brusca, 1973, 64, 68, fig. (Gulf of California).
microceraeus (Schmarda, 1859) Stummer-Traunfels, 1933, 3487, 3541, 3544, figs (type-specimen re-examd.); Marcus, 1950, 87 (note).
micronesianus Hyman, 1955*c*, 78, fig. (Marianas & Gilbert Is, Pacific Ocean).
micropapillosus Kato, 1934*b*, 123, 130, figs (Japan); 1944, 299; Marcus, 1950, 87 (note); du B.-R. Marcus, 1955*a*, 42 (subgen. *Acanthozoon*).
miniatus (Schmarda, 1859) Stummer-Traunfels, 1933, 3487, 3539, 3544, figs (type-specimen re-examd.), 3566, fig. (Indonesia); Marcus, 1950, 87 (note).
montereyensis Hyman, 1953*a*, 370 (California); Morris *et al.*, 1980, 60, fig. (coloured) (California).
mopsus Marcus, 1952, 5, 91, 116, figs (Brazil); du B.-R. Marcus, 1955*b*, 286; du B.-R. Marcus & Marcus, 1968, 75 (Caribbean & Florida).
muelleri (Delle Chiaje, 1829) Lang, 1884, 545; Carus, 1885, 154 (descr.); Marcus, 1950, 87 (note).
niger (Stimpson, 1857) Stummer-Traunfels, 1933, 3566, fig. (Indonesia; syn. *Proceros hancockanus*); Kato, 1944, 298 (*'nigrus'*).
nigrocinctus (Schmarda, 1859) Lang, 1884, 547; Stummer-Traunfels, 1933, 3487, 3537, 3538, figs (type-specimen re-examd.); Marcus, 1950, 87 (note).
nigromarginatus Yeri & Kaburaki, 1918*a*, 432, 440 (diagn.; Japan); 1918*b*, 2, 39, 49, figs (Japan); Kato, 1944, 299; Marcus, 1950, 87 (note).
nipponicus (Kato, 1944, 300, figs (Japan); Marcus, 1952, 93 (note).
papilionis (Kelaart, 1858) Lang, 1884, 546 (*'papilio'*); Kaburaki, 1923*a*, 646, fig. (Philippines; to *Prostheceraeus*); Palombi, 1938, 355, figs (synonymy; South Africa); Marcus, 1950, 87 (note); du B.-R. Marcus, 1955*a*, 42 (?subgen. *Acanthozoon*); Day *et al.*, 1971, 19 (False Bay, S. Africa).
paradoxus Bock, 1927*a*, 17 (Fiji); Marcus, 1950, 87 (note).
pardalis Verrill, 1900, 596, figs (Bermuda); Marcus, 1950, 87 (note).
periphaeus Bock, 1913, 255, figs (W. Australia); Marcus, 1950, 87 (note).
perviolaceus Hyman, 1959*a*, 566 (synonymy; Palau, Is); 1959*b*, 7, figs (California).
philippinensis Kaburaki, 1923*b*, 635, 645, fig. (Philippines).
pius Kato, 1938*a*, 560, 570, figs (Japan); 1944, 300; Marcus, 1950, 87 (note).
pleurostictus Bock, 1913, 257, figs (Malagasy); Stummer-Traunfels, 1933, 3573, fig.; Marcus, 1950, 87 (note).
punctatus Laidlaw, 1902, 296, fig. (Maldive Is, Indian Ocean); Marcus, 1950, 87 (note).
purpureus (Kelaart, 1858) Hyman, 1959*a*, 565.
regalis Haddon, *in* Laidlaw, 1903*a*, 3, 10 (Torres Straits, Australia); Marcus, 1950, 87 (note).
reticulatus Yeri & Kaburaki, 1918a, 432, 439 (diagn.; Japan); 1918*b*, 35, 49, figs (Japan); Kato, 1934*b*, 133 (Japan); 1944, 299; Dawydoff, 1952, 81 (Vietnam); Marcus, 1950, 87 (note); Prudhoe, 1982*a*, 222, 226 (Tasmania & South Australia); 1982*b*, 380 (descr.; Tasmania & S. Australia).
rubellus Laidlaw, 1903*c*, 302, 314, fig. (Malaysia); Marcus, 1950, 88 (note).
rubrocinctus (Schmarda, 1859) Stummer-Traunfels, 1933, 3487, 3537, 3538, figs (type-specimen re-examd.); Marcus, 1950, 88 (note).
rubrotentaculatus Kaburaki, 1923*b*, 635, 643, figs (Philippines); Dawydoff, 1952, 81 (Vietnam); Marcus, 1950, 88 (note).
sagamianus Kato, 1937*f*, 347., 362, figs (Japan); 1939*a*, 76, figs (Japan); 1944, 300; Marcus, 1950, 88 (note).
splendidus Stummer-Traunfels, 1933, 3487 (nom. nov. pro *superbus* Lang *nec* Schmarda); Marcus, 1950, 88 (note); Hyman, 1955*b*, 137 (Puerto Rico).
striatus (Kelaart, 1858) Lang, 1884, 546; Marcus, 1950, 88 (note).
striatus (Schmarda, 1859) Stummer-Traunfels, 1933, 3487, 3540, 3544, figs (type-specimen re-examd.); Marcus, 1950, 88, 115 (renamed *strigosus*); Hyman, 1959*a*, 566 (syn. *gratus*).
strigosus Marcus, 1950, 88, 115 (nom. nov. pro *striatus* Schmarda *nec* Kelaart); Hyman, 1959*a*, 566 (s.o. *gratus*).
superbus (Schmarda, 1859) Stummer-Traunfels, 1933, 3487, 3541, 3544, figs (type-specimen re-examd.), 3565, fig. (Indonesia); Marcus, 1950, 88 (note).
superbus Lang, 1884, 49, 540, fig. (Bay of Naples); Carus, 1885, 154 (descr.); Plehn, 1896*a*, 171 (Galapagos Is); Verrill, 1900, 596, fig. (Bermuda); Stummer-Traunfels, 1933, 3487 (renamed *Ps.*

splendidus); Hyman, 1939*d*, 19 (Bermuda); Dawydoff, 1952, 81 (Vietnam); Marcus, 1950, 88 (note).

susakiensis Kato, 1934*b*, 123, 131, figs (Japan); 1944, 302 (to *Eurylepta*); Marcus, 1950, 88 (note).

texanus Hyman, 1955*e*, 264, fig. (Texas); du B.-R. Marcus & Marcus, 1968, 75 (Caribbean & Florida).

tigrinus Laidlaw, 1902, 297, fig. (Laccadive Is, Indian Ocean); Marcus, 1950, 88 (note).

tomiokaensis Kato, 1938*a*, 560, 568, figs (Japan); 1944, 300.; Marcus, 1950, 88 (note).

tristriatus Hyman, 1959*a*, 576, fig. (Caroline Is, Pacific Ocean).

undulata (Kelaart, 1858) Marcus, 1950, 88.

velutinus (Blanchard, 1847) Lang, 1884, 49, 538, figs (Italy); Carus, 1885, 154 (descr.); Plehn, 1896*b*, 4, 9, fig. (Gulf Stream, 41°N, 56°W); Pruvot, 1897, 20 (Gulf of Lion, S. France); Lo Bianco, 1899, 479 (Naples); Palombi, 1928, 605 (descr.; Suez Canal); Steinböck, 1933, 19 (Gulf of Trieste); Marcus, 1950, 88 (note).

velutinus var. *violaceus* (Schmarda, 1859) Lang, 1884, 540; Boutan, 1892, 179 (Port Tewjik, Red Sea).

vinosum Meixner, 1907*a*, 171 (Somalia); 1907*b*, 470, figs; Marcus, 1950, 88 (note).

violaceus (Kelaart, 1858) Hyman, 1959*a*, 566.

violaceus (Schmarda, 1859) Stummer-Traunfels, 1933, 3487, 3539, 3544, figs (type-specimen re-examd.); Hyman, 1959*a*, 566 (renamed *Ps. perviolaceus*); Marcus, 1950, 88 (note).

virescens Hyman, 1959*a*, 569 (nom. nov. pro *viridis* Schmarda *nec* Kelaart).

viridis (Kelaart, 1858) Hyman, 1959*a*, 569.

viridis (Schmarda, 1859) Stummer-Traunfels, 1933, 3487, 3543, 3544, figs (type-specimen re-examd.); Marcus, 1950, 88 (note); Hyman, 1959*a*, 569 (renamed *Ps. virescens*).

yessoensis Kato, 1937*c*, 37, figs (Japan); 1937*e*, 131 (Japan); 1944, 300; Marcus, 1950, 88 (note).

zebra (Leuckart, 1828) Lang, 1884, 545; Marcus, 1950, 88 (note).

zeylandicus (Kelaart, 1958) Lang, 1884, 546; Marcus, 1950, 88 (note).

species Laidlaw, 1903*a*, 11 (Torres Straits, Australia).

species Laidlaw, 1903*a*, 11 (Rotuma I., Pacific Ocean).

species Laidlaw, 1903*c*, 315 (Malaysia; near *concinnus*).

species Laidlaw, 1904*b*, 135 (Ceylon).

species Crozier, 1917, 725, figs (Bermuda); Hyman, 1939*d*, 17 (to *Ps. crozieri*).

species Stummer-Traunfels, 1933, 3565, Pl. 1, fig. 1 (Indonesia).

species Stummer-Traunfels, 1933, 3565, Pl. 1; fig. 4 (Zanzibar).

species Stummer-Traunfels, 1933, 3565, Pl. 1, fig. 5 (Indonesia).

species Stummer-Traunfels, 1933, 3565, Pl. 1, figs 7–8 (Indonesia).

species Stummer-Traunfels, 1933, 3566, Pl. 1, fig. 12 (Indonesia).

species Stummer-Traunfels, 1933, 3566, Pl. 1, figs 13–14 (Indonesia).

species Stummer-Traunfels, 1933, 3566, Pl. 1, fig. 15 (Indonesia).

species Stummer-Traunfels, 1933, 3566, Pl. 1, fig. 16 (Philippines).

species Stummer-Traunfels, 1933, 3566, Pl. 1, fig. 17 (Indonesia).

species Stummer-Traunfels, 1933, 3566, Pl. 1, fig. 18 (Indonesia; poss, identical with *Eurylepta japonica* Stimpson).

species Stummer-Traunfels, 1933, 3566, Pl. 1, fig. 19 (Indonesia).

species Stummer-Traunfels, 1933, 3566, Pl. 1, fig. 20 (Indonesia).

species Villadolid & Villaluz, 1938, 395, figs (Philippines).

species Steinbeck & Ricketts, 1941, 337 (Lower California).

species MacGinitie & MacGinitie, 1949, 151, fig. (California); Hyman, 1953*a*, 370.

species Marcus, 1949, 87 (Brazil).

species Dawydoff, 1952, 82 (Vietnam).

species Corrêa, 1958, 81 (Brazil).

species Morton & Miller, 1968, 172, figs (New Zealand).

species Brusca, 1973, 70 (Gulf of California).

PSEUDOCEROTIDAE Poche, 1926, 100 (emendation of Pseudoceridae and Pseudoceratidae); Prudhoe, 1982*c*, 54 (defined).

PSEUDOSTYLOCHUS Yeri & Kaburaki, 1918*a*, 437 (features); 1918*b*, 25 (features) (type: *takeshitai*); Kato, 1939*b*, 147 (two groups); Hyman, 1953*a*, 358 (defined).

aino Kato, 1937*e*, 129, figs (Japan); 1939*a*, 75 (descr.; Japan); 1939*b*, 147 (Japan); 1944, 291; Okada *et al.*, 1971, 62 (Japan).

bellus Hyman, 1959*c*, 12, figs (N.S.W., Australia).

burchami (Heath & McGregor, 1913) Hyman, 1953*a*, 358, figs (syn. *Discosolenia washintonensis*; Puget Sound to Mexican waters; neotype); Brusca, 1973, 68 (Gulf of California). [See p. 114]

edurus Kato, 1938*b*, 577, 584, figs (Japan); 1939*a*, 76 (descr.; Japan); 1944, 291.

elongatus Kato, 1937*a*, 211, 218, figs (Japan); 1937*b*, 238 (descr.; Korea); 1944, 290.

fulvus Yeri & Kaburaki, 1918*a*, 432, 438 (diagn.; Japan); 1918*b*, 2, 28, 48, figs (Japan); Kato, 1944, 290 (Japan).

fuscoviridis Kato, 1934*b*, 123, 127, figs (Japan); 1944, 290.

intermedius Kato, 1939*b*, 141, 146, figs (Japan); 1944, 291.

longipenis Kato, 1937*b*, 233, 237, figs (Korea); 1944, 290.

maculatus Kato, 1938*b*, 577, 585, figs (Japan); 1944, 291.

meridialis Kato, 1938*a*, 559, 566, figs (Japan); 1944, 291.

nationalis Kato, 1939*a*, 65, 73, figs (Japan); 1944, 291.

notoensis Kato, 1944, 291, figs (Japan).

obscurus (Stimpson, 1857) Yeri & Kaburaki, 1918*a*, 432, 438 (diagn.; Japan); 1918*b*, 30, 48, figs (Japan); Kaburaki, 1923*a*, 193, figs (Japan); Kato, 1937*c*, 37 (*P. obscurus* of Kaburaki, 1923*a* = *P. stimpsoni*); 1937*e*, 131 (Japan); 1938*a*, 565 (Japan); 1938*b*, 584 (Japan); 1939*a*, 74 (descr.; Japan); 1940, 538 (development); 1944, 289; Yang, 1974, 7 (development; Korea).

okudai Kato, 1937*b*, 233, 236, figs (Korea); 1937*e*, 130, fig. (Japan); 1944, 290.

ostreophagus Hyman, 1955*a*, 4, figs (Puget Sound & Japan); Carl & Guiguet, 1958, 81 (British Columbia); Woelke, 1957, 62 (Puget Sound).

sadoensis Kato, 1944, 292, figs (Japan).

stimpsoni Kato, 1937*c*, 35, figs (Japan); 1937*e*, 130 (Japan); 1944, 290; Okada *et al.*, 1971, 62 (Japan).

takeshitai Yeri & Kaburaki, 1918*a*, 432, 438 (diagn.; Japan); 1918*b*, 2, 26, 48, figs (Japan); Kato, 1939*a*, 74, figs (Japan); 1944, 290.

PUCELIS Marcus, 1947, 99, 127, 165 (type: *evelinae*).

evelinae Marcus, 1947, 99, 128, 165, figs (Brazil); du Bois-Reymond Marcus ,1955*b*, 285.

PULCHRIPLANA Palombi, 1938, 329, 342, 350 (type: *insignis*).

insignis Palombi, 1938, 329, 342, figs (South Africa).

SCHEMATOMMATA Bock, 1913, 55, 166 (section of Acotylea); Poche, 1926, 99 (= superfam. Planocerides).

SCHEMATOMMATIDEA de Beauchamp, 1961, 60 (superfamily); Marcus & Marcus, 1966, 321 (superfamily), 327 (key to families).

SCHMARDEA Diesing, 1862, 493, 546 (type: *rubrocincta*); Lang, 1884, 550 (?s.o. *Yungia*).

rubrocincta (Schmarda, 1859) Diesing, 1862, 546 (to *Schmardea*); Lang, 1884, 550 (to *Yungia*).

SEMONIA Plehn, 1896*a*, 140, 157, 159 (type: *maculata*); 1896*c*, 329; Benham, 1901, 31; Laidlaw, 1903*c*, 309 (allied to *Discocelis*); Bock, 1913, 61 (diagn.); Marcus, 1950, 76; Hyman, 1955*c*, 70.

maculata Plehn, 1896*a*, 140, 157, figs (Indonesia); 1896*c*, 329, figs ('n.sp.', type-locality correctly Amboina); 1898, 146 (Indonesia); Bock, 1913, 62.

penangensis Laidlaw, 1903*c*, 302, 308, fig. (Malaysia); Bock, 1913, 62. [See p. 80]

SHELFORDIA Stummer-Traunfels, 1902, 160 (type: *borneensis*); Laidlaw, 1903*e*, 14; Bock, 1913, 147 (diagn.); 344 (corrigendum) (s.o. *Limnostylochus*).

amara Kaburaki, 1918, 188, fig. (Thailand).

annadalei Kaburaki, 1918, 185, figs (Thailand).

borneensis Stummer-Traunfels, 1902, 160 (Borneo); Shelford, 1904, 18; Bock, 1913, 147; 1923*c*, 346 (to *Limnostylochus*).

SIMPLICIPLANA Kaburaki, 1923*b*, 635, 648, (type: *marginata*).

marginata Kaburaki, 1923*b*, 635, 648, figs (Philippines).

SPHYNGICEPS Collingwood, 1876, 86, 87, 90 (type: *lacteus*); Lang, 1884, 538 (s.o. *Pseudoceros*).

lacteus Collingwood, 1876, 87, 90, figs (Singapore); Lang, 1884, 548 (to *Pseudoceros*).

SPINICIRRUS Hyman, 1953*a*, 350 (type: *inequalis*).

inequalis Hyman, 1953*a*, 350, figs (Gulf of California).

species Taneja, 1978, 127 (no. loc.).

STYLOCHID (undetermined genus & species) Prudhoe, 1977, 587, fig. (W. Australia).

STYLOCHIDAE Stimpson, 1857, 22; Lang, 1884, 433; Meixner, 1907*b*, 394, 444 (defined; geogr.

distribution); Bock, 1913, 110 (defined; key to genera); Yeri & Kaburaki, 1920, 595 (defined), 596 (key to genera); Bock, 1923*c*, 348 (key to genera); 1925*a*, 34 (position of genital pores); 1925*b*, 154 (arrangement of eyes), 162 (male complex), 169, 171 (key to genera), 173 (geogr. distribution); Bresslau, 1928–33, 285 (defined); Freeman, 1933, 112; Stummer-Traunfels, 1933, 3567, 3585, 3586; Hyman, 1939*b*, 129 (defined); 1940*a*, 459 (defined); 1940*b*, 16; 1953*a*, 280 (defined); 1959*c*, 3 (defined); de Beauchamp, 1961, 57 (generic constituents).

STYLOCHIDEA Diesing, 1862, 493, 562; Lang, 1884, 433, 553 (*ex parte* =Euryleptidae, *e.p.*= Planoceridae).

STYLOCHIDES Poche, 1926, 98 (superfam. of Acotylea =Craspedommata Bock, 1913); Nicoll, 1936, 74 (emend to Stylochoidea).

STYLOCHIDIUM Poche, 1926, 99 ('sedis incertae Stylochides').

STYLOCHINAE Laidlaw, 1903*e*, 12; 1904*a*, 2 (subfam. of Stylochidae; diagn.); Meixner, 1907*b*, 385, 394 (subfam. of Planoceridae: key to genera); Bresslau, 1928–33, 286.

STYLOCHOIDEA Nicoll, 1936, 74 (emendation for Stylochides Poche); Prudhoe, 1982*c*, 27.

STYLOCHOCESTIDAE Bock, 1913, 248; Bresslau, 1928–33, 289; Stummer-Traunfels, 1933, 3567, 3586.

STYLOCHOCESTINAE Laidlaw, 1903*e*, 12 (subfam. of Stylochidae); Meixner, 1907*b*, 391.

STYLOCHOCESTUS Laidlaw, 1903*e*, 12, 1904*b*, 131 (type: *gracilis*); Bock, 1913, 249.

gracilis Laidlaw, 1904*b*, 131, fig. (Ceylon); Meixner, 1907*b*, 391, fig. (sections of type-specimen re-examd.); Bock, 1913, 249.

STYLOCHOIDES Hallez, 1907, 1, 2 (type: *albus*); 1913, 23, 34, 41 (diagn.); Bock, 1913, 276.

albus (Hallez, 1905) Hallez, 1907, 1, 2, 5, figs (Antarctica); 1913, 1, 23, figs (Antarctica); Bock, 1913, 276, fig. (Tierra del Fuego; syns. *Nuchenceros orcadensis* & *Cotylocera michaelseni*); 1922, 17, 18 (synonymy); 1931, 282, figs (several localities in south polar regions); Westblad, 1952, 10 (Tierra del Fuego & South Georgia; referred to as *'Stylostomoides albus'* on p. 5); Hyman, 1958, 285, fig. (Kerguelen I.).

STYLOCHOIDIDAE Bock, 1913, 276; Stummer-Traunfels, 1933, 3576, 3587; Poche, 1926, 100 (syn. Laidlawiinae); Marcus, 1947, 141 (=Laidlawiidae).

STYLOCHOPLANA Stimpson, 1857, 22 (type: *maculata*); Lang, 1884, 429, 434, 455 (diagn.); Carus, 1885, 150; Hallez, 1893, 148, 1894, 202 (defined); Gamble, 1893*a*, 497; 1896, 19; Benham, 1901, 31; Haswell, 1907*b*, 471; Bock, 1913, 172 (diagn.; genus divided into three groups); Heath & McGregor, 1913, 463; Bock, 1924*a*, 2 (subgenera *Stylochoplana* (=group A) & *Stylochoplanoides*); Freeman, 1933, 116; Palombi, 1939*b*, 108 (table of spp. in group B of Bock); Hyman, 1939*b*, 139; 1940*a*, 467; 1953*a*, 301 (defined); Marcus, 1947, 120 (notes); 1949, 76; du B.-R. Marcus & Marcus, 1968, 22 (systematic notes).

aberrans Kato, 1944, 274, figs (Japan).

affinis Palombi, 1940, 109, 117, figs (in *Teredo* tube, Zaire).

agilis Lang, 1884, 111, 456, figs (Bay of Naples); Carus, 1885, 150 (descr.); Graff, 1886, 342 (Lesina); Vaillant, 1889, 654 (NW. France); Laidlaw, 1903*e*, 13 (?=*maculata*); Sabassow, 1905, 489 (S. France); Bock, 1913, 172, 178 (descr.); 1924*a*, 2 (to *Stylochoplana* (*Stylochoplana*)); Bresslau, 1928–33, 287, fig.; Stummer-Traunfels, 1933, 3571, fig.; Levetzow, 1939, 780 (Keil Bay, Germany); Pinto, 1947, 94, fig. (Portugal); Prudhoe, 1982*c*, 46, fig. (Britain).

amica Kato, 1937*a*, 211, 213, figs (Japan); 1944, 274.

angusta (Verrill, 1893) Palombi, 1928, 590, figs (Suez Canal); 1939*a*, 109 (*angusta* of Palombi, 1928, s.o. *suesensis*); 1939*b*, 107, 109 (*angusta* of Palombi, 1928, renamed *suesensis*); Hyman, 1939*b*, 139, figs (New England; type-specimens re-examd.); 1940*a*, 467 (descr.; synonymy); 1940*b*, 14 (prob. not indigenous of Atlantic coast of N. America); 1952, 196 (to *Zygantroplana*); Marcus, 1947, 110, figs (Brazil); Corrêa, 1949, 175 (to *Zygantroplana*), 202 (*S. angusta* of Marcus, 1947, renamed *Zygantroplana henriettae*).

aulica Marcus, 1947, 99, 114, 163, figs (Brazil); du B.-R. Marcus, 1955*b*, 284.

bayeri du Bois-Reymond Marcus & Marcus, 1968, 2, 28, figs (Florida).

californica Woodworth, 1894, 50, figs (pelagic in Gulf of California, 26°48'N, 110°45'20"W); Bock, 1913, 180 (diagn.); Heath & McGregor, 1913, 457 [the mention of this name by these authors is probably a typographical error for *Stylochoplana gracilis*]; Hyman, 1953*a*, 308, 314 (to *Parviplana*).

caraibica Palombi, 1924*a*, 36 (Caribbean); 1924*b*, 17, figs (17°12'N, 81°21'W – 13°55'N, 77°43'W); du B.-R. Marcus & Marcus, 1968, 22 (not a *Stylochoplana*).

challengeri (Graff, 1892) Prudhoe, 1950*b*, 713, fig. (type-specimens re-examd.; New Guinea).
chilensis (Schmarda, 1859) Stummer-Traunfels, 1933, 3486, 3494, 3496, figs (type-specimens re-examd.; syn. *Leptoplana lanceolata*).
chloranota (Boone, 1929) du B.-R. Marcus & Marcus, 1968, 23.
clara Kato, 1937*f*, 347, 357, figs (Japan); 1944, 274; Sandô, 1964, 29 (Japan); Kikuchi, 1966, 68 (Japan); 1968, 167 (Japan).
conoceraea (Schmarda, 1859) Meixner, 1907*b*, 398 (to *Conoceros*); Stummer-Traunfels, 1933, 3487, 3563 (type-specimen re-examd.).
diaphana (Stummer-Traunfels, 1933) du B.-R. Marcus & Marcus, 1968, 24.
divae Marcus, 1947, 99, 112, 163, figs (Brazil); du B.-R. Marcus, 1955*b*, 284.
elegans, (Kelaart, 1858) Collingwood, 1876, 88, 98, fig.; Lang, 1884, 465.
evelinae Marcus, 1952, 5, 83, 114, figs (Brazil); du B.-R. Marcus, 1955*b*, 284.
fasciata (Schmarda, 1859) Lang, 1884, 462; Bock, 1913, 223; Stummer-Traunfels, 1933, 3487, 3550 (to *Styloplanocera*).
floridana Pearse, 1938, 77, fig. (Florida); Pearse & Littler, 1938; 237, fig. (North Carolina); Hyman, 1940*a*, 478 (to *Gnesioceros*); du B.-R. Marcus & Marcus, 1968, 48 (s.o. *Gnesioceros sargassicola*).
folium (Grube, 1840) Stimpson, 1857, 22; Diesing, 1862, 568 (to *Stylochus*); Lang, 1884, 460 (*Planocera*).
genicotyla Palombi, 1939*b*, 96, 101, 109, figs (Rio de Oro, NW. Africa); Marcus, 1949, 50 (a *Leptoplana*); du B.-R. Marcus & Marcus, 1968, 22 (to *Leptoplana*).
gracilis Heath & McGregor, 1913, 463, figs (California) [on p. 457 the authors refer to *Stylochoplana californica*, but it is probable that this should have been *Stylochoplana gracilis*]; Hyman, 1953*a*, 303, figs (California).
graffii (Laidlaw, 1906) Bock, 1913, 172, 179 (diagn.; Cape Verde Is); Palombi, 1939*b*, 108 (*'graffi'*).
hancocki Hyman, 1953*a*, 307, figs (California).
heathi Boone, 1929, 35, figs (California); Ricketts & Calvin, 1948, 178, 300 (California); Hyman, 1953*a*, 308, 319 (s.o. *Notoplana inquieta* (Heath & McGregor)); Stasek, 1966, 9 (location of type-specimen).
inquilina Hyman, 1950, 55, fig. (Hawaii); Poulter, 1975, 334.
lactoalba (Verrill, 1900) Bock, 1913, 173, 179 (*'lacteolba'*); Palombi, 1939*b*, 108 (*Leptoplana bacteoalba [sic]* Verrill & *L. bacteoalba [sic]* var. *tincta* Verrill s.o. *Stylochoplana pallida*); Marcus, 1947, 118, 164 (not synonymous with *pallida*); Hyman, 1939*d*, 6 (to *Notoplana*).
lactoalba var. *tincta* (Verrill, 1900) Bock, 1913, 179 (*'lacteoalba'*).
leptalea Marcus, 1947, 99, 118, 164, figs (Brazil); 1948, 177, 195, figs (Brazil); du B.-R. Marcus, 1955*b*, 285; du B.-R. Marcus & Marcus, 1968, 24, figs (Brazil, Caribbean & Florida).
limnoriae (Hyman, 1953) du B.-R. Marcus & Marcus, 1968, 24.
longastyletta Freeman, 1933, 111, 119, figs (Puget Sound); Hyman, 1953*a*, 308, 325 (to *Notoplana*).
longipenis Hyman, 1953*a*, 305, figs (Mexico & California).
lynca du Bois-Reymond Marcus, 1958, 403, figs (Brazil).
maculata (Quatrefages, 1845) Stimpson, 1857, 22; Diesing, 1862, 568 (*Stylochus*); Uljanin, 1870, 106 (Black Sea); Lang, 1884, 459 (descr.; synonymy); Koehler, 1885, 14, 56 (Channel Is, U.K.); Vaillant, 1889, 654, fig.; Bergendal, 1890, 326 (Sweden); Hallez, 1893, 148 (NW. France; synonymy); Verrill, 1893, 468 (to *Heterostylochus*); Gamble, 1893*a*, 497 (descr.); 1894, 324 (Irish Sea); Herdman, 1894, 324 (Irish Sea); Hallez, 1894, 202, figs (NW. France); Gamble, 1896, 18, 19, fig. (Britain); Pruvot, 1897, 21 (NW. & S. France); Jameson, 1897, 174 (Irish Sea); Gamble, 1900, 812 (Ireland); Laidlaw, 1903*a*, 7 (note); 1903*e*, 13 (?=*agilis*); Jacubowa, 1909, 1, 8 (*S. maculata* of Uljanin, 1870, s.o. *S. taurica*); Bock, 1913, 172, 173, figs (Sweden); Southern, *in* Farran, 1915, 35 (Ireland); Bock, 1924*a*, 2 (to *Stylochoplana* (*Stylochoplana*)); Hadenfelt, 1929, 596, figs (nervous system); Remane, 1929, 73 (ecology in Keil Bay, Germany); Southern, 1936, 70 (Ireland); Moore, 1937, 62 (Irish Sea); Hyman, 1953*a*, 301 (*maculata* designated type-species of genus); Laverack & Blackler, 1974, 32 (St Andrews Bay, Scotland); Hendelberg, 1974*b*, 13, figs (Sweden); Galleni, 1976*a*, 85, figs (Tuscany coast, Italy); Galleni & Puccinelli, 1981, 33 (Tuscany coast); Prudhoe, 1982*c*, 44, fig. (Britain).
meleagrina (Kelaart, 1858) Collingwood, 1876, 88, 98, fig.; Lang, 1884, 613 (*Planaria*).
minuta Hyman, 1959*a*, 549, figs (Caroline Is, Pacific Ocean).
nadiae (Melouk, 1941) du B.-R. Marcus & Marcus, 1968, 23.
nationalis Plehn, 1896*b*, 5, figs (42°N, 55°W, off S. Labrador); Laidlaw, 1903*e*, 3, (? type of new

genus); Bock, 1913, 207 (to *Notoplana*).

oculifera (Girard, 1853) Pearse & Walker, 1939, 18, fig.; Hyman, 1940*a*, 478 (s.o. *Gnesioceros floridana*).

pallida (Quatrefages, 1845) Bock, 1913, 172, 179 (diagn.); Palombi, 1928, 589 (synonymy; Suez Canal); Stummer-Traunfels, 1933, 3571, fig.; Palombi, 1939*b*, 108 (synonymy); Pinto, 1947, 95 fig. (Portugal); Reidl, 1953, 122 (Sicily, as *Leptoplana pallida*).

palmula (Quatrefages, 1845) Lang, 1884, 127, 457, figs (Italy); Carus, 1885, 150 (descr.); Bock, 1913, 172, 178 (diagn.); 1924*a*, 2 (*Stylochoplana* (*Stylochoplana*)); Bresslau, 1933, 285, fig.

panamensis (Plehn, 1896*a*) Bock, 1913, 173, 179 (diagn.); Hyman, 1953*a*, 301, figs (Mexico & California); Brusca, 1973, 67 (Gulf of California).

parasitica Kato, 1935*a*, 123, figs (Japan); 1940, 538, 556, fig. (development); 1944, 273.

parva Palombi, 1939*b*, 96, 104, 109, figs (Rio de Oro, NW. Africa).

plehni Bock, 1913, 173, 180 (nom. nov. pro *Leptoplana californica* Plehn: diagn.); Steinbeck & Ricketts, 1941, 337 (Lower California); Hyman, 1953*a*, 305 (to *Notoplana; S. plehni* of Steinbeck & Ricketts is *Zygantroplana stylifera*).

pusilla (*Stylochoplanoides*) Bock, 1924*a*, 2, figs (Japan); Kato, 1934*b*, 124 (descr.; Japan); 1939*b*, 144 (Japan); 1944, 273.

reishi Hyman, 1959*b*, 5, figs (California).

reticulata (Stimpson, 1855) Stimpson, 1857, 22, 29 (descr.); Lang, 1884, 445 (to *Planocera*).

robusta (Palombi, 1928) du B.-R. Marcus & Marcus, 1968, 31.

sargassicola (Mertens, 1832) Graff, 1892*a*, 146 (Atlantic – 39°–44°N, 39°–47°W; descr.); 1892*b*, 207, figs (9°21′N, 18°25′W; 5°48′N, 14°20′W; Madeira, Sargasso Sea & off northern coast New Guinea); Plehn, 1896*b*, 3, 10 (42°N, 55°W (off Newfoundland); descr.); Laidlaw, 1906, 707 (Cape Verde Is); Bock, 1913, 233 (to *Pelagoplana*); Prudhoe, 1950*b*, 710 (Graff's form not identical with Mertens'); Cheng & Lewin, 1975, 518 (Baja California); Hyman, 1939*b*, 11 (to *Gnesioceros*).

selenopsis Marcus, 1947, 99, 116, 163, figs (Brazil); 1949, 74, figs (Brazil); du B.-R. Marcus, 1955*b*, 284.

siamensis Palombi, 1924*a*, 36 (Gulf of Siam); 1924*b*, 15, figs (9°55′N, 101°5′E, off Malaysian coast); du B.-R. Marcus & Marcus, 1968, 22 (not a *Stylochoplana*).

snadda du Bois-Reymond Marcus & Marcus, 1968, 2, 30, figs (Caribbean & Florida).

suesensis Palombi, 1939*b*, 107, 109 (nom. nov. pro *S. angusta* of Palombi, 1928; fig. 176 & fig. 180 must be transposed).

suoensis Kato, 1943*b*, 69, 70, fig. (Taiwan); 1944, 274 (Taiwan); du B.-R. Marcus & Marcus, 1968, 29 (diagn.).

taiwanica Kato, 1943*b*, 69, 70, figs (Taiwan); 1944, 274 (Taiwan).

tarda (Graff, 1878) Lang, 1884, 462; Carus, 1885, 150 (descr.); Micoletzky, 1910, 178 (Trieste); Bock, 1913, 178 (? = *palmula*); 1924*a*, 2 (to *S.* (*Stylochoplana*)).

taurica Jacubowa, 1909, 1, 8, figs (nom. nov. pro *Stylochus maculata* of Uljanin, 1870; Black Sea); Bock, 1913, 172, 178 (diagn.); 1924*a*, 2 (to *S.* (*Stylochoplana*)); Marinescu, 1971, 45 (Rumania).

tenera Stimpson, 1857, 22 ('*tenuis*'), 29 (pelagic, Atlantic Ocean, between 20° & 30°N); Diesing, 1862, 568 (to *Stylochus*); Lang, 1884, 461; Graff, 1892*b*, 199 (? = *Planocera pellucida*); Bock, 1913, 173, 223 (? = *Planocera pellucida*); 1931, 277 (s.o. *Planocera pellucida*); Kato, 1944, 283 (s.o. *pellucida*).

tenuis Stimpson, 1857, 22 [err. pro *tenera*]; Diesing, 1862, 568 (s.o. *Stylochus tener*).

tenuis Palombi, 1936, 13, figs (Still Bay, South Africa); 1939*a*, 126, figs (S. Africa); 1939*b*, 108.

utunomii Kato, 1943*b*, 69, 72, figs (Taiwan); 1944, 274 (Taiwan).

vesiculata Palombi, 1940, 113, figs (Angola); de Beauchamp, 1951*a*, 77, figs (Angola).

viridis Freeman, 1933, 111, 118, figs (Puget Sound); Hyman, 1953*a*, 308, 334 (to *Phylloplana*); Marcus & Marcus, 1966, 330 (not a *Phylloplana*, but a *Stylochoplana*).

walsergia du Bois-Reymond Marcus & Marcus, 1968, 2, 25, figs (Brazil).

wyona du Bois-Reymond Marcus & Marcus, 1968, 2, 32, figs (Caribbean).

species Laidlaw, 1902, 287 (Laccadive Is, Indian Ocean).

species Day *et al.*, 1970, 19 (False Bay, South Africa).

species Morton & Miller, 1968, 172, fig. (New Zealand).

(STYLOCHOPLANA) Bock, 1924, 2, 18 (subgen. of *Stylochoplana*, = Group A of Bock, 1913); 1913, 172.

STYLOCHOPLANINAE Meixner, 1907*b*, 386, 394, 447 (subfamily of Planoceridae); Hyman, 1953*a*, 343.

(STYLOCHOPLANOIDES) Bock, 1924*a*, 2 (subgen. of *Stylochoplana*; type: *S. pusilla*); Marcus, 1947, 115.

STYLOCHOPSIS Stimpson, 1857, 22 (type: *conglomeratus* – by page-precedence); Goette, 1882*b*, 190 (development); Diesing, 1862, 570 (s.o. *Stylochus*); Lang, 1884, 446 (s.o. *Stylochus*); Girard, 1893, 193; Verrill, 1893, 462 (s.o. *Stylochus*); Meixner, 1907*b*, 397.

conglomeratus Stimpson, 1857, 22, 29 (Japan); Diesing, 1862, 570 (to *Stylochus*); Lang, 1884, 454 (*Stylochus*); Meixner, 1907*b*, 397, 405.

lateralis [err. pro *littoralis* Verrill] Girard, 1893, 194.

limosus Stimpson, 1857, 22, 30 (Japan); Diesing, 1862, 570 (to *Stylochus*); Lang, 1884, 453 (*Stylochus*); Meixner, 1907*b*, 397, 405.

littoralis Verrill, 1873, 325, 332, 632, fig. (Massachusetts); Lang, 1884, 453 (to *Stylochus*); Verrill, 1893, 467 (s.o. *Eustylochus ellipticus*); Girard, 1893, 194 (*'lateralis'*); Laidlaw, 1903*b*, 107 (s.o. *Planocera elliptica*); Meixner, 1907*b*, 397, 405; Hyman, 1939, 130 (s.o. *Stylochus ellipticus*); 1940*a*, 459 (s.o. *ellipticus*).

malayensis Collingwood, 1876, 88, 94, fig. (Borneo); Lang, 1884, 567 (s.o. *Prostheceraeus hancockanus*); Meixner, 1907*b*, 398 (not a stylochid); Bock, 1913, 258 (to *Pseudoceros*); Kaburaki, 1923*a*, 639 (s.o. *Pseudoceros hancockanus*).

pilidium Goette, 1881, 189 (nom. nov. pro *Planaria neapolitana* Goette, 1878); 1882*a*, 1, figs (early development); Lang, 1884, 449 (to *Stylochus*); Meixner, 1907*b*, 398, 425 (*Stylochus*).

ponticus Metschnikoff, 1877, 1 (development; Black Sea); Meixner, 1907*b*, 398 (not a stylochid).

zebra Verrill, 1882, 371 (Massachusetts); 1893, 463, figs (to *Stylochus*); 1884, 666 (footnote 'n.sp.'); Girard, 1893, 193 (descr.); Meixner, 1907*b*, 398, 433.

STYLOCHUS Ehrenberg, 1831, 54 (type: *suesensis*); Quatrefages, 1845, 143; Diesing, 1850, 215 (diagn.); Girard, 1850, 252 (s.o. *Planocera*); Stimpson, 1857, 22; Schmarda, 1859, 14, 33; Diesing, 1862, 494, 564; Lang, 1884, 429, 446 (defined); Carus, 1885, 149; Gamble, 1896, 19; Benham, 1901, 31; Laidlaw, 1903*d*, 107 (table of spp.); Haswell, 1907*a*, 471; Meixner, 1907*b*, 395, 416 (defined, key to spp.); Bock, 1913, 128 (defined); Yeri & Kaburaki, 1918*a*, 434 (features); 1918*b*, 46 (features); Bock, 1925*b*, 171, 176 (defined; geographical distribution); Pearse, 1938, 72; Hyman, 1939*b*, 129, 131 (syn. *Eustylochus*); 1940*a*, 459 (defined); 1953*a*, 280 (defined), 289 (systematic note); Marcus, 1947, 105 (list of spp.); du B.-R. Marcus & Marcus, 1968, 5 (systematic note on subgenera).

albus Hallez, 1905*b*, 124 (Antarctica); 1907, 2, 5, figs (to *Stylochoides*); Meixner, 1907*b*, 439 (descr.).

alexandrinus Steinböck, 1937, 1, figs (Egypt); du B.-R. Marcus & Marcus, 1968, 6 (subgen. *Stylochus*); Galleni, 1976*b*, 15 (Tuscany coast); Galleni *et al.*, 1976, 62 (Tuscany coast); Galleni & Puccinelli, 1981, 33 (Tuscany coast).

amphibolus Schmarda, 1859, 34, figs (Ceylon); Diesing, 1862, 566; Lang, 1884, 444 (to *Planocera*); Meixner, 1907*b*, 402 (=*Planocera* (?) *amphibolus*); Stummer-Traunfels, 1933, 3487, 3552 (type-specimen re-examd.; s.o. *Paraplanocera oligoglena*); Kato, 1944, 286 (s.o. *P. oligoglena*).

aomori Kato, 1937*d*, 39, figs (Japan); 1939*b*, 142 (Japan); 1940, 561 (development); 1944, 262; du B.-R. Marcus & Marcus, 1968, 6 (subgen. *Imogene*); Okada *et al.*, 1971, 62 (Japan).

arenosus Willey, 1897, 154, 155 (New Britain); Jacubowa, 1906, 128 (type-specimen re-examd.); Meixner, 1907*b*, 417 (descr.); du Bois-Reymond Marcus & Marcus, 1968, 6 (subgen. *Stylochus*).

argus Czerniavsky, 1881, 221 (Black Sea); Lang, 1884, 454; Laidlaw, 1903*d*, 107; Meixner, 1907*b*, 405.

atentaculatus Hyman, 1953*a*, 283, figs (Oregon to Lower California); du B.-R. Marcus & Marcus, 1968, 6 (subgen. *Stylochus*).

bermudensis Verrill, 1903, 43 (Bermuda); Meixner, 1907*b*, 428 (descr.); Bock, 1925*b*, 176.

californicus Hyman, 1953*a*, 285, figs (California; syn. *Cryptophallus magnus* of MacGinitie & MacGinitie, 1949); 1955*a*, 9 (specimen of MacGinitie & MacGinitie is *Kaburakia excelsa*); du B.-R. Marcus & Marcus, 1968, 6, 9 (subgen. *Stylochus*).

castaneus Palombi, 1939*b*, 95, 98, figs (Senegal, W. Africa; syn. *S. neapolitanus* ? of Laidlaw, 1906; as host of tetraphyllidean cestode larva); du B.-R. Marcus & Marcus, 1968, 6 (subgen. *Stylochus*).

catus du Bois-Reymond Marcus, 1958, 401, figs (Brazil); du B.-R. Marcus & Marcus, 1968, 6, 11, figs (subgen. *Imogine*).

ceylanicus Laidlaw, 1904*b*, 130 (Ceylon); Meixner, 1907*b*, 426, fig. (type-specimen re-examd.); Bock, 1925*b*, 176; du B.-R. Marcus & Marcus, 1968, 6 (subgen. *Imogine*).

cinereus Willey, 1897, 154, 155 (New Britain); Jacubowa, 1906, 126 (type-specimen re-examd.); Meixner, 1907*b*, 418 (descr.).
conglomeratus (Stimpson, 1857) Diesing, 1862, 570; Lang, 1884, 454; Laidlaw, 1903*d*, 107; Meixner, 1907*b*, 405.
conoceraeus Schmarda, 1859) Diesing, 1862, 568; Lang, 1884, 446 (to *Conoceros*); Meixner, 1907*b*, 398 (a *Stylochoplana*).
corniculatus (Dalyell, 1853) Diesing, 1862, 571; Lang, 1884, 459 (s.o. *Stylochoplana maculata*); Meixner, 1907*b*, 402 (= ?*Stylochoplana maculata*).
corniculatus Stimpson, 1855*a*, 381 (China); 1857, 22, 29 (descr.; Hong Kong); Diesing, 1862, 565; Lang, 1884, 464; Meixner, 1907*b*, 402, 405.
coseirensis Bock, 1925*b*, 123, 176 (nom. nov. pro *St. reticulatus* of Meyer); du B.-R. Marcus & Marcus, 1968, 6 (subgen. *Stylochus*).
crassus Verrill, 1893, 466 (37°27′N, 73°33′W, off Maryland coast); Laidlaw, 1903*d*, 107 (not a *Stylochus*); Meixner, 1907*b*, 404 (presumed to be a *Cycloporus*); Hyman, 1940*a*, 491 (indeterminable); du B.-R. Marcus & Marcus, 1968, 7 (?an *Opisthogenia*).
crassus Bock, 1931, 262, 263, figs (Atlantic Ocean, 30°S, 15°W); Stummer-Traunfels, 1933, 3591, fig., 3594, fig.; Marcus, 1947, 106, 162 (renamed *sixteni*).
dictyotus Schmarda, 1859, 33, fig. (Jamaica); Diesing, 1862, 566; Lang, 1884, 443 (to *Planocera*); Meixner, 1907*b*, 402 (= *Planocera* (?) *dictyotus*); Hyman, 1955*b*, 148 (? = *Stylochus megalops*).
djiboutiensis Meixner, 1907*a*, 165 (Somalia); 1907*b*, 419, figs; Bock, 1925*b*, 176; du B.-R. Marcus & Marcus, 1968, 6 (subgen. *Stylochus*).
ellipticus (Girard, 1850) Kingsley, 1901, 165 (Maine); Meixner, 1907*b*, 399, 429; Allee, 1923*a*, 176 (Massachusetts); 1923*b*, 218 (Massachusetts); Hyman, 1939*b*, 130, figs (synonymy; Connecticut to Maine); 1940*a*, 459, fig.; 1940*b*, 17 (descr.; Texas to Prince Edward I); 1951, 164, fig. (New England); 1954*c*, 301 (Gulf of Mexico); Provenzano, 1961, 83 (attacking oysters, Massachusetts); Bahr & Hillman, 1966, 323, figs (among oysters dredged in St Mary's River, Maryland; chromosome number); du B.-R. Marcus & Marcus, 1968, 6 (subgen. *Imogine*); Linkletter *et al.*, 1977, 11 (New Brunswick); Overstreet, 1978, 10 (northern Gulf of Mexico).
exiguus Hyman, 1953*a*, 287, figs (California); du B.-R. Marcus & Marcus, 1968, 6 (subgen. *Imogine*).
fasciatus Schmarda, 1859, 33, figs (Jamaica); Diesing, 1862, 566; Lang, 1884, 462 (to *Stylochoplana*); Meixner, 1907*b*, 402 (= *Stylochoplana* (?) *fasciata*); Stummer-Traunfels, 1933, 3487, 3550, figs (type-specimen re-examd.; to *Styloplanocera*).
ferox Bock, 1925*b*, 175 (Japan; [MS name]).
flevensis Hofker, 1930, 206, figs (Netherlands); Levetzow, 1943, 194 (imported from America; development); de Beauchamp, 1961, 55; du B.-R. Marcus & Marcus, 1968, 6 (subgen. *Stylochus*).
floridanus Pearse, 1938, 71, fig. (Florida); Hyman, 1940*a*, 464 (type-specimen re-examd., s.o. *Stylochus oculiferus*).
folium Grube, 1840, 51, fig. (Sicily); Örsted, 1844*a*, 48 (to *Planocera*); Diesing, 1850, 216; 1862, 568; Stimpson, 1857, 4 (to *Stylochoplana*); Lang, 1884, 440 (*Planocera*); Meixner, 1907*b*, 401 (*Planocera*).
franciscanus Hyman, 1953a, 280, figs (California); du B.-R. Marcus & Marcus, 1968, 6 (subgen. *Stylochus*).
frontalis Verrill, 1893, 465, fig. (from bottom of whaling vessel, Massachusetts); Laidlaw, 1903*d*, 107; Meixner, 1907*b*, 427 (descr.); Bock, 1925*b*, 176; Pearse & Walker, 1939, 17, fig.; Hyman, 1940*a*, 461 (descr.; synonymy); 1940*b*, 14 (prob. not indigenous to Atlantic coast of N. America); 1954*c*, 301 (Gulf of Mexico); du B.-R. Marcus & Marcus, 1968, 6, 7, fig. (Caribbean & Florida; subgen. *Stylochus*).
gaimardi (Blainville, 1828) Schmarda, 1859, 33; Lang, 1884, 436 (= *Planocera gaimardi*); Meixner, 1907*b*, 403 (= *Planocera pellucida*).
hamanensis Kato, 1944, 263, figs (Japan); du B.-R. Marcus & Marcus, 1968, 6 (subgen. *Imogine*).
heteroglenus Schmarda, 1859, 34, figs (Jamaica); Diesing, 1862, 569; Lang, 1884, 444 (to *Planocera*); Meixner, 1907*b*, 402 (*Planocera* (?)); Stummer-Traunfels, 1933, 3486, 3490, 3558 (s.o. *Stylochus megalops*).
hyalinus Bock, 1913, 51, 136, figs (Thailand); 1925*b*, 176; du B.-R. Marcus & Marcus, 1968, 6 (subgen. *Imogine*).
ijimai Yeri & Kaburaki, 1918*a*, 431, 434 (diagn.; Japan); 1918*b*, 2, 6, 46, figs (Japan); Bock, 1925*b*, 177; Kato, 1934*b*, 124 (Japan); 1938*b*, 578 (Japan); 1944, 262; Utinomi, 1956, 29, fig. (coloured);

(Japan); du B.-R. Marcus & Marcus, 1968, 6 (subgen. *Imogine*); Yang, 1974, 7 (Korea; development).

inimicus Palombi, 1931*b*, 219, figs (commensal of *Ostrea virginica*, Florida); Pearse, 1938, 69, figs (Florida); Pearse & Walker, 1938, 607, figs (on oyster-beds, Florida; morphology, life-history, physiology & ecology); Hyman, 1940*a*, 461 (s.o. *Stylochus frontalis*).

insolitus Hyman, 1935*a*, 289, figs (California); du B.-R. Marcus & Marcus, 1968, 6 (subgen. *Stylochus*).

isifer du Bois-Reymond Marcus, 1955*a*, 37, figs (Brazil); du B.-R. Marcus & Marcus, 1968, 6 (subgen. *Stylochus*).

izuensis Kato, 1944, 265, figs (Japan); du B.-R. Marcus & Marcus, 1968, 6 (subgen. *Imogine*).

limosus (Stimpson, 1857) Diesing, 1862, 570; Lang, 1884, 453; Laidlaw, 1903*d*, 107; Meixner, 1907*b*, 405.

linteus Müller, 1854, 75, fig. (larval form; Mediterranean); Diesing, 1862, 571; Graff, 1882, 22 (*'luteus'*); Lang, 1884, 465 (*'luteus'*); Carus, 1885, 150 (descr.); Meixner, 1907*b*, 404; Micoletzky, 1910, 178 (Trieste); Kato, 1940, 559 (*'luteus'*).

littoralis (Verrill, 1873) Lang, 1884, 453; Verrill, 1893, 467 (s.o. *Eustylochus ellipticus*); Laidlaw, 1903*d*, 107 (= *Planocera elliptica*); Meixner, 1907*b*, 428, fig. (Connecticut); Bock, 1925*b*, 176; Hyman, 1939*b*, 130 (s.o. *Stylochus ellipticus*); 1940*a*, 459 (s.o. *ellipticus*).

luteus [pro *linteus*] Graff, 1882, 22; Lang, 1884, 465.

maculatus Quatrefages, 1845, 144, figs (NW. France); Diesing, 1850, 217; Stimpson, 1857, 22 (to *Stylochoplana*); Diesing, 1862, 568; Claparède, 1863, 20, figs (Normandy); 1864, 464 (ontogeny); Uljanin, 1870, 160; Lang, 1884, 459 (*Stylochoplana*); Verrill, 1893, 467 (to *Heterostylochus*); Meixner, 1907*b*, 401 (*Heterostylochus*); Jacubowa, 1909, 8 (*Stylochus maculatus* of Uljanin, 1870, renamed *Stylochoplana taurica*).

marginatus (Stimpson, 1857) Diesing, 1862, 569; Lang, 1884, 445 (to *Planocera*); Meixner, 1907*b*, 403 (*Planocera*(?)).

marmoreus Bock, 1925*b*, 98, 176, figs (Indonesia); du B.-R. Marcus & Marcus, 1968, 6 (subgen. *Imogine*).

martae Marcus, 1947, 99, 104, 162, figs (Brazil); du B.-R. Marcus & Marcus, 1968, 6 (subgen. *Stylochus*); du B.-R. Marcus, 1955*b*, 283.

mediterraneus (*Imogine*) Galleni, 1976*b*, 15 (Tuscany coast); Galleni *et al.*, 1976, 62 (attacking oysters and mussels); Galleni & Puccinelli, 1981, 33 (Tuscany coast); Lanfranchi *et al.*, 1981, 267 (larva in Ligurian Sea).

megalops (Schmarda, 1859) Stummer-Traunfels, 1933, 3486, 3488, 3558, figs (Jamaica; type-specimen re-examd.); Hyman, 1955*b*, 120, figs (Puerto Rico); du B.-R. Marcus & Marcus, 1968, 6 (subgen. *Imogine*); Espinosa, 1981, 1 (attacking oysters; Cuba).

meixneri Bock, 1925*b*, 123, 176 (nom. nov. pro *Stylochus reticulatus* Meixner *nec* Stimpson); du B.-R. Marcus & Marcus, 1968, 6 (subgen. *Stylochus*).

mertensi Diesing, 1850, 216 (nom. nov. pro *Planaria sargassicola* Mertens); 1862, 572 (to *Gnesioceros*); Moseley, 1877, 23; Lang, 1884, 454 (s.o. *Stylochus sargassicola*); Meixner, 1907*b*, 402 (= *Stylochus sargassicola*).

minimus Palombi, 1940, 109, 111, figs (Loango, Congo Rep.); du B.-R. Marcus & Marcus, 1968, 6 (subgen. *Imogine*).

miyadii Kato, 1944, 264, figs (Japan); du B.-R. Marcus & Marcus, 1968, 6 (subgen. *Imogine*).

neapolitanus (Delle Chiaje, 1841) Lang, 1884, 53, 447, 449, figs (Mediterranean); Carus, 1885, 149 (descr.); Lo Bianco, 1888, 400 (Naples); 1899*a*, 479 (Naples); Gérard, 1901, 143, fig. (ovogenesis; *'napolitanus'*); Laidlaw, 1903*d*, 107; 1906, 707. (Cape Verde Is); Meixner, 1907*b*, 422, figs; Bock, 1925*b*, 176; Bresslau, 1930, 142, fig.; Stummer-Traunfels, 1933, 3591, fig.; Palombi, 1939*b*, 98 (*S. neapolitanus*? of Laidlaw, 1906, s.o. *S. castaneus*); du B.-R. Marcus & Marcus, 1968, 6 (subgen. *Stylochus*); Arndt, 1943, 2 (Naples).

nebulosus (Girard, 1853) Meixner, 1907*b*, 399, 406, 431, figs (Connecticut); Bock, 1925*b*, 176; Hyman, 1940*a*, 459 (s.o. *S. ellipticus*).

obscurus Stimpson, 1857, 22, 29 (Yesso, Japan); Diesing, 1862, 566; Lang, 1884, 464; Meixner, 1907*b*, 405; Yeri & Kaburaki, 1918*a*, 438 (to *Pseudostylochus*); 1918*b*, 30, figs (to *Pseudostylochus*).

oculiferus (Girard, 1853), Diesing, 1862, 570; Lang, 1884, 446 (*Imogine*); Meixner, 1907*b*, 405; Hyman, 1940*a*, 464, fig. (syn. S. *floridanus*); 1955*b*, 122 (Bahamas); du B.-R. Marcus & Marcus,

1968, 6, 7, 9, figs (Caribbean; subgen. *Imogine*).

oligochlaenus [err. pro *oligoglenus*] Grube, 1868, 46 (Samao); Graff, 1892*b*, 205; Meixner, 1907*b*, 403.

oligoglenus Schmarda, 1859, 34, figs (Ceylon); Diesing, 1862, 567; Grube, 1868, 46 (Samoa); Lang, 1884, 444 (to *Planocera*); Meixner, 1907*b*, 402 (*Planocera* (?)); Stummer-Traunfels, 1933, 3487, 3552 (type-specimen re-examd.; to *Paraplanocera*).

orientalis Bock, 1913, 36, 128, figs (Thailand, Taiwan and Western Australia); Kato, 1944, 261; du B.-R. Marcus & Marcus, 1968, 6 (subgen. *Imogine*).

orientalis var. *splendida* Bock, 1913, 132, figs (Thailand); Kato, 1944, 261; du B.-R. Marcus & Marcus, 1968, 6.

oxyceraeus Schmarda, 1859, 35, figs (Ceylon); Diesing, 1862, 567; Lang, 1884, 445 (to *Planocera*); Meixner, 1907*b*, 403 (*Planocera* (?)); Stummer-Traunfels, 1933, 3488, 3558, figs (type-specimen re-examd.; s.o. *Callioplana marginata*); Kato, 1944, 288 (s.o. *Callioplana marginata*).

palmula Quatrefages, 1845, 143, figs (Sicily); Diesing, 1850, 217; 1862, 569; Lang, 1884, 127, 457, figs (to *Stylochoplana*); Meixner, 1907*b*, 401 (*Stylochoplana*).

papillosus Diesing, 1836, 316 (Trieste); Grube, 1840, 56 (to *Thysanozoon*); Diesing, 1850, 212 (s.o. *Thysanozoon diesingii*); Lang, 1884, 525 (s.o. *Thysanozoon brocchii*).

pelagicus Moseley, 1877, 24, figs (9°21′N, 18°25′W & 5°48′N, 14°20′W); Lang, 1884, 439 (to *Planocera*); Graff, 1892*b*, 207 (s.o. *Stylochoplana sargassicola*, but on p. 211 Graff states that Moseley may have been dealing with two species, as the latter's fig. of the eye arrangement corresponds with that of *Planocera pellucida*); Meixner, 1907*b*, 404 (s.o. *Stylochoplana sargassicola*); Bock, 1913, 233 (s.o. *Pelagoplana sargassicola*); Hyman, 1939*b*, 146 (s.o. *Gnesioceros sargassicola*).

pellucidus (Mertens, 1832) Ehrenberg, 1836, 67; Diesing, 1850, 216; Claparède, 1861, 75; Moseley, 1877, 23; Lang, 1884, 437 (s.o. *Planocera pellucida*); Meixner, 1907*b*, 401 (*Planocera*); Kato, 1944, 283 (*Planocera*).

pilidium (Goette, 1881) Lang, 1884, 321, 449; Carus, 1885, 150 (descr.); Lo Bianco, 1888, 400 (Naples); Plehn, 1896*a*, 171 (Chile, on ship from Italy); Steiner, 1898, 53; Lo Bianco, 1899*a*, 479 (Naples); Gérard, 1901, 143, figs (*'pelidium'*; ovogenesis); Laidlaw, 1903*d*, 107; Meixner, 1907*b*, 417, 425 (descr.); Bock, 1925*b*, 176, 178; Bytinski-Salz, 1935, 1, figs (biological notes; Rovigno, Gulf of Venice); Dawydoff, 1952, 81 (Vietnam); du B.-R. Marcus & Marcus, 1968, 6 (subgen. *Stylochus*); Galleni *et al.*, 1976, 62 (*S. pilidium* of Bytinski-Salz, 1935, s.o. *S.* (*I.*) *mediterraneus*); Rzhepishevskji, 1979, 23 (Black Sea).

plessisii Lang, 1884, 54, 450, figs (west coast of southern Italy); Carus, 1885, 150 (descr.); Laidlaw, 1903*d*, 107; Meixner, 1907*b*, 434 (descr.); Bock, 1925*b*, 176; Stummer-Traunfels, 1933, 3571, fig. (descr.); du B.-R. Marcus & Marcus, 1968, 6 (subgen. *Stylochus*); Riedl, 1953, 136 (Sicily).

pulcher Hyman, 1939*e*, 153; 1940*a*, 462, figs (North Carolina); du B.-R. Marcus & Marcus, 1968, 6 (subgen. *Imogine*).

pusillus Bock, 1913, 139, figs (Hong Kong); 1925*b*, 177; du B.-R. Marcus & Marcus, 1968, 6 (subgen. *Stylochus*).

refertus du Bois-Reymond Marcus, 1965, 129, figs (Brazil); du B.-R. Marcus & Marcus, 1968, 6 (subgen. *Imogine*); Narchi, 1975, 280 (Brazil).

reticulatus Stimpson, 1855*a*, 381 (Ryu Kyu Is, Japan); 1857, 22, 29 (to *Stylochoplana*); Diesing, 1862, 569; Lang, 1884, 445 (to *Planocera*); Meixner, 1907*b*, 435 (Somalia); Yeri & Kaburaki, 1918*b*, 22 (Meixner form different from that of Stimpson); Meyer, 1922, 145 (Kossier, Red Sea); Bock, 1925*b*, 123, 176 (*reticulatus* of Meyer, 1922, renamed *Stylochus coseirensis*).

roseus Sars, in Jensen, 1878, 75, figs (Scandinavia); Lang, 1884, 589 (to *Stylostomum*); Meixner, 1907*b*, 404 (*Stylostomum*); Bock, 1913, 270 (s.o. *Stylostomum ellipse*).

rutilus Yeri & Kaburaki, 1918*a*, 431, 434 (diagn.; Japan); 1918*b*, 2, 5, 46, figs (Japan); Bock, 1925*b*, 177; Kato, 1944, 262; du B.-R. Marcus & Marcus, 1968, 6 (subgen. *Imogine*).

salmoneus Meixner, 1907*a*, 165 (Somalia); 1907*b*, 420, figs; Bock, 1925*b*, 176; du B.-R. Marcus & Marcus, 1968, 6 (subgen. *Stylochus*).

sargassicola (Mertens, 1832) Ehrenberg, 1836, 67; Claparède, 1861, 75; Lang, 1884, 454 (synonymy); Graff, 1892*a*, 146 (to *Stylochoplana*); Meixner, 1907*b*, 401 (*Stylochoplana*); Bock, 1913, 233 (to *Pelagoplana*).

sixteni Marcus, 1947, 106, 162 (nom. nov. pro *crassus* Bock, 1931, *nec* Verrill, 1892); du B.-R. Marcus & Marcus, 1968, 6 (subgen. *Stylochus*).

speciosus Kato, 1937*f*, 347, figs (Japan); 1944, 262; du B.-R. Marcus & Marcus, 1968, 6 (subgenus *Imogine*).

suesensis Ehrenberg, 1831, 36, 37, 54, figs (Red Sea); Örsted, 1844*a*, 27, 48 (to *Planocera*); Diesing, 1850, 215; Stimpson, 1857, 22; Diesing, 1862, 565; Lang, 1884, 451 (synonymy); Laidlaw, 1903*d*, 106 (descr.; East Africa); Meixner, 1907*b*, 421 (descr.; synonymy); Palombi, 1928, 582, figs (synonymy; Suez Canal, Suez & Gulf of Suez); du B.-R. Marcus & Marcus, 1968, 6 (subgen. *Stylochus*).

tardus Graff, 1878, 460 (Trieste); Stossich, 1882, 225 (Trieste); Lang, 1884, 462 (to *Stylochoplana*); Meixner, 1907*b*, 404 (*Stylochoplana tarda*); Bock, 1913, 178 (? = *Stylochoplana palmula*).

tauricus Jacubowa, 1909, 2, 6, figs (Black Sea); Bock, 1925*b*, 176; du B.-R. Marcus & Marcus, 1968, 6 (subgen. *Stylochus*); Marinescu, 1971, 44 (external features; Rumania).

tenax Palombi, 1936, 1, 4, figs (Florida); Pearse, 1938, 71 (s.o. *Stylochus inimicus*); Hyman, 1940*a*, 461 (s.o. *Stylochus frontalis*).

tener (Stimpson, 1857) Diesing, 1862, 568; Lang, 1884, 461 (to *Stylochoplana tenera*); Meixner, 1907*b*, 402 (? = *Planocera pellucida*).

ticus Marcus, 1952, 5, 79, 114, figs (Brazil); du Bois-Reymond Marcus, 1955*b*, 282; du B.-R. Marcus & Marcus, 1968, 6, 11 (Brazil; subgen. *Imogine*).

tripartitus Hyman, 1953*a*, 282, figs (Oregon to Lower California); du B.-R. Marcus & Marcus, 1968, 6 (subgen. *Imogine*); Morris *et al.*, 1980, 77, fig. (coloured) (California).

truncatus (Schmarda, 1859) Diesing, 1862, 567; Stummer-Traunfels, 1933, 3562 (s.o. *Notoplana dubia*).

uniporus Kato, 1940, 561, figs (development; Japan); 1944, 262, figs. ('sp. nov.'); du B.-R. Marcus & Marcus, 1968, 6 (subgen. *Imogine*).

vesiculata Jacubowa, 1909, 2, 4, figs (Black Sea); Bock, 1925*b*, 176.

vigilax Laidlaw, 1904*a*, 2 (Thursday I., Torres Str., Australia); Meixner, 1907*b*, 438, fig. (type-specimen re-examd.); Bock, 1925*b*, 177, du B.-R. Marcus & Marcus, 1968, 6 (subgen. *Stylochus*).

zanzibaricus Laidlaw, 1903*d*, 105, fig. (Zanzibar); Meixner, 1907*b*, 416, 425, fig. (type-specimen re-examd.; poss. identical with *S. neapolitanus*); Bock, 1925*b*, 176; Skerman, 1960, 610 (New Zealand); du B.-R. Marcus & Marcus, 1968, 6 (subgen. *Stylochus*).

zebra (Verrill, 1882) Verrill, 1893, 463, figs (Massachusetts & Connecticut); Meixner, 1907*b*, 433 (descr.); Sumner *et al.*, 1913, 579 (Massachusetts); Allee, 1923*a*, 176 (Massachusetts); Bock, 1925*b*, 176; Pearse, 1938, 72 (Massachusetts): Pearse & Littler, 1938, 237, fig. (North Carolina); Pearse & Walker, 1939, 16, figs (Massachusetts); Hyman, 1939*b*, 129, fig. (New England); 1940*a*, 462 (descr.); 1940*b*, 17 (descr.); Pearse, 1949, 26 (N. Carolina); du B.-R. Marcus & Marcus, 1968, 6 (subgen. *Imogine*); Thomas, 1970, 219 (Massachusetts); Lytwyn & McDermott, 1976, 365 (symbionts of hermit-crabs, Florida & from North Carolina to Massachusetts; direct devel.).

species Grube, 1840, 52 (Sicily); Lang, 1879, 470 (= *Planocera graffii*).

species Minot, 1877*a*, 491 (Trieste); Lang, 1884, 466; Micoletzky, 1910, 178 (Trieste).

species Jacubowa, 1909, 1 (syn. *Endocelis ovata*).

species Bock, 1923*b*, 9 (Japan).

species Bock, 1925*b*, 174 (Japan).

species Bock, 1927*a*, 11 ('n.sp.'; Japan).

species Steinbeck & Ricketts, 1941, 336 (Lower California).

(?) species Steinbeck & Ricketts, 1941, 366 (Lower California).

species Day *et al.*, 1970, 19 (False Bay, S. Africa).

STYLOPLANOCERA Bock, 1913, 53, 233, 239 (type: *fasciata*); Hyman, 1955*b*, 135 (defined).

fasciata (Schmarda, 1859) Stummer-Traunfels, 1933, 3487, 3488, 3550, figs (type-specimen re-examd.; syn. *papillifera*); Hyman, 1955*b*, 135, fig. (Puerto Rico); du B.-R. Marcus & Marcus, 1968, 51 (from various localities in the Caribbean).

papillifera Bock, 1913, 233, figs (Barbados, Jamaica & St Croix, Caribbean); Bresslau, 1928–33, 144, fig.; Stummer-Traunfels, 1933, 3488, 3552 (s.o. *fasciata*); Hyman, 1955*b*, 135 (s.o. *fasciata*).

STYLOSTOMA [err. pro *Stylostomum*] Hallez, 1893, 178.

STYLOSTOMUM Lang, 1884, xi, 584 (type: *ellipse*); Carus, 1885, 157; Hallez, 1893, 178 ('*Stylostoma*'); 1894, 230 (defined); Gamble, 1893*a*, 511; 1896, 19; Benham, 1901, 31; Hallez, 1905*b*, 126; 1907, 9; Heath & McGregor, 1913, 476; Hyman, 1953*a*, 376 (defined); Prudhoe, 1982*c*, 66 (defined).

antarcticum Hallez, 1905*b*, 126 (Bay of Carthage, Antarctica); 1907, 3, 10, figs.

californicum Heath & McGregor, 1913, 455, 458 [appears to be an error for *S. lentum*].
ellipse (Dalyell, 1853) Lang, 1884, 588; Gamble, 1893*a*, 511 (poss. = *St. variabile*); Bock, 1913, 270 (synonymy; Sweden, Norway, Falkland Is, Tierra del Fuego & Cape Town); Steinböck, 1932, 334, 337 (distribution); 1933, 20 (Gulf of Trieste); Bresslau, 1928–33, 241, fig.; Stummer-Traunfels, 1933, 3575, fig.; Westblad, 1952, 9 (Falkland Is, South Georgia & Tierra del Fuego); Crothers, 1966, 22 (Wales); Laverack & Blackler, 1974, 32 (St. Andrews Bay, Scotland); Hendelberg, 1974*b*, 15, figs (Sweden); Galleni & Puccinelli, 1981, 42 (Britain); Prudhoe, 1982*c*, 68, fig. (U.K.).
felinum Marcus, 1954*a*, 70, figs (Chile).
frigidum Bock, 1931, 261, 279, figs (Kerguelen I., Indian Ocean); Marcus, 1954*a*, 68, figs (Chile); Westblad, 1952, 10 (Falkland Is).
fulvum Giard, 1888, 496 (NW. France); Hallez, 1894, 224, 225 (= *Cycloporus maculatus*).
hozawai Kato, 1939*b*, 142, 150, figs (Japan).
lentum Heath & McGregor, 1913, 476, figs (California); Hyman, 1953*a*, 376.
maculatum Kato, 1944, 305, figs (Japan).
punctatum Hallez, 1905*b*, 126 (Bay of Carthage, Antarctica); 1907, 2, 10, figs.
roseum (Sars, *in* Jensen, 1878) Lang, 1884, 589; Bock, 1913, 270 (s.o. *St. ellipse*).
rusticum Giard, 1888, 496 (NW. France); Hallez, 1894, 224.
sanguineum Hallez, 1893, 180, 197, figs (NW. France); 1894, 233 (NW. France); Bock, 1913, 270, 273 (?s.o. *St. ellipse*).
sanjuania Holleman, 1972, 409, figs (Puget Sound).
variabile Lang, 1884, 73, 584, figs (Gulf of Naples); 588 (? = *ellipse*); Carus, 1885, 157 (descr.); Lo Bianco, 1888, 400 (Naples); Vaillant, 1889, 656 (NW. France); Bergendal, 1893*a*, 237 (Sweden); Gamble, 1893*a*, 511, fig. (Britain); 1893*b*, 47, 1893*c*, 171, figs (Irish Sea); Hallez, 1893, 178, 197, figs (NW. France); 1894, 230, figs (NW. France); Gamble, 1896, 19, 22 (Britain); Plehn, 1896*a*, 172 (Patagonia); Lo Bianco, 1899*a*, 479 (Naples); Gamble, 1900, 813 (Ireland); Micoletzky, 1910, 180 (Trieste); Bock, 1913, 270 (s.o. *St. ellipse*); Southern, *in* Farran, 1915, 36 (Ireland); Southern, 1936, 72 (Ireland); Eales, 1939, 55, fig.; Bassindale & Barrett, 1957, 251 (Wales).
SUSAKIA Kato, 1934*b*, 123, 125 (type: *badiomaculata*).
badiomaculata Kato, 1934*a*, 123, 125, figs (Japan); 1944, 278 (s.o. *Discoplana gigas*); Marcus, 1947, 129 (congeneric, but not identical, with *Discoplana gigas*).

TAENIOPLANA Hyman, 1944*b*, 73 (type: *teredini*).
teredini Hyman, 1944*b*, 73, figs (in empty burrows of *Teredo*, Hawaii); Edmondson, 1945, 222, figs (experimental evidence to support claim that *T. 'teredinia'* preys on *Teredo*); Riser, 1970, 553 (in teredinid burrows, Florida & Israel); 1974, 521 (Hawaii, Florida, Israel & Panama, ?New Guinea).
TAENIOPLANIDAE Marcus & Marcus, 1966, 321, 322.
TERGIPES Risso, 1818, 373 (type: *brochi*).
brochi Risso, 1818, 373 (Mediterranean); 1826, 264 (to *Planaria brocchi*); Grube, 1840, 55 (a *Thysanozoon*); Lang, 1884, 525 (*Thysanozoon*); Quatrefages, 1845, 140 (to *Eolidiceros brocchii*).
dicquemari Risso, 1818, 373 (Mediterranean); 1826, 263 (to *Planaria*); Lang, 1884, 550 (to *Yungia*).
THALAMOPLANA Laidlaw, 1903*e*, 5, 11 (defined); 1904*b*, 132 (type: *herdmani*); Bock, 1913, 60 (s.o. *Discocelis*); Marcus, 1950, 75, 76 (valid genus).
herdmani Laidlaw, 1904*b*, 132, figs (Ceylon); Bock, 1913, 61 (to *Discocelis*); 1925*b*, 147 (*Discocelis*).
THALATTOPLANA Bock, 1925*a*, 24 (Bonin Island) (Pericelididae). [MS name; no specific name mentioned]; [see also Bock, 1922, 6, 30.]
THEAMA Marcus, 1949, 7, 72, 103 (type: *evelinae*).
evelinae Marcus, 1949, 7, 72, 103, figs (Brazil); du Bois-Reymond Marcus, 1955*b*, 283.
occidua Sopott-Ehlers & Schmidt, 1975, 198, figs (Galapagos Is).
THEAMATIDAE Marcus, 1949, 71, 103.
THYSANOPLANA Plehn, 1896*a*, 141, 162, 165 (type: *indica*); 1896*c*, 332, 333 ('n.g.'); Benham, 1901, 31; Laidlaw, 1902, 304 (s.o. *Thysanozoon*); 1903*a*, 2 (? = *Thysanozoon*).
indica Plehn, 1896*a*, 162, figs, (Java); 1896*c*, 329, 334, figs ('n.sp.'; type-locality correctly Amboina [Ambon I.]).
marginata Plehn, 1896*a*, 165, figs (Java); 1896*c*, 329, 334, figs 'n.sp.'; type-locality correctly Amboina).
THYSANOZOON Grube, 1840, 54 (type: *diesingii*); Örsted, 1844*a*, 46 (list of spp.); Diesing, 1850, 211 (defined); Grube, 1855, 137; Stimpson, 1857, 19; Schmarda, 1859, 14, 29; Diesing, 1862,

493, 555; Lang, 1884, 430, 524; Carus, 1885, 154; Girard, 1893, 195; Stummer-Traunfels, 1895, 690 (revision); Gamble, 1896, 19; Benham, 1901, 31; Laidlaw, 1902, 304, 305 (note; syn. *Thysanoplana*); Yeri & Kaburaki, 1918*a*, 439 (features); 1918*b*, 48 (defined); Hyman, 1940*a*, 484; 1952, 192 (remarks on Floridan spp.); 1953*a*, 362 (defined); Marcus, 1949, 83 (list of spp.); Prudhoe, 1982*c*, 55 (defined).

alderi Collingwood, 1876, 87, 88, figs (Labuan I., Malaysia); Lang, 1884, 537; Stummer-Traunfels, 1895, 704, figs (Ambon (Amboina), Indonesia).

allmani Collingwood, 1876, 87, 89, figs (Singapore); Lang, 1884, 538; Stummer-Traunfels, 1895, 706, figs (Ceylon); Laidlaw, 1903*c*, 302.

aucklandica Cheeseman, 1883, 213 (New Zealand); Lang, 1884, 617 (s.o. *brocchii*); Stummer-Traunfels, 1895, 148 (emended to *aucklandicum*); Marcus, 1949, 84 (note); Prudhoe, 1977, 601 (W. Australia).

aurantiacum (Delle Chiaje, 1822) Örsted, 1844*a*, 47; Diesing, 1850, 214; 1862, 558; Lang, 1884, 548 (to *Yungia*).

auropunctatum Kelaart, *in* Collingwood, 1876, 87, 54, figs (Ceylon); Lang, 1884, 537 (s.o. *Th. verrucosum*); Stummer-Traunfels, 1895, 701 (descr.; Indonesia); Laidlaw, 1903*c*, 302, 314 (Malaysia); Bock, 1913, 252, figs (W. Australia); Kaburaki, 1923*b*, 638, figs (Philippines; syn. *verrucosum*); Dawydoff, 1952, 81 (Vietnam).

australe Stimpson, 1855*b*, 389 (N.S.W., Australia); 1857, 20, 25 (descr.); Diesing, 1862, 556; Lang, 1884, 536; Whitelegge, 1890, 206 (N.S.W.); Marcus, 1949, 85 (note).

boehmigi Stummer-Traunfels, 1895, 699, 710, figs (Indonesia).

brocchii (Risso, 1818) Örsted, 1844*a*, 47; Diesing, 1850, 213 (Nice, Toulouse & Naples); Grube, 1855, 140, figs (Italy); Graeffe, 1860, 53 (Italy); Stossich, 1882, 225 (Trieste); Lang, 1884, 525, figs (Italy); Carus, 1885, 154 (descr.); Graff, 1886, 342 (Adriatic); Lo Bianco, 1888, 400 (Naples); Gamble, 1896, 18, 29, figs; Stricht, 1897*a*, 484; 1897*b*, 92; 1898, 367; Pruvot, 1897, 20 (Mediterranean); Lo Bianco, 1899, 479 (Naples); Plehn, 1899, 448 (Laysan I.); Benham, 1901, 34, figs; Monti, 1901, 1 (biology); Sabussow, 1905, 489 (S. France); Micoletzky, 1910, 179 (Trieste); Yeri & Kaburaki, 1918*a*, 439 (diagn.; Japan); 1918*b*, 34, 48, figs (Japan); Bresslau, 1928–33, 52, 81, figs; Palombi, 1928, 604, figs (synonymy; Port Said, Egypt); 1930, 2 (Italy); Steinböck, 1933, 19 (Gulf of Trieste); Stummer-Traunfels, 1933, 3566, 3573, 3578, 3592, figs (Trieste); Kato, 1934*b*, 133 (Japan); Pearse, 1938, 85 (descr.; Florida); Palombi, 1939*a*, 135 (S. Africa); 1939*b*, 110 (Rio de Oro, W. Africa); Hyman, 1940*a*, 485; Arndt, 1943, 2 (Naples); Levetzow, 1943, 190 (biol. notes); Kato, 1944, 294, figs (synonymy; distribution); Dieuzeide & Goëau-Brissonière, 1951, 31 (Algeria); Hyman, 1952, 197 (Florida); Marcus, 1949, 83 (note); Dawydoff, 1952, 81 (*'brooki* Grube'; Vietnam); Levetzow, 1939, 780 (Naples; regeneration); Utinomi, 1956, 29, fig. (coloured) (Japan); Powell, 1947, 15, fig. (New Zealand); Morton & Miller, 1968, 172, fig. (New Zealand); du B.-R. Marcus & Marcus, 1968, 66 (Brazil; Curaçao & Florida; syn. *Th. lagidium*); Kensler, 1964, 961 (S. France); Day *et al.*, 1970, 19 (False Bay, S. Africa); Miller & Batt, 1973, 70, 102, fig. (New Zealand); Lanfranchi *et al.*, 1981, 267 (larva in Ligurian Sea). Prudhoe, 1982*c*, 55, fig. (U.K.).

brocchii var. *cruciatum* (Schmarda, 1859) Laidlaw, 1906, 713 (descr.; Cape Verde Is, W. Africa).

brocchii var. *nigrum* (Girard, 1852*b*) Lang, 1884, 535; Hyman, 1939*d*, 15 (*nigrum* a distinct species).

brocchii var. *papillosum* (Sars, *in* Jensen, 1878) Lang, 1884, 536; Stummer-Traunfels, 1895, 700 (*papillosum* a distinct species).

brocchii var. *tentaculatum* (Gray, *in* Pease, 1860) Lang, 1884, 536; Stummer-Traunfels, 1895, 700 (*tentaculatum* a distinct species).

californicum Hyman, 1953*a*, 362, figs (California).

cruciatum (Eolidiceros) Schmarda, 1859, 30, figs (Australia & New Zealand); Diesing, 1862, 557; Hutton, 1879, 315 (New Zealand); Lang, 1884, 526 (?s.o. *Th. brocchii*); Whitelegge, 1890, 206 (Australia); Stummer-Traunfels, 1895, 714, (South Seas); Laidlaw, 1906, 713 (?var. of *brocchii*); Stummer-Traunfels, 1933, 3487, 3549, 3550, figs (type-specimen re-examd.); Marcus, 1949, 85 (note).

dicquemaris (Risso, 1818) Örsted, 1844*a*, 47; Diesing, 1850, 212; 1862, 555; Lang, 1884, 525 (s.o. *brocchii*).

diesingii Grube, 1840, 54, fig. (Sicily); Delle Chiaje, 1841, pt. V, p. 112 (*'Tisanozoon diesingi'* s.o. *Planaria tuberculata*); Örsted, 1844*a*, 47; Diesing, 1850, 212 (synonymy); 1862, 555; Mueller, 1852, 27 (nematocysts); Grube, 1855, 143; Schmarda, 1859, 29 (Ceylon); Schmidt, 1878, 153; Selenka, 1881*b*, 329 (Naples); Lang, 1881*b*, 87 (descr.); 1884, 525 (s.o. *brocchii*); Girard, 1893, 195

(*'diesingeri'*).
discoideum Schmarda, 1859, xiii, 29, figs (Ceylon); Lang, 1884, 537; Stummer-Traunfels, 1895, 716, figs (Baui I., Tanzania); Palombi, 1938, 354 (S. Africa); Marcus, 1949, 85 (note).
distinctum Stummer-Traunfels, 1895, 692, 721, figs (Indonesia); Marcus, 1949, 85 (note).
flavum (Delle Chiaje, 1822) Örsted, 1844*a*, 47; Diesing, 1850, 214, 1862, 558; Lang, 1884, 548 (s.o. *Yungia aurantiaca*).
flavotuberculatum Hyman, 1939*d*, 16, fig, (Bermuda); Marcus, 1949, 84 (note).
fockei Diesing, 1850, 213 (nom. nov. pro *Planaria ornata* Focke *in litt.*); 1862, 556; Stossich, 1882, 225 (Trieste); Lang, 1884, 526 (s.o. *brocchii*).
griseum Verrill, 1903, 41, fig. (Bermuda); Marcus, 1949, 84 (note).
hawaiiensis Hyman, 1960, 308, fig. (Hawaii).
huttoni Kirk, 1882, 267 (New Zealand); Lang, 1884, 617 (s.o. *brocchii*); Stummer-Traunfels, 1895, 700; Marcus, 1949, 85 (note).
japonicum Kato, 1944, 297, figs (Japan).
lagidium Marcus, 1949, 7, 81, 104, figs (Brazil); 1952, 90, 115 (descr.; Brazil); du B.-R. Marcus, 1955*b*, 285; du B.-R. Marcus & Marcus, 1968, 66 (s.o. *brocchii*).
langi Stummer-Traunfels, 1895, 700, 719, figs (Indonesia); Marcus, 1949, 85 (note).
minutum Stummer-Traunfels, 1895, 700, 718, figs (Japan); Marcus, 1949, 85 (note).
muelleri (Delle Chiaje, 1829) Örsted, 1844*a*, 47; Diesing, 1850, 215; 1862, 558; Lang, 1884, 545 (to *Pseudoceros*).
nigrum Girard, 1852*b*, 137 (Florida); Diesing, 1862, 558; Lang, 1884, 535 (var. of *brocchii*); Girard, 1893, 195 (descr.); Verrill, 1903, 41 (descr.; Bermuda); Hyman, 1939*d*, 15, fig. (Bermuda); 1940*a*, 484 (descr.; Florida & Bermuda); 1951, fig. (Bermuda); 1952, 196 (Florida); 1955*b*, 137 (Bahamas); 1955*e*, 263 (Texas); Marcus, 1949, 84 (note); du B.-R. Marcus & Marcus, 1968, 69, figs (Caribbean).
obscurum Stummer-Traunfels, 1895, 699, 712, figs (Indonesia).
ovale (Eolidiceros) Schmarda, 1859, 29, fig. (Ceylon); Diesing, 1862, 557; Lang, 1884, 526, (s.o. *brocchii*); Stummer-Traunfels, 1933, 3487, 3549, 3550, figs (type-specimen re-examd.); Marcus, 1949, 85 (note).
panormus (Quatrefages, 1845) Diesing, 1850, 213; Lang, 1884, 526 (s.o. *brocchii*).
papillosum Grube, 1840, 56 (Mediterranean); Lang, 1884, 525 (s.o. *brocchii*).
papillosum Sars, *in* Jensen, 1878, 79, figs (Norway); Lang, 1884, 536 (var. of *brocchii*); Stummer-Traunfels, 1895, 700; Bock, 1913, 263 (to *Cycloporus*).
plehni Laidlaw, 1902, 294, figs (Laccadive Is, Indian Ocean).
sandiegense Hyman, 1953*a*, 363 (nom. nov. pro *Thysanozoon* sp. of Johnson & Snook, 1927; California); Morris *et al.*, 1980, 78, fig. (coloured) (California).
semperi Stummer-Traunfels, 1895, 699, 709, figs (Philippines).
skottsbergi Bock, 1923*c*, 342, 358, figs (Juan Fernandez Is, Pacific Ocean); Marcus, 1949, 85 (note); Prudhoe, 1977, 602 fig. (W. Australia); 1982*a*, 226 (S. Australia); 1982*b*, 381 (S. Australia).
tentaculatum (Gray, *in* Pease, 1860) Diesing, 1862, 556; Lang, 1884, 536 (var. of *brocchii*); Stummer-Traunfels, 1895, 700; Marcus, 1949, 85 (note); Poulter, 1975, 337 (Hawaii).
tuberculatum (Delle Chiaje, 1829) Grube, 1840, 55; Örsted, 1844*a*, 47; Diesing, 1850, 212; 1862, 555; Stimpson, 1857, 2; Graeffe, 1860, 53; Claparède, 1867, 6 (Naples); Lang, 1884, 525 (s.o. *brocchii*).
verrucosum Grube, 1868, 46 (Samoa); Lang, 1884, 537 (syn. *auropunctatum*); Stummer-Traunfels, 1895, 700 (not synonymous with *auropunctatum*); Kaburaki, 1923*b*, 638 (s.o. *auropunctatum*); Marcus, 1949, 85 (note).
violaceum (Delle Chiaje, 1822) Örsted, 1844*a*, 47; Diesing, 1850, 214; 1862, 557; Lang, 1884, 563 (to *Prostheceraeus*).
'vulgaris Kato' Palombi, 1939*a*, 135; 1939*b*, 110 (corrected to *Prosthiostomum vulgaris* Kato); Kato, 1944, 295, figs (Japan); [Kato uses the name *'Thysanozoon vulgaris'* on the erroneous assumption that Palombi proposed a new name for *Thysanozoon brocchii* of Yeri & Kaburaki, 1918*a*].
species Schultze, 1854, 222; Lang, 1884, 526 (s.o. *brocchii*).
species Moseley, 1877, 29, figs (adult & larva, Philippines & off New Guinea, open sea, about 140 miles north of Pt d'Urville); Lang, 1884, 530 (s.o. *brocchii*).
species Koningsberger, 1891, 83 (Naples).
species Willey, 1897, 153 (New Guinea).

species Andrews, 1892, 75 (Jamaica).
species Johnston & Snook, 1927, 118, fig. (California); Hyman, 1953*a*, 363 (=*sandiegense*).
species Dakin, *et al.*, 1948, 218 (Australia).
species Ricketts & Calvin, 1948, 100, 300 (California).
species Hyman, 1952, 197 (?=*brocchii*; Florida).
species Dakin, 1953, 144 (eastern coasts of Australia).

TRACHYPLANA Stimpson, 1857, 22, 30 (type: *tuberculosa*); Diesing, 1862, 494, 572.
tuberculosa Stimpson, 1857, 22, 30 (Oshima, Japan); Diesing, 1862, 572; Lang, 1884, 464.

TRAUNFELSIA Laidlaw, 1906, 714, 718 (type: *elongata*).
elongata Laidlaw, 1906, 714, figs (Cape Verde Is, W. Africa); Bresslau, 1928–33, 70, fig.; Stummer-Traunfels, 1933, 3583, fig.
species Dawydoff, 1952, 82 (Vietnam).
species Sopott-Ehlers & Schmidt, 1975, 215, fig. (Galapagos Is).

TRIADOMMA Marcus, 1947, 99, 107, 162 (type: *evelinae*); 1949, 70 (definition modified).
curvum Marcus, 1949, 7, 70, 103, figs (Brazil); du B.-R. Marcus, 1955*b*, 283.
evelinae Marcus, 1947, 99, 107, figs (Brazil); du B.-R. Marcus, 1955*b*, 283.

TRICELIS (Ehrenberg) Quatrefages, 1845, 131; Schmarda, 1859, 13.
fasciatus Quatrefages, 1845, 131, fig. (Sicily); Diesing, 1850, 189 (emended to *fasciata*); 1862, 225; Lang, 1884, 516 (s.o. *Cestoplana rubrocincta*); Kato, 1944, 293 (s.o. *C. rubrocincta*).

TRIGONOPORINAE Laidlaw, 1903*c*, 12 (subfam. of Stylochidae); Meixner, 1907*b*, 391.

TRIGONOPORUS Lang, 1884, viii, 429, 502 (type: *cephalophthalmus*); Carus, 1885, 152; Verrill, 1893, 486; Gamble, 1896, 19; Benham, 1901, 31; Bock, 1913, 68 (diagn.); Kato, 1944, 261 (syn. *Bergendalia*).
anomala (Laidlaw, 1903) Kato, 1944, 261
cephalophthalmus Lang, 1884, 154, 503, fig. (Italy); Carus, 1885, 152 (descr.); Stummer-Traunfels, 1933, 3571, fig.; Bresslau, 1928–33; 285, fig.
dendriticus Verrill, 1893, 491, figs (Massachusetts); Bock, 1925*a*, 19 (note); Pearse & Walker, 1939, 16; Hyman, 1940*a*, 54 (type-specimen re-examd.; s.o. *Discocelides ellipsoides*).
diversus (Yeri & Kaburaki, 1918*a*) Kato, 1944, 261.
folium (Verrill, 1873) Verrill, 1893, 487, figs (Massachusetts & Maine); Laidlaw, 1903*e*, 14 (systematic note); Sumner *et al.*, 1913, 579 (Massachusetts); Bock, 1925*a*, 19 (note); Pearse & Walker, 1939, 16, fig.; Hyman, 1940*a*, 454 (s.o. *Discocelides ellipsoides*).
microps Verrill, 1903, 45, fig. (Bermuda); Bock, 1913, 68; Hyman, 1939*d*, 14, figs (to *Cestoplana*).
mirabilis (Kato, 1938) Kato, 1944, 261.

TRIPYLOCELIS Haswell, 1907*b*, 466, 468 (type: *typica*); Bock, 1913, 221 (diagn.).
typica Haswell, 1907*b*, 466, figs. (N.S.W., Australia); Bock, 1913, 221; Dakin, *et al.*, 1948, 218 (N.S.W.); Pope, 1943, 246 (N.S.W.); Prudhoe, 1982*a*, 226 (S. Australia); 1982*b*, 370, fig. (S. Australia).

TURBELLARES Poche, 1926, 95 (Klasse of Platodes [=Platyhelminthes]).

TURBELLARIES DENDROCELES Claparède, 1861, 137.

TURBELLARIA DENDROCOELA Ehrenberg: Diesing, 1850, 179 (=Turbellariea); Stimpson, 1857, 19 (families & genera); Diesing, 1862, 487 (subord. of Turbellaria).

TYPHLOCOLAX Stimpson, 1857, 21; Diesing, 1862, 521 (s.o. *Typhlolepta*); Girard, 1893, 202 (s.o. *Typhlolepta*).
acuminatus Stimpson, 1857, 21, 26 (parasitic in *Chirodota* sp., Bering Strait); Diesing, 1862, 523 (to *Typhlolepta*); Lang, 1884, 612.
acutus (Girard, 1854) Stimpson, 1857, 21; Lang, 1884, 612 (*Typhlolepta*).
marinus (Örsted, 1843) Stimpson, 1857, 21.

TYPHLOLEPTA Örsted, 1843, 548 (type: *coeca*); 1844*a*, 46, 50; Diesing, 1850, 200 (diagn.); Stimpson, 1857, 21; Schmarda, 1859, 13, 16; Diesing, 1862, 492, 521; Girard, 1893, 202.
acuminata (Stimpson, 1857) Diesing, 1862, 523; Lang, 1884, 612 (*Typhlocolax*); Girard, 1893, 203 (descr.).
acuta Girard, *in* Stimpson, 1853, 27 (New Brunswick, Canada); Stimpson, 1857, 3 (to *Typhlocolax*); Diesing, 1862, 523; Lang, 1884, 611; Girard, 1893, 203 (descr.).
bilobata (Leuckart, 1828) Diesing, 1862, 522.
byerleyana Collingwood, 1876, 88, 92, fig. (off W. coast of Borneo) Lang, 1884, 616; Laidlaw, 1902, 291 (to *Pericelis*).

coeca Örsted, 1843, 570 (Scandinavia); 1844*a*, 50; 1844*b*, 79 (Oresund, Denmark/Sweden); Diesing, 1850, 200; Stimpson, 1857, 2 ('*caeca*'); Diesing, 1862, 521; Lang, 1884, 608; Bergendal, 1893*a*, 241 (? = *Cryptocelides loveni*).
extensa Le Conte, 1851, 319 (Panama); Stimpson, 1857, 21; Diesing, 1862, 522; Lang, 1884, 611.
marina (Örsted, 1843) Diesing, 1862, 523.
opaca Schmarda, 1859, 16, figs (Table Bay, S. Africa); Diesing, 1862, 522; Lang, 1884, 614; Stummer-Traunfels, 1933, 3485 (a triclad).
retusa (Viviani, 1805) Diesing, 1850, 200; 1862, 523; Schmarda, 1859, 16; Lang, 1884, 606 (*Planaria*); Carus, 1885, 158.
rubrocincta (Grube, 1840) Stimpson, 1857, 21.
stimpsoni Diesing, 1862, 522 (nom. nov. pro *Cryptocoelum opacum* Stimps.); Lang, 1884, 612 (s.o. *Cryptocoelum opacum*).
species Graff, 1883, 7 (Adriatic).
TYPHLOLEPTIDAE Stimpson, 1857, 21.
TYPHLOLEPTIDEA Diesing, 1862, 492, 521 (family name).
TYSANOZOON [err. pro *Thysanozoon*] Schmarda, 1859, 29; Graeffe, 1860, 53.
TYPHLOPLANA Ehrenberg, *in* Hemprich & Ehrenberg, 1931, sign. *a* 2.
marina Örsted, 1843, 565; Stimpson, 1857, 21 (to *Typhlolepta*); Diesing, 1862, 523 (*Typhlolepta*).

WOODWORTHIA Laidlaw, 1903*e*, 10 (diagn.); 1904*b*, 128 (type: *insignis*); Meixner, 1907*b*, 442; Barbour, 1912, 187 (renamed *Idioplanoides*); Bock, 1913, 142 (defined).
atlantica Bock, 1913, 142, figs (St Thomas I., W. Indies); 1925*a*, 23 (to *Idioplanoides*).
insignis Laidlaw, 1904*b*, 128, figs (Ceylon); Meixner, 1907*b*, 442, fig. (type-specimen re-examd.); Barbour, 1912, 187 (to *Idioplanoides*).

YUNGIA Lang, 1884, vii, 431, 548 (type: *aurantiaca*); Carus, 1885, 154; Gamble, 1896, 19; Benham, 1901, 31.
aurantiaca (Delle Chiaje, 1822) Lang, 1884, 148, 548, fig. (Naples; synonymy); Carus, 1885, 154 (descr.); Lo Bianco, 1888, 401 (Naples); Gourret, 1890, 324 (S. France); Pruvot, 1897, 21 (S. France); Lo Bianco, 1899*a*, 480 (Naples); Benham, 1901, 34, fig.; Sabussow, 1905, 489 ('*Jungia*'; S. France); Micoletzky, 1910, 179 (Trieste); Arndt, 1943, 2 (Naples); Pastorino & Canu, 1965, 1 (Ligurian Sea).
dicquemari (Risso, 1818) Lang, 1884, 550 (Mediterranean).
miniata (Schmarda, 1859) Lang, 1884, 551; Stummer-Traunfels, 1933, 3487 (to *Pseudoceros*).
rubrocincta (Schmarda, 1859) Lang, 1884, 550; Stummer-Traunfels, 1933, 3487 (to *Pseudoceros*).
sasakii Kaburaki, 1923*a*, 196, figs (Japan); Kato, 1937*e*, 131 (Japan); 1944, 302; Okada *et al.*, 1971, 62 (Japan).
teffi Dawydoff, 1952, 81 (not descr.; Vietnam).

ZYGANTROPLANA Laidlaw, 1906, 709, 711 (type: *verrilli*); Bock, 1913, 222 (defined); Marcus, 1949, 76 (systematic note); Corrêa, 1949, 173 (systematics & biology); Hyman, 1953*a*, 308 (defined).
angusta (Verrill, 1893) Corrêa, 1949, 175; Hyman, 1952, 196.
clepeasta Kato, 1944, 278, figs (Japan).
henriettae Corrêa, 1949, 173, 176, figs (nom. nov. pro *Stylochoplana angusta* of Marcus, 1947); Hyman, 1952, 196 (prob. a geographical variant of *S. angusta* (Verrill).); du B.-R. Marcus, 1955*b*, 283.
plesia Corrêa, 1949, 173, 200, figs (Brazil); du B.-R. Marcus, 1955*b*, 283.
stylifera Hyman, 1953*a*, 308, figs (Gulf of California – *Stylochoplana plehni* of Steinbeck & Ricketts, 1941).
verrilli Laidlaw, 1906, 709, figs (Cape Verde Is, W. Africa); Bock, 1913, 222.
yrsa du Bois-Reymond Marcus & Marcus, 1968, 2, 19, figs (various Caribbean areas).

Alphabetical list of specific and infraspecific names used in Polycladida

aberrans Bock, 1925*b*, 97 — *Ilyplana.*
aberrans Kato, 1944, 274 — *Stylochoplana.*
acticola Boone, 1929, 38 — *Leptoplana*; Hyman, 1953*a*, 321 (*Notoplana*).
actinotrocha Dawydoff, 1940, 460 — *Lobophora.*
acuminatus Stimpson, 1857, 21 — *Typhlocolax*; Diesing, 1862, 523 (*Typhlolepta*).
acuta Girard, *in* Stimpson, 1853, 27 — *Typhlolepta*; Stimpson, 1857, 21 (*Typhlocolax*).
acuta Stimpson, 1855*a*, 381 — *Leptoplana*; Stimpson, 1857, 21 (*Elasmodes*).
adhaerens Bock, 1925*a*, 3 — *Euprosthiostomum.*
aegypticus Melouk, 1940, 125 — *Cryptophallus.*
affine Stimpson, 1857, 22 — *Prosthiostomum*; Diesing, 1862, 539 (*Leptoplana*).
affinis Kelaart, *in* Collingwood, 1876, 87 — *Eurylepta*; Stummer-Traunfels, 1933, 3566 (*Pseudoceros*).
affinis Palombi, 1924*a*, 35 — *Polyposthides.*
affinis Palombi, 1940, 109 — *Stylochoplana.*
agilis Lang, 1884, 111 — *Stylochoplana.*
aino Kato, 1937*e*, 129 — *Pseudostylochus.*
akkashiensis Kato, 1937*e*, 124 — *Mirostylochus.*
alba Kelaart, 1858, 139 — *Penula*; Diesing, 1862, *in* Lang, 1884, 213 (*Leptoplana*).
alba Lang, *in* Schmidtlein, 1880, 172 — *Leptoplana*; Lang, 1884, 101 (*Cryptocelis*).
albatrossi Hyman, 1955*d*, 14 — *Crassiplana.*
albicornis Stimpson, 1857, 20 — *Proceros*; Lang, 1884, 564 (*Prostheceraeus*); Kato, 1944, 298 (*Pseudoceros*).
albocinctus Lang, 1884, 557 — *Prostheceraeus.*
albomarginatus Hyman, 1959*c*, 14 — *Pseudoceros.*
albonigra Dawydoff, 1940, 457 — *Lobophora.*
albopapillosus Hyman, 1959*a*, 583 — *Acanthozoon.*
albopunctatus Prudhoe, 1977, 600 — *Acanthozoon.*
albus Hallez, 1905*b*, 124 — *Stylochus*; Hallez, 1907, 1 (*Stylochoides*).
albus Freeman, 1933, 111 — *Oligocladus*; Hyman, 1953*a*, 378 (*Acerostisa*).
alcha du Bois-Reymond Marcus & Marcus, 1968, 33 — *Notoplanides*; [*supra* to *Stylochoplana*].
alcinoi Schmidt, 1861, 5 — *Leptoplana*; Bock, 1913, 187 (*Notoplana*).
alderi Collingwood, 1876, 87 — *Thysanozoon.*
alexandrinus Steinböck, 1937, 1 — *Stylochus.*
allmani Collingwood, 1876, 89 — *Thysanozoon.*
amakusaensis Kato, 1936*a*, 17 — *Cryptocelis.*
amara Kaburaki, 1918, 188 — *Shelfordia*; Bock, 1925*b*, 177 (*Limnostylochus*).
americana Hyman, 1940*a*, 473 — *Digynopora.*
amica Kato, 1937*a*, 211 — *Stylochoplana.*
amphibolus Schmarda, 1859, 34 — *Stylochus*; Lang, 1884, 44 (*Planocera*).
angusta Verrill, 1893, 485 — *Leptoplana*; Palombi, 1928, 590 (*Stylochoplana*); Corrêa, 1949, 175 (*Zygantroplana*).
angustum Bock, 1913, 282 — *Prosthiostomum.*
annandalei Kaburaki, 1918, 185 — *Shelfordia*; Bock, 1925*c*, 177 (*Limnostylochus*).
annula du Bois-Reymond Marcus & Marcus, 1968, 43 — *Notoplana.*
anomala Laidlaw, 1903*c*, 302 — *Bergendalia*; Kato, 1944, 261 (*Trigonoporus*).
anomalus Haswell, 1907*b*, 481 — *Prostheceraeus.*
antarcticum Hallez, 1905*b*, 126 — *Stylostomum.*
antillarum Hyman, 1955*b*, 116 — *Adenoplana*; du Bois-Reymond Marcus & Marcus, 1968, 62 (*Boninia*).
aomori Kato, 1937*d*, 39 — *Stylochus.*

arctica Plehn, 1896*a*, 140 — *Acelis*; Bock, 1913, 70 (*Plehnia*).
arctica Hyman, 1953*a*, 379 — *Acerotisa*.
arctum Quatrefages, 1845, 135 — *Prosthiostomum*; Diesing, 1850, 196 (*Leptoplana*).
arenicola Hallez, 1893, 150 — *Cryptocelis*.
arenosus Willey, 1897, 154 — *Stylochus*.
argus Quatrefages, 1845, 137 — *Proceros*; Diesing, 1850, 209 (*Eurylepta*); Schmarda, 1859, 30 (*Prostheceraeus*).
argus Czerniavsky, 1881, 221 — *Stylochus*.
argus Laidlaw, 1903*c*, 302 — *Latocestus*.
armata Kelaart, 1858, 135 — *Planaria*; Collingwood, 1876, 87 (*Acanthozoon*); Lang, 1884, 545 (*Pseudoceros*); du Bois-Reymond Marcus, 1955*a*, 42 (*Pseudoceros* (*Acanthozoon*).)
armata Laidlaw, 1902, 282 — *Planocera*.
asamusiensis Kato, 1939*b*, 141 — *Pseudoceros*.
asiaticum Kato, 1937*b*, 233 — *Prosthiostomum*.
astis Bock, 1913, 125 — *Parastylochus*.
asymmetrica Hyman, 1953*a*, 359 — *Monosolenia*.
atentaculatus Hyman, 1953*a*, 283 — *Stylochus*.
ater Hyman, 1959*a*, 571 — *Pseudoceros*.
atlantica Bock, 1913, 207 — *Notoplana*.
atlantica Bock, 1913, 142 — *Woodworthia*; Bock, 1925*b*, 177 (*Idioplanoides*).
atlantica Hyman, 1940*a*, 479 — *Planctoplanella*.
atlanticus Plehn, 1896*a*, 140 — *Latocestus*.
atomata Müller, 1776, 282 — *Planaria*; Örsted, 1843, 569 (*Leptoplana*); Bock, 1913, 195 (*Notoplana*).
atraviridis Kelaart, *in* Collingwood, 1876, 87 — *Eurylepta*; Hyman, 1959*a*, 565 (*Pseudoceros*).
atropurpureus Kato, 1834*b*, 123 — *Pseudoceros*.
aucklandica Cheeseman, 1883, 213 — *Thysanozoon*.
aulica Marcus, 1947, 99 — *Stylochoplana*.
aurantiaca Delle Chiaje, 1830, pl. 78 — *Planaria*; Örsted, 1844*a*, 47 (*Thysanozoon*); Lang, 1879, 459 (*Proceros*); Lang, 1884, 148 (*Yungia*).
aurantiaca Collingwood, 1876, 88 — *Leptoplana*; Laidlaw, 1903*c*, 302 (*Prosthiostomum*).
aurantiaca Heath & McGregor, 1913, 458 — *Eurylepta*.
auratum Kato, 1937*f*, 347 — *Prosthiostomum*.
aurea Kelaart, 1858, 137 — *Planaria*; Diesing, 1862, 562 (*Planocera*).
aureolineata Verrill, 1903, 42 — *Pseudoceros*.
aureus Hallez, 1911*b*, 141 — *Enterogonimus*.
auriculata Müller, 1788*b*, 37 — *Planaria*; Diesing, 1850, 211 (*Eurylepta*).
aurita Claparède, 1861, 144 — *Eurylepta*; Diesing, 1863, 175 (*Proceros*); Lang, 1884, 583 (*Oligocladus*).
auropunctatum Kelaart, *in* Collingwood, 1876, 87 — *Thysanozoon*.
aurora Laidlaw, 1903*d*, 102 — *Paraplanocera*.
australe Stimpson, 1855*b*, 389 — *Thysanozoon*.
australiensis Woodworth, 1898, 63 — *Idioplana*.
australiensis [err. pro *austalis*] Bock, 1925*b*, 143 — *Notoplana*.
australiensis Prudhoe, 1982*a*, 223 — *Ancoratheca*.
australis Schmarda, 1859, 21 — *Polycelis*; Diesing, 1862, 529 (*Leptoplana*); Stummer-Traunfels, 1933, 3486 (*Notoplana*).
australis Schmarda, forma *huina* Marcus, 1954*a*, 56 — *Notoplana*.
australis Laidlaw, 1904*a*, 3 — *Leptoplana*; Bock, 1913, 205 (*Notoplana*).
australis Haswell, 1907*b*, 479 — *Cestoplana*.
australis Hyman, 1959*c*, 1 — *Discocelis*.
australis Prudhoe, 1982*a*, 222 — *Cycloporus*.
awaensis Yeri & Kaburaki, 1918*a*, 432 — *Prosthiostomum*.

bacteroalba [err. pro *lactoalba*] *Palombi, 1939b*, 108 — *Leptoplana*.
badiomaculata Kato, 1934*b*, 123 — *Susakia*.
badius Stimpson, 1855*b*, 389 — *Dionchus*; Diesing, 1862, 528 (*Leptoplana*).
baeckstroemi Bock, 1923*c*, 342 — *Aceros*; Marcus, 1947, 141 (*Acerotisa*).

bahamensis Bock, 1913, 187 — *Notoplana.*
baiae Hyman, 1939*a*, 153 — *Acerotisa.*
bajae Hyman, 1953*a*, 365 — *Pseudoceros.*
bartschi Kaburaki, 1923*b*, 635 — *Cryptophallus.*
bayeri Hyman, 1959*a*, 578 — *Nymphozoon.*
bayeri du Bois-Reymond Marcus & Marcus, 1968, 28 — *Stylochoplana.*
bedfordii Laidlaw, 1903*c*, 302 — *Pseudoceros* [*bedfordi*, p. 314].
bella Bock, 1922, 1 — *Chromoplana.*
bellostriatus Hyman, 1953*a*, 373 — *Prostheceraeus.*
bellum Kato, 1939*a*, 65 — *Prosthiostomum.*
bellus Hyman, 1959*c*, 12 — *Pseudostylochus.*
bermudensis Verrill, 1903, 43 — *Stylochus.*
bicolor Verrill, 1903, 42 — *Pseudoceros.*
bilobata Leuckart, 1828, 11 — *Planaria*; Schmarda, 1859, 24 (*Centrostomum*); Diesing, 1862, 522 (*Typhlolepta*).
bimarginatum Meixner, 1907*a*, 169 — *Pseudoceros.*
binoculata Verrill, 1903, 43 — *Discocelis*; Hyman, 1939*c*, 3 (*Notoplana*).
bitentaculata [err. pro *bituberculata*] Leuckart, *in* Diesing, 1850, 215.
bituberculata Leuckart, 1828, 13 — *Planaria*; Schmarda, 1859, 33 (*Planocera*).
bituna Marcus, 1947, 100 — *Acerotisa.*
bocki de Beauchamp, 1951*b*, 239 — *Polyphalloplana.*
boehmigi Stummer-Traunfels, 1895, 699 — *Thysanozoon.*
borealis Dieising, 1862, 524 — *Diopis.*
borneensis Stummer-Traunfels, 1902, 160 — *Shelfordia*; Bock, 1923*c*, 346 (*Limnostylochus*),
brasiliensis Palombi, 1924*a*, 37 — *Euryleptides.*
brasiliensis Hyman, 1955*d*, 12 — *Latocestus.*
brochi Risso, 1918, 373 — *Tergipes*; Örsted, 1844*a*, 47 (*Thysanozoon*); Quatrefages, 1845, 140 (*Eolidiceros*).
brocchi Risso, 1826, 164 — *Planaria*; Örsted, 1844*a*, 47 (*Thysanozoon*).
brocchii Quatrefages, 1845, 140 — *Eolidiceros*; Lang, 1884, 525 (*Thysanozoon*).
brocchii var. *cruciatum* Laidlaw, 1906, 713 — *Thysanozoon.*
brocchii var. *nigrum* Lang, 1884, 535 — *Thysanozoon.*
brocchii var. *papillosum* Lang, 1884, 536 — *Thysanozoon.*
brocchii var. *tentaculatum* Lang, 1884, 536 — *Thysanozoon.*
brunnea Cheeseman, 1883, 214 — *Leptoplana.*
burchami Heath & McGregor, 1913, 457 — *Planocera*; Hyman, 1953*a*, 358 (*Pseudostylochus*). [See p. 114]
buskii Collingwood, 1876, 87 — *Proceros*; Lang, 1884, 547 (*Pseudoceros*).
byerleyana Collingwood, 1876, 88 — *Typhlolepta*; Laidlaw, 1902, 291 (*Pericelis*).

caeca Hyman, 1953*a*, 277 — *Plehnia.*
caeca var *oculifera* Hyman, 1953*a*, 279 — *Plehnia.*
caecus Plehn, 1897, 90 — *Polyporus.*
caelata Sopott-Ehlers & Schmidt, 1975, 208 — *Mucroplana.*
caeruleocinctus Hyman, 1959*a*, 569 — *Pseudoceros.*
caeruleopunctatus Palombi, 1928, 579 — *Pseudoceros.*
caledonica Jacubowa, 1906, 141 — *Mesocela.*
californica Woodworth, 1894, 50 — *Stylochoplana*; Hyman, 1953*a*, 314 (*Parviplana*).
californica Plehn, 1897, 93 — *Leptoplana.*
californica Heath & McGregor, 1913, 455 — *Planocera*; Hyman, 1953*a*, 346 (*Alloioplana*).
californica Hyman, 1953*a*, 344 — *Hoploplana.*
californica Hyman, 1953*a*, 379 — *Acerotisa.*
californica Hyman, 1959*b*, 11 — *Eurylepta.*
californicum Heath & McGregor, 1913, 455 — *Stylostomum.*
californicum Hyman, 1953*a*, 362 — *Thysanozoon.*
californicus Hyman, 1953*a*, 285 — *Stylochus.*

callizona Marcus, 1947, 99 — *Alleena.*
canadensis Hyman, 1953*a*, 368 — *Pseudoceros.*
capense Bock, 1931, 261 — *Prosthiostomum.*
capensis Schmarda, 1859, 22 — *Polycelis;* Diesing, 1862, 530 (*Leptoplana*).
capensis Palombi, 1938, 329 — *Leptostylochus.*
caraibica Palombi, 1924*a*, 35 — *Polyposthides.*
caraibica Palombi, 1924*a*, 36 — *Stylochoplana.*
cardinalis Haswell, 1907*b*, 480 — *Pseudoceros.*
cardiosora Schmarda, 1859, 28 — *Eurylepta;* Diesing, 1862, 522 (*Proceros*); Lang, 1884, 546 (*Pseudoceros*).
caribbeana Hyman, 1939*c*, 2 — *Notoplana.*
caribbeanus Prudhoe, 1944, 322 — *Latocestus.*
carolinensis Hyman, 1939*e*, 153 — *Euplana.*
castaneus Palombi, 1939*b*, 95 — *Stylochus.*
cata du Bois-Reymond Marcus & Marcus, 1968, 59 — *Pericelis.*
catus du Bois-Reymond Marcus, 1958, 401 — *Stylochus.*
cavicola Heath & McGregor, 1913, 458 — *Euryleptodes.*
caymanensis Prudhoe, 1944, 326 — *Paraboninia.*
celeris Freeman, 1933, 111 — *Notoplana.*
celerrima Haswell, 1907*b*, 475 — *Echinoplana.*
cephalophthalmus Lang, 1884, 154 — *Trigonoporus.*
ceratommata Palombi, 1936, 1 — *Planocerodes.*
cerebralis Kelaart, 1858, 135 — *Planaria;* Collingwood, 1876, 87 (*Eurylepta*); Lang, 1884, 546 (*Pseudoceros*).
ceylanica Laidlaw, 1902, 302 — *Cestoplana.*
ceylanicus Laidlaw, 1904*b*, 130 — *Stylochus.*
challengeri Graff, 1892*b*, 190 — *Planctoplana;* Prudhoe, 1950*b*, 714 (*Stylochoplana*).
chierchiae Plehn, 1896*a*, 155 — *Leptoplana;* Bock, 1913, 211 (*Notoplana*).
chilensis Schmarda, 1859, 17 — *Leptoplana;* Stummer-Traunfels, 1933, 3486 (*Stylochoplana*).
chloranota Boone, 1929, 43 — *Phylloplana;* Hyman, 1953*a*, 310 (*Leptoplana*); du Bois-Reymond Marcus & Marcus, 1968, 23 (*Stylochoplana*).
chloreus Marcus, 1949, 7 — *Pesudoceros.*
cinereus Willey, 1897, 154 — *Stylochus.*
cinereus Palombi, 1931*a*, 3 — *Pseudoceros.*
clara Kato, 1937*f*, 347 — *Stylochoplana.*
clavicornis Schmarda, 1859, 32 — *Prostheceraeus;* Stummer-Traunfels, 1933, 3487 (*Pseudoceros*).
clepeasta Kato, 1944, 278 — *Zygantroplana.*
clippertoni Hyman, 1939*c*, 4 — *Euplana.*
coccinea Stimpson, 1857, 20 — *Eurylepta;* Kato, 1944, 298 (*Pseudoceros*).
coeca Örsted, 1843, 570 — *Typhlolepta;* Bergendal, 1893*a*, 241 (*Cryptocelides*).
colemani Prudhoe, 1977, 598 — *Pseudoceros.*
collaris Stimpson, 1855*a*, 381 — *Leptoplana;* Stimpson, 1857, 22 (*Prosthiostomum*).
collingwoodii Laidlaw, 1903*c*, 302 — *Pseudoceros.*
collingwoodi Laidlaw, 1903*c*, 314 — *Pseudoceros.*
colobocentroti Bock, 1925*a*, 4 — *Ceratoplana.*
colobocentroti var. *hawaiiensis* Bock, 1925*a*, 6 — *Ceratoplana.*
compacta Lang, 1884, 249 — *Cryptocelis.*
concinnus Collingwood, 1876, 87 — *Proceros;* Kaburaki, 1923*b*, 635 (*Pseudoceros*).
concolor Meixner, 1907*a*, 167 — *Leptoplana;* Bock, 1913, 221 (*Discoplana*); Hyman, 1954*b*, 333 (*Euplana*).
conglomeratus Stimpson, 1857, 22 — *Stylochopsis;* Diesing, 1862, 570 (*Stylochus*).
conoceraea Schmards, 1859, 35 — *Imogine;* Diesing, 1862, 568 (*Stylochus*); Lang, 1884, 446 (*Conoceros*); Stummer-Traunfels, 1933, 3487 (*Stylochoplana*).
constipatum Stimpson, 1857, 22 — *Prosthiostomum;* Diesing, 1862, 537 (*Leptoplana*).
cooperi Laidlaw, 1902, 301 — *Prosthiostomum.*
corallicola Woodworth, 1898, 64 — *Diposthus.*
corallophilus Hyman, 1954*a*, 223 — *Pseudoceros.*

corniculata Dalyell, 1853, 101 — *Planaria*; Leuckart, 1859, 183 (*Planocera*); Diesing, 1862, 571 (*Stylochus*).
corniculatus Stimpson, 1855*a*, 381 — *Stylochus.*
cornuta Müller, 1776, 221 — *Planaria*; Ehrenberg, 1831, 56 (*Eurylepta*); Schmarda, 1859, 30 (*Prostheceraeus*).
cornuta var. *melobesiarum* Lang, 1884, 73 — *Eurylepta.*
cornuta var. *wandeli* Hallez, 1907, 2 — *Eurylepta.*
coseirensis Bock, 1925*b*, 123 — *Stylochus.*
cotylifera Meixner, 1907*a*, 167 — *Notoplana.*
crassiusculum Stimpson, 1857, 22 — *Prosthiostomum*; Diesing, 1862, 537 (*Leptoplana*).
crassus Verrill, 1893, 466 — *Stylochus.*
crassus Bock, 1931, 262 — *Stylochus.*
cribrarium Stimpson, 1857, 22 — *Prosthiostomum*; Diesing, 1862, 537 (*Leptoplana*).
cristatus Quatrefages, 1845, 139 — *Proceros*; Diesing, 1850, 210 (*Eurylepta*). Schmarda, 1859, 14 (*Prostheceraeus*).
crosslandi Laidlaw, 1903*d*, 100 — *Planocera.*
crozieri Hyman, 1939*d*, 17 — *Pseudoceros.*
cruciatum Schmarda, 1859, 30 — *Thysanozoon* (*Eolidiceros*).
cuneata Sopott-Ehlers & Schmidt, 1975, 210 — *Cestoplana.*
cuneiformis Prudhoe, 1982*b*, 362 — *Candimboides.*
cupida Kato, 1938*b*, 577 — *Hoploplana.*
curvum Marcus, 1949, 7 — *Triadomma.*
cyclops Verrill, 1903, 44 — *Discocelis*; Hyman, 1939*d*, 19 (*Prosthiostomum*).
cynarium Marcus, 1950, 5 — *Prosthiostomum.*

dalyellii Johnston, 1865, 7 — *Eurylepta.*
deanna Kato, 1939*b*, 141 — *Hoploplana.*
deilogyna Hyman, 1959*a*, 562 — *Asolenia.*
delicata Plehn, 1896*a*, 140 — *Allioioplana.*
delicata Jacubowa, 1906, 135 — *Leptocera.*
delicata Yeri & Kaburaki, 1918*a*, 432 — *Notoplana.*
delicatula Stimpson, 1857, 22 — *Leptoplana.*
delicatum Palombi, 1939*a*, 123– *Prosthiostomum*; de Bois-Reymond Marcus & Marcus, 1968, 88 (*Lurymare*).
delmaris Hyman, 1953*a*, 337 — *Macginitiella.*
dendriticus Verrill, 1893, 491 — *Trigonoporus.*
devisii Woodworth, 1898, 63 — *Pseudoceros.*
diaphana Stummer-Traunfels, 1933, 3531 — *Leptoplana*; du Bois-Reymond Marcus & Marcus, 1968, 24 (*Stylochoplana*).
dicquemari Risso, 1818, 373 — *Tergipes*; Risso, 1826, 263 (*Planaria*); Örsted, 1844*a*, 47 (*Thysanozoon*); Lang, 1884, 550 (*Yungia*).
dicquemaris var. *verrucosa* Delle Chiaje, 1841, 132 — *Planaria.*
dictyotus Schmarda, 1859, 33 — *Stylochus*; Lang, 1884, 443 (*Planocera*).
diesingii Grube, 1840, 54 — *Thysanozoon.*
dimidiatus Graff, *in* Saville-Kent, 1893, 362 — *Pseudoceros.*
discoidea Willey, 1897, 156 — *Planocera.*
discoideum Schmarda, 1859, 29 — *Thysanozoon.*
discus Le Conte, 1851, 319 — *Elasmodes*; Diesing, 1862, 527 (*Leptoplana*).
discus Willey, 1897, 155 — *Planocera*; Bock, 1913, 246 (*Paraplanocera*).
distincta Prudhoe, 1982*b*, 367 — *Notoplana.*
distinctum Stummer-Traunfels, 1895, 692 — *Thysanozoon.*
divae Marcus, 1947, 99 — *Stylochoplana.*
divae Marcus, 1948, 111 — *Notoplana.*
divae Marcus, 1949, 7 — *Pentaplana.*
divae Marcus, 1949, 7 — *Candimba.*
divae Marcus, 1950, 5 — *Hoploplana.*
divae du Bois-Reymond Marcus & Marcus, 1968, 64 — *Boninia.*

diversa Yeri & Kaburaki, 1918*a*, 431 — *Bergendalia*; Kato, 1944, 261 (*Trigonoporus*).
djiboutiensis Meixner, 1907*a*, 165 — *Stylochus*.
dominicanus Hyman, 1955*b*, 132 — *Crassandros*.
dohrnii Lang, 1884, 274 — *Prosthiostomum*.
droebachensis Örsted, 1845, 415 — *Leptoplana*.
drygalskii Bock, 1931, 261 — *Prosthiostomum*; du Bois-Reymond, Marcus & Marcus, 1968, 88 (*Lurymare*).
dubia Blainville, 1826, 218 — *Planaria*.
dubia Laidlaw, 1903*d*, 103 — *Disparoplana*.
dubium Schmarda, 1859, 25 — *Centrostomum*; Lang, 1884, 499 (*Leptoplana*); Stummer-Traunfels, 1933, 3487 (*Notoplana*).
dulcis Kelaart, 1858, 137 — *Planaria*; Collingwood, 1876, 87 (*Eurylepta*); Hyman, 1959*a*, 565 (*Pseudoceros*).

edmondsi Prudhoe, 1982*b*, 373 — *Planocera*.
edurus Kato, 1938*b*, 577 — *Pseudostylochus*.
electrina Pennant, 1777, 43 — *Doris*.
elegans, Kelaart, 1858, 136 — *Planaria*; Diesing, 1862, 562 (*Planocera*); Collingwood, 1876, 88 (*Stylochoplana*).
elegans Laidlaw, 1902, 298 — *Prosthiostomum*.
elioti Laidlaw, 1903*d*, 109 — *Haploplana*.
ellipsis Dalyell, 1853, 101 — *Planaria*; Leuckart, 1859, 183 (*Polycelis*); Diesing, 1862, 542 (*Leptoplana*); Lang, 1884, 588 (*Stylostomum*).
ellipsoides Girard, *in* Stimpson, 1853, 27 — *Leptoplana*; Hyman, 1939*e*, 153 (*Discocelides*); Hyman, 1952, 195 (*Plehnia*).
elliptica Girard, 1850*a*, 398 — *Planocera*; Verrill, 1893, 467 (*Eustylochus*); Kingsley, 1901, 165 (*Stylochus*).
elongata Laidlaw, 1906, 714 — *Traunfelsia*.
elongata Yeri & Kaburaki, 1918*a*, 432 — *Neoplanocera*.
elongatum Quatrefages, 1845, 136 — *Prosthiostomum*; Diesing, 1850, 196 (*Leptoplana*).
elongatus Bock, 1925*b*, 98 — *Leptostylochus*.
elongatus Kato, 1937*a*, 211 — *Pseudostylochus*.
emarginatum Leuckart, 1863, 169 — *Prosthiostomum*.
equiheni Hallez, 1888, 104 — *Cryptocelis*.
ernesti Hyman, 1953*a*, 296 — *Marcusia*; Hyman, 1955*e*, 263 (*Pericelis*).
erythrotaenia Schmarda, 1859, 21 — *Polycelis*; Diesing, 1862, 529 (*Leptoplana*); Stummer-Traunfels, 1933, 3486 (*Notoplana*).
euscopa Marcus, 1952, 5 — *Nonatona*.
evansi [pro *evansii*] Bock, 1913, 187 — *Notoplana*.
evansii Laidlaw, 1903*c*, 301 — *Notoplana*.
evelinae Marcus, 1947, 99 — *Triadomma*.
evelinae Marcus, 1947, 99 — *Pucelis*.
evelinae Marcus, 1949, 7 — *Theama*.
evelinae Marcus, 1949, 7 — *Enchiridium*.
evelinae Marcus, 1950, 5 — *Adenoplana*.
evelinae Marcus, 1950, 5 — *Pseudoceros*.
evelinae Marcus, 1952, 5 — *Stylochoplana*.
evelinae Marcus, 1954*b*, 476 — *Callioplana*.
excelsa Bock, 1925*b*, 98 — *Kaburakia*.
exiguum Hyman, 1959*a*, 587 — *Prosthiostomum*.
exiguus Hyman, 1953*a*, 287 — *Stylochus*.
eximius Kato, 1937*f*, 347 — *Cryptophallus*.
exoptatus Kato, 1938*b*, 577 — *Pseudoceros*.
extensa Le Conte, 1851, 319 — *Typhlolepta*.

fallax Quatrefages, 1845, 135 — *Polycelis*; Diesing, 1850, 198 (*Leptoplana*); Bock, 1913, 204 (*Notoplana*).

faraglionensis Lang, 1884, 219 — *Cestoplana.*
fasciatus Quatrefages, 1845, 131 — *Tricelis.*
fasciatus Schmarda, 1859, 33 — *Stylochus*; Lang, 1884, 462 (*Stylochoplana*); Stummer-Traunfels, 1933, 3487 (*Styloplanocera*).
favis Sopott-Ehlers & Schmidt, 1975, 195 — *Amyris.*
felinum Marcus, 1954*a*, 70 — *Stylostomum.*
ferox Bock, 1925*b*, 175 — *Stylochus.*
ferruginea Schmarda, 1859, 22 — *Polycelis*; Diesing, 1862, 530 (*Leptoplana*); Stummer-Traunfels, 1933, 3486 (*Notoplana*).
ferrugineus Hyman, 1959*a*, 571 — *Pseudoceros.*
filiformis Plehn, 1896*a*, 140 — *Diplopharyngeata.*
filiformis Laidlaw, 1903*d*, 110 — *Cestoplana.*
flava Delle Chiaje, 1822, pl. 36, fig. 11 — *Planaria*; Örsted, 1844*a*, 47 (*Thysanozoon*).
flavofusca Dawydoff, 1940, 465 — *Lobophora.*
flavomaculatus Graff, *in* Saville-Kent, 1893, 362 — *Prostheceraeus*; Graff, 1904–05, 1888 (*Pseudoceros*) [?typographical error].
flavomarginata Ehrenberg, 1831, 56 — *Eurylepta*; Lang, 1884, 563 (*Prostheceraeus*).
flavomarginatus — Laidlaw, 1902 ,298 — *Pseudoceros.*
flavotuberculatum Hyman, 1939*d*, 16 — *Thysanozoon.*
flevensis Hofker, 1930, 206 — *Stylochus.*
flexilis Dalyell, 1814, 5 — *Planaria*; Diesing, 1850, 194 (*Leptoplana*); Stimpson, 1857, 21 (*Elasmodes*).
floridana Pearse, 1938, 77 — *Stylochoplana*; Hyman, 1940*a*, 478 (*Gnesioceros*).
floridanus Pearse, 1938, 71 — *Stylochus.*
floridanus Pearse, 1938, 88 — *Oligoclado.*
floridanus Hyman, 1955*b*, 138 — *Prostheceraeus.*
fockei Diesing, 1850, 213 — *Thysanozoon.*
folium Grube, 1840, 51 — *Stylochus*; Örsted, 1844*a*, 48 (*Planocera*); Stimpson, 1857, 22 (*Stylochoplana*).
folium Verrill, 1873, 487 — *Leptoplana*; Verrill, 1893, 487 (*Trigonoporus*).
formosa Darwin, 1844, 247 — *Planaria*; Diesing, 1850, 199 (*Leptoplana*).
formosum Kato, 1943*b*, 69 — *Prosthiostomum.*
franciscanus Hyman, 1953*a*, 280 — *Stylochus.*
frigidum Bock, 1931, 261 — *Stylostomum.*
fritillata Hyman, 1959*a*, 557 — *Paraplanocera.*
frontalis Verrill, 1893, 465 — *Stylochus.*
fulminata Stimpson, 1855*a*, 380 — *Eurylepta*; Kato, 1944, 298 (*Pseudoceros*).
fulva Kelaart, 1858, 139 — *Penula*; Lang, 1884, 298 (*Pseudoceros*).
fulva Kato, 1944, 260 — *Discocelis.*
fulvogriseus Hyman, 1959*a*, 573 — *Pseudoceros.*
fulvolimbata Grube, 1868, 46 — *Eurylepta.*
fulvopunctatus Yeri & Kaburaki, 1920, 591 — *Neostylochus.*
fulvum Giard, 1888, 496 — *Stylostomum.*
fulvus Yeri & Kaburaki, 1918*a*, 432 — *Pseudostylochus.*
furva Bock, 1913, 112 — *Meixneria.*
fusca Stimpson, 1857, 22 — *Leptoplana.*
fusca Kelaart, 1858, 136 — *Planaria*; Collingwood, 1876, 87 (*Eurylepta*); Hyman, 1959*a*, 565 (*Pseudoceros*).
fuscogriseus Hyman, 1959*a*, 575 — *Pseudoceros.*
fuscopunctatus Prudhoe, 1977, 586 — *Pseudoceros.*
fuscoviridis Kato, 1934*b*, 123 — *Pseudostylochus.*

gabriellae Marcus, 1949, 7 — *Prosthiostomum*; du Bois-Reymond Marcus & Marcus, 1968, 89 (*Lurymare*).
gabriellae Marcus, 1950, 5 — *Cycloporus.*
gaimardi Blainville, 1828, 578 — *Planocera*; Ehrenberg, 1831, 54 (*Planoceros*); Schmarda, 1859, 33 (*Stylochus*).
galapagensis Hyman, 1953*b*, 183 — *Latocestus.*

gamblei Laidlaw, 1902, 297 — *Pseudoceros*; Marcus, 1950, 85 (*Dicteros*).
gardineri Laidlaw, 1902, 296 — *Pseudoceros*.
gardineri Laidlaw, 1904*b*, 134 — *Leptoplana*; Bock, 1913, 211 (*Notoplana*).
gargantua Dawydoff, 1940, 461 — *Lobophora*.
gaymardi [err. pro *gaimardi*] Ehrenberg, 1831, 54 — *Planoceros*.
genicotyla Palombi, 1939*b*, 96 — *Stylochoplana*.
giesbrechtii Lang, 1884, 558 — *Prostheceraeus*.
gigantea Dawydoff, 1940, 461 — *Lobophora*.
gigas Leuckart, 1928, 11 — *Planaria*.
gigas Schmarda, 1859, 17 — *Leptoplana*; Diesing, 1862. 544 (*Centrostomum*); Stummer-Traunfels, 1933, 3486 (*Discoplana*); Hyman, 1955*c*, 76 (*Euplana*).
gilchristi Jacubowa, 1906, 153 — *Planocera*.
gilvum Marcus, 1950, 5 — *Prosthiostomum*.
glandulata Jacubowa, 1909, 2 — *Cryptocelis*.
gloriosa Kato, 1938*a*, 559 — *Kaburakia*.
gracile Girard, 1850*b*, 251 — *Prosthiostomum*; Stimpson, 1857, 21 (*Elasmodes*); Diesing, 1862, 541 (*Leptoplana*); Girard, 1893, 198 (*Euplana*).
gracilis Laidlaw, 1904*b*, 131 — *Stylochocestus*.
gracilis Heath & McGregor, 1913, 463 — *Stylochoplana*.
gracilis Kato, 1934*a*, 374 — *Leptostylochus*.
graffi [pro *graffii* Laidlaw] Bock, 1913, 172 — *Stylochoplana*.
graffi Heath & McGregor, 1913, 458 — *Anciliplana*.
graffii Lang, 1879, 470 — *Planocera*.
graffii Laidlaw, 1906, 708 — *Leptoplana*; Bock, 1913, 172 (*Stylochoplana*).
grande Stimpson, 1857, 22 — *Prosthiostomum*; Diesing, 1862, 539 (*Leptoplana*).
gratus Kato, 1937*a*, 211 — *Pseudoceros*.
gravieri Meixner, 1907*a*, 170 — *Pseudoceros*.
grisea Pearse, 1938, 67 — *Discocelis*.
griseum Verrill, 1903, 41 — *Discocelis*.
griseum Hyman, 1959*a*, 592 — *Prosthiostomum*.
griseus Hyman, 1959*b*, 10 — *Pseudoceros*.
grubei Graff, 1892*b*, 190 — *Planocera*; Bock, 1913, 225 (*Hoploplana*).
guttatomarginatus Stimpson, 1855*a*, 380 — *Eurylepta*; Kato, 1944, 298 (*Pseudoceros*).

habroptilus Hyman, 1959*a*, 569 — *Pseudoceros*.
haddoni Laidlaw, 1903*a*, 3 — *Pseudoceros*.
haloglena Schmarda, 1859, 21 — *Polycelis*; Diesing, 1862, 528 (*Leptoplana*); Stummer-Traunfels, 1933, 3486 (*Notoplana*).
hamanensis Kato, 1944, 263 — *Stylochus*.
hamatum Schmidt, 1861, 9 — *Prosthiostomum*; Diesing, 1862, 538 (*Leptoplana*).
hancockanus Collingwood, 1876, 87 — *Proceros*; Lang, 1884, 567 (*Prostheceraeus*); Kaburaki, 1923*b*, 635 (*Pseudoceros*).
hancocki Hyman, 1953*a*, 307 — *Stylochoplana*.
hancockianus [pro *honcockanus*] Laidlaw, 1903*c*, 301 (*Proceros*); Laidlaw, 1903*c*, 302 (*Pseudoceros*).
hawaiiensis Heath, 1907, 145 — *Planocera*.
hawaiiensis Hyman, 1960, 308 — *Thysanozoon*.
heathi Boone, 1929, 35 — *Stylochoplana*.
heda Kato, 1944, 285 — *Planocera*.
henrietta Corrêa, 1949, 173 — *Zygantroplana*
herberti Kirk, 1882, 267 — *Eurylepta*.
herdmani Laidlaw, 1904*b*, 132 — *Thalamoplana*; Bock, 1913, 61 (*Discocelis*).
heteroglenus Schmarda, 1859, 34 — *Stylochus*; Lang, 1884, 444 (*Planocera*).
hewatti Hyman, 1955*b*, 123 — *Indistylochus*.
hispidus du Bois-Reymond Marcus & Marcus, 1955*a*, 39 — *Pseudoceros* (*Acanthozoon*); Hyman, 1959*a*, 586 (*Acanthozoon*).
horrida Sopott-Ehlers & Schmidt, 1975, 204 — *Euplanina*.
hozawai Kato, 1939*b*, 150 — *Stylostomum*.

humilis Stimpson, 1857, 22 — *Leptoplana;* Yeri & Kaburaki, 1918*a*, 432 (*Notoplana*).
hummelincki du Bois-Reymond Marcus & Marcus, 1968, 53 — *Amyris.*
huttoni Kirk, 1882, 267 — *Thysanozoon.*
hyalina Ehrenberg, 1831, 56 — *Leptoplana.*
hyalinus Bock, 1913, 51 — *Stylochus.*
hymanae Marcus, 1947, 99 — *Euplana.*
hymanae Poulter, 1974, 93 — *Pericelis.*

igiliensis Galleni, 1974, 395 — *Notoplana.*
ijimai Yeri & Kaburaki, 1918*a*, 431 — *Stylochus.*
ijimai Bock, 1923*a*, 15 — *Cryptocelis.*
inarmata Bock, 1931, 261 — *Notoplanella.*
incisa Darwin, 1844, 248 — *Planaria;* Diesing, 1850, 200 (*Centrostomum*).
inconspicua Gray, *in* Pease, 1860, 37 — *Peasia;* Diesing, 1862, 536 (*Leptoplana*).
inconspicuus Lang, 1884, 589 — *Aceros;* Marcus, 1947, 141 (*Acerotisa*).
indica Plehn, 1896*a*, 162 — *Thysanoplana.*
inequalis Hyman, 1953*a*, 350 — *Spinicirrus.*
inimicus Palombi, 1931*b*, 219 — *Stylochus.*
inquieta Heath & McGregor, 1913, 456 — *Leptoplana;* Stummer-Traunfels, 1933, 3504 (*Discoplana*); Marcus, 1947, 131 (*Euplana*); Hyman, 1953*a*, 319 (*Notoplana*).
inquieta Freeman, 1933, 111 — *Notoplana.*
inquilina Wheeler, 1894, 195 — *Planocera;* Bock, 1913, 228 (*Hoploplana*).
inquilina subsp. *thaisana* (Pearse) Hyman, 1940*a*, 477 — *Hoploplana.*
inquilina Hyman, 1950, 55 — *Stylochoplana.*
inquilina Hyman, 1955*a*, 1 — *Notoplana.*
insignis Lang, 1884, 236 — *Planocera;* Laidlaw, 1902, 303 (*Hoploplana*).
insignis Laidlaw, 1904*b*, 128 — *Woodworthia;* Barbour, 1912, 187 (*Idioplanoides*).
insignis Palombi, 1938, 329 — *Pulchriplana.*
insolitus Hyman, 1953*a*, 289 — *Stylochus.*
insularis Hyman, 1939*c*, 1 — *Notoplana.*
insularis Hyman, 1944*a*, 7 — *Comprostatum.*
insularis Hyman, 1953*a*, 375 — *Euryleptodes.*
insularis Hyman, 1953*b*, 186 — *Cryptocelis.*
insularis Hyman, 1955*c*, 66 — *Discocelis.*
intermedius Kato, 1939*b*, 141 — *Pseudostylochus.*
interrupta Stimpson, 1855*a*, 380 — *Eurylepta;* Kato, 1944, 298 (*Pseudoceros*).
inversiporus Minot, 1877*a*, 451 — *Mesodiscus.*
irrorata Gray, *in* Pease, 1860, 38 — *Peasia;* Diesing, 1862, 536 (*Leptoplana*).
isifer du Bois-Reymond Marcus, 1955*a*, 37 — *Stylochus.*
izuensis Kato, 1944, 265 — *Stylochus.*
izuensis Kato, 1944, 301 — *Pseudoceros.*

jaltense Czerniavsky, 1881, 220 — *Centrostomum;* Lang, 1884, 500 (*Leptoplana*).
japonica Stimpson, 1857, 20 — *Eurylepta;* Lang, 1884, 565 (*Prostheceraeus*); Kato, 1944, 298 (*Pseudoceros*).
japonica Yeri & Kaburaki, 1918*a*, 431 — *Discocelis.*
japonica Bock, 1923*a*, 1 — *Plehnia;* Hyman, 1953*a*, 279 (*Paraplehnia*).
japonica Kato, 1937*a*, 211 — *Notoplana.*
japonicum Kato, 1943*b*, 69 — *Enchiridium.*
japonicum Kato, 1944, 297 — *Thysanozoon.*
japonicus Kato, 1944, 267 — *Cryptophallus.*
japonicus Kato, 1944, 305 — *Cycloporus.*
johnstoni Haswell, 1907*b*, 469 — *Diplosolenia.*

karimatensis Palombi, 1924*a*, 34 — *Polyposthides.*
katoi Poulter, 1975, 323 — *Prosthiostomum (Lurymare).*
kelaartii Collingwood, 1876, 87 — *Eurylepta;* Lang, 1884, 568 (*Prostheceraeus*); Laidlaw, 1903*c*, 302 (*Pseudoceros*).

kentii Graff, *in* Saville-Kent, 1893, 362 — *Pseudoceros.*
komaii Kato, 1944, 311 — *Prosthiostomum.*
koreana Kato, 1937*b*, 233 — *Notoplana.*
kuekenthalii Plehn, 1896*a*, 149 — *Leptoplana*; Bock, 1913, 202 (*Notoplana*).

lactea Stimpson, 1857, 22 — *Pachyplana*; Diesing, 1862, 531 (*Leptoplana*); Lang, 1884, 470 (*Discocelis*).
lactea [?err. pro alba] Lang, 1884, 473 — *Cryptocelis.*
lactea Laidlaw, 1903*d*, 107 — *Phylloplana.*
lactea Kato, 1937*a*, 211 — *Cestoplana.*
lacteoalba [pro *lactoalba*] Laidlaw, 1903*a*, 308 — *Leptoplana*; Bock, 1913, 173 (*Stylochoplana*).
lacteus Collingwood, 1876, 87 — *Sphyngiceps*; Lang, 1884, 548 (*Pseudoceros*).
lactoalba Verrill, 1900, 595 — *Leptoplana*; Hyman, 1939*d*, 6 (*Notoplana*).
lactoalba var. *tincta* Verrill, 1903, 46 — *Leptoplana*; Bock, 1913, 179 (*Stylochoplana*).
laetum Kato, 1938*b*, 578 — *Prosthiostomum.*
laevigata [pro *levigatus* Quatrefages] Diesing, 1850, 198 — *Leptoplana.*
lagidium Marcus, 1949, 7 — *Thysanozoon.*
laidlawi Jacubowa, 1906, 115 — *Paraplanocera.*
lanceolata Schmarda, 1859, 19 — *Leptoplana.*
langi Bergendal, 1893*a*, 241 — *Discocelides.*
langi Stummer-Traunfels, 1895, 700 — *Thysanozoon.*
langi Heath & McGregor, 1913, 458 — *Aceros*; Marcus, 1947, 141 (*Acerotisa*).
langi [pro *langii* Laidlaw] Bock, 1913, 246 — *Paraplanocera.*
langii Laidlaw, 1902, 286 — *Planocera*; Laidlaw, 1903*a*, 3 (*Paraplanocera*).
lapunda du Bois-Reymond Marcus & Marcus, 1968, 46 — *Notoplana.*
lateralis [err. pro *littoralis* Verrill] Girard, 1893, 194 — *Stylochopsis.*
latissimus Schmarda, 1859, 31 — *Prostheceraeus*; Stummer-Traunfels, 1933, 3487 (*Pseudoceros*).
latocelis Hyman, 1953*a*, 382 — *Prosthiostomum.*
lentum Heath & McGregor, 1913, 476 — *Stylostomum.*
leoparda Freeman, 1933, 111 — *Eurylepta.*
lepida Heath & McGregor, 1913, 458 — *Licheniplana* or *Lichenoplana*; du Bois-Reymond Marcus, 1955*a*, 42 (*Pseudoceros* (*Acanthozoon*)).
leptalea Marcus, 1947, 99 — *Stylochoplana.*
leptostictus Bock, 1913, 256 — *Pseudoceros.*
leuca Marcus, 1947, 100 — *Acerotisa.*
levigatus Quatrefages, 1845, 134 — *Polycelis*; Diesing, 1850, 198 (*Leptoplana*).
levis Hyman, 1953*a*, 293 — *Mexistylochus*; Hyman, 1955*a*, 9 (*Ommatoplana*).
libera Kato, 1939*a*, 65 — *Notoplana.*
lichenoides Mertens, 1832, 4 — *Planaria*; Ehrenberg, 1936, 67 (*Discocelis*); Örsted, 1844*a*, 49 (*Leptoplana*); Diesing, 1850, 199 (*Centrostomum*).
lilianae du Bois-Reymond Marcus & Marcus, 1968, 13 — *Cryptocelis.*
limbata Leuckart, 1828, 11 — *Planaria*; Örsted, 1844*a*, 50 (*Eurylepta*); Diesing, 1862, 554 (*Proceros*); Lang, 1884, 544 (*Pseudoceros*).
limbatus Haswell, 1907*b*, 480 — *Pseudoceros.*
limnoriae Hyman, 1953*a*, 313 — *Leptoplana*; du Bois-Reymond Marcus & Marcus, 1968, 24 (*Stylochoplana*).
limosus Stimpson, 1857, 22 — *Stylochopsis*; Diesing, 1862, 570 (*Stylochus*).
lineata Bock, 1922, 20 — *Amyella.*
lineata Dawydoff, 1940, 464 — *Lobophora.*
lineatum Meixner, 1907*a*, 172 — *Prosthiostomum.*
lineoliger Blanchard, 1849, 610 — *Polycelis.*
linteus Müller, 1854, 610 — *Stylochus*; Graff, 1882, 22 ('*luteus*').
liparis Marcus, 1950, 86 — *Pseudoceros.*
litoralis Bock, 1913, 259 — *Pseudoceros.*
litoricola Heath & McGregor, 1913, 458 — *Phylloplana*; Hyman, 1953*a*, *336* (*Freemania*).
littoralis Verrill, 1873, 325 — *Stylochopsis*; Lang, 1884, 453 (*Stylochus*).
littoralis [author?] Morgan, 1905, 187 — *Leptoplana.*

littoralis Kato, 1937*f*, 347 — *Cryptocelis.*
lividus Prudhoe, 1982*a*, 222 — *Pseudoceros.*
lobatum Heath, 1928, 187 — *Graffizoon.*
lobatum Pearse, 1938, 91 — *Prosthiostomum.*
lobianchii Lang, 1879, 486 — *Proceros*; Lang, 1884, 78 (*Eurylepta*).
lobianchoi [cf. *lobianchii*] Lang, 1884, 578 — *Eurylepta.*
longastyletta Freeman, 1933, 111 — *Stylochoplana*; Hyman, 1953*a*, 325 (*Notoplana*).
longicrumena Prudhoe, 1982*a*, 225 — *Notoplana.*
longiducta Hyman, 1959*c*, 10 — *Notoplana.*
longipenis Kato, 1937*b*, 233 — *Pseudostylochus.*
longipenis Kato, 1943*c*, 79 — *Discoplana*; Hyman, 1953*a*, 333 (*Euplana*).
longipenis Hyman, 1953*a*, 305 — *Stylochoplana.*
longisaccata Hyman, 1959*c*, 8 — *Notoplana.*
loveni Bergendal, 1890, 327 — *Cryptocelides.*
luracola Smith, 1961, 69 — *Hoploplana.*
lutea [pro *luteola*] Örsted, 1844*a*, 49 — *Leptoplana.*
luteola Delle Chiaje, 1822, p. 35, fig. 28 — *Planaria*; Örsted, 1844*a*, 49 (*Leptoplana* (?) *lutea*).
luteomarginatus Yeri & Kaburaki, 1918*a*, 432 — *Pseudoceros.*
luteus [pro *linteus* Müller] Graff, 1882, 22 — *Stylochus.*
luteus Plehn, 1897, 94 — *Amblyceraeus*; Hyman, 1953*a*, 366 (*Pseudoceros*).
lynca du Bois-Reymond Marcus, 1958, 403 — *Stylochoplana.*
lyrosora Schmarda, 1859, 24 — *Polycelis*; Diesing, 1862, 535 (*Leptoplana*).

macrorhyncha Schmarda, 1859, 23 — *Polycelis*; Diesing, 1862, 531 (*Leptoplana*); Stummer-Traunfels, 1933, 3486 (*Prosthiostomum*).
macrosora Schmarda, 1859, 18 — *Leptoplana.*
maculata Dalyell, 1853, 104 — *Planaria.*
maculata Gray, *in* Pease, 1860, 38 — *Peasia*; Diesing, 1862, 548 (*Eurylepta*); Lang, 1884, 547 (*Pseudoceros*).
maculata Plehn, 1896*a*, 140 — *Semonia.*
maculatum Haswell, 1907*b*, 482 — *Prosthiostomum.*
maculatum Kato, 1944, 305 — *Stylostomum.*
maculatus Quatrefages, 1845, 144 — *Stylochus*; Stimpson, 1857, 22 (*Stylochoplana*); Verrill, 1893, 468 (footnote) (*Heterostylochus*).
maculatus Hallez, 1893, 171 — *Cycloporus.*
maculatus Hallez, 1905*b*, 125 — *Aceros*; Hallez, 1913, 38 (*Leptoteredra*).
maculatus Kato, 1938*b*, 577 — *Pseudostylochus.*
maculosa Stimpson, 1857, 22 — *Leptoplana.*
maculosa Verrill, 1893, 495 — *Eurylepta*; Hyman, 1952, 197 (*Prostheceraeus*).
maculosus Pearse, 1938, 85 — *Pseudoceros*; du Bois-Reymond Marcus, 1955*a*, 42 (*Pseudoceros* (*Acanthozoon*).)
magnus Freeman, 1933, 111 — *Cryptophallus.*
malayana Laidlaw, 1903*c*, 302 — *Leptoplana*; Bock, 1913, 221 (*Discoplana*).
malayensis Collingwood, 1876, 88 — *Stylochopsis*; Bock, 1913, 258 (*Pseudoceros*).
maldivensis Laidlaw, 1902, 290 — *Cestoplana*; Bock, 1913, 64 (*Latocestus*).
marginata Stimpson, 1857, 22 — *Callioplana*; Diesing, 1862, 569 (*Stylochus*); Lang, 1884, 445 (*Planocera*).
marginata Plehn, 1896*a*, 165 — *Thysanoplana.*
marginata Meyer, 1922, 138 — *Paraplanocera.*
marginata Kaburaki, 1923*b*, 635 — *Simpliciplana.*
marginatus Meixner, 1907*a*, 168 — *Latocestus.*
marina Örsted, 1843, 565 — *Typhloplana*; Stimpson, 1857, 21 (*Typhlocolax*); Diesing, 1862, 523 (*Typhlolepta*).
marina Kato, 1938*a*, 559 — *Cestoplana.*
marmoratum Yeri & Kaburaki, 1918*a*, 432 — *Prosthiostomum.*
marmoratus Stimpson, 1857, 22 — *Diplonchus.*
marmoratus, Plehn, 1898, 145 — *Pseudoceros.*

marmoreus Bock, 1925*b*, 98 — *Stylochus.*
martae Marcus, 1947, 99 — *Stylochus.*
martae Marcus, 1948, 111 — *Notoplana.*
matarazzoi Marcus, 1950, 5 — *Prosthiostomum*; du Bois-Reymond Marcus & Marcus, 1968, 89 (*Lurymare*).
maximus Lang, 1884, 49 — *Pseudoceros.*
mediterraneus Galleni, 1976*b*, 15 — *Stylochus (Imogine).*
medvedica Marcus, 1952, 5 — *Phaenocelis.*
megala Marcus, 1952, 5 — *Notoplana.*
megalops Schmarda, 1859, 15 — *Dicelis*; Diesing, 1862, 523 (*Diopsis*); Stummer-Traunfels, 1933, 3486 (*Stylochus*).
meixneri Bock, 1925*b*, 123 — *Stylochus.*
meleagrina Kelaart, 1858, 137 — *Planaria*; Collingwood, 1876, 88 (*Stylochoplana*); Kaburaki, 1923*b*, 635 (*Prostheceraeus*).
melobesiarum Schmidtlein, 1880, 172 — *Proceros.*
memoralis Kato, 1938*a*, 560 — *Pseudoceros.*
meridialis Kato, 1938*a*, 559 — *Pseudostylochus.*
meridianus Ritter-Záhony, 1907, 5 — *Aceros*; Marcus, 1947, 141 (*Acerotisa*).
meridionalis Pearse, 1938, 73 — *Eustylochus.*
meridionalis Prudhoe, 1982*a*, 22 — *Cestoplana.*
mertensi Diesing, 1850, 216 — *Stylochus*; Diesing, 1862, 572 (*Gnesioceros*).
mertensii Claparède, 1861, 79 — *Centrostomum*; Lang, 1884, 499 (*Leptoplana*).
mexicana Hyman, 1953*a*, 299 — *Phaenocelis.*
mexicana Hyman, 1953*a*, 275 — *Alleena.*
mexicana Hyman, 1955*a*, 9 — *Ommatoplana.*
mexicanus Hyman, 1953*a*, 363 — *Pseudoceros.*
michaelseni Ritter-Zahony, 1907, 3 — *Cotylocera.*
micheli Marcus, 1949, 7 — *Notoplana.*
microceraeus Schmarda, 1859, 31 — *Prostheceraeus*; Stummer-Traunfels, 1933, 3487 (*Pseudoceros*).
micronesiana Hyman, 1959*a*, 551 — *Notoplana.*
micronesianus Hyman, 1955*c*, 78 — *Pseudoceros.*
micropapillosus Kato, 1934*b*, 123 — *Pseudoceros*; du Bois-Reymond Marcus, 1955*a*, 42 (*Pseudoceros (Acanthozoon)*); Hyman, 1959*a*, 583 (*Acanthozoon*).
microps Verrill, 1903, 45 — *Trigonoporus*; Bock, 1913, 68 (*Latocestus*); Hyman, 1939*d*, 14 (*Cestoplana*).
microsora Schmarda, 1859, 22 — *Polycelis*; Diesing, 1862, 529 (*Leptoplana*); Stummer-Traunfels, 1933, 3486 (*Notoplana*).
milcum du Bois-Reymond Marcus & Marcus, 1968, 87 — *Prosthiostomum.*
miniata Schmarda, 1859, 27 — *Eurylepta*; Diesing, 1862, 554 (*Proceros*); Lang, 1884, 551 (*Yungia*); Stummer-Traunfels, 1933, 3487 (*Pseudoceros*).
minimus Palombi, 1940, 109 — *Stylochus.*
minuta Hyman, 1959*a*, 549 — *Stylochoplana.*
minutum Stummer-Traunfels, 1895, 700 — *Thysanozoon.*
mira Bock, 1926*a*, 133 — *Apidioplana.*
mirabilis Bock, 1923*b*, 1 — *Boninia.*
mirabilis Kato, 1938*b*, 577 — *Bergendalia*; Kato, 1944, 261 (*Trigonoporus*).
misakiensis Yeri & Kaburaki, 1918*a*, 432 — *Paraplanocera.*
misakiensis du Bois-Reymond Marcus & Marcus, 1968, 77 — *Cycloporus.*
mitsuii Kato, 1944, 269 — *Ilyplanoides.*
miyadii Kato, 1944, 264 — *Stylochus.*
modestus Quatrefages, 1845, 133 — *Polycelis*; Diesing, 1850, 195 (*Leptoplana*). Stimpson, 1857, 21 (*Elasmodes*).
molle Freeman, 1930, 334 — *Prosthiostomum*; Hyman, 1953*a*, 386 (*Euprosthiostomum*).
monosora Schmarda, 1859, 16 — *Leptoplana*; Stummer-Traunfels, 1933, 3486 (*Prosthiostomum*).
montereyensis Hyman, 1953*a*, 370 — *Pseudoceros.*
montiporae Poulter, 1975, 317 — *Prosthiostomum (Lurymare).*
mopsus Marcus, 1952, 5 — *Pseudoceros.*

mortenseni Bock, 1913, 40 — *Notoplana.*
mortenseni Marcus, 1948, 111 — *Euprosthiostomum.*
moseleyi Lang, 1884, 500 — *Leptoplana.*
moseleyi Lang, 1884, 116 — *Prostheceraeus.*
muelleri Audouin, *in* Savigny, 1827, 247 — *Planaria*; Schmarda, 1859, 33 (*Planocera*).
muelleri Delle Chiaje, 1829, 179 — *Planaria*; Örsted, 1844*a*, 47 (*Thysanozoon*); Lang, 1884, 545 (*Pseudoceros*).
multicelis Hyman, 1953*a*, 384 — *Prosthiostomum.*
multicelis Hyman, 1955*b*, 139 — *Acerotisa.*
multitentaculata Kato, 1940, 559 — *Planocera.*
muscularis Hyman, 1955*b*, 125 — *Anandroplana.*
mutabilis Verrill, 1873, 746 — *Polycelis*; Verrill, 1893, 493 (*Discocelis*); Hyman, 1939*e*, 153 (*Coronadena*).

nadiae Melouk, 1941, 41 — *Leptoplana*; du Bois-Reymond Marcus & Marcus, 1968, 23 (*Stylochoplana*).
napolitanus [pro *neapolitanus*] Gerard, 1901, 143 — *Stylochus.*
natans Freeman, 1933, 111 — *Notoplana.*
natantis MS. name Hyman, 1953*a*, 328 — *Notoplana.*
nationalis Plehn, 1896*b*, 5 — *Stylochoplana*; Bock, 1913, 187 (*Notoplana*).
nationalis Plehn, 1896*b*, 6 — *Leptoplana.*
nationalis Plehn, 1896*b*, 7 — *Aceros*; Marcus, 1947, 141 (*Acerotisa*).
nationalis Plehn, 1896*b*, 8 — *Prosthiostomum.*
nationalis Kato, 1939*a*, 65 — *Pseudostylochus.*
neapolitana Delle Chiaje, 1841, 133 — *Planaria*; Lang, 1884, 53 (*Stylochus*).
nebulosa Girard, 1853, 367 — *Planocera*; Verrill, 1893, 471 (*Planocera* (*Planoceropsis*)); Meixner, 1907*b*, 399 (*Stylochus*).
nematoideum Le Conte, 1851, 319 — *Glossostoma.*
nemoralis [pro *memoralis* Kato] Nicoll, 1939, 73 — *Pseudoceros.*
neptis du Bois-Reymond Marcus, 1955*a*, 42 — *Eurylepta.*
nesidensis Delle Chiaje, 1822, pl. 41, fig. 1 — *Planaria.*
newtoni Willey, 1898, 203 — *Heteroplana.*
nexa Sopott-Ehlers & Schmidt, 1975, 213 — *Cestoplana.*
nigra Stimpson, 1857, 20 ('*niger*'), 26 — *Eurylepta*; Lang, 1884, 565 (*Prostheceraeus*); Stummer-Traunfels, 1933, 3566 (*Pseudoceros*).
nigricornis Schmarda, 1859, 31 — *Prosthecereaus.*
nigripunctata Örsted, 1843, 569 — *Leptoplana.*
nigrocincta Schmarda, 1859, 26 — *Eurylepta*; Diesing, 1862, 551 (*Proceros*); Lang, 1884, 547 (*Pseudoceros*).
nigromarginatus Yeri & Kaburaki, 1918*a*, 432 — *Pseudoceros.*
nigropapillosus Hyman, 1959*a*, 581 — *Acanthozoon.*
nigropunctata [pro *nigripunctata*] Diesing, 1850, 198 — *Leptoplana.*
nigrum Girard, 1852*b*, 137 — *Thysanozoon.*
nipponicus Kato, 1944, 300 — *Pseudoceros.*
norfolkensis Palombi, 1924*a*, 35 — *Metaposthia.*
notabilis Darwin, 1844, 249 — *Diplanaria*; Diesing, 1862, 542 (*Leptoplana*).
notoensis Kato, 1944, 291 — *Pseudostylochus.*
notoensis Kato, 1944, 310 — *Prosthiostomum.*
notulata Bosc, 1802, 254 — *Planaria*; Hyman, 1939*d*, 21 (*Acerotisa*).
novacambrensis Hyman, 1959*c*, 3 — *Leptostylochus.*
nozakensis Kato, 1944, 311 — *Prosthiostomum.*

oblonga Stimpson, 1857, 22 — *Leptoplana.*
oblongus Stimpson, 1855*b*, 389 — *Dioncus.*
obovata Schmarda, 1859, 20 — *Polycelis*; Diesing, 1862, 528 (*Leptoplana*); Stummer-Traunfels, 1933, 3486 (*Adenoplana*).
obscura Stimpson, 1855*a*, 381 — *Leptoplana*; Stimpson, 1857, 22 (*Prosthiostomum*).
obscurum Stummer-Traunfels, 1895, 699 — *Thysanozoon.*

obscurus Stimpson, 1857, 22 — *Stylochus*; Yeri & Kaburaki, 1918*a*, 432 (*Pseudostylochus*).
obtusus Collingwood, 1876, 88 — *Elasmodes*; Laidlaw, 1903*c*, 302 (*Leptoplana*).
occidentalis Hyman, 1953*a*, 294 — *Cryptocelis*.
occidua Sopott-Ehlers & Schmidt, 1975, 198 — *Theama*.
oceanica Darwin, 1844, 246 — *Planaria*; Diesing, 1850, 211 (*Eurylepta*); Stimpson, 1857, 20 (*Nautiloplana*); Schmarda, *in* Diesing, 1862, 546 (*Carenoceraeus*).
oceanica Hyman, 1953*b*, 191 — *Aquaplana*.
oceanica Hyman, 1955*c*, 72 — *Ommatoplana*.
ocellata Kelaart, 1858, 138 — *Penula*; Collingwood, 1876, 88 (*Centrostomum*).
ocellatus Marcus, 1947, 99 — *Latocestus*.
oculifera Girard, 1853, 367 — *Imogine*; Diesing, 1862, 570 (*Stylochus*); Pearse & Walker, 1939, 18 (*Stylochoplana*).
ohshimai Kato, 1938*a*, 560 — *Amakusaplana*.
okadai Kato, 1944, 287 — *Apidioplana*.
okudai Kato, 1937*b*, 233 — *Pseudostylochus*.
oligochlaenus [pro *oligoglenus*] Grube, 1868, 45, 46 — *Stylochus*.
oligoglenus Schmarda, 1859, 34 — *Stylochus*; Lang, 1884, 444 (*Planocera*); Stummer-Traunfels, 1933, 3487 (*Paraplanocera*).
oosora Schmarda, 1859, 22 — *Polycelis*; Diesing, 1862, 530 (*Leptoplana*); Stummer-Traunfels, 1933, 3486 (*Indiplana*).
opaca Schmarda, 1859, 16 — *Typhlolepta*.
opacum Stimpson, 1857, 21 — *Cryptocoelum*.
ophryoglena Schmarda, 1859, 20 — *Polycelis*; Diesing, 1862, 526 (*Leptoplana*).
opisthopharynx Palombi, 1928, 579 — *Notoplanides*.
opisthoporus Bock, 1913, 161 — *Emprosthopharynx*.
orbicularis Schmarda, 1859, 20 — *Polycelis*; Diesing, 1862, 527 (*Leptoplana*); Stummer-Traunfels, 1933, 3486 (*Enterogonia*).
orbicularis pigrans Hyman, 1959*c*, 5 — *Enterogonia*.
orcadensis Gemmill & Leiper, 1907, 823 — *Nuchenceros*.
orientalis Bock, 1913, 36 — *Stylochus*.
orientalis var. *splendida* Bock, 1913, 132 — *Stylochus*.
orientalis Kato, 1939*b*, 142 — *Cryptocelis*.
ornata Focke, *in* Diesing, 1850, 213 — *Planaria*.
ornata Yeri & Kaburaki, 1918*a*, 432 — *Hoploplana*.
ornata Marcus, 1947. 99 — *Itannia*.
ornata forma *murna* du Bois-Reymond Marcus, 1957, 174 — *Itannia*.
ostreae Kato, 1937*f*, 347 — *Prosthiostomum*.
ostreophagus Hyman, 1955*a*, 4 — *Pseudostylochus*.
otophora Schmarda, 1859, 18 — *Leptoplana*; Stummer-Traunfels, 1933, 3486 (*Notoplana*).
ovale Schmarda, 1859, 29 — *Thysanozoon (Eolidiceros)*.
ovale Simonetta & Delle Cave, 1978, 45 — *Platyendron*.
ovalis Bock, 1913, 212 — *Notoplana*.
ovata Schmankewitsch, 1873, 275 — *Endocelis*.
ovatus Kato, 1937*f*, 347 — *Leptostylochus*.
oxyceraeus Schmarda, 1859, 35 — *Stylochus*; Lang, 1884, 445 (*Planocera*).

pacifica Kato, 1939*a*, 65 — *Plehnia*; Hyman, 1953*a*, 279 (*Paraplehnia*).
pacifica Kato, 1943*c*, 79 — *Idioplana*.
pacifica Kato, 1944, 271 — *Amemiyaia*.
pacifica Hyman, 1954*b*, 333 — *Planocera*.
pacifica Hyman, 1959*a*, 555 — *Aquaplana*.
pacificola Plehn, 1896*a*, 140 — *Leptoplana*; Bock, 1913, 220 (*Discoplana*); Hyman, 1953*a*, 332 (*Euplana*).
pacificus Laidlaw, 1903*a*, 3 — *Latocestus*.
pacificus Jacubowa, 1906, 144 — *Dicteros*.
pacificus Bock, 1923*c*, 342 — *Neostylochus*; Prudhoe, 1982*b*, 363 (*Ancoratheca*).
pakium du Bois-Reymond Marcus & Marcus, 1968, 91 — *Euprosthiostomum*.

palaoensis Kato, 1943*c*, 79 — *Notoplana.*
pallidum Laidlaw, 1903*c*, 302 — *Prosthiostomum.*
pallidus Quatrefages, 1845, 133 — *Polycelis*; Diesing, 1850, 195 (*Leptoplana*); Stimpson, 1857, 21 (*Elasmodes*); Bock, 1913, 172 (*Stylochoplana*).
palmula Quatrefages, 1845, 143 — *Stylochus*; Lang, 1884, 127 (*Stylochoplana*).
palta Marcus, 1954*a*, 65 — *Notoplana.*
panamensis Woodworth, 1894, 51 — *Prostheceraeus.*
panamensis Plehn, 1896*a*, 151 — *Leptoplana*; Bock, 1913, 173 (*Stylochoplana*).
pannulus Heath & McGregor, 1913, 458 — *Euryleptodes.*
panormus Quatrefages, 1845, 142 — *Eolidiceros*; Diesing, 1850, 213 (*Thysanozoon*); Stimpson, 1857, 20 (*Planeolis*).
pantherina Grube, 1868, 46 — *Eurylepta.*
papilionis Kelaart, 1858, 136 — *Planaria*; Collingwood, 1876, 87 (*Acanthozoon*); Lang, 1884, 546 (*Pseudoceros*); Kaburaki, 1923*b*, 35 (*Prostheceraeus*); du Bois-Reymond Marcus, 1955*a*, 42 (*Pseudoceros* (*Acanthozoon*)).
papillifera Bock, 1913, 233 — *Styloplanocera.*
papillosa Lang, 1884, 236 — *Planocera*; Bock, 1913, 225 (*Hoploplana*).
papillosum Grube, 1840, 56 — *Thysanozoon.*
papillosum Sars, *in* Jensen, 1878, 79 — *Thysanozoon*; Bock, 1913, 262 (*Cycloporus*).
papillosus Diesing, 1836, 316 — *Stylochus*; Grube, 1840, 56 (*Thysanozoon*).
papillosus Lang, 1884, 568 — *Cycloporus.*
papillosus Lang var. *laevigatus* Lang, 1884, 570 — *Cycloporus.*
papillosus Lang var. *misakiensis* Kato, 1938*a*, 571 — *Cycloporus.*
paradoxa Bock, 1913, 214 — *Copidoplana.*
paradoxus Bock, 1927*a*, 17 — *Pseudoceros.*
parasitica Kato, 1935*a*, 123 — *Stylochoplana.*
parcus Bock, 1925*a*, 31 — *Discostylochus.*
pardalis Verrill, 1900, 596 — *Pseudoceros.*
pardalis Laidlaw, 1902, 287 — *Leptoplana.*
parva Palombi, 1939*b*, 96 — *Stylochoplana.*
parvicelis Hyman, 1939*c*, 6 — *Prosthiostomum.*
parvula Palombi, 1924*a*, 37 — *Notoplana.*
parvus Pearse, 1938, 81 — *Conjuguterus.*
patellarum Stimpson, 1855*b*, 389 — *Leptoplana*; Palombi, 1939*a*, 123 (*Notoplana*).
patellensis Collingwood, 1876, 88 — *Leptoplana.*
pegnis [pro *segnis*] Nicoll, 1934, 76 (*Notoplana*).
pelagicus Moseley, 1877, 24 — *Stylochus*; Lang, 1884, 236 (*Planocera*).
peleca du Bois-Reymond Marcus & Marcus, 1968, 2 — *Phaenocelis.*
pelidium [pro *pilidium*] Gérard, 1901, 143 — *Stylochus.*
pellucida Mertens, 1832, 8 — *Planaria*; Ehrenberg, 1836, 67 (*Stylochus*); Örsted, 1844*a*, 48 (*Planocera*); Diesing, 1862, 571 (*Gnesioceros*).
pellucida Grube, 1840, 53 — *Leptoplana*; Lang, 1884, 605 (*Prosthiostomum*).
pellucida Bosc, 1803, of Lang, 1884, 6 — *Planaria.*
pellucida Pearse, 1938, 90 — *Acerotisa*; Hyman, 1939*e*, 153 (*Enantia*).
penangensis Laidlaw, 1903*c*, 302 — *Semonia.* [See p. 80]
periommatum Bock, 1913, 287 — *Enchiridium.*
periphaeus Bock, 1913, 255 — *Pseudoceros.*
perviolaceus Hyman, 1959*a*, 566 — *Pseudoceros.*
philippinensis Kaburaki, 1923*b*, 635 — *Pseudoceros.*
phyllulus Heath & McGregor, 1913, 458 — *Euryleptodes.*
pigrans Haswell, 1907*a*, 644 — *Enterogonia.*
pigrans novaezealandiae Bock, 1925*b*, 142 — *Enterogonia.*
pilidium Goette, 1881, 189 — *Stylochopsis*; Lang, 1884, 321 (*Stylochus*).
piscatoria Marcus, 1947, 99 — *Acerotisa.*
pius Kato, 1938*a*, 560 — *Pseudoceros.*
plana Strøm, 1768, 365 — *Hirudo.*
platae Hyman, 1955*d*, 9 — *Adenoplana.*

plecta Marcus, 1947, 99 — *Notoplana.*
plehni Laidlaw, 1902, 294 — *Thysanozoon.*
plehni Laidlaw, 1906, 711 — *Latocestus.*
plehni, Bock, 1913, 173 — *Stylochoplana.*
plesia — Corrêa, 1949, 173 — *Zygantroplana.*
plessisii Lang, 1884, 54 — *Stylochus.*
pleurostictus Bock, 1913, 257 — *Pseudoceros.*
polycyclium Schmarda, 1859, 24 — *Centrostomum;* Lang, 1884, 498 (*Leptoplana*).
polygenia Palombi, 1938, 329 — *Laidlawia.*
polypora Meyer, 1922, 138 — *Cestoplana.*
polysorum Schmarda, 1859, 25 — *Centrostomum;* Lang, 1884, 499 (*Leptoplana*); Stummer-Traunfels, 1933, 3486 (*Leptostylochus*).
ponticus Metschnikoff, 1877, 1 — *Stylochopsis.*
popeae Hyman, 1959*c*, 15 — *Diposthus.*
portoricensis Hyman, 1955*b*, 127 — *Anandroplana.* [See p. 78]
praetexta Ehrenberg, 1831, 56 — *Eurylepta.*
profunda Kato, 1937*f*, 347 — *Planocera.*
promiscua Plehn, 1896*a*, 140 — *Plagiotata.*
prytherci Pearse & Littler, 1938, 239 — *Hymania.*
pseudolimax Lang, 1884, 559 — *Prostheceraeus.*
pulcher Hyman, 1939*e*, 153 — *Stylochus.*
pulchra Örsted, 1845, 415 — *Eurylepta.*
pulchrum Bock, 1913, 285 — *Prosthiostomum.*
puma Marcus, 1954*a*, 59 — *Notoplana.*
punctata Müller, 1776, 223 — *Planaria.*
punctata Stimpson, 1857, 22 — *Leptoplana.*
punctata Kelaart, 1858, 138 — *Penula;* Collingwood, 1876, 88 (*Centrostomum*).
punctata Diesing, of Lang, 1884, 616 — *Leptoplana.*
punctata Kaburaki, 1923*a*, 199 — *Eurylepta.*
punctatum Hallez, 1905*b*, 126 — *Stylostomum.*
punctatum Hyman, 1953*a*, 386 — *Enchiridium.*
punctatus Laidlaw, 1902, 226 — *Pseudoceros.*
purpurea Kelaart, 1858, 136 — *Planaria;* Collingwood, 1876, 87 (*Eurylepta*); Hyman, 1959*a*, 565 (*Pseudoceros*).
purpurea Schmarda, 1859, 18 — *Leptoplana;* Stummer-Traunfels, 1933, 3486 (*Phaenocelis*).
purpurea Yeri & Kaburaki, 1918*a*, 432 — *Planocera.*
purum Kato, 1937*f*, 347 — *Prosthiostomum;* du Bois-Reymond Marcus & Marcus, 1968, 88 (*Lurymare*); Poulter, 1975, 317 (*Prosthiostomum* (*Lurymare*).)
pusilla Bock, 1924*a*, 2 — *Stylochoplana (Stylochoplanoides).*
pusilla Kato, 1938*a*, 559 — *Discocelis.*
pusillus Bock, 1913, 139 — *Stylochus.*

queruca du Bois-Reymond Marcus & Marcus, 1968, 44 — *Notoplana.*

rabita du Bois-Reymond Marcus & Marcus, 1968, 35 — *Candimba;* Prudhoe, 1982*b*, 371 (*Candimboides*).
raffaelei Ranzi, 1928, 3 — *Cestoplana.*
rainieri Belloc, 1961, 297 — *Planocera.*
rasae Prudhoe, 1968, 408 — *Emprosthopharynx.*
refertus du Bois-Reymond Marcus, 1965, 129 — *Stylochus.*
regalis Hadden, *in* Laidlaw, 1903*a*, 3 — *Pseudoceros.*
reishi Hyman, 1959*b*, 5 — *Stylochoplana.*
reticulata Gray, *in* Pease, 1860, 37 — *Peasia;* Diesing, 1862, 561 (*Planocera*).
reticulatus Stimpson, 1855*a*, 381 — *Stylochus;* Stimpson, 1857, 22 (*Stylochoplana*); Lang, 1884, 445 (*Planocera*).
reticulatus Yeri & Kaburaki, 1918*a*, 432 — *Pseudoceros.*
retusa Viviani, 1805, 5 — *Planaria;* Diesing, 1850, 200 (*Typhlolepta*).
rickettsi Hyman, 1953*a*, 300 — *Longiprostatum.*

robusta Palombi, 1928, 579 — *Notoplana*; du Bois-Reymond Marcus & Marcus, 1968, 31 (*Stylochoplana*).
rosea Prudhoe, 1977, 588 — *Hoploplana*.
roseimaculata Blanchard, 1849, 72 — *Polycelis*.
roseus Sars, *in* Jensen, 1878, 75 — *Stylochus*; Lang, 1884, 589 (*Stylostomum*).
roseus Lang, 1884, 562 — *Prostheceraeus*.
rotumanensis Laidlaw, 1903*a*, 3 — *Paraplanocera*.
rubellus Laidlaw, 1903*c*, 302 — *Pseudoceros*.
rubra Dawydoff, 1940, 462 — *Lobophora*.
rubra Kato, 1944, 281 — *Hoploplana*.
rubrifasciata Kato, 1937*f*, 347 — *Paraplanocera*.
rubrocincta Schmarda, 1859, 26 — *Eurylepta*; Diesing, 1862, 546 (*Schmardea*); Lang, 1884, 550 (*Yungia*); Stummer-Traunfels, 1933, 3487 (*Pseudoceros*).
rubrocinctum Grube, 1840, 56 — *Orthostomum*; Stimpson, 1857, 21 (*Typhlolepta*); Lang, 1884, 49 (*Cestoplana*).
rubropunctatum Yeri & Kaburaki, 1918*a*, 432 — *Prosthiostomum*.
rubrotentaculatus Kaburaki, 1923*b*, 635 — *Pseudoceros*.
rugosa Hyman, 1959*a*, 585 — *Acerotisa*.
rupicola Heath & McGregor, 1913, 455 — *Leptoplana*; Hyman, 1939*a*, 437 (*Notoplana*).
russoi Palombi, 1939*a*, 123 — *Prosthiostomum*; du Bois-Reymond Marcus & Marcus, 1968, 89 (*Lurymare*).
rusticum Giard, 1888, 496 — *Stylostomum*.
rutilus Yeri & Kaburaki, 1918*a*, 431 — *Stylochus*.

sadoensis Kato, 1944, 292 — *Pseudostylochus*.
sadoensis Kato, 1944, 309 — *Prosthiostomum*.
saga Corrêa, 1958, 81 — *Chromyella*.
sagamianus Kato, 1937*f*, 347 — *Pseudoceros*.
salar Marcus, 1949, 7 — *Cestoplana*.
salmoneus Meixner, 1907*a*, 165 — *Stylochus*.
samoensis Palombi, 1924*a*, 33 — *Cryptocelides*.
sancum du Bois-Reymond Marcus, 1965, 131 — *Prosthiostomum*.
sandiegensis Boone, 1929, 33 — *Planocera*; Hyman, 1953*a*, 349 (*Alloioplana*).
sandiegensis Hyman, 1953*a*, 363 — *Thysanozoon*.
sanguinea Freeman, 1933, 111 — *Notoplana*.
sanguineum Hallez, 1893, 180 — *Stylostomum*.
sanguinolentus Quatrefages, 1845, 138 — *Proceros*; Diesing, 1850, 209 (*Eurylepta*); Lang, 1884, 127 (*Oligocladus*).
sanjuania Freeman, 1933, 111 — *Notoplana*.
sanjuania Holleman, 1972, 409 — *Stylostomum*.
sanpedrensis Freeman, 1930, 337 — *Notoplana*.
sargassicola Mertens, 1832, 11 — *Planaria*; Ehrenberg, 1836, 67 (*Stylochus*); Örsted, 1844*a*, 48 (*Planocera*); Diesing, 1850, 216 (renamed *Stylochus mertensi*); Graff, 1892*a*, 146 (*Stylochoplana*); Bock, 1913, 233 (*Pelagoplana*); Hyman, 1939*b*, 146 (*Gnesioceros*).
sargassicola var. *lata* Hyman, 1939*d*, 13 — *Gnesioceros*.
sasakii Kaburaki, 1923*a*, 196 — *Yungia*.
sawayai Marcus, 1947, 99 — *Notoplana*.
saxicola Heath & McGregor, 1913, 456 — *Leptoplana*; Hyman, 1939*a*, 437 (*Notoplana*).
schauinslandi Plehn, 1899, 449 — *Microcelis*.
schizoporellae Hallez, 1893, 155 — *Leptoplana*.
schizoporellae Kato, 1944, 280 — *Hoploplana*.
schlosseri Giard, 1873, 488 — *Planaria*.
schoenbornii Stimpson, 1857, 22 — *Leptoplana*.
sciophila Boone, 1929, 40 — *Leptoplana*; Hyman, 1953*a*, 323 (*Notoplana*).
segnis Freeman, 1933, 111 — *Notoplana*.
selenopsis Marcus, 1947, 99 — *Stylochoplana*.
semperi Stummer-Traunfels, 1895, 699 — *Thysanozoon*.

septentrionalis Kato, 1937*e*, 127 — *Notoplana.*
serica Kato, 1938*a*, 559 — *Notoplana.*
siamensis Palombi, 1924*a*, 36 — *Stylochoplana.*
similis Bergendal, 1892, 551 — *Polypostia*; Graff, 1896, 93 (*Polyposthia*).
similis Bock, 1927*a*, 87 — *Apidioplana.*
'*similis* Stimpson' Dawydoff, 1952, 81 — *Notoplana.*
simrothi Graff, 1892*b*, 190 — *Planocera.*
singulare Laidlaw, 1904*b*, 135 — *Prosthiostomum.*
singularis Hyman, 1953*a*, 341 — *Diplandros.*
siphunculus Delle Chiaje, 1822, pl. 35, fig. 26 — *Planaria*; Lang, 1884, 595 (*Prosthiostomum*).
sixteni Marcus, 1947, 106 — *Stylochus.*
skottsbergi Bock, 1923*c*, 342 — *Thysanozoon.*
snadda du Bois-Reymond Marcus & Marcus, 1968, 30 — *Stylochoplana.*
sondaicus Bock, 1925*b*, 120 — *Cryptophallus.*
sonorum Kato, 1938*a*, 560 — *Prosthiostomum.*
sophia Kato, 1939*a*, 65 — *Notoplana.*
sparsa Stimpson, 1855*a*, 381 — *Leptoplana*; Stimpson, 1857, 22 (*Prosthiostomum*).
speciosus Kato, 1937*f*, 347 — *Stylochus.*
spinifera Graff, 1890, 1 — *Enantia.*
splendidus Stummer-Traunfels, 1933, 3487 — *Pseudoceros.*
steueri Steinböck, 1937, 1 — *Cirroposthia*; Marcus, 1947, 133 (*Neoplanocera*).
stilifera Bock, 1923*c*, 341 — *Notoplana.*
stiliferum Bock, 1913, 152 — *Aprostatum.*
stimpsoni Diesing, 1862, 522 — *Typhlolepta.*
stimpsoni Diesing, 1862, 528 — *Leptoplana.*
stimpsoni Kato, 1937*c*, 35 — *Pseudostylochus.*
striata Kelaart, 1858, 137 — *Planaria*; Collingwood, 1876, 87 (*Eurylepta*); Lang, 1884, 546 (*Pseudoceros*).
striata Schmarda, 1859, 17 — *Leptoplana.*
striata Schmarda, 1859, 27 — *Eurylepta*; Diesing, 1862, 551 (*Proceros*); Stummer-Traunfels, 1933, 3487 (*Pseudoceros*).
strigosus Marcus, 1950, 88 — *Pseudoceros.*
stylifera Hyman, 1953*a*, 308 — *Zygantroplana.*
stylostomoides Gemmill & Leiper, 1907, 819 — *Aceros*; Bock, 1922, 15 (*Leptoteredra*); Marcus, 1947, 141 (*Acerotisa*).
subauriculata Johnston, 1836, 16 — *Planaria*; Diesing, 1850, 195 (*Leptoplana*).
subviridis Plehn, 1896*c*, 330 — *Leptoplana*; Bock, 1913, 220 (*Discoplana*).
suesensis Ehrenberg, 1831, 54 — *Stylochus*; Örsted, 1844*a*, 48 (*Planocera*).
suesensis Palombi, 1939*b*, 107 — *Stylochoplana.*
suoensis Kato, 1943*b*, 69 — *Stylochoplana.*
superba Schmarda, 1859, 28 — *Eurylepta*; Diesing, 1862, 552 (*Proceros*); Stummer-Traunfels, 1933, 3487 (*Pseudoceros*).
superbus Lang, 1884, 49 — *Pseudoceros.*
susakiensis Kato, 1934*b*, 123 — *Pseudoceros*; Kato, 1944, 302 (*Eurylepta*).
susakiensis Kato, 1944, 312 — *Prosthiostomum.*
suteri Jacubowa, 1906, 150 — *Leptoplana.*
syntoma Marcus, 1947, 99 — *Notoplana.*

taenia Schmarda, 1859, 24 — *Centrostomum*; Lang, 1884, 498 (*Leptoplana*).
taiwanica Kato, 1943*b*, 69 — *Stylochoplana.*
takeshitai Yeri & Kaburaki, 1918*a*, 432 — *Pseudostylochus.*
takewakii Kato, 1935*b*, 149 — *Discoplana*; Marcus, 1947, 131 (*Euplana*).
tardus Graff, 1878, 460 — *Stylochus*; Lang, 1884, 462 (*Stylochoplana*).
taurica Jacubowa, 1909, 1 — *Stylochoplana.*
tauricus Jacubowa, 1909, 2 — *Stylochus.*
tavoyensis Prudhoe, 1950*a*, 44 — *Notoplana.*
techa du Bois-Reymond Marcus, 1957, 174 — *Cestoplana.*

teffi Dawydoff, 1952, 81 — *Yungia.*
tenax Palombi, 1936, 1 — *Stylochus.*
tenebrosum Stimpson, 1857, 22 — *Prosthiostomum;* Diesing, 1862, 538 (*Leptoplana*).
tenellus Stimpson, 1857, 21 — *Elasmodes;* Diesing, 1862, 528 (*Leptoplana*).
tenera Stimpson, 1857, 24 — *Stylochoplana;* Diesing, 1862, 568 (*Stylochus*).
tentaculata Gray, *in* Pease, 1860, 37 — *Peasia;* Diesing, 1862, 556 (*Thysanozoon*).
tentaculata Palombi, 1928, 579 — *Opisthogenia.*
tentaculata Kato, 1943*a*, 48 — *Leptoteredra.*
tenuis [err. pro *tenera*] Stimpson, 1857, 22 — *Stylochoplana.*
tenuis Palombi, 1936, 1 — *Stylochoplana.*
teredini Hyman, 1944*b*, 73 — *Taenioplana.*
tergestinus Minot, 1877*a*, 451 — *Opisthoporus.*
terricola Schmarda, 1859, 30 — *Prostheceraeus.*
texanus Hyman, 1955*e*, 264 — *Pseudoceros.*
thaisana Pearse, 1938, 79 — *Hoploplana.*
thesea Kelaart, 1858, 136 — *Planaria;* Collingwood, 1876, 88 (*Planocera*).
ticus Marcus, 1952, 5 — *Stylochus.*
tigrinus Blanchard, 1847, 271 — *Polycelis;* Diesing, 1850, 195 (*Leptoplana*); Stimpson, 1857, 21 (*Elasmodes*); Lang, 1884, 71 (*Discocelis*).
tigrinus Laidlaw, 1902, 297 — *Pseudoceros.*
timida Health & McGregor, 1913, 455 — *Leptoplana.*
tincta Crozier, 1918, 379 — *Leptoplana.*
tipuca Marcus & Marcus, 1966, 327 — *Igluta.*
tomiokaensis Kato, 1938*a*, 560 — *Pseudoceros.*
trapezoglena Schmarda, 1859, 23 — *Polycelis;* Diesing, 1862, 531 (*Leptoplana*).
tremellaris Müller, 1774, 72 — *Fasciola;* Müller, 1776, 223 (*Planaria*); Örsted, 1843, 569 (*Leptoplana*).
tremellaris forma *mediterranea* Bock, 1913, 184 — *Leptoplana.*
tremellaris var. *taurica* Jacubowa, 1909, 21 — *Leptoplana.*
tremellaris Grube, 1840, 52 — *Planaria;* see Lang, 1884, 607.
tridentata Hyman, 1953*b*, 188 — *Planocera.*
trigonopora Herzig, 1905, 329 — *Laidlawia.*
trilineatum Yeri & Kaburaki, 1920, 595 — *Prosthiostomum.*
tripartitus Hyman, 1953*a*, 282 — *Stylochus.*
tripyla Hyman, 1953*a*, 339 — *Copidoplana.*
tristriatus Hyman, 1959*a*, 576 — *Pseudoceros.*
tropica Hyman, 1959*a*, 549 — *Plehnia.* [See p. 46]
tropicalis Hyman, 1954*b*, 331 — *Euplana;* Marcus & Marcus, 1966, 330 (*Phylloplana*).
trullaeformis Stimpson, 1855*a*, 381 — *Leptoplana.*
truncata Schmarda, 1859, 35 — *Imogine;* Diesing, 1862, 567 (*Stylochus*).
tuba Grube, 1871, 28 — *Leptoplana.*
tuberculata Delle Chiaje, 1822, pl. 35, fig. 29 — *Planaria;* Grube, 1840, 55 (*Thysanozoon*).
tuberculata Laidlaw, 1903*d*, 111 — *Ommatoplana.*
tuberculatus Schmidtlein, 1880, 172 — *Proceros.*
tuberculatus Lang, *in* Vaillant, 1889, 656 — *Cycloporus.*
tuberculatus Hyman, 1953*a*, 29 — *Mexistylochus;* Hyman, 1955*a*, 9 (renamed *Ommatoplana mexicana*).
tuberculosa Stimpson, 1857, 22 — *Trachyplana.*
turma Marcus, 1952, 5 — *Eurylepta.*
typhlus Bock, 1913, 273 — *Aceros;* Marcus, 1947, 141 (*Acerotisa*).
typica Haswell, 1907*b*, 466 — *Tripylocelis.*

ujara du Bois-Reymond Marcus & Marcus, 1968, 55 — *Amyris.*
uncinata Palombi, 1939*a*, 123 — *Planocera.*
undulata Kelaart, 1858, 137 — *Planaria;* Collingwood, 1876, 87 (*Eurylepta*); Lang, 1884, 552 (*Prostheceraeus*); Marcus, 1950, 88 (*Pseudoceros*).
uniporus Kato, 1940, 538 — *Stylochus.*
usaguia Smith, 1960, 385 — *Hoploplana.*

utarum Marcus, 1952, 5 — *Prosthiostomum*; du Bois-Reymond Marcus & Marcus, 1968, 89 (*Lurymare*).
utunomii Kato, 1943*b*, 69 — *Stylochoplana*.

vanhoffeni Bock, 1931, 261 — *Emprosthopharynx*.
variabile Lang, 1884, 73 — *Stylostomum*.
variabilis Girard, 1850*b*, 251 — *Polycelis*; Diesing, 1862, 542 (*Leptoplana*).
variegatus Kato, 1934*b*, 123 — *Cycloporus*.
velellae Lesson, 1830, 453 — *Planaria*.
velutinus Blanchard, 1847, 273 — *Proceros*; Diesing, 1850, 210 (*Eurylepta*); Lang, 1884, 49 (*Pseudoceros*).
velutinus var. *violaceus* Schmarda, Lang, 1884, 540 — *Pseudoceros*.
veneris [err pro *insularis*] Hyman, 1944*a*, fig. 16 — *Comprostatum*.
verrilli Laidlaw, 1906, 709 — *Zygantroplana*.
verrilli Hyman, 1939*b*, 146 — *Gnesioceros*.
verrucosa Delle Chiaje, 1831, 180 — *Planaria*.
verrucosum Grube, 1868, 46 — *Thysanozoon*.
vesiculata Hyman, 1939*a*, 434 — *Leptoplana*.
vesiculata Palombi, 1940*a*, 109 — *Stylochoplana*.
vesiculatus Jacubowa, 1909, 2 — *Stylochus*.
vigilax Laidlaw, 1904*a*, 2 — *Stylochus*.
villosa Lang, 1884, 62 — *Planocera*; Bock, 1913, 225 (*Hoploplana*).
vinosum Meixner, 1907*a*, 171 — *Pseudoceros*.
violacea Delle Chiaje, 1830, pl. 108 — *Planaria*; Örsted, 1844*a*, 47 (*Thysanozoon*); Lang, 1884, 563 (*Prostheceraeus*).
violacea Kelaart, 1858, 135 — *Planaria*; Collingwood, 1876, 87 (*Eurylepta*); Hyman, 1959*a*, 566 (*Pseudoceros*).
violacea Schmarda, 1859, 25 — *Eurylepta*; Diesing, 1862, 553 (*Proceros*); Stummer-Traunfels, 1933, 3487 (*Pseudoceros*).
virescens Hyman, 1959*a*, 569 — *Pseudoceros*.
virgae Sopott-Ehlers & Schmidt, 1975, 202 — *Copidoplana*.
viridis Kalaart, 1858, 135 — *Planaria*; Collingwood, 1876, 87 (*Eurylepta*); Lang, 1884, 567 (*Prostheceraeus*); Hyman, 1959*a*, 569 (*Pseudoceros*).
viridis Schmarda, 1859, 32 — *Prostheceraeus*; Stummer-Traunfels, 1933, 3487 (*Pseudoceros*).
viridis Bock, 1913, 64 — *Latocestus*.
viridis Bock, 1925*a*, 4 — *Neostylochus*.
viridis Freeman, 1933, 111 — *Stylochoplana*; Hyman, 1953*a*, 334 (*Phylloplana*).
virilis Lang, 1884, 62 — *Anonymus*.
virilis Verrill, 1893, 478 — *Leptoplana*; Bock, 1913, 187 (*Notoplana*).
viscosum Palombi, 1936, 1 — *Euprosthiostomum*
vitrea Lang, 1884, 71 — *Leptoplana*; Bock, 1913, 207 (*Notoplana*).
vittata Montagu, 1813, 25 — *Planaria*; Diesing, 1850, 209 (*Eurylepta*); Lang, 1884, 49 (*Prostheceraeus*).
vulgaris Kato, 1938*a*, 560 — *Prosthiostomum*.
vulgaris Palombi, 1939*a*, 135 — *Thysanozoon*.

wagurensis Kato, 1944, 309 — *Prosthiostomum*.
wahlbergi Bock, 1913, 120 — *Cryptophallus*.
walsergia du Bois-Reymond Marcus & Marcus, 1968, 25 — *Stylochoplana*.
washingtonensis Freeman, 1933, 111 — *Discosolenia*.
whartoni Pearse, 1938, 83 — *Oculoplana*; Hyman, 1940*a*, 458 (*Latocestus*).
willeyi Jacubowa, 1906, 131 — *Notoplana*.
woodworthi Laidlaw, 1903*c*, 302 — *Asthenoceros*.
wyona du Bois-Reymond Marcus & Marcus, 1968, 32 — *Stylochoplana*.

yatsui Kato, 1937*f*, 347 — *Discostylochus*.
yerii Kato, 1937*f*, 347 — *Prosthiostomum*.
yessoensis Kato, 1937*c*, 37 — *Pseudoceros*.
yrsa du Bois-Reymond Marcus & Marcus, 1968, 19 — *Zygantroplana*.

zanzibaricus Laidlaw, 1903*d*, 105 — *Stylochus.*
zebra Leuckart, 1828, 11 — *Planaria*; Diesing, 1850, 211 (*Eurylepta*); Diesing, 1862, 554 (*Proceros*); Lang, 1884, 544 (*Pseudoceros*).
zebra Verrill, 1882, 371 — *Stylochopsis*; Verrill, 1893, 463 (*Stylochus*).
zebra Hyman, 1955*e*, 266 — *Prostheceraeus.*
zeylanica Kelaart, 1858, 138 — *Planaria*; Collingwood, 1876, 87 (*Eurylepta*); Lang, 1884, 546 (*Pseudoceros*).

Subject index

PHYSIOLOGY

PIGMENTATION

PREDATION

REGENERATION

REPRODUCTION & DEVELOPMENT

References

Allee, W.C. 1923*a*. Studies in marine ecology. I. The distribution of common littoral invertebrates of the Woods Hole region. *Biol. Bull. mar. biol. Lab., Woods Hole* **44:** 167–191.

—1923*b*. Studies in marine ecology. III. Some physical factors related to the distribution of littoral invertebrates. *Biol. Bull. mar. biol. Lab., Woods Hole* **44:** 205–253.

Anderson, D.T. 1977. The embryonic and larval development of the turbellarian *Notoplana australis* (Schmarda, 1859) (Polycladida: Leptoplanidae). *Aust. J. mar. freshw. Res.* **28:** 303–310.

Andrews, E.A. 1892. Notes on the fauna of Jamaica. *John Hopkins Univ. Circ.* **11** No. 97: 72–77.

Arndt, W. 1943. Polycladen und maricole Tricladen als Giftträger. *Mem. Estud. Mus. Zool. Univ. Coimbra* No. 148: 1–15.

Audouin, V. 1827. *In:* Savigny, *Description de l'Egypte [etc.]* (2nd ed.), Paris **22:** 248. Histoire naturelle. Planches de Zool. **2,** pl. 5, figs. 6–7.

Ax, P. 1963. Relationships and phylogeny of the Turbellaria. *In:* E.C. Dougherty (Ed.) *The Lower Metazoa*, pp. 191–224. Univ. California Press, Berkeley & Los Angeles.

Bahr, L.M. & Hillman, R.E. 1966. Chromosome number of *Stylochus ellipticus* (Girard). *Trans. Am. microsc. Soc.* **85:** 323–324.

Barbour, T. 1912. Two preoccupied names. *Proc. biol. Soc. Washington* **25:** 187.

Bassindale, R. & Barrett, J.H. 1957. The Dale Fort marine fauna. *Proc. Bristol nat. Soc.* **29:** 213–226.

Bates, M. 1956. Ifalik, lonely paradise of the South Seas. *Nat. geogr. Mag.* **109:** 547–571.

Bedini, C. & Papi, F. 1974. Fine structure of the turbellarian epidermis. *In:* N.W. Riser & M. P. Morse (Eds), *Biology of the Turbellaria.* pp. 108–147. L.H. Hyman Mem. Vol. McGraw-Hill, New York: (McGraw-Hill series in the Invertebrates).

Belloc, G. 1961. Note préliminaire sur un turbellarié du genre *Planocera* de Blainville, capturé a Villefranche-sur-Mer, par S.A.S. Le Prince Rainier III, Prince Souverain de Monaco. *Rapp. Comm. int. Mer. Medit.* **16:** 297.

Benazzi, M. & Benazzi-Lentati, G. 1976. *Animal cytogenetics. I. Platyhelminthes.* G. Bornträger, Berlin-Stuttgart. 182 pp.

Bendl, W.E. 1909. Der "Ductus genito-intestinalis" der Platyhelminthen. *Zool. Anz.* **34:** 294–299.

Beneden, P.J. van 1861. Recherches sur la faune littorale de Belgique. Turbellariés. *Mem. Acad. roy. Sci. Belg.* **32:** 56 pp.

Benham, W.B. 1901. The Platyhelmia, Mesozoa and Nemertini. *In:* E. Ray Lankester (Ed.), *A Treatise on Zoology* Pt. IV: 6–42. Black, London.

Bennett, I. 1971. *The Great Barrier Reef.* Lansdowne Press, Melbourne. 183 pp.

Bergendal, D. 1890. Studien über nordische Turbellarien und Nemertinen. Vorläufige Mittheilung. *Öfv. K. VetenskAkad. Förh. Stockh.* No. 6: 323–328.

—1892. Några anmärkningar om Sveriges Tricladen. *Öfv. K. VetenskAkad, Förh. Stockh.* **49:** 539–557.

—1893*a*. Quelques observations sur *Cryptocelides Loveni* mihi. (Note préliminaire). *Rev. biol. N. Fr.* **5:** 237–241.

—1893*b*. *Polypostia similis* nov. gen., nov. spec. (Polyclade acotylé pourvu de nombreux appareils copulateurs mâles). *Rev. biol. N. Fr.* **5:** 366–368.

—1893*c*. *Polypostia similis* n.g., n.sp. en Acotyl Polyklad med Många Hanliga Parningapparater. (Eine acotyle Polycladide mit zahlreichen männlichen Begattungsapparaten.) *Acta Univ. lund.* **29** [*Kgl. Fysiogr. Sällsk. Lund Handl.* **6**]: 1–29 [German summary 27–29.]

—1893*d*. Einiger Bermerkungen über *Cryptocelides Loveni* mihi. *Acta Univ. lund.* **29:** [*Kgl. Fysiogr. Sällsk. Lund Handl.* **7**] 1–7.

—1902. Über die Polycladengattung *Polypostia* Bgdl. *Verh. Intern. Zool. Congr. Berlin, 1902:* pp. 63 & 750.

Blainville, D. de, 1826 & 1828. Dictionnaire des Sciences naturelles. Art. Planaire. **41:** 204–218 Paris, 1826: Art. Vers **57:** 578–579, Paris, 1828, pl. 40.

Blanchard, E. 1847. Recherches sur l'organisation der vers. *Annls. Sci. nat.* (3) Zool. **8:** 119–149, 271–275.

—1849. Planarianos. *In:* Cl. Gay; *Historica fisica y politica de Chile.* Zool. 3, **8:** 69–72. Atlas zoologico. Anelides Fol. lám. 3, F.1. Paris.

Bock, S. 1913. Studien über Polycladen. *Zool. Bidr. Uppsala* **2:** 29–344.

—1922. Two new cotylean genera of polyclads from Japan and remarks on some other cotyleans. *Ark. Zool.* **14:** No. 13, 1–31.

—1923*a* Two new acotylean polyclads from Japan. *Ark. Zool.* **15:** No. 17, 1–39.

—1923*b*. *Boninia,* a new polyclad genus from the Pacific. *Nova Acta R. Soc. upsal.* (4) **6:** No. 3, 1–32.

—1923*c*. Polycladen aus Juan Fernandez. *Nat. Hist. Juan Fernandez and Easter Island,* Uppsala, **3** (Zool.): 341–372.

—1924. Eine neue *Stylochoplana* aus Japan. *Ark. Zool.* **16:** No. 7, 1–24.

—1925*a*. Papers from Dr. Th. Mortensen's Pacific Expedition 1914–16. XXV. Planarians, Pts. I–III. *Vidensk. Medd. dansk. naturh. Foren.* **79:** 1–84.

—1925*b*. Papers from Dr. Th. Mortensen's Pacific Expedition 1914–16. XXVII. Planarians, Pt. IV. New stylochids. *Vidensk. Medd. dansk. naturh. Foren.* **79:** 97–184.

—1926*a*. Eine Polyclade mit muskulösen Drüsenorganen rings um dem Körper. *Zool. Anz.* **66:** 133–138.

—1926*b*. *Polyporus caecus* Plehn. *Zool. Anz.* **67:** 293–302.

—1927*a*. *Apidioplana.* Eine Polycladengattung mit muskulösen Drüsenorganen. *Göteborgs K. Vetensk.-o. vitterSamh.* (4) **30:** No. 1, 1–116.

—1927*b*. Ductus genito-intestinalis in the polyclads. *Ark. Zool.* **19:** No. 14, 1–15.

—1931. Die Polycladen. *Dt. Südpol. Exped. 1901–1903.* XX, Zool. **12:** 259–304.

Boone, E.S. 1929. Five new polyclads from the Californian coast. *Ann. Mag. nat. Hist.* (10) **3:** 33–46.

Borcea, L. 1927. Données sommaires sur la faune de la mer noire (littoral de Roumanie). *Annls Univ. Jassy* **14:** 536–581.

Bosc, L.A.G. 1802. Histoire naturelle der vers, contenant leur description et leurs moeurs. I: 324 pp. Paris. (Des Planaires pp. 248–262) [2nd Ed. 1830].

—1803. Article "Planaire." *In: Nouveau Dictionnaire d'Histoire naturelle.* **18:** 61–63. Paris.

Boutan, L. 1892. Voyage dans la Mer Rouge. *Rev. biol. N. Fr.* **4:** 173–183.

Boyer, B.C. 1972. Ultrastructural studies of differentiation in the oocyte of the polyclad turbellarian, *Prostheceraeus floridanus. J. Morph.* **136:** 273–296.

Brammertz, W. 1913. Morphologie des Glykogens während der Eibildung und Embryonalenwicklung von Wirbellosen. *Arch. Zellforsch.* **11:** 389–412.

Bresslau, E. 1928–33. Turbellaria. *Kükenthal & Krumbach, Handb. d. Zool.* **2** (1), Lief.1 (1928); 52–112; Lief.9 (1930); 113–192; Lief.16 (1933): 193–293, 314–319.

Briggs, J.C. 1974. *Marine Zoogeography.* McGraw-Hill, New York. 475 pp. (McGraw-Hill series in population biology).

Bruce, J.R. 1948. Additions to faunal records, 1941–46. (Supplement to "The marine fauna of the Isle of Man" 1937). *Rep. mar. biol. Stn. Port Erin* 1945–47 (Nos. 58–60); 39–58.

Brusca, R.C. 1973. Platyhelminths (Flatworms). In: *A Handbook to the common intertidal invertebrates of the Gulf of California,* pp. 63–70. Univ. Arizona Press.

Bush, L. 1964. Phylum Platyhelminthes, Class Turbellaria. *In:* R.I. Smith (Ed.), *Keys to marine invertebrates of the Woods Hole region;* Contrib. No. 2: 30–39. Marine Biol. Lab., Woods Hole.

Bytinski-Salz, H. 1935. Un policlado (*Stylochus pilidium* Lang) dannoso ai parchi ostricoli. *Thalassia* **2:** 1–24.

Carl, G.C. & Guiguet, C.J. 1958. Alien animals in British Columbia. *Handb. B.C. Prov. Mus.* No. 14: 81–83.

Carus, J.V. 1885. Vermes Polycladidea: In his *Prodromus Faunae Mediterraneae* [etc.] Stuttgart. 148–148.

Cheeseman, F.F. 1883. On two new planarians from Auckland Harbour. *Trans. & Proc. N.Z. Inst.* **15:** 213–214.

Cheng, L. & Lewin, R.A. 1975. Flatworms afloat. *Nature, Lond.* **258:** 518–519.

Chien, P. & Koopowitz, H. 1972. The ultrastructure of neuro-muscular systems in *Notoplana acticola,* a free-living polyclad flatworm. *Z. Zellforsch. mikrosk. Anat.* **133:** 277–288.

Child, C.M. 1904*a*. Studies on regulation. IV. Some experimental modifications in form regulation of *Leptoplana* sp. *J. exp. Zool.* **1:** 95–133.

—1904*b*. Studies on regulation. V. The relation between central nervous system and regeneration in *Leptoplana*: posterior regeneration. *J. exp. Zool.* **1:** 463.

—1904*c*. Studies on regulation. VI. The relation between the central nervous system and regeneration in *Leptoplana*: anterior and lateral regeneration. *J. exp. Zool.* **1:** 513–557.
—1905*a*. Studies on regulation. VII. Further experiments on form-regulation in *Leptoplana. J. exp. Zool.* **2:** 253–285.
—1905*b*. Studies on regulation. VIII. Functional regulation and regeneration in *Cestoplana. Arch. Entw. Mech. Org.* **20:** 261–294.
—1905*c*. Studies on regulation. IX. The positions and proportions of parts during regulation in *Cestoplana* in the presence of the cephalic ganglia. *Arch. Entw. Mech. Org.* **20:** 48–75.
—1905*d*. Studies on regulation. X. The positions and proportions of parts during regulation in *Cestoplana* in the absence of the cephalic ganglia. *Arch. Entw. Mech. Org.* **20:** 157–186.
—1906. Contribution towards a theory of regulation. I. The significance of the different methods of regulation in Turbellaria. *Arch. Entw. Mech. Org.* **20:** 380–426.
—1907. Studies on regulation. XI. Functional regulation in the intestine of *Cestoplana. J. exp. Zool.* **4:** 357–398.
—1910. The central nervous system as a factor in the regeneration of polyclad Turbellaria. *Biol. Bull. mar. biol. Lab., Woods Hole* **19:** 333–338.
Ching, H.L. 1977. Redescription of *Eurylepta leoparda* Freeman, 1933 (Turbellaria; Polycladida) a predator of the ascidian *Corella willmeriana* Herdman, 1898. *Can. J. Zool.* **55:** 338–342.
—1978. Redescription of a marine flatworm *Pseudoceros canadensis* Hyman, 1953 (Polycladida: Cotylea). *Can. J. Zool.* **56:** 1372–76.
Christensen, D.J. 1971. Early development and chromosome number of the polyclad flatworm *Euplana gracilis. Trans. Am. microsc. Soc.* **90:** 457–463.
Chumley, J. 1918. *The fauna of the Clyde sea area.* University Press, Glasgow. pp. vi + 200.
Chun, C. 1882. Die Verwandischaftsbeziehungen zwischen Würmern und Cölenteraten. *Biol. Zbl.* **2:** 5–16.
Claparède, E. 1861. Études anatomiques sur les annélides, turbellariés, opalines et grégarines observés dans le Hébrides. *Mem. Soc. phys. Hist. nat.* **16:** 71–164.
—1863. Beobachtungen über Anatomie und Entwickelungsgeschichte wirbelloser Thiere an der Küste von Normandie angestellt. Engelmann, Leipzig. pp. viii + 120.
—1867. De la structure des annelides [etc.] *Arch. Sci. phys. nat.* **30:** 6.
Clark, R.B. & Milne, A. 1955. The sublittoral fauna of two sandy bays on the Isle of Cumbrae, Firth of Clyde. *J. mar. biol. Assoc. U.K.* **34:** 161–180.
Colgan, N. 1907. A note on *Leptoplana tremellaris. Irish Nat.* **16:** 323.
Collingwood, C. 1876. On thirty-one species of marine planarians, collected partly by the late Dr. Kelaart, F.L.S., at Trincomalee, and partly by Dr. Collingwood, F.L.S., in the eastern seas. *Trans. Linn. Soc. London* (2) **1** (Zool.), pt. 3: 83–98.
Corrêa, D.D. 1949. Sôbre o gênere *Zygantroplana. Bolm. Fac. Filos. Ciênc. Univ. S. Paulo* **99** (Zool. No. 14); 173–218.
—1958. A new polyclad from Brazil. *Bolm. Inst. Oceanogr. S. Paulo* **7:** 81–86.
Crothers, J.H. 1966. (Ed.) Dale Fort marine fauna. *Fld. Stud. London* **2** (Suppl.): xxiv + 169.
Crozier, W.J. 1917. On the pigmentation of a polyclad. *Proc. Am. Acad. Arts Sci.* **52:** 723–730.
—1918. On the method of progression in polyclads. *Proc. nat. Acad. Sci. Wash.* **4:** 379–381.
Czerniavsky, V. 1881. Materialia ad Zoographiam ponticam comparatum. Fasc. III. Vermes. *Bull. Soc. Nat.,* Moscow, **55,** No. 4, 213–363.
Dakin, W.J. 1953. *Australian seashores.* Angus & Robertson, London: xii + 379 pp.
Dakin, W.J., Bennett, I. & Pope, E. 1948. A study of certain aspects of the ecology of the intertidal fauna of the New South Wales coast. *Aust. J. Sci. Res. Melbourne* **1B** 2: 176–230.
Dalyell, J.G. 1814. Observations on some interesting phenomena in animal physiology exhibited by several species of planariae, pp. 5–23. Constable, Edinburgh.
—1853. The powers of the creator displayed in the creation; or, observations on life amidst the various forms of the humbler tribes of animated nature: with practical comments and illustrations. **2:** xiii + 359 pp. van Voorst, London.
Danglade, E. 1920. The flatworm as an enemy of the Florida oyster. *Rep. U.S. Commnr Fish. 1918.* Appendix 5: pp. 1–8.
Darwin, C. 1844. Brief descriptions of several terrestrial planariae and of some remarkable marine species, with an account of their habits. *Ann. Mag. nat. Hist.* **14:** 241–251.

Dawydoff, C.N. 1930. Les larves des polyclades des côtes d'Annam. *C.r. hebd. Seanc. Acad. Sci. Paris* **190:** 74–75.

—1940. Les formes larvaires de polyclads et de némertines du plancton indochinois. *Bull. biol. Fr.-Belg.* **74:** 443–496.

—1952. Contribution à l'étude des invertebrés de la faune marine benthique de l'Indochine. *Bull. biol. Fr.-Belg.* (Suppl.) No. 39: 1–158.

Day, J.H., Field, J.R. & Penrith, H.J. 1970. The benthic fauna and fishes of False Bay, South Africa. *Trans. roy. Soc. S. Afr.* **39:** 1–108.

de Beauchamp, P. 1951*a*. *Turbellariés de l'Angola (récoltes de M.A. de Barres Machado). Publ. Cult. Companhia Diam. Angola.* No. 11: 75–84.

—1951*b*. Turbellariés polyclades du Maroc et de Mauritanie. (Première Note). *Bull. Soc. Sci. nat. Maroc.* **39:** (1949): 239–249.

—1961. Classe des Turbellariés. Turbellaria (Ehrenberg, 1831). *In:* P.P. Grasse, *Traite de Zoologie; anatomie, systematique, biologie.* **4,** fasc. 1. (Masson et Cie: 35–212, 887–890.)Paris.

Delle Chiaje, S. 1822–1831. Memorie sulla storia notomia degli animali senza vertebre del regno di Napoli. **1822** Pls.; **1823** vol. I: 1–84; **1824** vol. 1: 1–184; **1825–27** vol. II; **1828** vol. III; **1830** vol. IV; 1–116; **1831** vol. IV: 117–214. Fratelli Fernandes, Napoli.

—1841–1844. *Descrizione e notomia degli animali invertebrati della Sicilia citeriore osservati vivi negli anni 1822–1830.* Parts 1–8. Batteli & Co., Naples.

den Hartog, C. *see* Hartog, C. den.

Deton, W. 1909. L'Étape synaptique dans l'ovogénèse du *Thysanozoon brocchii.* La Cellule **25:** 131–148.

Dicquemare, J.F. 1781. Lists des extraits du portefeuille de M. l'Abbe D. La pellicule animee. *Rozier et Mongez' J. Phys., Paris* **17:** 141–142.

Diesing, C.M. 1836. Helminthologische Beiträge. *Nova Acta Acad. Leopoldina* **18:** 316.

—1850. Systema Helminthum. **I:** 679 pp. Vindobonae.

—**K.M.** 1862. Revision der Turbellarien. Abtheilung: Dendrocoelen. *Sber. Akad. Wiss. Wien* **44,** Abth. I: 485–578.

—1863. Nachtrage zur Revision der Turbellarien. *Sber. Akad. Wiss. Wien* **46:** Abth. I: 173–181.

Dieuzeide, R. & Goëau-Brissonnière, W. 1951. Les prairies de zosteres naines et de cymodocées ("mattes") aux environs d'Alger. *Bull. Stn. Aquic. Pêche Castiglione* (N.S.) No. 3: 5–53.

Domenici, L., Galleni, L. & Gremigni, V. 1974. Primi reporti ultrastrutturali e citochimici sulla ovogenesi di *Notoplana alcinoi* (Turbellaria: Polycladida). *Boll. Zool.* **41:** 480–481.

—1975. Electron microscopical and cytochemical study of egg-shell globules in *Notoplana alcinoi* (Turbellaria: Polycladida). *J. Submicr. Cytol.* **7:** 239–247.

du Bois-Reymond Marcus, Eveline. 1955*a*. On Turbellaria and *Polygordius* from the Brazilian coast. *Bol. Fac. Filos. Ciênc. Univ. S. Paulo* No. 207 (Zool. No. 20): 19–65.

—1955*b*. Chave dos Polycladida do litoral de São Paulo. *Bol. Fac. Filos. Cienc. Univ. S. Paulo* Zool. No. 19: 281–288.

—1957. On Turbellaria. *Anu. Acad. Brasil. Cienc.* **29:** 153–191.

—1958. On South American Turbellaria. *Anu. Acad. Brasil. Cienc.* **30:** 391–417.

—1965. Drei neue neotropische Turbellarien. *Sitz. Ges. naturf. Freunde Berl.* (N.F.) **5:** 129–135.

—1970. On Brazilian Turbellaria. *Am. Zool.* **10:** 546 [Abstract].

— **& Marcus, Ernst.** 1968. Polycladida from Curaçao and faunistically related regions. *Stud. Fauna Curaçao* **26:** 1–106.

Dujardin, F. 1845. *Histoire naturelle des helminthes ou vers intestinaux.* Librarie encyclopédique de Robert, Paris. 654 pp.

Eales, N.B. 1939. *The littoral fauna of Great Britain.* A handbook for collectors. Cambridge University Press. xvii + 301 pp.

Edmonson, C.H. 1945. Natural enemies of shipworms of Hawaii. *Trans. Am. microsc. Soc.* **64:** 220–224.

Ehrenberg, C.G. 1831. *Symbolae physicae. Animali evertebrata [etc.]* Berolini.

—1836. In: *Hemprich & Ehrenberg:* Die Acalephen des Rothen Meeres und der Organisms der Medusen der Ostsee erläutert und auf Systematik angewandt. *Abh. Berl. Akad. (1835): 67.*

Emery, C. 1905. Proposta di una partizione generale dei metazoi. [With discussion of phylogeny of Polycladida]. *Real. Accad. Bologna* **8:** 61–75.

Espinosa, J. 1981. *Stylochus megalops* (Platyhelminthes: Turbellaria), nuevo depredator del ostion en Cuba. *Poeyana* (Inst. zool. Acad. Cie. Cuba) No. 228: 1–5.
Euzet, L. & Poujol, M. 1963. La faune associée a *Mercierella enigmata* Fauvel (Annélide Serpulidae) dans quelques stations des environs de Sète. *Rapp. P.-v. Reun. Cons. perm. int. Explor. Mer.* **17:** 833–842.
Farran, G.P. 1915. Results of a biological survey of Blacksod Bay, Co. Mayo. *Dublin Fish. Sci. Invest.* No. 3: 1–72.
Fischer, E. 1929. Recherches de bionomie et d'oceanographie littorales sur la Rance et la littoral de la Manche. *Annls. Inst. oceanogr. Monaco* (N.S.) **5:** 211–429.
Fischer, P.H. & Duval, M. 1926. Rôle de la tension superficielle dans certaine reptation d'un turbellarié marin (*Leptoplana tremellaris* Oersted). *Annls. Physiol. Physiochem. Biol.* **2:** 1–6.
Fleming, J. 1823. Gleanings of natural history, gathered on the coast of Scotland during a voyage in 1821. *Edinb. phil. J.* **8:** 294–303.
Fleming, K.M.G. 1936. The natural history of Barra, Outer Hebrides. Turbellaria and Nematomorpha. *Proc. phys. Soc. Edinb.* **22:** 264–265.
Forbes, E. & Goodsir, J. 1840. Notice of zoological researches in Orkney and Shetland during the month of June 1839. *Rep. Br. Ass. Advmt. Sci.* 9th Meet. (1839). Trans. of Sects. p. 79.
Francotte, P. 1883. *In:* P.J. van Beneden (Ed.), Compte rendu sommaire des recherches entreprises à la station biologique d-Ostende pendant les mois d'été 1883. *Bull. Acad. r. Belg.* (3) **6:** 465–468.
—1894. *In:* P.J. van Beneden (Ed.), Quelques essais d'embryologie pathologique experimentale. *Bull. Acad. r. Belg.* (3) **27:** 382–391.
—1897. *In:* P.J. van Beneden (Ed.), Recherches sur la maturation, la fecondation et la segmentation chez les polyclades. *Mem. Acad. r. Belg.* **55:** 1–72.
—1898. Recherches sur la maturation, la fecondation et la segmentation chez les polyclades. *Arch. Zool. exp. gen.* (3) **6:** 189–298.
Franzen, A. 1956. On spermiogenesis, morphology of the spermatozoon and biology of fertilization among invertebrates. *Zool. Bidr. Uppsala* **31:** 355–480.
Freeman, D. 1930. Three polyclads from the region of Point Firmin, San Pedro, California. *Trans. Am. microsc. Soc.* **49:** 334–341.
—1933. The polyclads of the San Juan region of Puget Sound. *Trans. Am. microsc. Soc.* **52:** 107–146.
Frey, H. & Leuckart, R. 1847. *Beiträge zur Kenntniss wirbelloser Thiere mit besonderer Berücksichtigung der Fauna des norddeutschen Meeres.* Vieweg & Son, Braunschweig. 170 pp.
Gallagher, M. 1976. Specimens collected by the Rafos Masirah Expedition 1976. *In:* Curry *et al.*, The Royal Air Force Ornithological Society's *Expedition to Masirah Island, 6–26 Oct., 1976:* pp. 36–37.
Galleni, L. 1974. Polycladi delle coste Toscane. I. *Notoplana igiliensis* n.sp., nuovo leptoplanide (Polycladida. Acotylea) dell'isole del Giglio. *Cah. Biol. mar.* **15:** 395–402. (English summary).
—1976*a*. *Stylochoplana maculata* (Quatrefages) un leptoplanide (Polycladida, Acotylea) nuovo per le coste Italiane. *Annali Univ. Ferrara* (N.S.) (Seg. Ecol) **1:** 85–92. (English summary).
—1976*b*. Polyclads of the Tuscan coast. II. *Stylochus alexandrinus* Steinböck and *Stylochus mediterraneus* n.sp. from the rocky shores near Pisa and Livorno. *Boll. Zool.* **43:** 15–25.
—1978*a*. Il genere *Echinoplana* Haswell (Polycladida: Acotylea). *Boll. Zool.* **45:** 214–215.
—1978*b*. Polycladi delle coste toscane. III. *Echinoplana celerrima* Haswell, planoceride nuovo per il mediterraneo e note sul *Echinoplana. Atti Soc. tosc. Sci. nat.* Mem. ser. B, **85:** 139–148. (English summary).
Galleni, L., Mannocci, M., Salghetti, U. & Tongiorgi, P. 1976. Prime osservazioni sull' etoecologie di *Stylochus (Imogine) mediterraneus,* polyclade predatore di mitili. *Mem. biol. mar. Oceangr.* **6** (Suppl.): 62–64. (English translation).
Galleni, L., Ferrero, E., Salghetti, U., Tongiorgi, P. & Salvadego, P. 1977. Ulteriori osservazione sulla predazione di *Stylochus mediterraneus* (Turbellaria, Polycladida) sui mitili e suo orientamente chemiotatico. *Atti. Congr. Soc. ital. Biol. mar.*, Ischia 19–22 Maggio, pp. 259–261.
Galleni, L. & Puccinelli, I. 1975. Karyology of *Notoplana igiliensis* Galleni (Polycladida – Acotylea). *Caryologia* **28:** 375–387.
—1977*a*. Cariologie di polycladi e tricladi maricola della Gran Bretagna. *Atti. IX Congr. Soc. ital. Biol. mar.*, Ischia 19–22 Maggio: 263–264.
—1977*b*. Karyometric analysis in *Thysanozoon brocchii* (Risso) (Turbellaria Polycladida). *Rend. Cl. Sci. fis. mat. nat.* (8) **63:** 436–439.
—1981. Karyological observation on polyclads. *Hydrobiol.* **84:** 31–44.

Gamble, F.W. 1893*a*. Contributions to a knowledge of British marine Turbellaria. *Quart. J. microsc. Sci.* **34:** 433–528.
—1893*b*. The Turbellaria of Plymouth Sound and the neighbourhood. *J. mar. biol. Soc. U.K.* **3:** 30–47.
—1893*c*. Report on the Turbellaria of the L.M.B.C. district. *Proc. Liverpool biol. Soc.* **7:** 148–174.
—1894. The marine zoology of the Irish Sea. *Rept. 64th Meet. Brit. Ass. Adv. Sci.*, Oxford, 1894. p. 324.
—1896. Platyhelminthes and Mesozoa. *In*: S.F. Harmer & A.E. Shipley (Eds), *Cambridge Nat. Hist.* **II:** 1–96. Macmillan & Co., London.
—1900. Report on the Turbellaria. *In*: The fauna and flora of Valencia Harbour on east coast of Ireland. *Proc. roy. Irish Acad.* (3) **5:** 812–814.

Gemmill, J.F. & Leiper, R.T. 1907. Turbellaria of the Scottish Antarctic Expedition. *Trans. roy. Soc. Edinb.* **45,** Pt. III: 819–827.

Gérard, O. 1901. L'Ovocyte de premier ordre du *Prostheceraeus vittatus* avec quelques observations relative à la maturation chez trois autres polyclades. *La Cellule* **18:** 139–248.

Gerzeli, G. 1960. Ricerche istochimiche sulla ovogenesi nei policladi. *Ist. Lomb. Acc. Sci. e Lett.* **94:** 195–204.

Giard, A. 1873. Contributions à l'histoire naturelle des synascidies. *Arch. Zool. exp. gen.* **2:** 481–514.
—1877. Sur les Orthonectida, class nouvelle d'animaux parasites des echinodermes et des turbellaries. *C.r. hebd. Seanc. Acad. Sci. Paris* **85:** 812–814.
—1888. Le laboratoire de Wimereux en 1888 (Recherches fauniques). *Bull. Sci. France Belg.* **19:** 492–513.
—1894*a*. Contributions à la faune du Pas-de-Calais et de la Manche. *C.r. Séanc. Soc. Biol.* (10) **1:** 245–247.
—1894*b*. À propos d'une note de M. Francotte sur quelques essais d'embryologie pathologique expérimentale. *C.r. Seanc. Soc. Biol.* (10) **1:** 385–387.

Girard, C.F. 1850*a*. On the embryology of planariae. *Proc. Am. Ass. Adv. Sci.* (2nd Meeting, Cambridge, Mass., Aug. 1849): 398–402.
—1850*b*. Descriptions of several new species of marine planariae from the coast of Massachusetts. *Proc. Boston Soc. nat. Hist.* **3:** 251–256.
—1851*a*. On the development of *Planocera elliptica*. *Proc. Boston Soc. nat. Hist.* **3:** 348.
—1851*b*. Essay on the classification of nemertes and planariae: preceeded by some general considerations on the primary divisions of the Animal Kingdom. *Proc. Am. Ass. Adv. Sci.* (4th Meeting, New Haven, Conn., Aug. 1850): 258–273. [See also *Am. J. Sci.* (2) **11:** 41–53.]
—1851*c*. Die Planarien und Nemertinen Nordamerikas. *Keller und Tiedemann's Nordamerik. Monatsber. Natur. Heilkunde* **2:** 1–5.
—1852*a*. Sur l'embryogénie des planaires. *Bull. Soc. Sci. nat. Neuchatel* **2:** 300–308.
—1852*b*. Descriptions of a new *Planaria* and a new *Nemertes* from the coast of Florida. *Proc. Boston Soc. nat. Hist.* **4:** 137.
—1853. Descriptions of new nemerteans and planarians from the coast of the Carolinas. *Proc. Acad. nat. Sci. Philadelphia* **6:** 365–367.
—1854*a*. Researches upon nemerteans and planarians. I. Embryonic development of *Planocera elliptica*. *J. Acad. nat. Sci. Philadelphia* **2:** 307–325.
—1854*b*. *In*: William Stimpson, Synopsis of the marine Invertebrata of Grand Manan: or the region about the mouth of the Bay of Fundy, New Brunswick. *Smithsonian Contr. Knowl.* **6** Art. 5: 1–67.
—1893. Recherches sur les planariéns et les némertiens de l'Amerique du Nord. *Annls. Sci. nat.* (7) Zool. **15:** 145–310.

Goette, A. 1878. Zur Entwickelungsgeschichte der Seeplanarien. *Zool. Anz.* **1:** 75–76.
—1881. Zur Entwickelungsgeschichte der Würmer. *Zool. Anz.* **4:** 189.
—1882*a*. Untersuchungen zur Entwickelungsgeschichte der Würmer. Beschreibenden Teil. I. Entwickelungsgeschichte von *Stylochopsis pilidium* n.sp. *Abhandlungen zur Entwickelungsgeschichte der Tiere* 1 Heft. Leipzig. pp. 1–58.
—1882*b*. Zur Entwickelungsgeschichte der marin Dendrocoelen. *Zool. Anz.* **5:** 190–194.
—1884. Untersuchungen zur Entwickelungsgeschichte der Würmer. Vergleichender. Teil. II. Über die Verwandtschaftabeziehungen der Würmer. *Abhandlungen zur Entwickelungsgeschichte der Tiere.* 2 Heft. Leipzig. pp. 50–215.

Götte, A. *see* Goette, A.

Gourret, P. 1890. Nouvelle contribution à la faune pélagique du Golfe de Marseille. *Arch. Biol. Paris* **10:** 324.
Graeffe, E. 1860. Beobachtungen über Radiaten und Würmer in Nizza. *Neue Denkschr. schweiz. naturf. Ges.* **17:** 1–59.
Graff. L. von 1878. Kurze Berichte über fortgesetzte Turbellarien-studien. *Z. wiss. Zool.* 30 (Suppl.): 457–465.
—1882. *Monographie der Turbellarien.* **I.** Rhabdocoelida. Engelmann, Leipzig. pp. xii + 442.
—1886. Turbellarien von Lesina. *Zool. Anz.* **9:** 338–342.
—1890. *Enantia spinifera* der Repräsentant einer neuen Polycladen-Familie. *Mitt. naturw. Ver. Steierm.* (1889): 1–16.
—1892*a*. Sur une planaire de la mer des sargasses (*Stylochoplana sargassicola* Mertens). *Bull. Soc. zool. Fr.* **17:** 146–147. [See also *Verh. dt. zool. Ges.* **2** (Vers.): 117–119.]
—1892*b*. Pelagischen Polycladen. *Z. wiss. Zool.* **55:** 189–219.
—1893. *In:* W. Saville-Kent, *The Great Barrier Reef of Australia; its products and potentialities,* p. 362. W. H. Allen, London.
—1903. Die Turbellarien als Parasiten und Wirte. *Festschr. Univ. Graz (1902),* pp. vi + 66.
—1904–05. Turbellaria [Literature] In: Bronn's *Kl. Ordnung Tierreichs,* Bd. **IV,** Abt. 1c, Lief. 63–71, pp. 1733–1940.
—1907. Turbellaria [Literature, contd.] *In*: Bronn's *Kl. Ordnung Tierreichs,* Bd. **IV,** Abt. 1c, Lief. 75–79, pp. 1986–2040.
—1912–17. Turbellaria [Literature, contd.] *In*: Bronn's *Kl. Ordnung. Tierreichs,* Bd. **IV:** 2601–3369.
Graw, W.L. & Darsie, L.D. 1918. Notes on flat worms at Laguna Beach. *J. Ent. Zool.* **10:** 68.
Gray, J.E. 1860. *In*: W.H. Pease, Descriptions of new species of planaridae collected in the Sandwich Islands. *Proc. zool. Soc. London* Pt. 28: 37–38.
Grube, A.E. 1840. *Actinien, Echinodermen und Würmer des adriatischen und Mittelmeers, nach eigenen Sammlungen beschreiben.* Königsberg. 92pp.
—1855. Bemerkungen über einige Helminthen und Meerwürmer. *Arch. Naturgesch.* **21:** 137–158.
—1864. *Die Insel Lussin and ihre Meeresfauna. Nach einem sechswochentlichen Aufenthalte geschildert.* Bresslau. pp. vi + 116.
—1868. Ueber Land und Seeplanarien. *Jber. schles. Ges. vaterl. Kult.* **45:** 45–46.
—1871. Ueber die Fauna des Baikalsee's, sowie über einige Hirudineen und Planarien andered Faunen. *In*: Bericht über die Thatigheit der naturwissenschaftlichen Section der Schlesischen Gesellschaft im Jahre 1871, erstattet von Grube und Römer. pp. 27–28.
Guerin-Meneville, F.E. 1844. *Iconographie du Règne animal de G. Cuvier.* Paris 1827–1844. Tom. **II.** Planches des animaux invertebres, Zoophytes. Pl. XI, figs 3–7; Tom. **IV,** Zoophytes, p. 14.
Hadenfeldt, D. 1929. Das Nervensystem von *Stylochoplana maculata* und *Notoplana atomata. Z. wiss. Zool.* **133:** 586–638.
Hadzi, J. 1944. *Turbelarijska Teorija Knidarijev.* [Turbellarien-Theorie der Knidarier.] *Slov. Akad. Znan. Um., Ljubljana.* 238 pp. (Slovenian, with German summary).
—1958. Zur Diskussion über die Abstammung der Eumetazoen. *Verh. dtsch. zool. Ges.* (**1952**): 169–179. (*Zool. Anz.,* Suppl. 21).
—1963. *The evolution of the Metazoa.* Pergamon Press, Oxford. 499pp.
Hallez, P. 1878*a*. Contributions à l'histoire des turbellariés. 1 re. Note: Sur le developpement des turbellariés. *Bull. scient. Dép. Nord.* (2) **1:** 193–195.
—1878*b*. Considérations au sujet de la segmentation des oeufs. *Bull. scient. Dép. Nord* (2) **1:** 227–229.
—1878*c*. Contributions à l'histoire des turbellariés. 4me Note. *Bull. scient. Dép. Nord* (2) **1:** 251–260.
—1878*d*. Considerations sur la determination des plans de segmentation dans l'embryogenie du *Leptoplana tremellaris. Bull. scient. Dép. Nord* (2) **1:** 264–266.
—1879*a*. Contributions à l'histoire des turbellariés. *Thesis,* Lille, 213 pp. (see also *Bull. scient. Dép. Nord* **11:** 325–338).
—1879*b*. Contributions à l'histoire naturelle des turbellariés. *Trav. Inst. zool. Lille* **2:** 213 pp.
—1888. Draguages effectués dans le Pas-de-Calais pendant les mois d'Aout et Septembre 1888. II. Les fonds côtiers. *Rev. biol. Nord. Fr.* **1:** 102–108.
—1892. Morphogénie générale et affinités des turbellariés. (Introduction à une embryologie comparie de ces animaux.) *Trav. Mém. Fac. Lille,* **2:** 29 pp.
—1893. Catalogue des turbellariés (Rhabdocoelides, Triclades et Polyclades) du Nord de la France et de la côte Boulonnaise. *Rev. biol. Nord Fr.* **5:** 135–158 and 165–197.

—1894. *Catalogue des Rhabdocelides, Triclades & Polyclades du Nord de la France.* 2nd ed. L. Danel, Lille. 239 pp.
—1899. Régénération et heteromorphose. *Revue scient. Paris* (4) **12:** 506–507.
—1900*a.* Rhabdocoeles (*Gyrator notops* Dugès), triclades (*Dendrocoelum lacteum* (Müller)), polyclades (*Leptoplana tremellaris* Oersted). In: *Zoologie descriptive. Anat.-Hist et Dissect. formes typiq. d'Invert.* **I:** 449–584. Paris.
—1900*b.* Régénération comparée chez les polyclades et les triclades. *C.r. Ass. fr. Avanc. Sci.* 28me Session, Pt. I (1899): 270–271.
—1900*c* Hétéromorphoses comparées chez les polyclads et les triclades. *C.r. Ass. fr. Avanc. Sci.* 28me Session, Pt. I (1899); 271.
—1905*a.* Notes fauniques. *Arch. Zool. exp. gén.* (4) **3,** Notes et Revue: xlvii–lii.
—1905*b.* Note préliminaire sur des polyclades recueillis dans l'expedition antarctique du "Francais". *Bull. Soc. zool. France* **30:** 124–127.
—1907. Polyclades et Triclades maricoles. *Expéd. Antarct. franc.* (1903–05), Vers, pp. 1–26.
—1911*a.* Sur les terminaisons nerveuses dans l'épiderme des planaires. *Arch. Zool. exp. gén.* (5) **7** (Notes et Revue, pp. xx–xxii.)
—1911*b.* Double fonction des ovaires de quelques polyclades. *C.r. hebd. Séanc. Acad. Sci. Paris* **153:** 141–143.
—1913. Polyclades et Triclades maricoles. *Deuxième Expéd. Antarct. frac.* (1908–10), pp. 1–70.
Harman, M.T. & Stebbing, F.M. 1928. The maturation and segmentation of the eggs of *Leptoplana* (sp.). *Publs. Puget Sound mar. biol. Stat.* **6:** 239–251.
Hartog, C. den, 1968. Marine triclads from the Plymouth area. *J. mar. biol. Ass. U.K.* **48:** 209–223.
—1977. Turbellaria from intertidal flats and salt-marshes in the estuaries of the south-western parts of the Netherlands. *Hydrobiologia* **52:** 29–32.
Harvey, W.H. 1857. *The Seaside Book* 4th edn. van Voorst, London. 8vo. p. 157.
Haswell, W.A. 1907*a.* A genito-intestinal canal in polyclads. *Zool. Anz.* **31:** 643–644.
—1907*b.* Observations on Australian polyclads. *Trans. linn. Soc. Lond.* (2) Zool. **9:** 465–485.
Heape, W. 1888. Preliminary report upon the fauna and flora of Plymouth Sound. *J. mar. biol. Ass. U.K.* **1:** 153–193.
Heath, H. 1907. A new turbellarian from Hawaii. *Proc. Acad. nat. Sci. Philadelphia* **59:** 145–148.
—1928. A sexually mature turbellarian resembling Müller's larva. *J. Morph.* **45:** 187–207.
— **& McGregor, E.A.** 1913. New polyclads from Monterey Bay, California. *Proc. Acad. nat. Sci. Philadelphia* **64:** 455–488.
Hendelberg, J. 1965. On different types of spermatozoa in Polycladida, Turbellaria. *Ark. Zool.* **18:** 287–304.
—1974*a.* Spermiogenesis, sperm morphology, and biology of fertilization in the Turbellaria. *In:* N.W. Riser & M.P. Morse (Eds) *Biology of the Turbellaria,* pp. 148–164. L.H. Hyman, Mem. Vol. McGraw-Hill, New York (McGraw-Hill series in the Invertebrates).
—1974*b.* Polyclader vid svenska västkusten. [Polyclads of the Swedish west coast.] *Zool. Rev.* **36:** 3–18. [English summary.]
Henley, C. 1974. Platyhelminthes (Turbellaria). *In*: A.C. Giese & J.S. Pearse (Eds), *Reproduction of marine invertebrates,* Vol. 1, *Acoelomate and pseudocoelomate metazoans,* pp. 267–343. Academic Press, New York & London.
Herdman, W.A. 1894. The marine zoology of the Irish Sea – second report of the committee. *Rept. Brit. Ass. Adv. Sci. Oxford,* p. 324.
Herzig, E.M. 1905. *Laidlawia trigonopora* n.g., n.sp. *Zool. Anz.* **29:** 329–332.
Hesse, R. 1897. Untersuchungen über die Organe der Lichtempfindung bei niederen Thieren. II. Die Augen der Plathelminthen, insonderheit der tricladen Turbellarien. *Z. wiss. Zool.* **62:** 527–582.
Hill, M.G. 1974. Report of the zoology section. *Rept. Trans. Soc. guernes* **19:** 263–264.
Hofker, J. 1930. Faunistische Beobachtungen in der Zuidersee während Trockenlegung. *Z. Morph. Okol. Tiere* **18:** 189–216.
Holleman, J.J. 1972. Marine turbellarians of the Pacific Coast. I. *Proc. biol. Soc. Washington* **85:** 405–412.
—1974. Order Polycladida. *In*: B.N. Kozloff (Ed.), *Keys to the marine invertebrates of Puget Sound, the San Juan Archipelago and adjacent regions,* pp. 28–33. Univ. of Washington Press, Seattle.
Hoyle, W.E. 1889. On the deep-water fauna of the Clyde sea-area. *J. Linn. Soc. Lond.* (Zool.) **20:** 458.

Hurley, A.C. 1975. The establishment of populations of *Balanus pacificus* Pilsbry (Cirripedia) and their elimination by predatory turbellarian. *J. Anim. Ecol.* **44:** 521–536.
—1976. The polyclad flatworm *Stylochus tripartitus* Hyman as a barnacle predator. *Crustaceana* **31:** 110–111.
Hutton, F.W. 1879. Catalogue of the hitherto described worms of New Zealand. *Trans. N.Z. Inst.* **11:** 314–327.
Hyman, L.H. 1938. Faunal notes. *Bull. Mt. Desert Is. biol. Lab.*: 24–25.
—1939*a*. New species of flatworms from North, Central and South America. *Proc. U.S. natn. Mus.* **86,** No. 3055: 419–439.
—1939*b*. Some polyclads of the New England coast, especially of the Woods Hole region. *Biol. Bull. mar. biol. Lab. Woods Hole* **76:** 127–152.
—1939*c*. Polyclad worms collected on the Presidential Cruise of 1938. *Smithsonian miscell. Collns.* **98,** No. 17: 1–13.
—1939*d*. Acoel and polyclad Turbellaria from Bermuda and the Sargassum. *Bull. Bingham Oceanogr. Coll.* **7,** Art. I: 1–26.
—1939*e*. Atlantic coast polyclads. *Anat. Rec. Philadelphia* **75** (Suppl.): 153. (Abstract).
—1939*f*. A new polyclad genus of the family Discocelidae, with some remarks on the family. *Vest. ceskol. zool. Spolnec. Praze* **6–7:** 237–246.
—1940*a*. The polyclad flatworms of the Atlantic coast of the United States and Canada. *Proc. U.S. natn. Mus.* **89:** 449–495.
—1940*b*. Revision of the work of Pearse and Walker on littoral polyclads on New England and adjacent parts of Canada. *Bull. Mt. Desert Is. biol. Lab.* (Research Abstrs. for 1939): 14–20.
—1944*a*. Marine Turbellaria from the Atlantic coast of North America. *Am. Mus. Novitates,* No. 1266: 1–15.
—1944*b*. A new Hawaiian polyclad flatworms associated with *Teredo*. *Occ. Pap. Bernice P. Bishop Mus.* **18:** 73–75.
—1950. A new Hawaiian polyclad, *Stylochoplana inquilina,* with commensal habits. *Occas. Pap. Bernice P. Bishop Mus.* **20:** 55–58.
—1951. *The Invertebrates:* Vol. II. Platyhelminthes and Rhynchocoela; the acoelomate Bilateria. McGraw-Hill, New York. vii + 572 pp.
—1952. Further notes on the turbellarian fauna of the Atlantic coast of the United States. *Biol. Bull. mar. biol. Lab. Woods Hole* **103:** 195–200.
—1953*a*. The polyclad flatworms of the Pacific coast of North America. *Bull. Amer. Mus. nat. Hist.* **100:** 265–392.
—1953*b*. Some polyclad flatworms from the Galapagos Islands. Allan Hancock Pacific Expedn. **15:** 183–210.
—1954*a*. The polyclad genus *Pseudoceros,* with special reference to the Indo-Pacific region. *Pacific Sci.* **8:** 219–225.
—1954*b*. Some polyclad flatworms from the Hawaiian Islands. *Pacific Sci.* **8:** 331–336.
—1954*c*. Free-living flatworms (Turbellaria) of the Gulf of Mexico. *Fishery Bull. Fish. Wildl. Serv. U.S.* **55:** 301–302.
—1954*d*. Key to the more common polyclads from Monterey Bay to Vancouver Island. *In*: Light, Smith, Pitelka & Weesner, *Intertidal Invertebrates of the Central Californian Coast,* 2nd ed., pp. 50–54. Univ. California Press, Berkeley & Los Angeles.
—1955*a*. The polyclad flatworms of the Pacific coast of North America: Additions and corrections. *Am. Mus. Novitates* No. 1704: 1–11.
—1955*b*. Some polyclad flatworms from the West Indies and Florida. *Proc. U.S. natn. Mus.* **104:** 115-150.
—1955*c*. Some Polyclad flatworms from Polynesia and Micronesia. *Proc. U.S. natn. Mus.* **105:** 65–82.
—1955*d*. Miscellaneous marine and terrestrial flatworms from South America. *Am. Mus. Novitates* No. 1742: 1–33.
—1955*e*. A further study of the polyclad flatworms of the West Indian region. *Bull. mar. Sci. Gulf Carib.* **5:** 259–268.
—1958. Turbellaria. *Rept. B.A.N.Z.A.R.E. Antarct. Res. Exped.* **6B,** Pt. 12: 277–290.
—1959*a*. A further study of Micronesian polyclad flatworms. *Proc. U.S. natn. Mus.* **108:** 543–597.
—1959*b*. Some Turbellaria from the coast of California. *Am. Mus. Novitates* No. 1943: 1–17.
—1959*c*. Some Australian polyclads (Turbellaria). *Rec. Austr. Mus.* **25:** 1–17.

—1960. Second report on Hawaiian polyclads. *Pacific Sci.* **14:** 308–309.
Jacubowa, L. 1906. Polycladen von Neu-Britannien und Neu-Caledonien. *Jena. Z. Naturw.* **41:** 113–158.
—1908. A new species of *Planocera (P. gilchristi)* from South Africa. *Trans. S. Afr. philos Soc.* **17:** 145–149.
—1909. [Die Polycladida der Bucht von Sevastopol]. *Mém. Acad. Sci. St. Petersburg* (8) **24:** 1–32. (In Russian.)
Jameson, H.L. 1897. Additional notes on the Turbellaria of the L.M.B.C. District. *Trans. Liverpool biol. Soc.* **11:** 163–181.
Jennings, J.B. 1970. Digestive physiology of the Turbellaria. *Am. Zool.* **10:** 549 (Abstract).
—1971. Parasitism and commensalism in the Turbellaria. *Adv. Parasit.* **9:** 1–32.
—1974. Digestive physiology of the Turbellaria. *In*: N.W. Riser & M.P. Morse (Eds), *Biology of the Turbellaria,* pp. 173–197. L.H. Hyman Mem. Vol. McGraw-Hill, New York (McGraw-Hill series in the Invertebrates).
Jensen, O.S. 1878. Turbellaria ad litora norvegiae occidentalis. *Bergens Mus. Skr.* Raekke I, No. 1: 1–98.
Johnson, M.E. & Snook, H.J. 1927. *Seashore animals of the Pacific coast.* Macmillan & Co. New York. xiv + 656 pp.
Johnston, G. 1832. Illustrations in British Zoology. 3. *Planaria cornuta. Mag. nat. Hist. Lond.* **5:** 344–346, 429 and 678.
—1836. Illustrations in British Zoology. 52. *Planaria subauriculata. Mag. nat. Hist. Lond.* **9:** 16–17.
—1846. An index to the British annelides. *Ann. Mag. nat. Hist.* **16** (Suppl.): 435–462.
—1865. *A catalogue of the British non-parasitical worms in the collection of the British Museum.* British Museum, London. 365 pp.
Jokiel, P.L. & Townsley, S.J. 1974. Biology of the polyclad *Prosthiostomum (Prosthiostomum)* sp., a new coral parasite from Hawaii. *Pacific Sci.* **28:** 368–373.
Jones, N.S. 1939. Some recent additions to the off-shore fauna of Port Erin. *Rept. mar. biol. Stn. Port Erin* **52:** 18–32.
Joubin, L. 1899. Recherches sur la faune des turbellariés des côtes de France. *C.r. Ass. fr. Avanc. Sci.* **2:** 520–579.
Juei-Un Shu & Yao-Sung Liu. 1981. Biological studies on the oyster predator *Stylochus inimicus. In: Reports on Fish Diseases Research (III) CAPD Fisheries series* No. 3: 39–51.
Kaburaki, T. 1918. Zoological results of a tour of the Far East. Brackish-water polyclads. *Mems. Asiatic Soc. Bengal* **6:** 183–192.
—1923*a*. Notes on Japanese polyclad turbellarians. *Annotnes zool. jap.* **10,** Art. 19: 191–201.
—1923*b*. The polyclad turbellarians from the Philippine Islands. *U.S. natn. Mus., Bull. 100,* **1,** pt. 10: 635–649.
Kaltenbach, Dr. 1915. Beitrag zur Kenntnis der Centrosomenbildung in *Thysanozoon brocchii. Arch. Zellforsch.* **13:** 525–529.
Karling, T.G. 1966. On nematocysts and similar structures in turbellarians. *Acta zool. fenn.* **116:** 1–28.
Kato, K. 1934*a*. *Leptostylochus gracilis,* a new polyclad turbellarian. *Proc. imp. Acad.* **10:** 374–377.
—1934*b*. Polyclad turbellarians from the neighborhood of the Mitsui Institute of Marine Biology. *Jap. J. Zool.* **6:** 123–138.
—1935*a*. *Stylochoplana parasitica* sp. nov., a polyclad parasitic in the pallial groove of the chiton. *Annotnes zool. jap.* **15:** 123–129.
—1935*b*. *Discoplana takewakii* sp.nov., a polyclad parasitic in the genital bursa of the ophiuran. *Annotnes zool. jap.* **15:** 149–156.
—1936*a*. A new polyclad turbellarian, *Cryptocelis amakusaensis,* from southern Japan. *Jap. J. zool.* **7:** 17–20.
—1936*b*. Notes on *Paraplanocera. Jap. J. Zool.* **7:** 21–31.
—1937*a*. Polyclads collected in Idu, Japan. *Jap. J. Zool.* **7:** 211–232.
—1937*b*. Polyclads from Korea. Jap. J. Zool. **7:** 233–240.
—1937*c*. Three polyclads from northern Japan. *Annotnes zool. jap.* **16:** 35–38.
—1937*d*. *Stylochus aomori,* a new polyclad from northern Japan. *Annotnes zool. jap.* **16:** 39–41.
—1937*e*. The fauna of Akkeshi Bay. V. Polycladida. *Annotnes Zool. jap.* **16:** 124–133.
—1937*f*. Thirteen new polyclads from Misaki. *Jap. J. Zool.* **7:** 347–371.

—1938*a*. Polyclads from Amakusa, southern Japan. *Jap J. Zool.* **7:** 559–576.

—1938*b*. Polyclads from Seto, middle Japan. *Jap. J. Zool.* **7:** 577–593.

—1938*c*. On a pelagic polyclad, *Planocera pellucida* (Mertens) from Japan. *Zool. Mag. (Japan)* **50:** 230–232.

—1939*a*. Polyclads in Onagawa and vicinity. *Sci. Rept. Tôhoku imp. Univ.* (4) Biol. **14:** 65–79.

—1939*b*. Report on the biological survey of Mutsu Bay. 34. The polyclads of Mutsu Bay. *Sci. Rept. Tôhoku imp. Univ.* (4) Biol. **14:** 141–153.

—1940. On the development of some Japanese polyclads. *Jap. J. Zool.* **8:** 537–573.

—1943*a*. A new polyclad with anus. *Bull. biogeogr. Soc. Japan* **13:** 47–53.

—1943*b*. Polyclads from Formosa. *Bull. biogeogr. Soc. Japan* **13:** 69–77.

—1943*c*. Polyclads from Palao. *Bull. biogeogr. Soc. Japan* **13:** 79–90.

—1944. Polycladida of Japan. *Sigenkagaku Kenkyusko (J. Res. Inst. nat. Resources)* **1:** 257–318.

—1948. The distribution of polyclad Turbellaria. *Proc. biogeogr. Soc. Tokyo* No. 1: 39–41.

—1968. Platyhelminthes. *In*: M. Kumé & K. Dan (Eds); translated by J.C. Dan, *Invertebrate Embryology*, pp. 125–143. NOLIT, Publ. House, Belgrade, Yugoslavia.

Keferstein, W. 1869. Beiträge zur Anatomie Entwicklungsgeschichte einiger Seeplanarien von St. Malo. *Abh. K. Ges. Wiss. Göttingen* **14:** 1–38.

Kelaart, E.F. 1858. Description of new and little known species of Ceylon nudibranchiate molluscs, and zoophytes. *J. Ceylon Branch Roy. Asiatic Soc.* **3** (1856–58): 84–139.

Kennel, J. von 1879. Die in Deutschland gefundenen Landplanarien *Rhynchodemus terrestris* O.F. Müller und *Geodesmus bilineatus* Mecznikoff. *Arb. zool. zootom. Inst. Würzburg* **5:** 120–160.

Kensler, C.B. 1964. The Mediterranean crevice habitat. *Vie Milieu* **15B:** 947–977.

—1965 [1966]. Distribution of crevice species along the Iberian Peninsula and northwest Africa. *Vie Milieu* **16B:** 851–887.

Kingsley, J.S. 1901. Preliminary catalogue of the marine Invertebrata of Casco Bay, Maine. *Proc. Portland Soc. nat. Hist.* **2,** pt. 5: 159–183.

Kikuchi, T. 1966. An ecological study on animal communities of the *Zostera marina* belt in Tomioka Bay, Amakushu. *Publs. Amakusa mar. biol. Lab.* **1:** 1–106.

—1968. Faunal list of the *Zostera marina* belts in Tomioka Bay, Amakusa, Kyushu. *Publs. Amakusa mar. biol. Lab.* **1:** 163–192.

Kirk, J.W. 1882. On some new marine planarians. *Trans. N.Z. Inst.* **14:** 267–268.

Klinckowström, A. von, 1896. Beiträge zur Kenntnis der Eireifung und Befruchtung bei *Prostheceraeus vittatus. Arch. microsc. mikr. Anat.* **48:** 587–605.

Knappert, B. 1865. Bijdragen tot de ontwikkelings-geschiedenis Zoetwater-Planariën. *Natuurk. Verh. Utrecht Genoot. Wet.* I Deel, 4 Stuk. 39 pp.

Koehler, R. 1885. Contribution a l'etude de la faune littorale des îles Anglo-Normandes (Jersey, Guernsey, Herm et Sark). *Annls. Sci. nat.* (6) Zool., **20:** 11–62. (Also *Bull. Soc. Nancy* (2) **18:** 76–82.) English translation in *Ann. & Mag. nat. Hist.* (5) **18:** 229, 290, 350.

Koningsberger, J.C. 1891. Over het watervaatstelsel bÿ de Polycladen. *Tijdschr. neder. dierk. Vereen.* (2) **3:** 83.

Koopowitz, H. 1970. Organization and physiology of flatworm nervous system. *Am. Zool.* **10:** 549 (Abstract).

—1974. Some aspects of the physiology and organization of the nerve plexus in polyclad flatworms. *In*: N.W. riser & M.P. Morse (Eds) *Biology of the Turbellaria*, pp. 198–212. L.H. Hyman Mem. Vol.; McGraw-Hill, New York (McGraw-Hill series in the Invertebrates).

—& Chien, P. 1974. Ultrastructure of the nervous plexus in flatworms. 1. Peripheral organization. *Cell Tissue Res.* **155:** 337–351.

—, Silver, D. & Rose, G. 1976. Primitive nervous systems: Control and recovery of feeding behavior in the polyclad flatworm, *Notoplana acticola. Biol. Bull. mar. biol. Lab. Woods Hole* **150:** 411–425.

Kowalewsky, A.O. 1870. Bemerkung über der Bau des Darm-Canals der dendrocölen Planaria. *Schrift. Gesellsche. Naturf. Kiew* **1:** 109–110.

—1880. Ueber *Coeloplana metschnikowii. Zool. Anz.* **3:** 140.

Laidlaw, F.F. 1902. The marine Turbellaria, with an account of the anatomy of some of the species. *Fauna & Geogr. Maldive & Laccadive Archipelagoes* **1:** 282–312.

—1903*a*. Notes on some marine Turbellaria from Torres Straits and the Pacific, with a description of new species. *Mem. Proc. Manchester lit. philos. Soc.* **47,** Art. 5: 1–12.

—1903*b*. On a land planarian from Hulule, Male Atoll, with a note of *Leptoplana pardalis* Laidlaw. *Fauna & Geogr. Maldive & Laccadive Archipelagoes* **2:** 579–580.

—1903*c*. On a collection of Turbellaria Polycladida from the Straits of Malacca. (Skeat Expedition, 1899–1900). *Proc. zool. Soc. Lond.* **1:** 301–318.

—1903*d*. On the marine fauna of Zanzibar and British East Africa, from collections made by Cyril Crossland in the years 1901 and 1902. – Turbellaria: Polycladida. *Proc. zool. Soc. Lond.* **2:** 99–113.

—1903*e*. Suggestions for a revision of the classification of the polyclad Turbellaria. *Mem. Proc. Manchester lit. philos. Soc.* **48:** Art. 4: 1–16.

—1904*a*. Notes on some polyclad Turbellaria in the British Museum. *Mem. Proc. Manchester lit. philos. Soc.* **48,** Art. 15: 1–6.

—1904*b*. Report on the polyclad Turbellaria collected by Professor Herdman at Ceylon, in 1902. *Roy. Soc. Rept. to Ceylon Governm. on Pearl Oyster Fisheries of Gulf of Manaar,* Pt. II: 127–136.

—1906. On the marine fauna of the Cape Verde Islands, from collections made in 1904 by Mr C. Crossland — The polyclad Turbellaria. *Proc. zool. Soc. Lond.* **2:** 705–719.

Landers, W.S. & Rhodes, E.W., Jr. 1970. Some factors influencing predation by the flatworm *Stylochus ellipticus* (Girard) on oysters. *Chesapeake Sci.* **11:** 55–60.

Lanfranchi, A., Bedini, C. & Ferrero, E. 1981. The ultrastructure of the eyes in larval and adult polyclads (Turbellaria). *Hydrobiologia* **84:** 267–275.

Lang, A. 1879. Untersuchungen zur vergleichenden Anatomie und Histologie des Nervensystems der Plathelminthen. I. Das Nervensystem der marinen Dendrocoelen. *Mitt. zool. Stn. Neapel.* **1:** 459–488.

—1881*a*. Les relations des platyelmes avec les coelentérés d'un côte et les hirundiniés de l'autre. *Arch. Biol. Paris* **2:** 533–552.

—1881*b*. Der Bau of *Gunda segmentata* und die Verwandtschaft der Plathelminthen mit Coelenteraten und Hirudineen. *Mitt. zool. Stn. Neapel,* **3:** 187–251.

—1881*c*. Sur mode particulier de copulation chez les vers marins dendrocèles ou polycladès. *Archs. Sci. phys. nat.* (3) **6:** 308–309.

—1884. Die Polycladen (Seeplanarien) des Golfes von Neapel und der angrenzenden Meeresabschnitte. Eine Monographie. *Fauna Flora Golfes v. Neapel,* Leipzig, **11:** ix + 688 pp.

—1903. Segmentation of *Discocelis tigrina.* Table. *Arch. Zool. exp. gen.* (3) **10:** No. 4. pl. xlii.

Lankester, E.R. 1866. Annelida and Turbellaria of Guernsey. *Ann. Mag. Nat. Hist.* (3) **17:** 388–390.

Laverack, M.S. & Blackler, M. 1974. *Fauna and flora of St. Andrews Bay.* Scottish Academic Press, Edinburgh & London.

Lawler, A.R. 1969. Occurrence of the polyclad *Coronadena mutabilis* (Verrill, 1873) in Virginia. *Chesapeake Sci.* **10:** 65–67.

Le Conte, L. 1851. Zoological notes. *Proc. Acad. nat. Sci. Philadelphia* **5:** 316–320.

Lenhoff, H.M. 1964. Rearing response of a marine flatworm to the lowering of light intensity. *Nature, Lond.* **201:** 841–842.

Lesson, M. 1830. *Autour du Monde, exécuté par ordre du Roi sur la corvette de la Majeste "La Coquille", pendant les annees 1822–1825.* Zoologie, **2:** pt. 1, p. 453.

Leuckart, F.S. 1828. *In*: Ed. Rüppell, *Atlas zu der Reise im nordlichen Afrika. Neue wirbellose Thiere des rothen Meers,* pp. 11 & 15. Bearbeitet von Dr. Ed. Rüppell & Dr F.S. Leuckart. Brönner, Frankfurt a.M.

Leuckart, R. 1856. Nachträge und Berichtigungen zu dem ersten Bande von J. Van der Hoeven's Handbuch der Zoologie. Eine systematisch Geordnete übersicht der hauptsächlichsten neueren Leistungen über die Zoologie der wirbellosen Thiere, pp. 107–111. [Anhang zum II Bande dieses Handbuches.] Voss, Leipzig.

—1859. Bericht über die wissenschaftlichen Leistungen in der Naturgeschichte de niederen Thiere während des Jahres 1858. Turbellarii. *Arch. Naturgesch.* **25,** II: 179–183.

—1863. Bericht über die wissenschaftlichen Leistungen in der Naturgeschichte der niederen Thiere während der Jahre 1861 und 1862. Turbellarii. *Arch. Naturgesch.* **29,** II: 163–175.

Levetzow, K.G. von, 1936. Beitrage zur Reizphysiologie der Polycladen Strudelwürmer. *Z. vergl. Physiol.* **23:** 721–726.

—1939. Die Regeneration der Polycladen Turbellarien. Wilhelm- *Roux Arch. EntwMech. Org.* **139:** 780–818.

—1943. Zur Biologie and Verdaunngsphysiologie der Polycladen Turbellarien. *Zool. Anz.* **141:** 189–196.

Levinsen, G.M.R. 1879. Bidrag til Kunskab om Grønlands Turbellarie-Fauna. *Vidensk. Meddr. dansk. naturh. Foren.* **1879–1880:** 165–204.

Linkletter, L.E., Lord, E.L. & Dadswell, M.J. 1977. *A checklist of marine fauna and flora of the Bay of Fundy.* St. Andrews, New Brunswick: Huntsman Mar. Lab. 68 pp.

Linnaeus, C. 1758. *Systema naturae [etc.]* Edit. decima, reformata **1:** 823 pp. Laurentii salvii, Holmiae.

Lo Bianco, S. 1888. Notizie biologiche riguardanti specialmente il periodo di maturità sessuale degli animali del Golfo di Napoli. *Mitt. zool. Stn. Neapel,* **8:** 385–440.

—1899*a*. Notizie biologiche riguardanti specialmente il periodo di maturità sessuale degli animali del Golfo di Napoli. *Mitt. zool. Stn. Neapel,* **13:** 48–573.

—1899*b*. The methods employed at the Naples Zoological Station for the preservation of marine animals. *Bull. U.S. Mus.* No. 30: 42 pp.

—1909. Notizie biologiche riguardanti specialmento il periodo di maturita sessuale degli animali del Golfo di Napoli. *Mitt. zool. Stn. Neapel.* **19:** 513–761.

Loeb, J. 1894. Beitrage zur Gehirnphysiologie der Würmer. *Pfleiger's Arch. Physiol.* **56:** 249–258.

—1905. *Studies in general physiology,* pp. 73, 77, 221, 287, 343, 352–356. University of Chicago.

Lytwyn, M.W. & McDermott, J.J. 1976. Incidence, reproduction and feeding of *Stylochus zebra,* a polyclad turbellarian symbiont of hermit crab. *Mar. Biol.* **38:** 365–372.

MacGinitie, G.E. & MacGinitie, N. 1949. *Natural history of marine animals.* McGraw-Hill, New York. xii + 473 pp.

Maitland, R.T. 1851. *Fauna Belgii septentrionalis. Descriptio systematica Animalium, Belgii septentrionalis adjectis synonymis nec non locis in quibus reperiuntur.* van der Hoek, Leyden. xxxviii + 234 pp.

Makino, S. 1951. Polycladida. In: *An Atlas of the chromosome numbers in animals.* Iowa State College Press, Ames, Iowa. p. 10.

Marcus, Ernst, 1947. Turbellários marinhos do Brasil. *Bol. Fac. Filos. Cienc. Univ. S. Paulo,* Zoologia No. 12: 99–215. (English summary).

—1948. Turbellaria do Brazil. *Bol. Fac. Filos. Cienc. Univ. S. Paulo,* Zool. No. 13: 111–243. (English summary).

—1949. Turbellaria brasileiro (7). *Bol. Fac. Filos Cienc. Univ. S. Paulo.* Zool. No. 14: 7–155. (English summary).

—1950. Turbellaria brasileiros (8). *Boll. Fac. Filos. Cienc. Univ. S. Paulo.* Zool. No. 15: 5–191. (English summary).

—1952. Turbellaria brasileiros (10). *Bol. Fac. Filos. Cienc. Univ. S. Paulo.* Zool. No. 17: 5–187. (English summary).

—1954*a*. Reports of the Lund University Chile Expedition 1948–1949. 11. Turbellaria. *Acta Univ. lund.,* N.F. **49,** No. 13: 1–115.

—1954*b*. Turbellaria brasileiros — XI. *Papeis avuls. Dep. Zool. S. Paulo* **11:** 419–489. (English summary).

Marcus, Eveline & Marcus, Ernst. 1951. Contributions to the natural history of Brazilian Turbellaria. *Comun. Zool. Mus. Hist. nat. Montevideo* **3,** No. 63: 1–25.

—1966. Systematische Übersicht der Polycladen. *Zool. Beitr.* **12:** 319–343.

Marenzeller, E. von, 1886. Poriferen, Anthozoen, Ctenophora und Würmer von Jan Mayen gesammelt von Dr F. Fischer. *Die internat. Polarforschung,* 1882–83. *Die Osterreich Polarstation Jan Mayen,* Wien, III, 16 pp. (*Kaiserlich Akad. Wissenschaften,* Vienna.)

Marinescu, A. 1971. Contributions à la connaissance des polyclades du littoral roumain de la Mar Noire. *Trav. Mus. hist. nat. "Gr. Antipa"* **11:** 41–47. (French summary).

Martin, C.H. 1908. The nematocysts of Turbellaria. *Quart. J. microsc. Sci.* N.S. (2) **52:** 261–270.

—1914. A note on the occurrence of nematocysts and similar structures in the various groups of the Animal Kingdom. *Biol. Zentbl.* **34:** 248–273.

Mast, S.O. 1911. Preliminary report on reactions to light in marine Turbellaria. *Yearbk. Carnegie Inst.,* Washington, **9:** 131–133.

McIntosh, W.C. 1874. On the invertebrate marine fauna and fishes of St. Andrews. *Ann. Mag. nat. Hist.* (4) **19:** 144–155.

—1875. *The marine invertebrates and fishes of St. Andrews.* Adam & Charles Black, Edinburgh. vi + 186 pp.

Meixner, A. 1907*a*. Polyclades recueilles par M. Ch. Gravier dans le golfe de Tadjourah en 1904, *Bull. Mus. natn. Hist. nat., Paris,* No. 2: 164–172.

—1907*b*. Polycladen von der Somaliküste, nebst einer Revision der Stylochinen. *Z. wiss. Zool.* **88:** 385–498.

Melouk, M.A. 1940. A new polyclad from the Red Sea, *Cryptophallus aegypticus* nov.spec. *Fouad I Univ., Cairo, Bull. Fac. Sci.* No. 22: 125–140.

—1941. *Leptoplana nadiae,* a new acotylean polyclad from Ghardaqa (Red Sea). *Fouad I Univ. Cairo, Bull. Fac. Sci.* No. 23: 41–49.

Mereschkowsky, K.S. 1879. Über einige neue Turbellarien des weissen Meeres. *Arch. Naturgesch.* **45:** Bd. 1: 35–55.

Mertens, H. 1832. Untersuchungen über den innern Bau verschiedener in der See lebender Planarien. *Mem. Acad. Sci. St. Petersbourg* (6) **2:** 3–17.

Metschnikoff, E. 1877. [Untersuchungen über die Entwickelung der Planarien.] *Zap. novoross. Obshch. Estest.* **5,** Heft. 1, 16 pp. (In Russian).

Mettrick, D.F. & Boddington, M.J. 1972. The chemical composition of some marine and freshwater turbellarians. *Carib. J. Sci.* **12:** 1–7.

Meyer, F. 1922. Polycladen von Koseir (Rotes Meer). (Kollektion Professor Klunzinger). *Arch. Naturgesch.* Abt. A, **87,** Heft 10: 138–158.

Micoletsky, H. 1910. Die Turbellarienfauna des Golfes von Triest. *Arb. zool. Inst. Univ. Wien* **18:** 167–182.

Miller, M. & Batt, G. 1973. *Reef and Beach Life of New Zealand.* W. Collins (N.Z.) Ltd, Auckland & London. 141 pp.

Milne-Edwards, H. 1859. *Lecons sur la physiologie et l'anatomie comparée de l'homme et des animaux faites a la Faculté des Sci. d. Paris* **5:** 455–458.

Minot, C.S. 1877*a*. Studien an Turbellarien. Beitrage zur Kenntnis der Plathelminthen. *Arb. zool.-zootom. Inst. Würzbrg* **3:** 405–472.

—1877*b*. On the classification of some of the lower worms. *Proc. Boston Soc.* **19:** 17–25.

Möbius, K. 1875. *Jahresbericht der Commission zur wissenschaftlichen Untersuchung der deutschen Meere in Kiel für die Jahre 1872, 1873.* [Vermes, p. 154.] Wiegandt, Hempel Parey, Berlin.

Montagu, G. 1813. Description of several new or rare animals, principally marine, found on the south coast of Devonshire. *Trans. Linn. Soc. Lond.* **11:** 1–26.

Monti, C.R. 1900. La rigenerazione nelle planairie marine. *Mem. R. Ist. Lombardo* **19:** 1–16.

Moore, A.R. 1933. On the rôle of the brain and cephalic nerves in the swimming and righting movements of the polyclad worm, *Planocera reticulata. Sci. Rep. Tohoku Univ.* (4) (iv Biol.): 193–200.

Moore, H.B. 1937. Marine fauna of the Isle of Man. *Proc. Liverpool biol. Soc.* **50:** 1–293.

Morgan, L.V. 1905. Incomplete anterior regeneration in the absence of the brain in *Leptoplana littoralis. Biol. Bull. mar. biol. Lab. Woods Hole* **9:** 187–193.

Morris, R.H., Abbot, D.P. & Haderlie, E.C. 1980. *Intertidal Invertebrates of California.* Stamford Univ. Press. 690 pp.

Morton, J. & Miller, M. 1968. *The New Zealand sea shore.* W. Collins (N.Z.) Ltd, Auckland & London. 638 pp.

Moseley, H.N. 1874. On the anatomy and histology of the land-planarians of Ceylon, with some account of their habits and a description of two new species, and with notes on the anatomy of some European aquatic species. *Phil. Trans. roy. Soc. London:* 105–171.

—1877. On *Stylochus pelagicus,* a new species of pelagic planarian, with notes on other pelagic species, on the larval forms of *Thysanozoon* and of a gymnosomatous pteropod. *Quart. J. microsc. Sci.* **17:** 23–34.

Mueller, M. 1852. *Observationes anatomicae de vermibus quibusdam maritimis.* Diss. inaug. Berolini: pp. 27–30. Schade.

Müller, J. 1850. Ueber eine eigenthumliche Würmlarve, aus der Classe der Turbellarien und aus der Familie der Planarien. *Arch. Anat. Physiol.* **1850:** 485–500.

—1854. Ueber verschiedene Formen von Seethieren. *Arch. Anat. Physiol.* **1854:** 69–98.

Müller, O.F. 1774. *Vermium terrestrium et fluviatilium, seu animalium infusoriorum, helminthicorum, et testaceorum, non marinorum, succincta historia.* Vol. I, Pt. 2, 72 pp. Havniae et Lipsiae.

—1776. *Zoologiae danicae prodromus, seu animalium Daniae et Norvegiae indigenarum characteres, nomina, et synonyma imprimis popularium.* pp. xxxii + 282. Havniae.

—1779. *Zoologia danica seu animalium Daniae et Norvegiae rariorum ac minus notorum descriptiones et historia.* **I:** pp. viii + 104. Havniae.

—1788*a*. *Zoologia danica seu animalium Daniae et Norvegiae rariorum ac minus notorum descriptiones et historia*. **I** (3rd ed.), pp. vi + 52. Havniae.

—1788*b*. *Zoologia danica seu animalium Daniae et Norvegiae rariorum ac minus notorum descriptiones et historia*. **II:** iv + pp. 56. Havniae.

Narchi, W. 1975. Levantamento de fauna macroscopica da bahia de Santos. *Cienc. Cult. S. Paulo* **23** (Suppl.); 280.

Nicoll, W. 1929. Vermes. *Zool. Rec. 1928 London* **65:** 46.

—1934. Vermes. *Zool. Rec. 1933 London* **70:** 76.

Okada, S., Igarashi, T. & Kobayashi, K. 1971. Invertebrates and fishes of Oskoro Bay and neighbouring area. *Bull. Plankt. Soc. Japan* **18:** 59–72. (In Japanese).

Olmsted, J.M.D. 1922*a*. The rôle of the nervous system in the regeneration of polyclad Turbellaria. *J. exp. Zool.* **36:** 49–56.

—1922*b*. The rôle of the nervous system in the locomotion of certain marine polyclads. *J. exp. Zool.* **36:** 57–66.

Örsted, A.D. 1843. Forsøg til en ny Classification af Planarierne (Planariea Dugés) grundet paa mikroskopiskanatomiske Undersøgelser. *Kroyer's Naturhist. Tidsskr.* (I) **4:** 519–582.

—1844*a*. *Entwurf einer systematischer Eintheilung und speciellen Beschreibung der Plattwürmer auf microscopische Untersuchungen gegründet*. Copenhagen. pp. viii + 96.

—1844*b*. *De regionibus marinis. Elementa topographiae historico naturalis freti*. Öresund. *Diss. inaug.* Havniae. pp. x + 88.

—1845. Fortegnelse over Dyr, samlede i Christianiafjord ved Drøbach fra 21–24 Juli 1844. *Kroyer's Naturhist. Tidsskr.* (2) **1:** 400–427.

Overstreet, R.M. 1978. *Marine maladies? Worms, Germs and other Symbionts from the northern Gulf of Mexico*. Mississippi–Alabama Sea Grant Constitution, Ocean Springs. 140 pp.

Palombi, A. 1924*a*. Diagnosi di nuove specie di polycladi della R.N. "Liguria". Note preliminaire. *Boll. Soc. nat. Napoli* **35** (Ser. II, XV) (1923): 33–37.

—1924*b*. Polycladi pelagici. Raccolte planktonische fatte dalla R.N. "Liguria" nel viaggio di circonnavigazione del 1903–05. *Pubbl. Ist. Studi sup. prat. Firenze* Sect. III (Sci. fis. nat.) **1:** 28 pp.

—1927. Notizie faunistiche sul Canale di Suez. *Ann. R. Liceo Scient. Avellino*: p. 94.

—1928. Report on the Turbellaria. Cambridge Expedition to Suez Canal, 1924. *Trans. zool. Soc. Lond.* **22:** 579–631.

—1929. Gli apparecchi copulatori della famiglia Polyposthiidae (Policladi Acotilei). Ricerche sistematiche e considerazioni sulle affinità dell'ordine dei polycladi. *Boll. Soc. nat. Napoli* **15** (Ser. II, Vol. II), Anno XLII (1928): 196–209.

—1930. *Notoplana alcinoi, Thysanozoon brocchii. Faune et Flore de la Mediterr.*, Paris, 4 pp.

—1931*a*. Turbellari della Nuova Guinea. *Mem. Mus. r. Hist. nat. Belg.* **2** Fasc. 8: 1–14. Brussels, Hors Ser. Res. Sci. Voyage Indes Orient, Neerland.

—1931*b*. *Stylochus inimicus* sp. nov. polyclade acotileo commensale di *Ostrea virginica* Gmelin delle coste della Florida. *Boll. Zool.*, Naples, **2:** 219–226.

—1936. Policladi liberi e commensali raccolti sulle coste del Sud Africa. Della Florida e del golfo di Napoli. *Arch. zool. Italiano* **23:** 1–45.

—1938. Turbellari del Sud Africa. Secondo contributo. *Arch. zool. Italiano* **25:** 329–383.

—1939*a*. Turbellari del Sud Africa. Policladi di East London. Terzo contributo. *Arch. zool. Italiano* **28:** 123–149.

—1939*b*. Turbellaria Polycladidea. *Mem. Mus. r. Hist. nat. Belg.* (2) **15:** 95–114.

—1940. Policladi delle coste occidentali dell' Africa. *Rev. Zool. Bot. Afric.* **33:** 109–121.

Panceri, P. 1875. Catalogo degli annelidi gefirei e turbellarie d'Italia. *Attia Soc. Ital. Sci. nat.* **18:** 246–247.

Pastorino, E. & Canu, S. 1965. Osservazioni intorno allo fauna marina bentonica di Camoglia Dintorni (Riviera Ligure di Levanti). *Doriana* **4,** No. 159: 1–9.

Patterson, J.T. & Wieman, H. 1912. The uterine spindle of the polyclad *Planocera inquilina*. *Biol. Bull. mar. biol. Lab. Woods Hole* **23:** 271–292.

Pearse, A.S. 1936. Zoological names, a list of phyla, classes and orders prepared for Section F., *Amer. Assoc. Adv. Sci.* 24 pp. Duke Univ. Press.

—1938. Polyclads of the east coast of North America. *Proc. U.S. natn. Mus.* **86:** 67–98.

—1949. Observations on flatworms and nemerteans collected at Beaufort, N.C. *Proc. U.S. natn. Mus.* **100:** No. 3255; 25–38.

— **& Littler, J.W.** 1938. Polyclads of Beaufort, N.C. *J. Elisha Mitchell sci. Soc.* **54:** 235–244.

— **& Walker, A.M.** 1939. Littoral polyclads from New England, Prince Edward Island, and Newfoundland. *Bull. Mt. Desert I. biol. Lab.:* 15–22.

— **& Wharton, G.W.** 1938. The oyster "leech", *Stylochus inimicus* Palombi, associated with oysters on the coasts of Florida. *Ecol. Monogr.* **8:** 605–655.

— **& Williams, L.G.** 1951. The biota of the reefs off the Carolinas. *J. Elisha Mitchell sci. Soc.* **67:** 133–161.

Pease, W.H. 1860. See Gray, J.E.

Pedersen, K.J. 1966. The organization of the connective tissue of *Discocelides langi* (Turbellaria, Polycladida). *Z. Zellforsch.* **71:** 94–117.

Pennant, T. 1777. *British Zoology* (4th ed.) **4:** viii + 154 pp.

Pinto, J. dos S. 1947. Breves apontamentos sôbre alguns Turbelários marinhos de Portugal. *Bol. Soc. portug. Ciênc. nat.* **15:** 91–96.

Plehn, M. 1896*a*. Neue Polycladen, gesammelt von Herrn Kapitän Chierchia bei der Erdumschiffung der Korvette Vettor Pisani, von Herrn Prof. Dr Kükenthal im nördlichen Eismeer und von Herrn Prof. Dr Semon in Java. *Jena Z. Naturw.* **30:** 137–176.

—1896*b*. Die Polycladen der Plankton-Expedition. *Ergebn. Plankton Expdn. Humboldt-Stiftung,* Kiel u. Leipzig, **2:** 1–14.

—1896*c*. Polycladen von Ambon. *In*: Semon, *Zoologische Forschungsreisen in Australien und dem Malayischen Archipel.* Denk. med. naturw. Ges. Jena **8** 5 (Syst.) Lief. 3: 327–334.

—1897. Drei neue Polycladen. *Jena Z. Naturw.* **31:** 90–99.

—1898. Polycladen von Ternate. *Abh. Senckenbg. naturf.* **24:** 145–146.

—1899. Ergebnisse einer Reise nach dem Pacific. (Schauinsland 1896–1897). Polycladen. *Zool. Jb.* (Syst.) **12:** 448–452.

Poche, F. 1926. Das System der Platodaria. *Arch. Naturg.* **A:** 1–458.

Pope, E.C. 1943. Animal and plant communities of the coastal rock-platform at Long Reef, New South Wales. *Proc. Linn. Soc. N.S.W.* **68:** 221–254.

Poulter, J.L. 1970. A new species of the genus *Pericelis,* a polyclad flatworm from Hawaii. *Am. Zool.* **10:** 553 (Abstract).

—1974. A new species of the genus *Pericelis,* a polyclad flatworm from Hawaii. *In*: N.W. Riser & M.P. Morse (Eds), *Biology of the Turbellaria,* pp. 93–107. L.H. Hyman Mem. Vol. McGraw-Hill, New York. (McGraw-Hill series in the Invertebrates).

—1975. Hawaiian polyclad flatworms. Prosthiostomids. *Pacific Sci.* **29:** 317–339.

Powell, A.W.B. 1947 Native animals of New Zealand. *Auckland Mus. Handb. Zool.* 96 pp. Unity Press, Auckland.

Prenant, M. 1919. Recherches sur les rhabdites des turbellariés. *Arch. Zool. exp. gen.* **58:** 219–250.

—1922. Recherches sur le parenchyme des platyhelminthes. Essai d'histologie comparée. *Arch. morph. gén. exp.* **5:** 1–174.

Procter, W. 1933. *Biological Survey Mount Desert Region Marine Fauna.* **5:** 402 pp. Wistar Institute of Anatomy and Biology, Philadelphia.

Provenzano, A.J., Jr. 1961. Effects of the flatworm *Stylochus ellipticus* (Girard) on oyster spat in two salt water ponds in Massachusetts. *Proc. natl. shellfish. Ass.* **50:** 83–88.

Prudhoe, S. 1944. On some polyclad turbellarians from the Cayman Islands. *Ann. Mag. nat. Hist.* (11) **11:** 322–334.

—1945*a*. On the species of the polyclad genus *Paraplanocera. Ann. Mag. nat. Hist.* (11) **12:** 195–202.

—1945*b*. Two notes on trematodes. *Ann. Mag. nat. Hist.* (11) **12:** 381–383.

—1950*a*. On some polyclad turbellarians from Burma. *Ann. Mag. nat. Hist.* (12) **3:** 41–50.

—1950*b*. On the taxonomy of two species of pelagic polyclad turbellarians. *Ann. Mag. nat. Hist.* (12) **3:** 710–716.

—1952. The "Manihine" Expedition to the Gulf of Aqaba 1948–1949. IV. Turbellaria: Polycladida. *Bull. Br. Mus. nat. Hist.* (Zool.) **1** (8): 175–179.

—1968. A new polyclad turbellarian associating with a hermit crab in the Hawaiian Islands. *Pacific Sci.* **22:** 408–411.

—1977. Some polyclad turbellarians new to the fauna of the Australian coasts. *Rec. Aust. Mus.* **31:** 586–604.

—1982*a*. Polyclad flatworms (Phylum Platyhelminthes). *In*: S.A. Shepherd & I.M. Thomas (Eds),

Marine invertebrates of southern Australia, Part I, pp. 220–227. *Handb. Flora & Fauna of South Australia.* Handbook Committee of the South Australian Government, Adelaide.

—1982*b*. Polyclad turbellarians from the southern coasts of Australia. *Rec. S. Aust. Mus.* **18:** 361–384.

—1982*c*. British polyclad turbellarians. *In*: D.M. Kermack & R.S.K. Barnes (Eds), *Synopses of the British Fauna* (NS.) No. 26: 27 pp. Linn. Soc. Lond. & Estuarine and Brackish-water Sci. Ass. Cambridge Univ. Press.

Pruvot, G. 1897. Essai sur les fonds et la faune de la Mache occidentale (côtes de Bretagne) comparés a ceux du golfe du Lion. Catalogue des invertebrés benthiques du golfe du Lion et de la Manche occidentale, avec leur habitat. *Arch. Zool. exp. gen.* (3) **5:** 617–660.

Purchon, R.D. 1948. Studies on the biology of the Bristol Channel. xvii. The littoral and sublittoral fauna of the northern shores, near Cardiff. *Proc. Bristol nat. Soc.* **27:** 285–310.

—1957. Studies on the biology of the Bristol Channel. xviii. The marine fauna of five stations on the northern shores of the Bristol Channel and Severn Estuary. *Proc. Bristol nat. Soc.* **29:** 213–226.

Quatrefages, A. de, 1845. Études sur les types inférieurs de l'embranchement des annelés: mémoire sur quelques planairées marines appartenant aux genres *Tricelis* (Ehr.), *Polycelis* (Ehr.), *Prosthiostomum* (Nob.), *Proceros* (Nob.), *Eolidiceros* (Nob.) et *Stylochus* (Ehr.). *Annls Sci. nat.* (3) Zool. **4:** 129–184.

Quoy, J.R.C. & Gaimard, P. 1833. Zoologie de la voyage de la corvette l'Astrolabe, exécuté par ordre du roi pendant les annees 1826–1829 sous le commandement de M. Dumont d'Urville. **4:** 326.

Randall, J.E. & Emery, A.R. 1971. On the resemblance of the young of the fishes *Platax pinnatus* and *Plectorhynchus chaetodontoides* to flatworms and nudibranchs. *Zoologica N.Y.* **56:** 115–117.

Ranzi, S. 1928. Nuovo turbellario policlade del Golfo di Napoli (*Cestoplana raffaelei* n.sp.). *Boll. Soc. nat. Napoli* **39** ((2) 19): 3–11.

Reisinger, E. 1923. Turbellaria. *In*: P. Schulte's *Biologie de Thiere Deutschlands* **4,** Lief.6: 1–64. Berlin.

—1925. Untersuchungen am Nervensystem der *Bothrioplana semperi* Braun. (Zugleich ein Beitrag zur Technik der vitalen Nervenfärbung und zur vergleichenden Anatomie des Plathelminthennervensystems.) *Z. Morph. Okol. Tiere* **5:** 119–149.

Remane, A. 1929. Die Polycladen der Keiler Förde (8 Beitrag zur Fauna der Kieler Bucht). *Schr. naturw. Ver. Schlesw.-Holst.* **19:** 73–79.

Retzius, G. 1906. Die Spermien der Turbellarien. *In*: Retzius, *Biologisch. Untersuchung.,* Stockholm (N.F.) **13:** 41–44.

Ricketts, E.F. & Calvin, J. 1939 [1948]. *Between Pacific tides. An account of the habits and habitats of some five hundred of the common, conspicuous sea-shore invertebrates of the Pacific coast between Sitka, Alaska and northern Mexico.* 2nd ed. Stanford Univ. Press.

—1952. *Between Pacific tides* [etc.] 3rd ed.

Riedl, R. 1953. Quantitative ökologie Methoden marines Turbellarienforschung. *Öst. Zool. Z.* **4:** 108–145.

—1959. Turbellarien aus submarinen Hohlen. I. Archoophora. Ergebnisse der Österreichischen Tyrrhenia-Expedition 1952, Teil VII. *Pubbl. Staz. zool. Napoli,* **30** (Suppl.); 178–208.

Riser, N.W. 1970. Biological studies on *Taenioplana teredini* Hyman, 1944. *Am. Zool.* **10:** 533 (Abstract).

—1974. Epilogue. *In*: N.W. Riser & M.P. Morse (Eds), *Biology of the Turbellaria,* pp. 517–524. L.H. Hyman Mem. Vol. McGraw-Hill, New York (McGraw-Hill series in the Invertebrates).

Risso, A. 1818. Sur quelques gasteropodes nouveaux, nudibranches et testibranches observes dans la mer de Nice. *J. Phys. Chim. Hist. nat.* **87:** 368–376.

—1826. *Histoire naturelle des principales productions de l'Europe meridionale et particulierement de celles des environs de Nice et des Alpes maritimes.* **5:** viii + 403. Paris.

Ritter-Záhony, R. von, 1907. Turbellarien: Polycladen. *In*: *Hamburger Magalhaensische Sammelreise,* Lfg.8, No. 1: 1–19.

Robertson, J.A. 1932. A cursory survey of the Bear Island trawling ground. *Rapp. Cons. Explor. Mer Copenhagen* **81:** 115–139.

Rzhepishevskji, I.K. 1979. Acorn barnacles eating [*sic*] away by *Stylochus pilidium. Biologiya Morya* **48:** 23–28. (Russian; English summary).

Sabussow, H.P. 1897. [Vorläufiger Bericht über die Turbellarien der Insel von Solowetzk.] *Protok. Zased. Obshch. Estest. Kazan,* 1896–1897 No. 167: 15 pp. (In Russian).

—1900. Beobachtungen über die Turbellarien der Insel von Solowetzk. *Trudy Obshch. Estest. imp. kazan. Univ.* **34:** 208 pp. (Russian; German summary).

—1905. Zur Kenntniss der Turbellarienfauna des Golfes von Villefrance s.m. *Zool. Anz.* **28:** 486–489.

Sandô, H. 1964. Faunal list of the *Zostera marina* region at Kugurizaka coastal waters, Aomore Bay. *Bull. mar. biol. Sta. Asemashi* **12:** 27–35.

Schechter, V. 1943. Two flatworms from the oyster-drilling snail, *Thais floridana haysae* Clench. *J. Parasit.* **29:** 362.

Schmankewitsch, W. 1873. Ueber die Wirbellosen Thiere der Limane bei Odessa. *Zap. novoross. Obshch. Estest.* **2:** 275–276, 278–280, 294. (In Russian).

Schmarda, L.K. 1859. *Neue wirbellose Thiere beobachtet und gesammelt auf einer Reise um die Erde 1853 bis 1857.* Bd. I: Turbellarien, Rotatorien und Anneliden. 1 Halfte: 66 pp. W. Engelmann, Leipzig.

Schmidt, O. 1861. Untersuchungen über Turbellarien von Corfu und Cephalonia, nebst Nachträgen zu früheren Arbeiten. *Z. wiss. Zool.* **11:** 1–30.

—1878. Die niederen Thiere. *In*: Brehm's *Thierleben*, 2nd Aufl., Bd. 10: 147–154. Verlag des Bibliographischen Instituts, Leipzig.

Schmidtlein, R. 1878. Beobachtungen über Trachtigkeits-und Eiablage-Perioden verschiedener Seethiere, Januar 1875 bis July 1878. *Mitt. zool. Stn. Neapel.* **1:** 124–136.

—1880. Vergleichende Übersicht über das Erscheinen grösser pelagischer Thiere und Bermerkungen über Fortpflaszungsverhaltnisse einiger Seethiere im Aquarium. *Mitt. zool. Stn. Neapel.* **2:** 162–175.

Schneider, K.C. 1902. Lehrbuch der vergleichenden Histologie. Jena. 293–309.

Schockaert, R. 1900. Nouvelle recherches sur la maturation de l'ovocyte de premier ordre du *Thysanozoon brocchii. Anat. Anz.* **18:** 30–33.

—1901. L'Ovogénèse chez le *Thysanozoon brocchii* (Premiere partie). *La Cellule* **18:** 35–137.

—1902. L'Ovogénèse chez le *Thysanozoon brocchii* (Deuxieme partie). *La Cellule* **20:** 101–177.

—1905. La fecondation et la segmentation chez le *Thysanozoon brocchii. La Cellule* **22:** 5–37.

Schultz, E. 1901. Über Regeneration bei Polycladen. *Zool. Anz.* **24:** 527–529.

—1902. Aus dem Gebiete der Regeneration. II. Über die Regeneration bei Turbellarien. *Z. wiss. Zool.* **72:** 1–30.

—1905. Études sur la régénération chex les Vers. *Protok. St. Petersbg. Obshch.* **34** (15): 1–136.

Schultze, M. 1851. Stäbchenformige Körper in den Haut der Turbellarien. *Tagesber. Fortschr. Nat. Heilk.* (Zool. II) No. 371: 137–141.

—1854. Bericht über einige im Herbst 1853 an der Küste des Mittelmeeres angestellte zootomische Untersuchungen. *Verh. phys.-med. Ges. Würzb.* **4:** 222–224.

Selenka, E. 1881*a*. Über eine eigenthümliche Art der Kernmetamorphose. *Biol. Zbl.* **1:** 492–497.

—1881*b*. Zur Entwicklungschichte der Seeplanarien. *Biol. Zbl.* **1:** 229.

—1881*c*. Die Keimblätter der Planarien. *Sber. phys.-med. Soc. Erlangen* **1881:** 4 pp.

—1881*d*. *Zoologische Studien.* II. Zur Entwicklungsgeschichte der Seeplanarien. Ein Beitrag zur Keimblätterlehre und Descendenztheorie. Engelmann, Leipzig. 44 pp.

Shelford, R. 1901–04. Report on the Sarawak Museum for *1900*, pp. 21–22 (Also reports for *1901*, p. 18 and for *1903*, p. 18.)

Simonette, A. & Delle Cave, L. 1978. Notes on new and strange Burgess Shale fossils (Middle Cambrian of British Columbia). *Atti Soc. tosc. Sci. nat. Mem.* (A) **85:** 45–49.

Skerman, T.M. 1960. Note on *Stylochus zanzibaricus* Laidlaw (Turbellaria, Polycladida), a suspected predator of barnacles in the Port of Auckland, New Zealand. *N.Z. J. Sci.* **3:** 610–614.

Smith, E.H. 1960. On a new polyclad commensal of Prosobranchs. *An. Acad. An. Acad. bras. Cienc.* **32:** 385–390.

—1961. A new commensal polyclad from Panama. *The Veliger* **4:** 69–70.

Sopott-Ehlers, B. & Schmidt, P. 1975. Interstielle Fauna von Galapagos. XIV. Polycladida (Turbellaria). *Mikrofauna Meeresbodens* **54:** 193–222.

Southern R. 1912. Platyhelmia. *Proc. r. Irish Acad.* (Clare Island Survey, pt. 56): pp. 1–18.

—1936. Turbellaria of Ireland. *Proc. r. Irish. Acad.* **43**(B): 43–72.

Stasek, C.R. 1966. Type specimens in the California Academy of Sciences, Department of Invertebrate Zoology. *Occ. Papers Calif. Acad. Sci.* No. 51: 38 pp.

Stauber, L.A. 1941. The polyclad *Hoploplana inquilina thaisana* Pearse, 1938, from the mantle-cavity of oyster drills. *J. Parasit.* **27:** 541–542.

Stead, D.G. 1907. Preliminary note on the wafer (*Leptoplana australis*), a species of dendrocoelous turbellarian worm, destructive to oysters. *Dept. Fish. N.S.W.*, Sydney, 6 pp.

Steinbeck, J. & Ricketts, E.F. 1941. *Sea of Cortez.* A leisurely journal of travel and research. With a scientific appendix comprising materials for a source book on the marine animals of the Panamic faunal province. (Polyclads pp. 335–338) Viking Press, New York.

Steinböck, O. 1931. Marine Turbellaria. *In: Zoology of the Faroes,* Copenhagen. **1:** No. 8. 26 pp.

—1932. Die Turbellarien des arktischen Gebietes. *In*: Römer u. Schaudinn, *Fauna Arctica* **6:** 295–342.

—1933. Die Turbellarienfauna der Umbegung von Rovigno. *Thalassia, Jena,* **1,** No. 5: 1–33.

—1937. The fisheries grounds near Alexandria. 14. Turbellaria. *Hydrobiol. Fish. Direct. Min. Comm. Ind. Egypt* Notes & Mem. No. 25: 1–15.

—1938. Marine Turbellaria. *In*: Fridriksson *et al.* (Eds), *Zoology of Iceland* **2,** Pt. 9: 1–26. Copenhagen.

Steiner, J. 1898. *Die Functionen des Centralnervensystems und ihre Phylogenese.* 3 Abth. Die Wirbellosen Thiere. Braunschweig. pp. 53–57.

Steopoe, I. 1931. Sur les parasomes des ovocytes de *Leptoplana tremellaris. C.r. Séanc. Soc. Biol.* **107:** 1187–1189.

—1933. Parasomii din ovocytile de *Leptoplana tremellaris* si *Prosthiostomum siphunculus. Bul. Soc. Nat. Romania* **4:** 21.

—1934. Observations cytologiques sur les cellules nerveuses de *Leptoplana tremellaris* et *Prosthiostomum siphunculus. C.r. Séanc. Soc. Biol.* **115:** 1315–1317.

Steven, D. 1938. The shore fauna of Amerdloq Fjord, West Greenland. *J. Anim. Ecol.* **7:** 53–70.

Stimpson, W. 1853. Synopsis of the marine Invertebrata of Grand Manan: or the region about the mouth of the Bay of Fundy, New Brunswick. *Smithsonian Contr. Knowl.* **4** Art. 5: 1–67.

—1855*a*. Descriptions of some of the new marine Invertebrata from the Chinese and Japanese Seas. *Proc. Acad. nat. Sci. Philadelphia* **7:** 375–384.

—1855*b*. Descriptions of some new marine Invertebrata. *Proc. Acad. nat. Sci. Philadelphia* **7:** 385–394.

—1857. Prodromus descriptionis animalium evertebratorum quae in expeditione ad oceanum, pacificum septentrionalem a Republica Federata missa, Johanne Rodgers Duce observavit et descripsit. Pars. I. Turbellaria Dendrocoela. *Proc. Acad. nat. Sci. Philadelphia* **9:** 19–31.

—1861. On the genus *Peasia. Amer. J. Sci. Arts* (2) **31:** 134.

Stone, G. & Koopowitz, H. 1976. Primitive nervous systems; electrophysiology of the pharynx of the polyclad flatworm, *Enchiridium punctatum. J. Exp. Biol.* **65** (3); 627–642.

Stossich, A. 1882. Prospetto della fauna del mare Adriatico. *Boll. Soc. Adriat. Sc. nat. Trieste* **7:** 168–247.

Strand, E. 1926. Miscellanea nomenclatoria zoologica. *Arch. Naturg.* **92:** 30–75.

Stricht, O. van der, 1894. De la figure achromatique de l'ovule en mitose chez le *Thysanozoon brocchii. Verh. Anat. Ges.* **8:** 223–232.

—1896*a*. Le premier amphiaster de l'ovule de *Thysanozoon brocchii.* Une figure mitosique peutelle retrograder? *Bibliogr. anat. par. 5 la dir. de M.A. Nicolás* **4:** 27–30.

—1896*b*. Anomalies lors de la formation de l'amphiaster de rebut. *Bibliogr. anat. par. 5 la dir. de M.A. Nicolás* **4:** 33–34.

—1897*a*. La maturation et la fécondation de l'oeuf de *Thysanozoon brocchii. Ass. franc. Adv. Sci. 25th. Session (Carthage 1896)* 2nd pt. Notes & Mem. pp. 484–489.

—1897*b*. Les ovocentres et les spermocentres de l'ovule de *Thysanozoon brocchii. Verh. Anat. Ges. II Vers in Gent.* pp. 92–99.

—1898. La formation des deux globules polaires et l'apparition des spermocentres dans l'oeuf de *Thysanozoon brocchii. Arch. Biol. Paris* **15:** 367–451.

—1899. Étude de plusiere anomalies intéressantes lors de la formation des globules polaires. Étude de la sphère attractive ovulaire à l'état pathologique dans les oocytes en voie de dégénérescence. *Livre Jub. Charles Van Bambecke.* pp. 225–270.

Strøm, H. 1768. Beskrivelse over Norske Insecter; Andet Stykke. *K. norsk. Vidensk. Selsk. Skr.* **4:** 313–371.

Studer, T. 1876. Ueber Seethiere aus dem antarctischen Meere. *Mitt. naturf. Ges. Bern*: 75–84.

—1879. Die Fauna von Kerguelensland. Verzeichniss der bis jetzt auf Kerguelensland beobachteten Thierspecies nebst kurzen Notezin über ihr Vorkommen und ihre zoogeographischen Begiehungen. *Arch. Naturgesch.* **45** I: 104–141.

Stummer-Traunfels, R. von. 1895. Tropische Polycladen. I. Das Genus *Thysanozoon* Grube. *Z. wiss. Zool.* **60:** 689–725.
—1902. Eine Süsswasserpolyclade aus Borneo. *Zool. Anz.* **26:** 159–161.
—1933. Polycladida (continued). *Bronn's Kl. Ordnung. Tierreichs,* **4,** Abt. 1c, Lief. 179: 3485–3596.
— **& Meixner, J.** 1930. Turbellaria. Polycladida, Literature V. *Bronn's Kl. Ordnung. Tierreichs,* **4,** Abt. 1c, Lief. 178: 3371–3484.
Sumner, F.B., Osburn, R.C. & Cole, L.J. 1913. A biological survey of the waters of Woods Hole and vicinity. Section III. A catalogue of the marine fauna of Woods Hole and vicinity. Class Turbellaria. *Bull. Bur. Fish. Washington* **30:** 579–582.
Surface, F.M. 1907. Note on the origin of the mesoderm of the polyclad, *Planocera inquilina* Wh. *Science,* New York (N.S.) **25:** 732.
—1908. The early development of a polyclad *Planocera inquilina* Wh. *Proc. Acad. nat. Sci. Philadelphia* **59:** 514–559.
Taneja, S.K. 1978. Acrosome formation in *Spinicirrus* sp. (Turbellaria: Polycladida). *Res. Bull. Panjab. Univ.* **25** (1974): 127–129.
Théel, H. 1907. Om utvecklingen af Sveriges Zoologiska Hafsstation Kristineberg och om djurlifvet i angränsande haf och fjordar. *Ark. Zool.* **4,** 5: 61.
Thomas, M.B. 1970. Transitions between hetical and protofibrillar configurations in doublet and single microtubules in spermatozoa of *Stylochus zebra* (Turbellaria: Polycladida). *Biol. Bull. mar. biol. Lab. Woods Hole* **138:** 219–234.
Thompson, T.E. 1965. Epidermal acid secretion in some marine polyclad Turbellaria. *Nature,* Lond. **206:** 954–955.
Thompson, W. 1840. Additions to the fauna of Ireland. *Ann. Mag. nat. Hist.* **5:** 245–257.
—1845. Additions to the fauna of Ireland, including descriptions of some apparently new species of Invertebrata. *Ann. Mag. nat. Hist.* **15:** 308–322.
—1846. Additions to the fauna of Ireland, including a few species unrecorded in that of Britain; with the description of a new *Glossiphonia. Ann. Mag. nat. Hist.* **18:** 392–394.
—1849. Additions to the fauna of Ireland. *Ann. Mag. nat. Hist.* (2) **3:** 351–357.
Thum, A.B. 1970. Reproductive ecology of *Notoplana acticola* (Boone) 1929 (Polyclad) on the central California coast. *Am. Zool.* **10:** 553 (Abstract).
—1974. Reproduction ecology of the polyclad turbellarian *Notoplana acticola* (Boone, 1929) on the central Californian coast. *In*: N.W. Riser & M.P. Morse (Eds), *Biology of the Turbellaria,* pp. 431–445. L.H. Hyman Mem. Vol. McGraw-Hill, New York (McGraw-Hill series in the Invertebrates).
Tseng-Jui Tu, 1939. Geschlichtlicher Überblick über das Studium der Turbellarien in Ostasien und Stand unserer Kenntnisse von diesen. *Zool. Jb.* (Syst.) **73:** 201–260.
Turner, R.L. 1946. Observations on the central nervous system of *Leptoplana acticola. J. Compar. Neurol.* **85:** 53–63.
Uljanin, W. 1870. Materialien zur Fauna des Schwarzen Meeres. *Prof. Freunde Naturw. Anthropol. Ethnogr.,* Moscow, **7:** 106–107.
Utinomi, H. 1956. *Coloured illustrations of sea-shore animals of Japan.* Osaka (Hoikusha). xvii + 167 pp. (Japanese). [Fauna & Flora of Japan No. 8.]
Vaillant, L. 1866. Sur le développement des *Polycelis laevigatus* Quartref. *L. Institute* **1,** Sect. 34: 183.
—1868. Remarques sur le développement d'une planaire dendrocoele le *Polycelis laevigatus* Quat. *Mem. Acad. Montpelier* **7:** 93–108.
—1889. Lombriciniens, hirudiniens, bdellomorphes, teretulariens et planariens. *In*: Quatrefages, *Histoire naturelle des anneles marins et d'eau douce* **3:** xii + 768 pp.
Valcurone, M. 1954. Richerches istochimiche sui granuli vitelline dei policladi. *Arch. Zool. ital.* **38:** 245–265.
Vanhöffen, E. 1897. Die Fauna und Flora Grönlands. *Grönl. Exped. Ges. Erdkunde Berlin, 1891–1893.* Berlin (Kuhl): p. 175.
Van Name, W.G. 1899. The maturation, fertilization and early development of the planarians. *Trans. Conn. Acad. Sci.* **10:** 263–300.
Vàtova, A. 1928. Compendis della flora e fauna del Mare Adriatico presso Rovigno con la distribuzione geografica delle species bentoniche. *Mem. Com. talassogr. Ital.* **143:** 154–174.
Verany, J.B. 1846. *Catalogo degli animali invertebrati del golfo di Genova e Nizza.* Genova. 30 pp.
Verrill, A.E. 1873. Report upon the invertebrate animals of Vineyard Sound and the adjacent waters,

with an account of the physical characters of the region. *Rep. U.S. Comm. Fish & Fisheries,* for 1871 and 1872: 295–778.
—1879. Check list of the marine Invertebrata of the Atlantic coast, from Cape Cod to the Gulf of St. Lawrence. *Rep. U.S. Comm. Fish. & Fisheries,* p. 13.
—1882. Notice of the remarkable marine fauna occupying the outer banks off the southern coast of New England, No. 7, and of some additions to the fauna of Vineyard Sound. *Am. J. Sci.* (3) **24:** 360–371. [Also *Rep. U.S. Comm. Fish. & Fisheries* for 1882 [1884]: 641–669.]
—1893. Marine planarians of New England. *Trans. Conn. Acad. Sci.* **8:** 459–520.
—1895. Supplement to the marine nemerteans and planarians of New England. *Trans. Conn. Acad. Sci.* **9:** 523–534.
—1900. Additions to the Turbellaria, Nemertina, and Annelida of the Bermudas, with revisions of some New England genera and species. *Trans. Conn. Acad. Sci.* **10:** 595–672.
—1903. Additions to the fauna of the Bermudas from the Yale Expedition of 1901, with notes on other species. *Trans. Conn. Acad. Sci.* **11:** 15–62.
—1907. *Zoology of Bermudas.* 2nd ed. Vol. 1. New Haven, Conn.
Vialli, M. 1934. Ricerche istochimiche sui granuli vitellini dei policladi. *Boll. Zool. Torino* **5:** 21–23.
Villadolid, D.V. & Villaluz, D.K. 1938. Animals destructive to oysters in Bacoor Bay, Luzon. *Philippine J. Sci.* **67:** 393–396.
Viviani, D. 1805. *Phosphorescentia maris. Quatuordecim lucescentium animalculorum novis speciebus illustrata ... Accedit novi cujusdam generis e molluscorum familia descriptio et anatomes.* 17 pp. Genuae.
von Graff, L. See Graff, L. von.
Vonck, E. 1932. Les vers marins du littoral belge. *Naturalistes belg.* **13:** 22–25, 47–50, 65–69.
Wagner, N. 1885. Die Wirbellosen des weissen Meeres. *Zool. Forsch. Küste Solowetzkischen Meerbasin, 1878, 1879 u. 1882,* Leipzig. **1:** 50, 60.
Watanabe, Y. & Child, C.M. 1933. The longitudinal gradient in *Stylochus ijimai*: with a critical discussion. *Physiol. Zool.* **6:** 542–591.
Wesenberg-Lund, E. 1928. Turbellarier og Nemertiner. *Medd. Grønland Kjobenhavn* **23** (Suppl.); 35–77.
Westblad, E. 1952. Turbellaria (excl. Kalyptorhynchia) of the Swedish South Polar Expedn. 1901–1903. *Further Zool. Res. Swed. Antarct. Exped.* **4** (8): 1–55.
—1955. Some Hydroidea and Turbellaria from western Norway with description of three new species of Turbellaria. *Univ. Bergen Ärbok.* **1954** No. 10: 1–12.
Wharton, G.W. 1938. Polyclad larvae. *J. Elisha Mitchell sci. Soc.* **54:** 196.
Wheeler, W.M. 1894. *Planocera inquilina,* a polyclad inhabiting the branchial chamber of *Sycotypus canaliculatus* Gill. *J. Morph.* **9:** 195–201.
Whitelegge, T. 1890. List of the marine and freshwater invertebrate fauna of Port Jackson and the neighbourhood. *J. roy. Soc. N.S.W.* **23:** 163–330.
Wilhelmi, J. 1913. Kulter und Natur an Meerestrande. Betractungen über die Verunreinigungen von Küsten durch Abwasse, mit einem einleitenden Abschnitt über die biologische Analyse des Süsswassers. *Naturw. Wschr.* 28 (n.s.12): 452–456, 470–473.
Willey, A. 1897. Letters from New Guinea on *Nautilus* and some other organisms. *Quart. J. microsc. Sci.* **39:** 153–159.
—1898. On *Heteroplana,* a new genus of planarians. *Quart. J. microsc. Sci.* **40:** 203–205.
Williams, G. 1954. Fauna of Strangford Lough and neighbouring coasts. *Proc. roy. Irish Acad.* **56** (B): 29–133.
Wilson, E. 1898. Considerations on cell-lineage and ancestral reminiscence, based on a re-examination of some points in the early development of annelids and polyclads. *Ann. New York Acad. Sci.* **11:** 63–67.
Woelke, C.E. 1957. Flatworm, *Pseudostylochus ostreophagus* Hyman, a predator of oysters. *Proc. natl. shellfish Ass.* **47:** 62–67.
Woodworth, W.McM. 1894. Report of the Turbellaria (Albatross Report IX). *Bull. Mus. comp. Zool. Harvard* **25:** 49–52.
—1898. Some planarians from the Great Barrier Reef of Australia. *Bull. Mus. comp. Zool. Harvard* **31:** 63–67.
Yang, Han-Choon. 1974. Spawning and larval developments of two species of polyclad worms

Stylochus ijimai Yeri & Kaburaki and *Pseudostylochus obscurus* (Stimpson). *Bull. Korean Fish. Soc.* **7:** 7–14 (English summary).
Yeri, M. & Kaburaki, T. 1918*a*. Bestimmungsschlussel für die japanischen Polycladen. *Annotnes zool. jap.* **9:** Pt. 4: 431–442.
—1918*b*. Description of some Japanese polyclad Turbellaria. *J. Coll. Sci. imp. Univ. Tokyo* **39** Art. 9: 1–54.
—1920. Notes on two new species of Japanese polyclads. *Annotnes zool. jap.* **9:** 591–598.
Young, R.T. 1912. The epithelium of Turbellaria. *J. Morph.* **23:** 255–268.
Záhony, R.R. von *see* Ritter-Záhony, R. von.